MOLECULAR MARKERS,
NATURAL HISTORY
AND EVOLUTION

MOLECULAR MARKERS, NATURAL HISTORY AND EVOLUTION

JOHN C. AVISE

CHAPMAN & HALL

I(T)P An International Thomson Publishing Company

New York • Albany • Bonn • Boston • Cincinnati • Detroit • London • Madrid • Melbourne •
Mexico City • Pacific Grove • Paris • San Francisco • Singapore • Tokyo • Toronto • Washington

Copyright © 1994 by Chapman & Hall

Printed in the United States of America

For more information contact:

Chapman & Hall
115 Fifth Avenue
New York, NY 10003

Chapman & Hall
2-6 Boundary Row
London SE1 8HN
England

Thomas Nelson Australia
102 Dodds Street
South Melbourne, 3205
Victoria, Australia

Chapman & Hall GmbH
Postfach 100 263
D-69442 Weinheim
Germany

Nelson Canada
1120 Birchmount Road
Scarborough, Ontario
Canada M1K 5G4

International Thomson Publishing Asia
221 Henderson Road #05-10
Henderson Building
Singapore 0315

International Thomson Editores
Campos Eliseos 385, Piso 7
Col. Polanco
11560 Mexico D.F.
Mexico

International Thomson Publishing - Japan
Hirakawacho-cho Kyowa Building, 3F
1-2-1 Hirakawacho-cho
Chiyoda-ku, 102 Tokyo
Japan

4 5 6 7 8 9 XXX 01 00 99 98 97 96

Library of Congress Cataloging-in-Publication Data

Avise, John C.
    Molecular markers, natural history & evolution/John C. Avise.
        P.   Cm.
    Includes bibliographical references (p.) and index.
    ISBN 0-412-03771-8 (hb); 0-412-03781-5 (pbk)
    1. Biochemical markers. 2. Molecular evolution. I. Title.
    QH438.4B55A95   1993   93-20406
    574.87'328--dc20   CIP

Visit Chapman & Hall on the Internet http://www.chaphall.com/chaphall.html

To order this or any other Chapman & Hall book, please contact **International Thomson Publishing, 7625 Empire Drive, Florence, KY 41042.** Phone (606) 525-6600 or 1-800-842-3636. Fax: (606) 525-7778. E-mail: order@chaphall.com.

For a complete listing of Chapman & Hall titles, send your request to **Chapman & Hall, Dept. BC, 115 Fifth Avenue, New York, NY 10003**

Contents

Preface

Part I. Background

Part II. Applications

Preface

I never cease to marvel that the DNA and protein markers magically appearing on laboratory gels and autoradiographs can reveal so many otherwise hidden facets about the world of nature. Can individual plants sometimes exist as genetic mosaics derived from multiple zygotes? Is reproduction in unicellular eukaryotes predominantly sexual or clonal? In the absence of recombinationally derived genetic variability, how long do evolutionary lineages persist in parthenogenetic all-female "species"? Do social groups within various species of insects, fishes, and other organisms consist of close relatives whose behavior might have evolved under the influence of kin selection? How often does intraspecific brood parasitism occur in birds, whereby females "dump" eggs in the nests of foster parents? When distinctive morphotypes exist among related fishes within a lake, do these reflect species' differences (as opposed to developmental switches within a species), and, if so, did the species flock arise in situ? Do migratory marine turtles return to natal sites for nesting? How often has carnivory evolved among plants? What are the evolutionary origins of mitochondrial DNA and chloroplast DNA within eukaryotic cells? Are transposable elements the phylogenetic precursors of retroviruses, or vice versa? Can plate tectonics and continental drift account for the presence of large flightless birds in Australia, South America, and Africa? Have demographic bottlenecks diminished genetic variability in some populations to the extent that they can no longer adapt to environmental challenges? How distinct genetically are the taxa currently afforded protection under the Endangered Species Act? These are but a small sample of the diverse problems addressed and answered (at least provisionally) through use of molecular genetic markers.

This treatment of molecular natural history and evolution is written at a level appropriate for the advanced undergraduate or graduate student, or for scientists in fields such as ecology, genetics, ethology, molecular biology, population biology, and conservation biology who may wish a readable introduction or refresher to the burgeoning application of molecular markers to problems in their disciplines. I hope to have captured and conveyed the genuine excitement that can be brought to such fields when genetic attributes with known patterns of inheritance are applied to organismal-level issues. I also hope to have provided a wellspring of research ideas for those entering the field. My goal has been to present material in a manner that is technically straightforward, without sacrificing the richness of underlying concepts and biological applications. For the reader, the only necessary prerequisites are an introductory knowledge of genetics and an acute interest in the natural biological world.

Why is a treatment of this topic necessary when numerous excellent texts in molecular evolution already are available? Recent publications in the field have focused on (a) proteins and DNA as primary objects of interest from a molecular perspective; (b) broad conceptual issues regarding patterns and processes of molecular evolution; (c) statistical or mathematical aspects of population genetics theory; (d) procedures of molecular data analysis; or (e) descriptions of laboratory methods. Some books have approached more closely what is attempted here, but are restricted either to topic, laboratory method, or taxonomic group. Hoelzel and Dover (1991a) and Hoelzel (1992) have produced fine introductions to molecular ecology, but to my knowledge, no extended available treatment explicitly addresses the multifaceted biological applications for molecular genetic markers from the perspectives of population biology, natural history, and organismal phylogeny.

The field of molecular evolution is at a stage where reflection on the past 30 years may provide useful historical perspective, as well as a springboard to the future. The mid-1960s witnessed an explosion of interest in molecular techniques with the seminal introduction of protein electrophoretic approaches to population genetics and evolutionary biology. In the late 1970s, attention shifted to methods of DNA analysis, primarily through restriction enzymes, and in the 1980s, mitochondrial DNA analyses and DNA fingerprinting approaches gained immense popularity. Recently, the introduction of PCR-mediated DNA sequencing has brought the field close to a final technical fruition by providing the first ready access to the "ultimate" genetic data, nucleotide sequences themselves. Nonetheless, it would be a mistake to conclude that direct sequence information invariably provides the preferred or most accessible pool of genetic markers for all biological applications. Several alternative assays retain tremendous power and utility and, because of ease, cost, amount of genetic information accessed, or simplicity of data interpretation, continue to be the techniques of choice for many evolutionary problems. Ecologists, as well as molecular geneticists, sometimes are unaware of the arguments for and against various molecular genetic approaches, and one major goal of this book is to clarify these positions.

In scientific advance, timing and context are all-important. Imagine for sake of argument that DNA sequencing methods had been widely employed for the past 30 years and that only recently had protein-electrophoretic approaches been introduced. No doubt a headlong rush into allozyme techniques would ensue, on justifiable rationales that (a) the methods are cost-effective and technically simple, (b) the variants revealed reflect independent Mendelian polymorphisms at several loci scattered around the genome (rather than as linked polymorphisms in a single stretch of DNA), and (c) the amino acid replacement substitutions uncovered by protein electrophoresis (as opposed to the silent base changes often revealed in DNA assays) might bring molecular evolutionists closer to the real "stuff" of adaptive evolution. To carry the argument farther, suppose that molecular genetic methods had been employed throughout the last century but that an entrepreneurial scientist finally ventured into the world of nature and discovered organismal phenotypes and behaviors. Finally, the interface of gene products with the environment would have been revealed! Imagine the sense of excitement and the research prospects!

These fanciful scenarios are raised to emphasize a point—molecular approaches carry immense popularity at the present time, but they nonetheless provide but one of many avenues toward the goal of understanding the natural histories and evolutionary biologies of organisms. Studies of morphology, ecology, and behavior undeniably have shaped the great majority of our perceptions about the natural world. Molecular approaches are especially exciting at this point in time because they open new empirical windows and novel insights on these traditional subjects.

In this book, I have attempted to identify and highlight select case histories where molecular methods have made significant contributions to natural history and evolutionary biology. The treatment is not intended to be exhaustive, because literally thousands of studies have utilized genetic markers. Rather, I have tried to choose classic, innovative, or otherwise interesting examples illustrative of the best that molecular methods have to offer. The literature is represented through 1992. Overall, I have attempted to retain a balanced taxonomic perspective that includes examples from plants, animals, and microbes, and indeed I hope that common threads will be evident that tie together the similar classes of biological questions that frequently apply to such disparate organisms.

This book is organized into two parts. Part I provides introductory material and background: the rationale for molecular approaches in natural history and evolution (Chapter 1); the history of molecular phylogenetics (Chapter 2); outlines of various laboratory methods and the nature of the genetic data each molecular method provides (Chapter 3); and descriptions of some of the interpretive tools of the trade, including molecular clock concepts and phylogenetic methods as applied to molecular data (Chapter 4).

Part II departs significantly from previous books in molecular evolution by emphasizing empirical examples of significant biological applications for molecular genetic markers. Topics are arranged along a phylogenetic hierarchy ranging from micro- to macro-evolutionary: the assessment of genetic identity/nonidentity and parentage (Chapter 5); kinship and intraspecific phylogeny (Chapter 6); speciation, hybridization, and introgression (Chapter 7); and assessment of the deeper phylogenetic structures in the evolutionary tree of life (Chapter 8). A concluding chapter deals with the relevance of molecular studies to conservation biology and to the preservation of genetic diversity (Chapter 9).

I would like to dedicate this book to my current and former graduate students and research technicians, without whom most of our work would not have been possible, nor nearly as much fun: Charles Aquadro, Marty Ball, Eldredge Bermingham, Brian Bowen, Robert Chapman, Michael Douglas, Matt Hare, Steve Karl, Lou Kessler, Trip Lamb, Joe Neigel, Bill Nelson, Jay Parker, John Patton, Joe Quattro, Carol Reeb, Nancy Saunders, Kim Scribner, DeEtte Walker, and Kurt Wollenberg. Recent graduate students were forced to read early drafts of several chapters, and they responded by providing numerous helpful criticisms. Drs. Fred Allendorf, Jim Hamrick and Linda Maxson read drafts of the entire book and suggested numerous improvements. Excellent criticisms of individual sections or chapters were kindly provided by Drs. Jeff Palmer, David Stock, Gary Olsen, and Carl Woese. Of course, I remain responsible for any errors or omissions. I am also indebted to my own advisors in graduate school—Robert K. Selander, Michael H. Smith, and Francisco J. Ayala—for getting me started; and to my wonderful colleagues at the University of Georgia—Wyatt Anderson, Jim Hamrick, Jonathan Arnold, Marjorie Asmussen, John McDonald, Mike Arnold, and others—for keeping me going. Much of this book was written during a sabbatical supported by the Sloan Foundation, and I am indebted to my host Denny Powers for providing a wonderful working environment at the Hopkins Marine Station of Stanford University. Over the years, my laboratory has been supported by grants primarily from the National Science Foundation and the National Geographic Society.

I especially want to dedicate this book to my wife Joan, with whom I have collaborated to put into practice the fitness concepts that we evolutionary biologists so often discuss. The result was the light of my life, Jennifer. Finally, I want to thank my parents, Dean and Edith, for unwavering support.

Part I

Background

1

Introduction

The stream of heredity makes phylogeny; in a sense, it is phylogeny. Complete genetic analysis would provide the most priceless data for the mapping of this stream.

G.G. Simpson, 1945

This book is about the study of phylogeny and natural history, late 20th-century style. Scientists now routinely utilize the genetic information in biological macromolecules—proteins and DNA—to address numerous aspects of the behaviors, life histories, and evolutionary relationships of organisms. When used to best effect, molecular data are integrated with information from such fields as ethology, field ecology, comparative morphology, systematics, and paleontology. These time-honored biological disciplines remain highly active today, but each has been enriched if not rejuvenated by contact with the relatively young but burgeoning field of molecular evolution.

Interest in molecular evolution can center in either of two areas: (a) characterization of the molecular basis of variation in particular genetic systems or (b) application of genetic data to questions in natural history and organismal evolution. The latter constitutes the primary focus of this book. However, molecular and organismal issues are intertwined. Knowledge of such molecular-level properties as the mode of genetic transmission and the nature of mutational differences underlying a polymorphism are critical to proper interpretation of molecular markers in a population context; and, patterns of variation and divergence in genetic markers can be highly informative regarding evolutionary forces impinging on molecular evolution.

3

In the not-too-distant future, it might become possible to isolate and characterize directly the genes contributing to variation in particular phenotypic features and adaptations. Such molecular-level analyses will revolutionize the study of phenotypic variability and its maintenance. But there is much more to evolution, and many topics in organismal biology already can be addressed critically by examining variation in "randomly chosen" DNAs or proteins—in other words, by employing molecular genetic markers. These subjects include organismal forensics, mating systems, population structure, gene flow, speciation, hybridization, and systematics, to name a few.

Phylogeny is evolutionary history—the topology of the proverbial tree of life. All organisms share certain features (most notably, nucleic acids as hereditary material) that suggest a single or monophyletic origin on earth, some three to four billion years ago. The pattern of phylogeny involves successive branching from this trunk of life, with organisms alive today representing tips of the twigs on the currently outermost branches. A complete picture of phylogeny requires knowledge of both the branching order (cladogenetic splitting of lineages) and of the branch lengths (anagenetic changes within lineages through time). The possibility of occasional genetic transfer between branches (reticulate evolution), perhaps mediated by interspecific hybridization, viral conveyance, or establishment of endosymbiotic associations among genomes, also must be considered.

Most studies that utilize molecular markers can be viewed as attempts to estimate phylogeny, broadly defined, at one or another hierarchical stage of evolutionary divergence (Fig. 1.1). Phylogenetic relationships thus can be assessed at levels ranging from extreme micro- to macro-evolutionary: (a) genetic identity versus nonidentity, as in clonally-reproducing organisms; (b) parentage (maternity and paternity); (c) extended kinship within a local group or population; (d) differentiation among geographic populations and subspecies; (e) differentiation among reproductively isolated species; and (f) phylogenetic structure at intermediate and great evolutionary depths in the tree of life. Different types of molecular assay provide genetic information ideally suited to different subsets of this hierarchy, and a continuing challenge is to develop and utilize molecular methods appropriate for a particular biological problem at hand.

It is also appropriate to orient this book around phylogeny because of the central importance of historical events in evolutionary biology. As noted by Hillis and Bull (1991), "Virtually all comparative studies of biological variation among species depend on a phylogenetic framework for interpretation." If Dobzhansky's (1973) famous dictum is correct, that "Nothing in biology makes sense except in the light of evolution," then it might be appended that "much in evolution makes more sense in the light of phylogeny." Brooks and McLennan (1991) provide an eloquent and extended treatment of the general role of phylogenetic analysis in ecology and ethology, as do Eldredge and Cracraft (1980) and Harvey and Pagel (1991). Molecular approaches will contribute to these

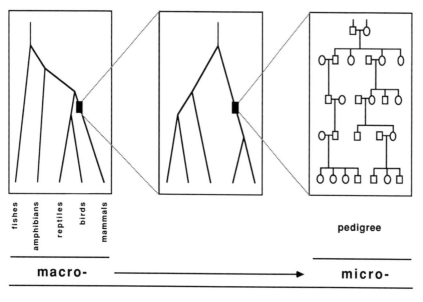

fishes
amphibians
reptiles
birds
mammals

pedigree

macro- ⟶ **micro-**

PHYLOGENY

Figure 1.1. The hierarchical nature of phylogenetic assessment (after Avise et al., 1987a).

phylogenetic enterprises because with evolutionary relationships properly sorted out at the molecular level, unprecedented opportunities emerge for understanding the evolutionary origins and histories of other organismal attributes.

WHY EMPLOY MOLECULAR GENETIC MARKERS?

In describing organismal relationships, the approaches previously available have rested on the data of comparative morphology, physiology, and other assayable phenotypic features. Molecular systematists also employ the comparative method, but the comparisons involve direct or indirect information on nucleic acid and protein sequences. Why are molecular data of special significance to phylogeny estimation?

Molecular Data are Genetic

This simple truism is of overriding significance. Because phylogeny is "the stream of heredity," only genetically transmitted traits are informative to phylogeny estimation. Not only are protein and DNA features heritable, but most molecular assays reveal variable character states whose particular genetic bases and modes of transmission can be specified explicitly. Thus, from knowledge of the amount and nature of genetic information assayed, statements of relative confidence can be placed on molecular-based phylogenetic conclusions.

This situation contrasts with the insecure knowledge concerning the genetic bases of most conventional characters employed in organismal systematics (Barlow, 1961; Boag, 1987). Seldom can the genes and alleles controlling particular morphological, physiological, or behavioral traits be specified, and, indeed, some such traits conventionally utilized in taxonomy may be influenced by nongenetic factors. For example, a significant fraction of the variance in morphometric features used to describe named subspecies of the red-winged blackbird (*Agelaius phoeniceus*) proved to be due to environmental rearing conditions: from eggs experimentally transplanted from one geographic population to another, the resulting progeny converged significantly on some of the morphological features of their foster parents (James, 1983). Developmental or phenotypic plasticity, whereby form is influenced directly by environment, is especially pronounced in many plants (Clausen et al., 1940). Of course, such ecotypic variation can be misleading if interpreted as providing genetic or systematic characters.

Molecular Methods Open the Entire Biological World for Genetic Scrutiny

Prior to the introduction of molecular approaches, genetic studies mostly were confined to the small handful of species that could be crossed under controlled laboratory conditions, such as Mendel's pea plants, the bacterium *Escherichia coli* and its phages, the corn *Zea mays,* the fruit fly *Drosophila melanogaster,* and the house mouse *Mus musculus.* From transmission patterns across generations, the genetic bases of assayable morphological or physiological traits in these species were deduced, but such analyses could hardly be expected to capture the full flavor of genetic diversity across the earth's broader biota (nor the richness of diversity among genetic elements within a genome). By contrast, various molecular assays provide direct structural evidence regarding essentially any genes or their protein products and can be applied to the genetics of any creatures, microbes to whales.

Molecular Methods Access a Nearly Unlimited Pool of Genetic Variability

Genomes are enormous in information content. For example, a typical mammalian genome consists of some three thousand million nucleotide pairs, in a composite linear sequence roughly 100 times longer than the total string of letter characters in an 18-volume World Book encyclopedia. Like an encyclopedia, each genome is a repository of information, coding not only the proteins and other cellular machinery of life but also retaining within its nucleotide sequence a record of evolutionary relationships to other genomes. Bacterial genomes range in size from about 0.6×10^6 to 13.2×10^6 base pairs (bp); protist genomes from 23×10^6 to 686×10^9 bp; and genomes of multicellular fungi, plants and animals from 8.8×10^6 to more than 300×10^9 bp (Cavalier-Smith, 1985; Li and Graur, 1991; Sparrow et al., 1972). Molecular assays involve sampling, often more or less at random, from such vast informational pools.

Genetic variability within most species is also tremendous. Consider, for example, the human gene pool, which in comparison to many other species is unexceptional if not low with regard to magnitude of variation (Li and Sadler, 1991). Although genome-wide estimates of nucleotide diversity still are somewhat uncertain, a conservative estimate is that homologous pairs of randomly drawn human DNA sequences may differ at 0.03% or more of nucleotide positions, on average (Ewens et al., 1981; Li and Sadler, 1992). The human genome sequencing project, perhaps to be completed within the next decade, promises to provide the complete three-billion-bp nucleotide sequence of one genome equivalent. Thus, if a second human genome were to be sequenced, it would likely differ from the first at nearly one million nucleotide sites! Nucleotide diversities in some other species have been estimated at 0.5–2.0% (Stephan and Langley, 1992), such that an average pair of chromosome sets drawn from such populations likely would differ at several million nucleotide positions.

Obtaining complete DNA sequences from more than a few model organisms will remain impractical for the foreseeable future [the first complete sequence of a single entire chromosome from any organism, the 315 kb chromosome III of the yeast *Saccharomyces cerevisiae,* was obtained in 1992 (Oliver et al., 1992)]. Complete genomic sequences will also remain quite unnecessary for purposes of providing genetic markers in population biology and evolution because available methods already can uncover ample variability for even the finest of diagnostic requirements. For example, over 2000 human DNA polymorphisms revealed by restriction endonuclease analysis had been cataloged by 1990 (Stephens et al., 1990a) and the list is growing rapidly (Weissenbach et al., 1992); nearly 100 human protein and blood group polymorphisms had been surveyed by that time among the major human races (Nei and Livshits, 1990). Assume for the sake of an extremely conservative argument that only 200 polymorphisms were present in the human genome, each with the minimum possible two alleles. Rules of Mendelian heredity show that the potential number of different human genotypes would then be an astronomical 3^{200}, or roughly 10, 000, 000, 000, 000, 000, 000, 000, 000, 000, 000, 000, 000, 000, 000, 000, 000, 000, 000, 000, 000, 000, 000, 000, 000, 000, 000, 000, 000, 000, 000, 000, 000. The total number of people estimated to be alive in the year 2000 is 6,000,000,000, and the total number of humans who have ever lived will be roughly 13,000,000,000 by then. Thus, the potential number of distinct human genotypes vastly exceeds the number of individuals who will ever inhabit the earth, and no humans past, present, or future (barring identical twins) are likely to be the same genetically. In human forensic practice, assays of even modest numbers of highly allelic Mendelian polymorphisms have proved sufficient to provide individual-specific DNA fingerprints that have withstood jurisprudential scrutiny (Chapter 5).

Molecular Data Can Distinguish Homology from Analogy

The central problem of phylogenetics always has been to distinguish the component of biological similarity due to descent from common ancestry (homology) from that due to convergence from different ancestors (analogy). Evolutionary classifications should reflect only true homologies, records of genealogical ancestry. But shared morphologies and behaviors, which until recently provided the major guides to classification, sometimes can evolve independently in unrelated species in response to common environmental challenges. For example, Old World and New World vultures share numerous adaptations for carrion-feeding existence (e.g., soaring food-search behavior, featherless face and head, powerful hooked bill) which formerly were thought to be indicative of a close evolutionary relationship and resulted in their taxonomic placement with other diurnal raptors (Falconiformes). However, recent data from DNA-DNA hybridization (as well as detailed reappraisals of morphology) showed conclusively that New World vultures are related more closely to storks (formerly Ciconiiformes) and have independently acquired adaptations for the carrion-feeding life-style (Sibley and Ahlquist, 1990).

In referring to DNA hybridization methods, the paleontologist Gould (1985) wrote "I do not fully understand why we are not proclaiming the message from the housetops . . . We finally have a method that can sort homology from analogy." Gould was referring to the fact that when species are assayed with respect to hundreds or thousands of molecular characteristics, any widespread and intricate similarities that appear are unlikely to have arisen by convergent evolution and, therefore, must reflect true phylogenetic descent. This is not to say that particular molecular characters, such as individual nucleotide sites, are free from homoplasy (convergences, parallelisms, or evolutionary reversals that muddy the historical record). Indeed, some molecular characteristics considered individually may be especially prone to homoplasy, due to a small number of interconvertible character states and to a sometimes rapid rate of change among them (e.g., each nucleotide position is characterized by only four assumable states—adenine, guanine, thymine, and cytosine). Other molecular features, such as duplications, deletions, or rearrangements of genetic material, may be relatively rare events for which hypotheses of monophyly can sometimes be justified. In generating molecular-based phylogenetic hypotheses, the nature of the underlying data must be taken into account.

Molecular Data Provide Common Yardsticks for Measuring Divergence

A singularly important aspect of molecular data is that they allow direct comparisons of relative levels of genetic differentiation among essentially any groups of organisms (Wheelis et al., 1992). Suppose, for example, that one wished to quantify evolutionary differentiation within a taxonomic family or genus of fishes as compared to that within a taxonomic counterpart of birds. The kinds of morphological features traditionally employed in fish systematics (e.g., number of lateral line scales, fin rays, gill rakers, and position of the swim bladder) clearly would be of little utility for direct comparison with avian systematic characters (e.g., plumage features, structure of the syrinx, or arrangement of toes on the feet), and perhaps for such reasons, traditional systematic practices and standards have developed separately among ichthyologists, ornithologists, and other organismal specialists. However, birds and fishes (as well as other animals, plants, and microbes) do share numerous molecular traits, such as genes and enzymes involved in central metabolic pathways for the respiration and synthesis of carbohydrates, fats, amino acids, and nucleic acids.

To introduce the concept of common yardsticks in molecular evolution, consider Figure 1.2 which summarizes reported levels of genetic divergence among congeneric species representing five vertebrate classes, as measured by the common denominator of protein electrophoretic divergence among similar suites of enzymes. One unanticipated generality to emerge is that by this criterion, avian congeners typically exhibit far less genetic differentiation than do many reptilian or amphibian species within a genus, despite considerable anatomical differentiation among birds (Wyles et al., 1983). Perhaps avian congeners are evolutionarily younger than many nonavian congeners, or perhaps avian proteins evolve more slowly on average than those of amphibians, for example (Avise, 1983a; Avise and Aquadro, 1982). In either event, these comparative genetic results raise exciting evolutionary issues that were not evident from morphological comparisons alone.

In another example of this "common yardstick" perspective, King and Wilson (1975) reviewed evidence that the assayed proteins and nucleic acids of humans and chimpanzees are only about as divergent as are those of morphologically similar sibling species of fruit flies and rodents, and much less divergent than those of typical amphibian congeners. They suggested that the traditional placement of humans and chimps in different taxonomic families (Hominidae and Pongidae) might have stemmed from morphological differences attributable to changes at a few regulatory genes acting during ontogeny. An alternative possibility is that the perceived morphological distinctiveness of humans from chimps and other primates may have been exaggerated for reasons of anthropocentric bias. In a fascinating paper entitled "Frog perspective on the morphological difference between humans and chimpanzees," Cherry et al.

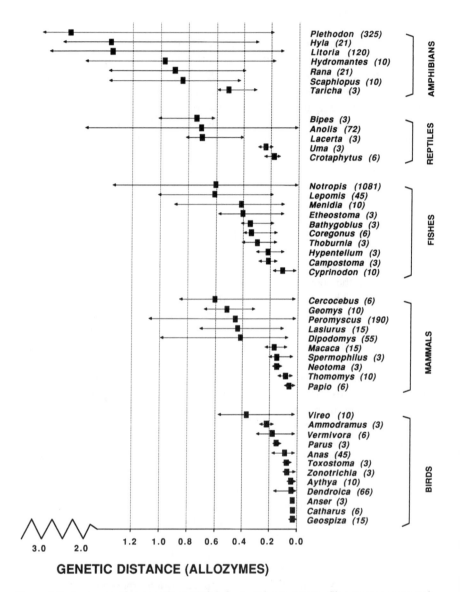

GENETIC DISTANCE (ALLOZYMES)

Figure 1.2. A comparative genetic yardstick for vertebrate genera. Shown are means and ranges of molecular differentiation, plotted on a common scale, among congeneric species within each of five vertebrate classes (after Avise and Aquadro, 1982). Genetic distances are in units of codon substitutions per locus, as assessed by multilocus protein electrophoresis. In parentheses are numbers of pairwise species comparisons for each genus.

(1978) employed the anatomical traits normally used to discriminate among frogs (eye–nostril distance, forearm length, toe length, etc.) to quantify the human–chimpanzee morphological separation. By these criteria, morphological divergence between the two primates was large even by frog standards (whereas the molecular divergence was not), a result interpreted as consistent with the postulate that morphological and molecular evolution can proceed at independent rates. It is ironic, but understandable, that this rare attempt to utilize a comparative yardstick approach in morphological evolution came from a research laboratory normally devoted to molecular genetics, where such comparisons come quite naturally.

Thus, one challenging question posed by molecular approaches is whether a genetic basis for taxonomy can be developed that for the first time will provide a universal standard for comparing all forms of life (Chapter 8). This is not to imply that the overall magnitude of genetic divergence between taxa is necessarily the only or even the best guide to phylogenetic relationships within groups (Chapters 2 and 4), but it certainly is a novel and important consideration for intergroup comparisons that simply falls outside the framework of traditional systematic practice.

Molecular Approaches Facilitate Mechanistic Appraisals of Evolution

Although traditional assessments of phenotype have been immensely informative in identifying spontaneous visible mutations and in elucidating the action of evolutionary forces including natural selection, the mechanistic (i.e., molecular) underpinnings of such attributes remained unknown. Modern methods such as DNA sequencing provide previously inaccessible information about the fundamental molecular basis of evolutionary change. For example, various morphological and physiological mutations in *Drosophila* now have been characterized at the molecular level and have proved attributable to specifiable point mutations in the coding regions of single-copy genes, mutations in flanking and nonflanking regulatory domains, insertions of transposable elements, and to "homeotic" mutations in special classes of repetitive segmentation loci (Lambert et al., 1988; Lewin, 1990). A growing enterprise in quantitative genetics involves use of DNA markers to physically map loci for polygenic traits (phenotypic characteristics influenced by variation at multiple genes) (e.g., Lander and Botstein, 1989; B. Martin et al., 1989). Scientists are still a long way from the goal of relating all phenotypic differences to particular DNA changes, but increasing numbers of genetic disorders in humans and a few other species are yielding to detailed molecular-level characterizations (e.g., Randall, 1991; Stine, 1989; Wallace et al., 1988).

Although these kinds of mechanistic appraisals fall somewhat outside the subject matter of this book, data of relevance to molecular biology frequently arise as a direct or indirect by-product of molecular phylogenetic analyses. For example, through phylogenetic reconstructions based on restriction-site maps for the alcohol dehydrogenase locus in *Drosophila melanogaster,* researchers have discovered that a mutation conferring a higher capacity to utilize and/or detoxify environmental alcohols arose recently in evolution, perhaps within the past few thousand to one million years (Aquadro et al., 1986; Stephens and Nei, 1985). From sequence analyses of introns at "*t*-loci" on chromosome 17 in house mice, particular chromosomal inversions influencing embryonic development were estimated to have originated about three million years ago, and to have accumulated recessive lethality factors that have been spread worldwide within the last 800,000 years (Morita et al., 1992). Detailed molecular analysis of an esterase locus in mosquitos has shown that the global distribution of an insecticide-resistance attribute most likely resulted from the migrational spread of a single mutation, rather than a high mutation rate to independent resistance alleles (Raymond et al., 1991); a similar molecular-based conclusion was reached for the global spread of a methicillin-resistance gene in the pathogenic bacterium *Staphylococcus aureus* (Kreiswirth, 1993). Much interest in human medicine now centers on the use of molecular markers to assess whether particular genetic disorders (such as phenylketonuria in Yemenite Jews, Huntington's chorea in Afrikaners, or the fragile X syndrome) are of monophyletic or polyphyletic mutational origin (Avigad et al., 1990; Diamond and Rotter, 1987; Hayden et al., 1980; Richards et al., 1992). In general, the mapping of phenotypic traits onto phylogenies estimated from molecular markers is revolutionizing understanding of the evolutionary origins of numerous organismal features (Chapter 8).

Molecular Approaches are Challenging and Exciting

Another appeal of molecular phylogenetics is the sheer intellectual challenge provided by this new discipline. Many of the discoveries in molecular biology clearly affect the practice of molecular phylogenetic assessment, and some relevant molecular-level phenomena taken for granted today were undreamed of even a few years ago. For example, nucleotide sequences among loci belonging to some multigene families often evolve in concert within a species, and thereby remain relatively homogeneous. This process of concerted evolution (Zimmer et al., 1980), first noted by Brown et al. (1972), apparently is due to the homogenizing effects of unequal crossing over among tandem repeats and to gene conversion events even among unlinked loci (Arnheim, 1983; Dover, 1982; Ohta, 1980, 1984; Smithies and Powers, 1986). Thus, the genes in multilocus families may not provide the independent bits of phylogenetic information formerly assumed (Ohno, 1970). To the extent that sequences of multicopy genes within a species evolve in concert subsequent to the duplications from which they arose, the usual distinction between orthology (sequence similarity tracing to a speciation event) versus paralogy (sequence similarity tracing to a gene duplication) loses some relevancy (Fig. 1.3).

Another example of the changing perspectives on phylogeny prompted by molecular methods involves the introduction of mitochondrial (mt) DNA methods to population biology in the late 1970s. Prior to that time, most biologists viewed intraspecific evolution primarily as a process of shifting allele frequencies, a perspective that fit well with the traditional language and framework of population genetics but that failed to accommodate adequately the phylogenetic component of population history (Avise et al., 1987a; Wilson et al., 1985). By providing the first accessible data on "gene genealogies" at the within-species level, mtDNA methods have forged an empirical and conceptual bridge between

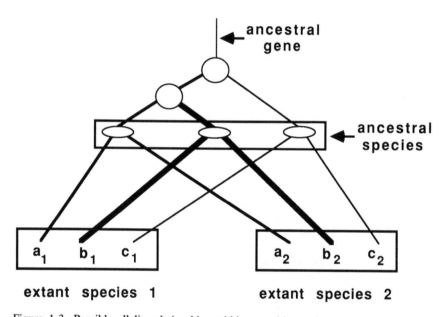

Figure 1.3. Possible allelic relationships within a multigene family. The two circles indicate gene duplication events from an ancestral locus, producing three extant genes a, b, and c. The three ellipses represent allelic separations, leading to the extant alleles a_1, b_1, and c_1 in species 1 and a_2, b_2, and c_2 in species 2. Genetic comparisons between a_1, and a_2, b_1 and b_2, or c_1 and c_2 are orthologous, whereas all other comparisons (e.g., between a_1 and b_2, a_1 and c_2, or a_1 and b_1) are paralogous. In the absence of concerted evolution, orthologous similarities should stem to times near the speciation event, whereas paralogous similarities date to the relevant gene duplication events (which could vastly predate the speciation times of the organisms compared). However, under strong concerted evolution, all or portions of a_1, b_1, and c_1 would appear related more closely to one another than to their respective allelic counterparts in species 2.

Table 1.1. Eleven unorthodox perspectives on evolution prompted by genetic findings on animal mitochondrial DNA.[a]

1. *Asexual Transmission* (Chapter 3)
   Cytoplasmic genes within sexually-reproducing species can exhibit uniparental, nonrecombining transmission.

2. *Population Hierarchy*
   Populations of mtDNA molecules inhabit each somatic and germ cell lineage within the individuals comprising organismal populations (Birky et al., 1989).

3. *Non-universal Code* (Chapter 8)
   MtDNA genetic codes sometimes differ from one another, and from the nuclear codes formerly thought to be universal.

4. *Conserved Function, Rapid Evolution* (Chapters 3 and 4)
   Considerations in addition to level of functional constraint must be required to explain the rapid pace of animal mtDNA evolution.

5. *Lack of Mobile Elements, Introns, Repetitive DNA*
   Genes with selfish motives gain no fitness advantage by becoming repetitive within an asexually-transmitted genome (Hickey, 1982).

6. *Endosymbiotic Origins* (Chapter 8)
   Eukaryotic organisms are genetic mosaics containing nuclear and organellar genomes with formerly independent evolutionary histories.

7. *Intergenomic Conflicts of Interest*
   Mutations with sex-limited detrimental effects on male fitness may accumulate disproportionately in maternally-transmitted genomes and lead to conflicts of interest between nuclear and cytoplasmic genes.

8. *Intergenomic Cooperation*
   Interactions between products of mitochondrial and nuclear genes also lead to expectations of functional coevolution.

9. *Matriarchal Phylogeny* (Chapter 6)
   Mutational differences among mtDNA haplotypes record the phylogenetic histories of female lineages within and among species.

10. *Gene Trees Versus Organismal Phylogenies* (Chapters 4, 6, and 7)
    Organismal pedigrees contain gene genealogies that can differ greatly from locus to locus.

11. *Degenerative Diseases*
    Genetic defects in mitochondrial oxidative phosphorylation provide a new paradigm for the study of aging and degenerative disease (Wallace, 1992).

[a]For elaboration, see the chapters or references listed, and also Avise (1991a).

the nominally rather separate disciplines of population genetics (micro-evolution) and traditional phylogenetics (mostly macro-evolutionary study) (Avise, 1989a). Indeed, a current intellectual challenge is how to come to grips with the true meaning of "organismal phylogeny," which in some sense must be viewed as an emergent property of the multitudinous genealogies for independent loci whose surviving alleles have trickled through an animal or plant pedigree under the vagaries of Mendelian genetic transmission (Chapters 4, 6, and 7). Overall, phylogenetic studies on mtDNA have stimulated a wide variety of novel perspectives on evolution (Table 1.1); similar claims could be made for some other molecular systems as well.

WHY NOT EMPLOY MOLECULAR GENETIC MARKERS?

Against these advantages of molecular genetic methods appear to stand only two major detractions: (a) considerable training normally is required of practitioners and (b) the monetary cost is high (but also quite variable across methods) relative to the standards of nonmolecular systematics (Weatherhead and Montgomerie, 1991). However, a fact sometimes overlooked is that the great majority of molecular-level assessments merely support earlier phylogenetic hypotheses based on morphology (e.g., Avise, 1974; Sibley and Ahlquist, 1990). Thus, a complete molecular reanalysis of the biological world is unnecessary. Molecular markers are used most intelligently when they address controversial areas or when they are employed to analyze problems in natural history and evolution that have proven beyond the purview of traditional nonmolecular observation.

2

History of Molecular Phylogenetics

*The field . . . —molecular evolution—does not have clas-
sification as its primary objective, relevant though molecular data may be to that end.*

R.K. Selander, 1982

From its inception in the latter half of this century, the field of molecular evolution has been dominated by a series of fundamental controversies about the nature and evolutionary significance of genetic variation. Underlying these debates (discussed later) were exciting conceptual issues that understandably captured the attention of molecular evolutionists. However, to a considerable degree these controversies also directed attention away from what often were perceived as more mundane applications of molecules as genetic markers. Thus, prior to the mid-1980s, with a few notable exceptions, most applications of molecular markers in areas of natural history or phylogenetic estimation were viewed as ancillary by-products of research programs whose main goals were to uncover genetic mechanisms or elucidate broad evolutionary processes. Only in the last few years has molecular phylogenetics per se begun to assume its appropriate status as an essential scientific discipline, both empirically rich and conceptually challenging. What were these broad controversies in evolutionary biology that so dominated attention, relegating molecular phylogenetics to the back burner? These historical debates are the primary subject of this chapter.

DEBATES AND DIVERSIONS FROM MOLECULAR SYSTEMATICS

Empiricism and the Classical-Balance Debate

CLASSICAL VERSUS BALANCE VIEWS OF GENOME STRUCTURE

One long-standing controversy concerned the magnitude of genetic variability characterizing animal and plant genomes. Evolution has been defined as a change in genetic composition of populations through time (Dobzhansky, 1937). Genetic variation is prerequisite. Understandably, a central challenge for empirical population genetics always has been to measure genetic variability, under the rationale that such quantification would enlighten understanding of the operation of natural selection, drift, or other evolutionary forces. Unfortunately, the variation routinely observable at the phenotypic level seldom can be characterized in population genetic terms (e.g., the numbers of variable loci and the alleles responsible). This problem of empirical insufficiency plagued population genetics throughout the first half of this century, as evidenced by the development of diametrically opposed scientific opinions about the magnitude of genetic variability in nature. Advocates of the "classical school" maintained that genetic variability in most species was low, such that conspecific individuals were homozygous for the same "wild-type" allele at nearly all genetic loci. Proponents of the "balance view" maintained that genetic variation was high, such that most loci were polymorphic and individuals typically heterozygous at a large fraction of genes (Fig. 2.1).

Numerous corollaries and ramifications stem from these opposing schools of thought (Lewontin, 1974). Under the classical view, natural selection was seen as a purifying agent, cleansing the genome of inevitable mutational variation. Deleterious recessive alleles in heterozygotes might temporarily escape elimination, but were prevented from reaching high frequencies in populations because of their negative fitness consequences when homozygous. The classicists did not deny adaptive evolution but felt that the process was attributable to rare, selectively advantageous mutations that quickly swept through a species to become the new wild-type alleles. Because little variability existed to be shuffled into new multilocus allelic combinations, recombination was viewed as a relatively insignificant process. Furthermore, any genetic differences discovered between populations or species must be of profound importance (because of the low within-population component of variation). Central to the classical school was the concept of genetic load (see Wallace, 1970, 1991)—the idea that genetic variability produced a heavy burden of diminished fitness, which in the extreme might even cause population extinction. This perception of genetic variation as a curse was summarized forcefully by Muller (1950), who predicted, from ge-

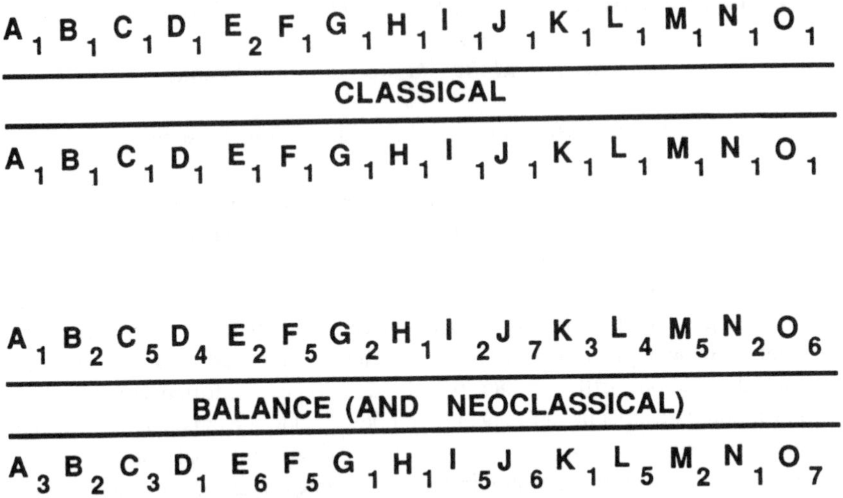

$$A_1 \quad B_1 \quad C_1 \quad D_1 \quad E_2 \quad F_1 \quad G_1 \quad H_1 \quad I_1 \quad J_1 \quad K_1 \quad L_1 \quad M_1 \quad N_1 \quad O_1$$

CLASSICAL

$$A_1 \quad B_1 \quad C_1 \quad D_1 \quad E_1 \quad F_1 \quad G_1 \quad H_1 \quad I_1 \quad J_1 \quad K_1 \quad L_1 \quad M_1 \quad N_1 \quad O_1$$

$$A_1 \quad B_2 \quad C_5 \quad D_4 \quad E_2 \quad F_5 \quad G_2 \quad H_1 \quad I_2 \quad J_7 \quad K_3 \quad L_4 \quad M_5 \quad N_2 \quad O_6$$

BALANCE (AND NEOCLASSICAL)

$$A_3 \quad B_2 \quad C_3 \quad D_1 \quad E_6 \quad F_5 \quad G_1 \quad H_1 \quad I_5 \quad J_6 \quad K_1 \quad L_5 \quad M_2 \quad N_1 \quad O_7$$

Figure 2.1. Classical versus balance views of genome structure. Shown for each case are two homologous chromosomes with genes A–O exhibiting alleles indicated by subscripts.

netic load calculations, that only 1 locus in 1000 would prove to be heterozygous in a typical human individual.

On the contrary, the balance school viewed natural selection as favoring genetic polymorphisms through balancing mechanisms such as fitness superiority of heterozygotes (Dobzhansky, 1955), variation in genotypic fitness among habitats, or frequency-dependent fitness advantage (Ayala and Campbell, 1974). Genetic variability was thought to be both ubiquitous and adaptively relevant. Deleterious alleles were not ruled out, but these were held in check by natural selection and contributed little to heterozygosity. Because of the high variability predicted for sexually reproducing species, no allele could properly be termed wild-type. Genetic recombination, therefore, assumed a greater significance than de novo mutation in producing interindividual fitness variation from one generation to the next. Furthermore, genetic differences among populations were perhaps of less import because of the large within-population component of overall variability. How much genetic variation was predicted under the balance view? Wallace (1958) raised a proposal that seemed extreme at the time, but not at all unreasonable today: "the proportion of heterozygosis among gene loci of representative individuals of a population tends towards 100 percent."

The balance hypothesis gained indirect support from several lines of evidence: (a) extensive phenotypic variation in most natural populations, which in a few well-studied species often proved to have a genetic basis and to be of adaptive relevance (e.g., Ford, 1964); (b) a genetic basis for many naturally occurring morphological variants and fitness characters in populations that could be ma-

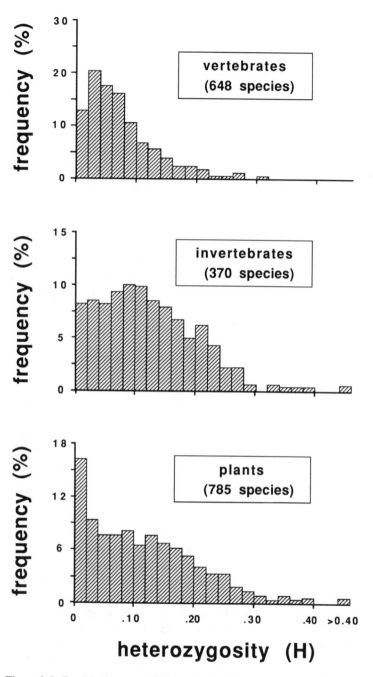

Figure 2.2. Empirical heterozygosity estimates from allozymes. Shown are frequency distributions of mean *H* per species derived from protein-electrophoretic surveys conducted on 1803 species of vertebrate and invertebrate animals (after Ward et al., 1992) and plants [data originally tabulated for a review by Hamrick and Godt (1989) and kindly provided by J. Hamrick]. An average of more than 20 loci was scored per study.

nipulated experimentally (e.g., by inbreeding, or by use of "common garden" experiments in which the fraction of phenotypic differentiation attributable to genetic variation could be estimated by controlling for environmental differences); and (c) the genetic response to artificial selection exhibited by many domesticated animals and plants for numerous traits (see the review in Ayala, 1982a). However, none of these or related observations allowed direct answers to the fundamental question: what fraction of genes is heterozygous in an individual and polymorphic in a population? An answer requires that variability be assessed at multiple independent loci, chosen without bias with respect to magnitude of variability. But this requirement introduces a Catch-22 for any genetic appraisal based on traditional Mendelian genetic approaches: Genes underlying a phenotypic trait can be identified only when they carry segregating polymorphisms. In other words, because genetic assignments for phenotypic features were inferred from the segregation patterns of allelic variants, invariant loci escaped detection and no accumulation of such data could provide an uncolored estimate of overall genetic variability. Other means were required to screen genetic variation more directly, and in a manner that allowed assay of an unbiased sample of polymorphic and monomorphic loci.

MOLECULAR INPUT TO THE DEBATE

A fundamental breakthrough occurred in 1966, when independent research laboratories published the first estimates of genetic variability based on multilocus protein electrophoresis (Harris, 1966; Johnson et al., 1966; Lewontin and Hubby, 1966). This method involves separation of non-denatured proteins by net charge under influence of an electric current, followed by application of histochemical stains to reveal the enzyme or other protein products of particular, specifiable genes (Chapter 3). Because proteins are revealed whether or not they vary, this approach provided the first valid attempt to obtain an unbiased estimate of genome variability at a reasonable number (usually 20–50) of genetic loci. The empirical results were clear—fruit fly and human genomes harbored a wealth of variation, with 30% or more of assayed genes polymorphic, and roughly 10% of loci heterozygous in a typical individual (Box 2.1). Over the next two decades, multilocus electrophoretic surveys conducted on hundreds of plant and animal species revealed levels of genetic variation that normally were high but also variable among species (Fig. 2.2; see the reviews in Hamrick and Godt, 1989; Nevo, 1978; Powell, 1975; Ward et al., 1992).

Protein electrophoretic techniques were not entirely new in 1966—indeed, crude methods had been available for nearly 30 years (see Brewer, 1970). Rather, the scientific impact of the landmark electrophoretic surveys lay primarily in the manner in which methods and data from the field of molecular biology were applied for the first time to long-standing issues in the previously separate

---

Box 2.1. Measures of Genetic Variability within a Population

For multilocus protein electrophoretic data, one useful measure of genetic diversity is population heterozygosity (H), defined as the mean percentage of loci heterozygous per individual (or equivalently, the mean percentage of individuals heterozygous per locus). Estimates of H can be obtained by direct count from a raw data matrix, the body of which consists of observed diploid genotypes as in the following hypothetical example involving eight allozyme loci (A–H) scored in each of five individuals:

				Locus (j)					
Individual (i)	A	B	C	D	E	F	G	H	h_i
1	aa	aa	aa	aa	a̲b̲	aa	aa	aa	0.125
2	bb	a̲b̲	a̲b̲	a̲b̲	bb	aa	a̲b̲	aa	0.500
3	cc	a̲c̲	b̲c̲	b̲d̲	dd	a̲d̲	c̲d̲	b̲c̲	0.750
4	aa	aa	aa	aa	aa	aa	aa	aa	0.000
5	cc	cc	cc	cc	cc	cc	cc	cc	0.000
h_j	0.0	0.4	0.4	0.4	0.2	0.2	0.4	0.2	$H = 0.275$

Here, diploid genotypes are indicated by lowercase letters (each letter representing an electrophoretic allele), and heterozygotes have been underlined. In this example, 11 of 40 assayed genotypes are heterozygous ($H = 0.275$). Equivalently, H may be interpreted as the mean of the row or column totals, which represent direct-count heterozygosities for single individuals (h_i) or single loci (h_j), respectively. Heterozygosities also may be estimated from observed frequencies of alleles (rather than genotypes), assuming the population is in Hardy–Weinberg proportions. Thus $h_j = 1 - \Sigma q_k^2$, where q_k is the frequency of the kth allele. Other common measures of population variability for allozyme data are the mean number of alleles per locus and the percentage of polymorphic loci (P), which is 0.6 in the above example. To avoid an expected positive correlation between P and sample size, a locus usually is considered polymorphic only if the frequency of the most common allele falls below an arbitrary cut-off, typically 0.99 or 0.95. In most protein-electrophoretic surveys, estimates of H occur in the range of 0.0–0.2 (Fig. 2.2) and estimates of P range from 0.0 to 0.80.

For DNA-level data (which typically involve restriction sites or sequences along a particular stretch of DNA that can be thought of as a "locus"), a useful statistic summarizing heterozygosity at the nucleotide level is nucleotide diversity (Nei and Li, 1979; Nei and Tajima, 1981), or the mean sequence divergence between haplotypes (alleles): $p = \Sigma f_i f_j p_{ij}$, where f_i and f_j are the frequencies of the ith and jth haplotypes in the population, and p_{ij} is the sequence divergence between these haplotypes. Another informative measure is haplotype diversity: $h = 1 - \Sigma f_i^2$. This measure is a DNA-level analogue of the h for protein-electrophoretic data because its calculation entails no assessment of the magnitude of genetic divergence between the alleles involved. Depending on the loci and species surveyed, nucleotide diversities within a population typically fall in the range 0.0005–0.020 (Stephan and Langley, 1992), and haplotype diversities sometimes approach 1.0 for rapidly evolving genomes such as animal mtDNA (Avise et al., 1989).

discipline of population genetics. After the mid-1960s, contacts between molecular and population genetics were only to expand, and at an ever-faster pace.

The data and conceptual orientations of the early protein electrophoretic (allozyme) studies were to exert overriding influence on research goals in population genetics for the next quarter century. Although this work provided important information on genetic variation in natural populations, it also inhibited a stronger molecular phylogenetic perspective in at least two ways. First, allozyme methods revealed for each locus allelic products (electromorphs) that represent qualitative multistate characters, the phylogenetic order of which cannot safely be inferred from the observable property, band mobility on the gel. Much of traditional population genetic theory, built around the seminal works of Fisher (1930), Haldane (1932), and Wright (1931), can be couched in the language of the expected frequency dynamics of such phylogenetically unordered alleles under the separate or joint evolutionary forces of natural selection, genetic drift, gene flow, mutation, and recombination. Protein electrophoretic approaches produced data that could be interpreted within the perspectives and language of traditional population genetics theory, but the net effect was to channel thought away from the phylogenetic perspectives which increasingly characterized other areas of evolutionary biology. Thus, 20 years after the protein-electrophoretic revolution began, Lewontin (1985) concluded, "population genetics is conceptually the study of gene lineages, [but] until now, the data to study such lineages have not really been available."

Second, by focusing on issues of genetic variation per se (rather than on the possible phylogenetic content of the molecular information), the seminal protein electrophoretic surveys launched an intensive and still-continuing quest to refine and interpret empirical estimates of molecular variability. The field of empirical population genetics, thus, became preoccupied with several nonphylogenetic questions, as described next.

RESEARCH PREOCCUPATIONS OF THE PROTEIN-ELECTROPHORETIC REVOLUTION

(a) How much protein variation remained hidden beyond the resolving power of conventional gel electrophoresis? To address this question of cryptic polymorphism, two general experimental protocols were followed. In "backward experiments" (Selander and Whittam, 1983), protein variants of known amino acid sequence were electrophoresed to determine the proportion of known alleles detectable (Ramshaw et al., 1979). In "forward experiments," assay conditions were varied in attempts to discriminate additional alleles within the electromorph classes identified in the original tests. Such assays involved the use of varied buffers or other electrophoretic running conditions, "gel-sieving" through acrylamide matrices of differing pore size, thermostability analyses, and miscellaneous other biochemical techniques. These approaches frequently uncovered hidden protein variants (particularly at loci that were polymorphic in the initial assays), and left the general impression that the original electrophoretic methods had revealed only the tip of the genetic variability iceberg (Aquadro and Avise, 1982a, 1982b; Ayala, 1982b; Bernstein et al., 1973; Bonhomme and Selander, 1978; Coyne, 1982; Johnson, 1976a, 1977; McDowell and Prakash, 1976; Milkman, 1976; Prakash, 1977). Unfortunately, data from some of these biochemical methods were difficult to interpret because the particular genetic bases of the polymorphisms seldom were verifiable directly from the assays. Primarily for this reason, none of these ''refined'' methods proved widely useful as a source of genetic markers for population biology.

(b) How representative of other genes were the variability estimates derived from protein electrophoretic loci? Because the availability (and cost) of histochemical stains were deciding criteria for inclusion of proteins in most electrophoretic surveys, dehydrogenases and other enzymes of the glycolytic pathway and citric acid cycle were represented disproportionately. An initial concern was whether variability at the loci encoding these proteins might be misrepresentative of other protein-coding genes. For a brief time in the late 1970s and early 1980s, attention was directed to the assay of abundant membrane-associated, ribosomal, and other structural proteins revealed by nonspecific protein stains, using newly introduced two-dimensional gel techniques (which separate proteins on the basis of charge by isoelectric focusing in the first dimension, and then by molecular weight in the second dimension) (O'Farrell, 1975). Results from several species indicated somewhat lower heterozygosities than had been estimated from the original protein electrophoretic surveys (Aquadro and Avise, 1981; Leigh Brown and Langley, 1979; Racine and Langley, 1980; Smith et al., 1980). However, any lingering thoughts that genomes might lack appreciable variation were dispelled conclusively with the explosion of data on restriction fragment-length polymorphisms (RFLPs) and DNA sequences that already had begun to revolutionize the field by the early 1980s. One legacy of the genomic variability debate is that an important branch of population genetics still focuses on global estimates of genomic variation per se as a means of attempting to understand the evolutionary forces governing genome structure (Gillespie, 1987; Kimura, 1991; Kreitman, 1987; Li, 1978; Ohta and Tachida, 1990).

(c) How was genetic variability related to organismal fitness? Considering the electrophoretic results in the historical context of the debate between the classical and balance schools, it is hardly surprising that many population geneticists turned empirical research efforts to the problem of how natural selection might maintain so much protein polymorphism. As described next, allozyme researchers attacked this problem in several ways.

EMPIRICAL APPROACHES TO THE VARIABILITY/FITNESS CONUNDRUM

In the ''multilocus'' approach, searches were launched for correlations between mean overall heterozygosity and organismal life history attributes or fitness components. One widely addressed issue was whether protein variability might be correlated with environmental heterogeneity (Hedrick, 1986; Levene, 1953; Soulé and Stewart, 1970). Some intriguing associations were reported. For example, Nevo and Shaw (1972) attributed low heterozygosity in burrowing mole rats to selection for homozygosity in the supposedly constant and narrow subterranean niche. Selander and Kaufman (1973a) suggested that genic heterozygosity generally was highest in small, immobile animals that perceive environments as coarse-grained patches of alternative habitat (Levins, 1968) and lowest in large mobile animals that perceive environments as fine grained. Smith and Fujio (1982) concluded that heterozygosities in marine fishes were correlated positively with degree of habitat specialization. Powell and Taylor (1979) summarized evidence that environmental heterogeneity in conjunction with habitat choice contributed to genotypic diversity, whereas Valentine and Ayala (1974) favored an environmental selection model consistent with an observed correlation in marine invertebrates between low genetic variability and temporal trophic resource stability (Ayala et al., 1975a; Valentine, 1976). Two influential studies using experimental cages of fruit flies reported significantly higher heterozygosities in populations maintained under variable as opposed to uniform environmental regimes (McDonald and Ayala, 1974; Powell, 1971).

On the other hand, Sage and Wolff (1986) suggested that differing levels of genic heterozygosity in large mammals were attributable not to varying environmental selection pressures per se, but rather to environment-dependent population histories and effects of genetic drift—species in glaciated regions tended to have lower variabilities than their counterparts in temperate and tropical regions, purportedly due to population bottlenecks accompanying serial recolonizations of northern latitudes following retreats of the Pleistocene glaciers. In general, unless species are normally at genetic equilibrium (which seems unlikely), the extant standing crop of genetic variation must be a function both of the genetic diversity originally available to a species (its phylogenetic legacy) and of additional processes such as selection, gene flow, and the mating system that govern how that available variation is partitioned within and among populations.

Positive correlations also were noted between genetic variability and particular life history attributes, such as short generation time, small maximum body size, and small egg size in bony fishes (Mitton and Lewis, 1989; but see Waples, 1991), and high fecundity, outcrossing mode of reproduction, pollination by wind, and long generation time in plants (Hamrick et al., 1979). Among conspecific organisms, correlations were reported between individual heterozygosity and a variety of phenotypic characters presumably associated with fitness (see the review in Mitton, 1993): exploratory behavior in mice (Garten, 1977); antler characteristics in deer (Scribner and Smith, 1990); shell shape in blue mussels (Mitton and Koehn, 1985); growth rate in fishes, salamanders, oysters, trees and other species (Ferguson, 1992; Garton et al., 1984; Koehn et al., 1988; Ledig et al., 1983; Mitton and Grant, 1984; Pierce and Mitton, 1982; Singh and Zouros, 1978); herbivory resistance in pines (Mopper et al., 1991); disease resistance in trout (Ferguson and Drahushchak, 1990); and developmental stability in many species, as supposedly evidenced by lower phenotypic variance between individuals (Lerner, 1954; Zink et al., 1985) or by lower "fluctuating asymmetry" (the difference between bilateral features) within individuals (Allendorf and Leary, 1986; Leary et al., 1985; Palmer and Strobeck, 1986; Van Valen, 1962). Of course many of these physiological and developmental characteristics likely are interrelated.

Another aspect of the multilocus approach involved searches for molecular or metabolic features correlated with heterozygosity. Among the examined factors arguably associated with genic variability were the following: molecular size of the enzyme (Eanes and Koehn, 1978a); quaternary structure (Solé-Cava and Thorpe, 1989; Ward, 1977; Zouros, 1976); level of intragenic recombination (Koehn and Eanes, 1976); physiological role in regulating flux through metabolic pathways (Johnson, 1976b); enzymatic action on intracellular versus extracellular substrates (Ayala and Powell, 1972a; Gillespie and Langley, 1974; Kojima et al., 1970); and others (reviews in Koehn and Eanes, 1978; Selander, 1976).

Several difficulties accompanied attempts to interpret such multilocus associations, beyond the obvious point that correlation by itself cannot prove causality. First, there likely is a reporting bias in favor of positive correlations, and the number of variables that can be examined essentially is limitless. Second, mean heterozygosity as estimated from a small number of protein loci may not accurately rank-order specimens within a population with respect to genome-wide variability (Chakraborty, 1981; Mitton and Pierce, 1980) [unless, perhaps, individuals vary dramatically along an outbred-inbred continuum, due to demographic cycles, fine demic structure, or mating behaviors (Mitton, 1993; Scribner, 1991; Smith et al., 1975; Smouse, 1986)]. This point led some authors to conclude that associations of individual heterozygosity with fitness were attributable not to differing levels of genome-wide variation, but rather to physiological advantages stemming from heterozygosity at the particular glucose-metabolizing or other enzymes under survey [or perhaps to tightly linked genes in the chromosomal blocks that they mark (Koehn et al., 1983; Mitton and Grant, 1984)]. Third, several of the associations with heterozygosity listed above involved weak trends for which exceptions could readily be cited, or alternative explanations advanced. For example, high genetic variability characterizes some species inhabiting proverbially "stable" environments such as the deep sea, and caves or other subterranean settings, and low genetic variability certainly can result from demographic population contractions in any environment (Avise and Selander, 1972).

Ward et al. (1992) recently updated the correlational approach to the study of allozyme heterozygosity. From a literature survey of over 1000 animal species, they concluded that approximately 21–34% of the variance in mean protein heterozygosity could be attributed to taxonomic effects (e.g., fishes tend to have low H values, amphibians the highest such values), and 41–52% of the variance was due to protein effects (including subunit size, subunit number, enzyme function, etc.). No comparable summary appears to be available for plants. Nonetheless, with regard to postulated molecular or physiological correlates of multilocus heterozygosity, the conclusion reached in 1976 by Selander still stands: ". . . notwithstanding the immense amount of effort expended in surveying variation in organisms in the last decade, the sample sizes of loci are generally inadequate for satisfactory analyses of the variation: molecular heterogeneity is too great." The numerous correlations involving heterozygosity listed above remain intriguing, but largely unexplained.

Frustration with such multilocus approaches to assessing natural selection's role in maintaining genetic variability led other researchers to the "vertical" or "single-locus" approach, wherein particular polymorphisms were studied at multiple levels ranging from biochemistry, physiology, and developmental expression to transmission patterns, population dynamics and ecological associations (Clarke, 1975; Koehn and Hilbish, 1987; McDonald, 1983). Intensive studies of several such model systems all uncovered convincing evidence for differences between allozyme genotypes upon which selection probably operates (Table 2.1). Unfortunately, relatively few polymorphisms have been analyzed so intensively. Furthermore, most of the genes studied had been identified a priori as likely candidates for natural selection, and thus the polymorphisms analyzed by the vertical approach probably comprise a biased sample with regard to the issue of selective maintenance.

Genetic Theory and the Neutralist–Selectionist Debate

The discovery of extensive molecular variation did not clinch the case for the philosophical perspective on genetic variation embodied in the balance school of thought, but instead stimulated development of an alternative explanation for molecular genetic variability that was to assume a prominent role in population genetics to the present time. Under the strict neutral mutation theory, alternative alleles confirm no differential fitness effects on their bearers. As summarized by Kimura (1991), "the great majority of evolutionary mutant substitutions at the molecular level are caused by random fixation, through sampling drift, of selectively neutral (i.e., selectively equivalent) mutants under continued mutation pressure." As applied to intraspecific molecular variability, neutrality theory predicts that polymorphisms are maintained by a balance between mutational input and random allelic extinction by genetic drift. Neutralists did not deny the existence of high molecular variability, but rather questioned its relevance to

Table 2.1 Examples of major research programs on allozyme polymorphisms that employed the "vertical" approach[a]

Protein	Organism	Evidence	Introductory references
Alcohol dehydrogenase	*Drosophila*	Kinetic differences between isozymes associated with differences in survivorship, developmental time, and environment	Aquadro et al., 1986; Clarke, 1975; van Delden, 1982.
α-glycerophosphate dehydrogenase	*Drosophila*	Kinetic differences correlated with flight metabolism, power output, and environmental temperature	Miller et al., 1975; O'Brien and MacIntyre, 1972; Oakeshott et al., 1982.
Carboxylesterase	*Drosophila*	Differences in enzyme activity associated with reproduction	Gilbert and Richmond, 1982; Richmond et al., 1980.
Glucose-6 phosphate dehydrogenase and 6-phosphogluconate dehydrogenase	*Drosophila*	Differences in metabolic flux associated with differences in fitness	Barnes and Laurie-Ahlberg, 1986; Cavener and Clegg, 1981; Hughes and Lucchesi, 1977.
Glucosephosphate isomerase	*Colias* butterflies	Kinetic differences correlated with mating success and survivorship	Watt, 1977; Watt et al., 1983, 1985.
Glutamate pyruvate transaminase	*Tigriopus* copepods	Enzyme activity differences associated with differential responses to hyperosmotic stress	Burton and Feldman, 1983.
Lactate dehydrogenase	*Fundulus* fishes	Differences in kinetic and other biochemical properties associated with differences in metabolism and fitness	DiMichele et al., 1986, 1991; Place and Powers, 1979, 1984.
Leucine aminopeptidase	*Mytilus* mollusks	Enzyme activity differences associated with osmoregulation and fitness	Hilbish and Koehn, 1985; Hilbish et al., 1982; Koehn and Immerman, 1981.

[a] In each case, kinetic differences demonstrated between allelic products have suggested that the polymorphisms may be maintained by natural selection.

organismal fitness. Because neutralists and classicists share the position that balancing selection plays little role in maintaining molecular polymorphism and that most selection is directional or "purifying" against deleterious alleles, neutrality theory also has been referred to as the neoclassical theory (Lewontin, 1974).

Several points should be made clear at the outset. First, neutralists do not suggest that most genes or allelic products are dispensable (of course they are not). Rather, they propose that different alleles are functionally equivalent such that organismal fitness is not a function of the particular genotypes possessed. Second, neutralists do not deny that many de novo mutations are deleterious and eliminated by purifying selection. Rather, the focus is on the supposed neutrality of segregating polymorphisms that escape selective elimination. Indeed, one cornerstone of neutrality theory is that nucleotide positions or genic regions that are functionally less constrained are those most likely to harbor neutral variation and to exhibit the most rapid pace of allelic substitution. Third, neutralists do not challenge the Darwinian mode of adaptive evolution for organismal morphologies and behaviors [although some important recent extensions of the neutrality theory do propose a significant role for genetic drift in organismal evolution as well (Kimura, 1990)]. Rather, neutrality theory developed in response to the intellectual challenge provided by the unexpectedly high levels of molecular variability observed.

Neutrality concepts were introduced in the late 1960s (Kimura, 1968a, 1968b), and gained immediate widespread attention due in part to a paper by King and Jukes (1969) provocatively titled: "Non-Darwinian evolution: random fixation of selectively neutral mutations." Indeed, the theory did challenge a prevailing approach of naively extending to molecular biology the neo-Darwinian views on the adaptive significance of nearly all organismal differences (see Gould and Lewontin, 1979). It is quite remarkable that within a decade, and continuing today, neutrality theory gained sufficient acceptance to be viewed widely as molecular evolution's gigantic "null hypothesis"—the simplest and most straightforward way to interpret molecular variability, and the hypothesis whose predictions were to be falsified before alternative proposals involving balancing selection could be entertained seriously. This is not to say that the selectionist-neutralist debate is fully resolved.

The neutrality school has strong roots in the quantitative tradition of theoretical population genetics developed earlier in the century (Fisher, 1930; Haldane, 1932; Wright, 1931). An elegant and elaborate theory predicts the amount of genetic variability within a given population as a function of mutation rate, gene flow (where applicable), and population size (Kimura and Ohta, 1971). Conspicuously absent from the calculations are selection coefficients, because alleles are assumed to be neutral. Under strict neutrality theory, molecular variability is a function of the neutral mutation rate and the evolutionary effective population

Box 2.2 Effective Population Size

Not all individuals in a population contribute gametes to the next generation with equal probability. This has led to the concept of effective population size (N_e), originally due to Wright (1931). The effective number of individuals refers to the size of an idealized population that would have the same genetic properties (such as the intergeneration variance in allele frequencies due to chance sampling error) as that observed for the real population. Usually, N_e is much smaller than N (the census size) for one or more of the following reasons:

(a) *Separate sexes:* In organisms with separate sexes, one gender may be more common than the other. Let N_m and N_f be the census numbers of males and females in such a population. Then the effective population size due to this disparity alone is $N_e = 4N_m N_f/(N_m + N_f)$. Unless $N_m = N_f$, this equation shows that N_e is less than the total census count ($N_m + N_f$).

(b) *Fluctuations in population size:* Most populations in nature probably fluctuate greatly in size, due to diseases, changes in habitat quality, predation, etc. The effective population size due to such fluctuations is equal to the harmonic mean of the breeding population sizes across generations. A harmonic mean is a function of the mean of reciprocals, or in this case $N_e = n/[\Sigma(1/N_i)]$, where N_i is the population size in the ith generation and n is the number of generations. A harmonic mean is closer to the smaller rather than to the larger of a series of numbers being averaged, so N_e can be much lower than most population censuses. A severe reduction in population size is called a "population bottleneck" and can greatly depress evolutionary N_e.

(c) *Combination of separate sexes and fluctuating population size:* If census population sizes of males and females are known across multiple generations, the joint effects of the above factors on N_e can be determined. For each generation, the census sizes of males and females are converted to an effective size for that generation, as in (a). Then the equation in (b) is employed to take the harmonic mean of the single-generation estimates.

(d) *Variation in progeny numbers:* Even in a nonfluctuating population with equal numbers of males and females, some individuals may leave many more progeny than others, creating a large variance across families. Only when offspring numbers follow a Poisson distribution with mean (and hence variance) of 2.0 per parent, does $N_e = N$. In more realistic situations, where the variance often exceeds the mean, N_e is smaller than the census breeding population size (Crow, 1954). Hedgecock et al. (1992) have argued that organisms with extremely high fecundities are particularly prone to gross disparities between N_e and N due to high variability in fertility across individuals; empirically, such disparities indeed have been reported for several aquatic and marine species such as shellfish (Hedgecock and Sly, 1990).

(e) *Other factors:* Various other factors also can reduce N_e relative to N. For example, in a species composed of many subpopulations each of which is subject to periodic extinction and recolonization, the species as a whole will have a far lower N_e than might have been predicted had only composite census sizes for particular generations been available (Maruyama and Kimura, 1980).

size, N_e (Box 2.2). For example, the heterozygosity expected for electrophoretically detectable alleles at equilibrium between mutation and genetic drift is given by

$$H = 1 - 1/(1 + 8N_e\mu)^{1/2}, \qquad (2.1)$$

where μ is the per locus per generation mutation rate to neutral alleles (Ohta and Kimura, 1973). Figure 2.3 plots this expected relationship between H and N_e for reasonable neutral mutation rates and also shows the range of allozyme heterozygosities empirically observed for numerous animal species with indicated population census sizes. Such comparisons should deal with species effective sizes because the theory involves equilibrium expectations over long-term evolution.

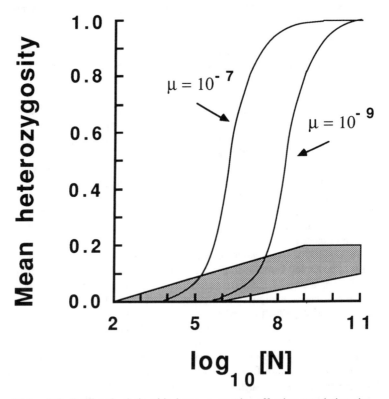

Figure 2.3. Predicted relationship between species effective population size and protein-electrophoretic heterozygosity under neutrality theory. Expectations for two neutral mutation rates (μ) are presented. Also shown (shaded area) are observed heterozygosities for numerous animal species as a function of current-day population size (N, shown logarithmically as powers of 10) (after Soulé, 1976).

Except perhaps for some of the least abundant species, observed values of H have proved to be much lower than neutrality theory predicts. This conclusion generally holds even when mutations are assumed to be mildly deleterious (Nei, 1983). One likely explanation for the relative paucity of genetic variation is that long-term effective population sizes for most species are vastly smaller than might otherwise be supposed from current-day census sizes.

This raises a remarkable irony about the neutralist–selectionist debate that stems from historical precedents of the classicist–balance controversy. When protein variation was uncovered in the seminal electrophoretic surveys, selectionists interpreted the observations as consistent with the balance view and sought (as described earlier) to discover the selective forces responsible for such extensive polymorphism. At the same time, neutralists were facing a dilemma of how to account for the *paucity* of polymorphism relative to neutrality expectations, given mutation rates and population sizes thought to characterize most species. The dearth of variation from the neutralist perspective extends to the level of some DNA sequences as well (Box 2.3). For example, with regard to mitochondrial DNA alleles (which are maternally inherited), the expected mean time to common ancestry under neutrality theory is approximately

$$G = N_{f(e)}, \qquad (2.2)$$

where G is the number of generations and $N_{f(e)}$ is the effective population size of females (Avise et al., 1988). Figure 2.4 plots values of $N_{f(e)}$ for a number of vertebrate and invertebrate species as estimated from observed mtDNA haplotype distances (Avise, 1992), using conventional evolutionary rate calibrations for the mtDNA molecule (Brown et al., 1979). Most observed values fall orders of magnitude below theoretical expectations based on neutrality theory and present-day census population sizes (N_f values). In other words, despite extensive genetic heterogeneity, mtDNA diversity typically is much lower than neutrality theory predicts. Either mtDNA evolution is slower than generally believed or evolutionary effective population sizes are vastly lower than are present-day population sizes for most species. In summarizing these types of observations, Nei and Graur (1984) concluded that ''. . . polymorphism is actually much lower than the neutral expectation and that if the bottleneck effect is not sufficient for explaining the observed level, the type of selection to be considered is not diversity-enhancing selection but diversity-reducing selection.'' The irony of this neutralist perspective in the history of the classical-balance debate still is not appreciated widely by many of the proponents of balancing selection.

Another important aspect of the neutral mutation theory concerns predictions about molecular evolutionary rate. Two aspects of rate must be distinguished carefully. *With regard to shifts in frequencies of preexisting alleles,* the rate of neutral evolution can be greater in small populations. Genetic drift refers to random changes in allele frequency due to sampling variation of gametes from

Box 2.3. Mean Times to Shared Allelic Ancestry

Another way to formulate neutrality theory regarding the association between genetic variability and population size is through consideration of the expected frequency distribution of times to common ancestry among alleles. Imagine an idealized population with nonoverlapping generations and large constant size N. Suppose further that in each generation, individuals contribute to a gamete pool from which $2N$ nuclear gametes are drawn at random (effectively with replacement) to produce individuals of the next generation. The probability that two gametes carry copies of the same allele from the prior generation is $1/2N$. This is also the probability that the time to common ancestry of two alleles is one generation ago ($G = 1$). The probability that a pair of alleles is not identical from the prior generation is $1 - 1/2N$. Thus, the probability that these latter alleles trace to an identical copy two generations ago is $(1 - 1/2N)(1/2N)$. From an extension of such reasoning, the probability that two randomly chosen alleles derive from a common ancestral allele that existed G generations ago is

$$f(G) = (1/2N)(1 - 1/2N)^{G-1}, \text{ or approximately } (1/2N)e^{-(G-1)/2N}.$$

These equations give the probability distribution of times to common ancestry in terms of the number of generations (Tajima, 1983). The distribution is geometric, with mean approximately $2N$. The mean time to shared haplotype ancestry for mtDNA genes can be derived similarly (Avise et al., 1988), but is only one-fourth as large as for nuclear genes, the difference being attributable to a twofold effect due to the haploid transmission of mtDNA and another twofold effect due to mtDNA's normal pattern of uniparental transmission through females. The above theory assumes that times to common ancestry for allelic pairs are independent. Therefore, in interpreting empirical data for any particular species against these expectations, caution must be exercised because the history of lineage coalescence within a real population imposes a severe correlation on the pairwise comparisons (Ball et al., 1990; Felsenstein, 1992; Hudson, 1990; Slatkin and Hudson, 1991).

generation to generation and is a special case of the more general phenomenon of sampling error, which is inversely related to sample size. However, *with regard to the origin and substitution of new alleles,* the rate of neutral evolution is independent of population size and depends only on the mutation rate to neutral alleles.

This latter conclusion can be demonstrated as follows. In a diploid population of size N, there are $2N$ allelic copies of each nuclear gene. In time, the descendents of only one of these copies will remain (i.e., is destined for fixation). The chance that any newly arisen neutral mutation will undergo random fixation is simply $1/2N$. On the other hand, the probability that a new neutral mutation arises in a population is $2N\mu$, where μ is again the mutation rate to neutral alleles. It follows that the rate of fixation of new neutral mutations is the product of the origination rate of mutations and their probabilities of fixation once present, or $2N\mu \times 1/2N = \mu$. In other words, the rate of substitution in evolution under strict

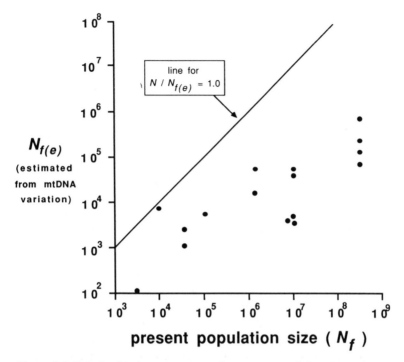

Figure 2.4. Relationship between current-day census population size and evolutionary effective population size, as estimated from empirical mtDNA nucleotide diversities for several marine species in the southeastern United States (after Avise, 1992). Both axes are in logarithmic scale.

neutrality equals the rate of mutation to neutral alleles. This simple conclusion is the theoretical basis for the neutrality prediction that biological macromolecules can provide standard "molecular clocks" (Chapter 4), irrespective of population size.

The selection–neutrality controversy has dominated both the theoretical and empirical sides of population genetics for the last 25 years (see the reviews in Avise, 1977a; Ayala, 1976a; Lewontin, 1974, 1991; Nei and Koehn, 1983), and the debate is not yet settled, for at least two major reasons. First, both selectionist and neutralist theories are immensely powerful constructs in the sense of being capable of explaining nearly any set of observations by appropriate alteration of parameters and assumptions. Because of the multitudinous ways in which natural selection can operate, falsification of all selectionist scenarios for a given data set is nearly impossible (indeed this was a primary motivation for the development of a quantitative neutrality theory that specifies expectations explicitly). But neutrality theory also can yield a nearly limitless array of predictions by varying parameters that are notoriously difficult to measure in real populations [N_e, μ, and selection intensities against alleles that may be slightly deleterious or "nearly neutral" (Ohta, 1992a)]. Furthermore, many apparent departures from neutrality expectations might be due to unknown historical factors that no doubt remove populations and species from the equilibrium conditions that commonly are assumed in most neutral models.

A second probable reason why the selection–neutrality debate continues without final resolution is the difficulty of defining exactly what is meant by natural selection. For example, is the phenomenon of "meiotic drive" (whereby certain alleles appear to "cheat" during meiosis by distorting Mendelian segregation ratios in their favor) to be viewed as a form of natural selection at the gametic level? In general, are "selfish genes" (Dawkins, 1989) that compete for transmission within an organismal lineage to be interpreted as evolving under the influence of natural selection? Holmquist (1989) argues that the molecules inside a cell form an interacting community rather like an ecosystem. This "molecular interplay" is quite different from what traditionally has been meant by natural selection at the organismal level (Ohta, 1992b). Perhaps the concept of natural selection should be broadened to encompass evolutionary effects at hierarchical levels both below (Dawkins, 1989) and above (Gould, 1980) the level of differential fitness among individuals.

In any event, the final answer likely lies somewhere between the polarized neutralist and selectionist camps. Certainly some loci are under strong balancing or other forms of natural selection, whereas the segregating alleles at many other loci must have negligible differential influence on organismal fitness. Knowledge about the evolutionary forces governing the dynamics of molecular markers probably is much more important for some applications (e.g., phylogeny reconstruction) than it is for others (e.g., clonal identification and parentage assessment).

Systematic Philosophy and the Phenetic–Cladistic Debate

Another area of evolutionary research that diverted attention from the application of molecules as genetic markers centered around conflicting philosophical approaches to systematics. In the 1960s, a heated debate developed (between the pheneticists and the cladists) that was to dominate attention in the field of systematics for more than two decades. The philosophical differences involved, relevant though they may be to the field of molecular evolution, initially centered around interpretations of morphologic and other traditional systematic characters.

Until the mid-1900s, the science of classifying organisms involved the operational approach of comparative morphological assessment, introduced in crude form by Carl Linnaeus two centuries earlier (Linnaeus, 1759). Typically, specialists devoted years of study to a particular group such as birds or beetles, and on the basis of accumulated experience and gestalt classified their creatures into a hierarchical taxonomy. This approach contributed greatly to a cataloging of the tremendous diversity of the natural world and resulted in most of the biological classifications still followed today. Potential difficulties of this approach stemmed from the lack of unifying or standardized classification methods (either conceptual or operational) with the following consequences: (a) the centering of systematic authority within a small number of researchers for each taxonomic group; (b) the lack of formalized procedures for corroboration or refutation of a proposed classification; (c) the absence of a uniform measure by which classifications for different taxonomic groups might meaningfully be compared; and (d) the lack of a clear philosophical orientation on precisely which aspects of evolution were reflected in a particular classification.

Explicit concern with these shortcomings of traditional systematic practice prompted the rise of numerical taxonomy, or the phenetic approach to systematics (Sokal and Sneath, 1963). Pheneticists proposed that organisms should be grouped and classified according to overall similarity (or its converse, distance), as measured by defined rules preferably using as many organismal traits as possible. Among the principles guiding numerical taxonomy are the following (Sneath and Sokal, 1973): (a) the best classifications usually result from the analyses based on the largest numbers of characters; (b) at least at the outset, every character is to be afforded equal weight; (c) classifications are based on quantitative measures of overall (phenetic) similarity or distance between the taxa (operational taxonomic units or OTUs) under comparison; and (d) patterns of character correlations can be used to recognize distinct taxa and to draw systematic inferences under certain assumptions about evolutionary pathways and mechanisms. Operationally, numerical taxonomy involves the application of quantitative methods for estimating phenetic similarity, examining character correlations, and grouping OTUs. Philosophically, "numerical taxonomy aims to develop methods that are objective, explicit, and repeatable. . . ." (Sneath and Sokal, 1973).

The development of numerical taxonomy provided a valuable service to science by opening for scrutiny traditional practices in systematics that had been needlessly opaque. Nonetheless, pheneticists were attacked on several fronts, most notably by cladists who proposed an alternative philosophy and protocol for phylogeny reconstruction and classification (Eldredge and Cracraft, 1980). Under tenets of the cladistic school, organismal relationships cannot be deduced from overall similarity, but *are* reflected in a particular subset of similarity attributable to synapomorphic or shared-derived traits (Box 2.4; Fig. 2.5). Cladists focus almost exclusively on assessing the branch-splitting component of evolutionary trees (cladogenesis) rather than branch lengths (accumulated change within lineages, or anagenesis). The ultimate goal is to develop organismal classifications based on correctly inferred cladogenetic histories.

As sometimes practiced, cladistic approaches themselves are not entirely im-

Box 2.4. Cladistic Terminology and Concepts

The following definitions are relevant to cladistic-phenetic discussions:

I. Classes of organismal resemblance:

    a. Phenetic similarity—overall resemblance between organisms.

    b. Patristic similarity—the component of overall similarity due to shared ancestry.

    c. Homoplastic similarity (homoplasy)—the component of overall similarity due to convergence from unrelated ancestors. [The term homoplasy also is used frequently to describe the "extra steps" implied in a phylogenetic network beyond those that distinguish taxa in the raw data matrix. In this usage, homoplasy may arise from convergence, parallelism, or evolutionary reversals in character states (Chapter 4)].

II. Classes of character state used to characterize organismal resemblance:

    a. Plesiomorph—an ancestral character state (one present in the common ancestor of the taxa under study).

    b. Symplesiomorph—an ancestral character state shared by two or more descendant taxa.

    c. Apomorph—a derived or newly evolved character state, not present in the common ancestor of the taxa under study.

    d. Synapomorph—a derived character state shared by two or more descendant taxa.

    e. Autapomorph—a character state unique to a single taxon.

III. Other relevant definitions

    a. Monophyletic group or clade—an evolutionary assemblage that includes a common ancestor and all of its descendents.

    b. Polyphyletic group—an artificial assemblage derived from two or more distinct ancestors.

    c. Paraphyletic group—an artificial assemblage that includes a common ancestor and some but not all of its descendents.

    d. Outgroup—a taxon phylogenetically outside the clade of interest.

    e. Sister taxa—taxa stemming from the same node in a phylogeny.

Phenetic resemblance may be due to patristic and/or homoplastic similarity. Patristic similarity may arise from symplesiomorphic and/or synapomorphic character states. Cladists attempt to distinguish between symplesiomorphic and synapomorphic similarity and to identify clades on the basis of synapomorphs only (Fig. 2.5). Pheneticists usually make no such attempts to distinguish sources of resemblance. Phylogenetic reconstructions based on either cladistic or phenetic principles can be compromised by extensive homoplasy.

Because cladists must distinguish symplesiomorphs from synapomorphs, much effort is devoted to the elucidation of evolutionary "polarities" (derived versus ancestral conditions) of character states. The following are among the criteria that have been used to suggest primitiveness for a character:

    a. Presence in fossils

    b. Commonness among an array of taxa

    c. Early appearance in ontogeny

    d. Presence in an outgroup

Criterion (d) is most widely employed now, as the others have proved misleading or incorrect in many instances (Stevens, 1980).

CHARACTERS AND STATES

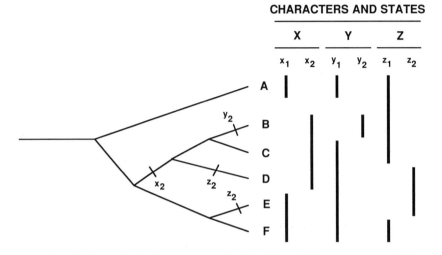

Figure 2.5. Philosophical rationale underlying Hennigian cladistic attempts to distinguish sources of similarity. Shown is the true (but unknown to the researcher) phylogeny for taxa A–F and the distribution of observed binary states for characters X, Y, and Z. Suppose that taxon A is known to be an outgroup for B–F, whose phylogeny is to be reconstructed. Character states x_1, y_1, and z_1, possessed by various taxa and the outgroup, are symplesiomorphs (shared ancestral states), and hence identify no clades. In particular, y_1 and z_1 could be positively misleading in amalgamating ingroup members (C–F and B, C, F, respectively) had their ancestral status gone unrecognized. Character state y_2 defines no multitaxon clade because it is an autapomorph, and z_2 could be misleading as a putative clade marker because it evolved in parallel (convergent) fashion in taxa D and E. Only x_2 is a valid synapomorph in this example, correctly identifying the true clade composed of taxa B–D.

mune from criticism. For example, a widely held belief is that ''one *true* synapomorphy is enough to define a unique genealogical relationship'' (Wiley, 1981). Although incorrect [in part because of the distinction between phylogenies for particular genes and organismal pedigrees (Chapter 4)], this perception sometimes has led to dogged advocacy for putative clades that receive support from only a few presumptive but favored synapomorphs. Thus, unless many characters are assayed (as advocated under the phenetic school), there is a potential danger in cladistics of the kinds of authoritarianism that plagued systematics earlier in the century and that prompted the original rise of numerical taxonomy.

Indeed, there seems little justification for the rancor of the cladistic attack on phenetics, for at least two reasons: (a) cladism owes a deep debt to numerical taxonomy for having first opened new discussions and viewpoints on traditional systematic practices, and (b) curiously, cladistic methods as applied to large data sets can come rather close to the procedures of numerical taxonomy (because for such data, character conflicts in clade delineation almost inevitably arise, thereby requiring some form of numerical tallying of putative synapomorphs). Thus, numerical cladistics and numerical phenetics are not as distinct operationally as they might at first appear, although cladistic methods do attempt to distinguish among sources of similarity and thereby account for specified character state distributions in terms of phylogenetic history.

The original "bible" of the cladistic school, published in 1950 by the German entomologist Willi Hennig, was translated to the English version *Phylogenetic Systematics* in 1966. In the last two decades, cladistic methods based on Hennig's insights have revolutionized systematic practice as applied to traditional taxonomic characters. Cladograms generated for many taxonomic groups now summarize phylogenetic reconstructions based on hypotheses concerning the polarities (ancestral versus derived conditions), transformations, and temporal orders of appearance of various morphological, physiological, or behavioral character states. Thus, one major strength of cladistic approaches involves the formulation of explicit, potentially testable hypotheses for the origin of particular character states (Buth, 1984; Patton and Avise, 1983). The cladistic school also has given rise to the important "vicariance" subdiscipline of biogeography (Chapter 8).

Few researchers now dispute the philosophical pillar of the cladistic school— that shared-derived traits are a solid basis for clade delineation. Rather, most questions center on operational issues: How reliably can synapomorphies be identified? What kinds of characters are best suited? How are character conflicts resolved when putative clades identified by different presumptive synapomorphs disagree? How is a phylogeny to be translated into a classification?

The relatively simple principles of Hennigian cladistics have attracted much devotion and prompted strident claims concerning the theory and practice of systematics. Future historians of science no doubt will address reasons for the heated polarization between the phenetic and cladistic camps [see Hull (1988) for an early perspective], but at least two important ramifications for molecular evolution have stemmed from this controversy. First, the debate itself unquestionably gave a renewed energy to morphology-based systematics. This development came at a time when some traditional systematists may have felt threatened by the rise and increasing dominance of molecular biology. One unfortunate and unnecessary consequence of this timing is that molecular and morphological approaches to systematics sometimes have been viewed in opposition, a perception with no valid basis. Second, the cladistic–phenetic war, although waged primarily in the context of morphology-based systematics, occasionally spilled over such that molecular phylogenetics also was caught in the crossfires. For example, strict Hennigian approaches cannot be applied to raw data consisting solely of numerical distance values between taxa, and as a result some cladists automatically discredited all such information derived from the important immunological and nucleic acid hybridization methods of molecular biology (Chapter 3). Attacks also were mounted against the widespread practice of summarizing molecular data using the UPGMA algorithm of numerical taxonomy, despite the fact that an important assumption of this phenetic procedure (constant evolutionary rate across phenogram branches) appeared to mesh well with the considerable independent evidence of a clocklike behavior for many biological macromolecules (Chapter 4). Furthermore, many molecular data, even those in the form of qualitative character states such as protein electromorphs or DNA sequences, are not particularly well suited for strict Hennigian cladistic analysis, in part because of the high risk of homoplasy at the level of individual electromorph or nucleotide character states (Straney, 1981).

Apparent conflicts among characters in clade delineation arise primarily from homoplasy (Box 2.4). To minimize the number of ad hoc hypotheses required to resolve character conflicts along a phylogeny, principles of "maximum parsimony" are employed widely today (Felsenstein, 1983; Sober, 1983). As applied to phylogenetic inference, parsimony algorithms operate by estimating evolutionary trees of minimum total length (i.e., trees that minimize the number of evolutionary transformations among character states required to explain a given data set). Although notions of parsimony have long been a part of general biological reasoning, developments in cladistic philosophy provided an important historical step in the further elaboration of parsimony approaches in phylogeny reconstruction. Thus, a close intellectual connection exists between numerical cladistics and numerical parsimony, both of which strive to resolve data conflicts in the simplest possible ways consistent with particular assumptions about the evolutionary process (Swofford and Olsen, 1990). Further discussion of maximum parsimony and other phylogenetic methods as applied to molecular data will be deferred to Chapter 4.

Phylogenetic Data and the Molecule–Morphology Debate

Overlapping to some degree with the phenetics–cladistics debate was another controversy about whether molecules or morphologies provide a better guide to phylogeny reconstruction. Particularly in the 1960s and early 1970s (in the early years of protein electrophoresis), many systematists trained in traditional organismal disciplines understandably viewed the new molecular data with considerable incredulity. As the field of molecular evolution grew, this overt skepticism softened somewhat to a fundamental disquietude over the relative merits of molecular versus morphological information in systematics. Nonetheless, the underlying tone of antagonism between molecular and morphological approaches persisted, and remains today in many circles. The question too often has been "Which of these data bases is superior?"

Perhaps we recently have entered a more mature era in systematic relations with the recognition that molecular and morphological data can be reciprocally informative, and indeed require each other's services (Hillis, 1987). For example, a new enterprise in molecular evolution may be termed "phylogenetic character mapping" (Chapter 8). This involves plotting the taxonomic distribution of morphological (or other) characters along a molecular-inferred phylogeny, the intent being to uncover the evolutionary histories of organismal attributes. Of course, phylogenetic mapping also may be conducted in the reverse direction; that is, by plotting the distributions of molecular characters along a morphology-inferred phylogeny. For example, several instances of horizontal gene transfer between otherwise unrelated organisms have been revealed under the compelling logic of this general approach (Chapter 8). Such enlightened analyses that attempt to capitalize on the comparative information content of multiple classes of data should lead eventually to a more contented marriage between molecular- and morphology-based systematics. Under this developing perspective, the interplay between alternative lines of evidence becomes of greater interest and significance than does either data source considered alone.

MOLECULAR PHYLOGENETICS

While these grand controversies of evolutionary biology were being played out, other researchers adopted a more pragmatic approach of simply applying protein and DNA markers to resolvable problems in natural history and evolution. Beginning as a subsidiary endeavor in molecular evolution, this previously neglected perspective has grown steadily and now occupies a position of central prominence, as this book will attest.

Advances in molecular systematics have proceeded as a series of waves, each initiated by the development of a new laboratory method. A typical pattern is as follows. A laboratory technique is introduced, and a flurry of evaluative activity follows. Methods that fail to meet advance billing (e.g., for reasons of technical difficulty, poor repeatability, or ambiguity in genetic data) are abandoned. Approaches that survive the initial screening are then incorporated first into research programs addressing the general conceptual issues of evolutionary biology described above. For example, concerns invariably arise about whether natural selection plays a role in maintenance of molecular polymorphisms revealed by the new technique under consideration, and numerous tests (involving observational or experimental data evaluated against the predictions of neutrality theory) are applied. Discussion also ensues about methods of analysis most appropriate for the new class of molecular data. In the meantime, genetic markers provided by each method are applied to particular problems in natural history or evolution where their use appears appropriate. Success in such endeavors stimulates further interest, and the more utilitarian of the molecular approaches wash over the field. Eventually, usually after a period of several years, the enthusiasm crests, and a new wave of interest in another method may begin to take shape. Typically, the earlier methods are not abandoned, but merely become incorporated into the growing pool of molecular techniques that find continued application in studies of natural history and evolution.

The first molecular approach employed widely in the field was protein electrophoresis as applied to allozyme and isozyme systems. "Allozymes" are protein variants of a genetic locus that behave in straightforward Mendelian fashion and, hence, are interpretable as simple allelic products of a gene. "Isozymes" are a broader class encompassing all protein variants observed on electrophoretic gels, including heteromeric products of multiple loci, posttranslational variants, and other protein alterations (Chapter 3). Allozyme methods were introduced in the mid-1960s, and for the next 10 years dominated molecular systematics (see the reviews in Avise, 1974, 1983b; Buth, 1984; Gottlieb, 1977; Whitt, 1983, 1987). Today, protein electrophoresis remains a popular method for generation of molecular markers.

The next widely employed technique in molecular systematics involved analyses of "restriction fragment-length polymorphisms" (RFLPs) in DNA. For both technical and conceptual reasons, mitochondrial (mt) DNA received a great bulk of the early attention. Mitochondrial approaches dominated molecular systematics during the late 1970s and 1980s (Avise et al., 1979a, 1979b; Brown and Wright, 1979; Brown et al., 1979; reviews in Avise, 1986, 1991a; Avise and Lansman, 1983; Birley and Croft, 1986; Harrison, 1989; Moritz et al., 1987; Palmer, 1990; Wilson et al., 1985), much as had allozyme studies a decade earlier, and strong interest in mtDNA markers continues today. In the middle and late 1980s, another wave of excitement attended RFLP analyses as applied to hypervariable nuclear DNA regions, in a class of procedures that because of diagnostic power became known as "DNA fingerprinting" methods (Burke, 1989; Hill, 1987; Jeffreys et al., 1985a,b, 1988a; Kirby, 1990).

The current wave of excitement taking shape in molecular phylogenetics began with the introduction of the polymerase chain reaction (PCR) for in vitro amplification of specific DNA fragments (Erlich and Arnheim, 1992; Erlich et al., 1991; Mullis, 1990; Mullis et al., 1986; Saiki et al., 1988; White et al., 1989). When coupled with the further development of amplification primers (Kocher et al., 1989), improved laboratory methods for sequence determination (Innis et al., 1988; Ruano et al., 1990; Scharf et al., 1986; Wrischnik et al., 1987), and development of appropriate methods for interpretation of haplotype data (Clark, 1990; Stephens et al., 1990b), PCR-based approaches permit increased direct access to the phylogenetic information content of DNA sequences from both nuclear and cytoplasmic genes. Furthermore, because the PCR can amplify particular DNA segments from tiny amounts of starting tissue [or even

Box 2.5. Abbreviated Chronology of Some Significant Historical Developments in the Application of Molecular Markers (for an extended history of genetic discoveries, see King and Stansfield, 1990).

1944   Avery, MacLeod, and McCarty provide experimental evidence that DNA and not protein is the genetic material.

1953   Watson and Crick propose a molecular model for the structure of DNA.

1955   Smithies uses starch-gel electrophoresis to identify protein polymorphisms.

1963   Margoliash determines amino acid sequences for cytochrome c in several taxa and generates the first phylogenetic tree for a specific gene product.

1966   Several independent researchers (see text) use electrophoretic methods and histochemical enzyme stains to assess levels of genetic variability in animal populations and humans.

1967   Sarich and Wilson provide an early application of protein immunological methods and discover a more recent shared ancestry for human and great apes than previously suspected.

1968   Kimura proposes the neutral theory of molecular evolution. Meselson and Yuan isolate and characterize the first specific restriction enzyme. Britten and Kohne use DNA hybridization methods to characterize animal genomes.

1971   Publication of the first periodical devoted explicitly to molecular evolution (*Journal of Molecular Evolution*).

1975   Southern describes a method for transferral of DNA fragments to nitrocellulose filters, hybridization to radioactive probes, and detection of fragments by autoradiography.

1977   Maxam and Gilbert, and Sanger, Nicklen, and Coulson describe laboratory methods for DNA sequencing.

1978   Maniatis and colleagues develop a procedure for gene isolation which involves construction and screening of cloned libraries of eukaryotic DNA.

1979   Avise, Lansman, and Shade, and Brown, George, and Wilson introduce mtDNA approaches to analyses of natural animal populations.

1981   Palmer and colleagues begin an important series of papers utilizing cpDNA for phylogenetic reconstructions in plants.

1985   Jeffreys, Wilson, and Thein develop the DNA fingerprinting technique and point out its potential for forensic science. Saiki and six colleagues report the enzymatic amplification of DNA using the polymerase chain reaction.

1989   Kocher and six colleagues report the discovery of conserved PCR primers that can be employed to amplify mtDNA segments from many species.

1992   Proliferation of periodicals devoted to evolutionary applications for molecular markers, for example, *Molecular Ecology* (Blackwell); *Molecular Phylogenetics and Evolution* (Academic Press); *Molecular Marine Biology and Biotechnology* (Blackwell).

from some well preserved fossils (Chapter 8)], it has extended molecular applications to a much wider biological arena (Arnheim et al., 1990).

In addition to these major approaches in molecular systematics, other powerful but less widespread methods have added significant contributions. Particularly important among these have been immunological comparisons of proteins, which provided some of the initial evidence for molecular clocks (Benjamin et al., 1984; Goodman, 1963; Sarich and Wilson, 1966, 1967; Wilson et al., 1977) and are still employed today (Maxson and Maxson, 1986, 1990), and DNA-DNA hybridization methods, which have been available for many years (Britten and Kohne, 1968; Doty et al., 1960) and have had special impact in the systematics of certain groups such as birds (Sibley and Ahlquist, 1990), insects (Caccone and Powell, 1987; Caccone et al., 1988a,b) and hominoid primates (Caccone and Powell, 1989; Sibley and Ahlquist, 1987).

Given the burgeoning interest today in molecular phylogenetics and molecular ecology, it is useful to remain cognizant of the remarkably shallow history of these scientific disciplines. An abbreviated chronology of some significant developments in the application of molecular markers is summarized in Box 2.5.

SUMMARY

1. The study of evolution from a molecular perspective is a fairly recent enterprise, dating in substantive form only to the latter half of the 20th century.

2. Several major controversies have dominated attention in molecular evolution and related fields. These include the classical-balance debate on the magnitude of genetic variation, the selection-neutrality debate on the adaptive significance of molecular variation, the phenetic-cladistic debate on procedures for interpreting molecular or other data in a systematics context, and the relative phylogenetic utility of molecular versus morphologic characters. The latter controversies include issues that are particularly relevant to some phylogenetic applications for molecular data.

3. To a considerable extent, these important debates diverted attention from some of the more utilitarian applications for protein and DNA markers in natural history and evolution. Only recently has a focus of primary research effort shifted to this latter arena.

4. Several waves of excitement in molecular evolution have followed introduction of new laboratory techniques. Among the most influential methods have been protein electrophoresis in the late 1960s and 1970s, RFLP analyses of mtDNA in the late 1970s and 1980s, DNA fingerprinting in the mid-to-late 1980s, and PCR-mediated DNA sequencing in the 1990s.

3

Molecular Tools

*Perhaps nowhere has the power of the scientific method
been more brilliantly demonstrated than in the development of procedures for the study of the chemistry of life.*

M.O. Dayhoff and R.V. Eck. 1968

There exists a wide variety of laboratory assays for revealing molecular genetic
markers. Detailed laboratory protocols can be found in the papers and manuals
listed at the end of this chapter, but in practice there is no substitute for hands-on
training "at the bench" under the guidance of an experienced practitioner.
Therefore, this chapter will merely outline the procedural steps of various molecular methods and will emphasize instead the nature of genetic information
produced by each of the several molecular techniques that has had major impact
in ecological and evolutionary studies.

PROTEIN ASSAYS

Protein Immunology

PRINCIPLES AND PROCEDURES

Immunological methods rely on the antigenic properties of proteins (Fig. 3.1).
When a protein from species A is injected into a suitable host such as a rabbit,
this antigen elicits production of antibodies with high specificity for antigenic

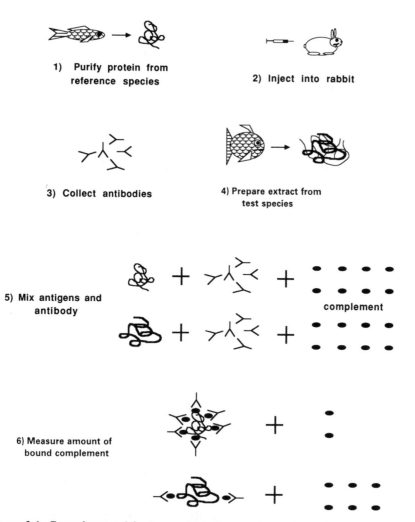

1) Purify protein from
   reference species

2) Inject into rabbit

3) Collect antibodies

4) Prepare extract from
   test species

5) Mix antigens and
   antibody

complement

6) Measure amount of
   bound complement

Figure 3.1. General protocol for immunological comparisons of proteins by microcomplement fixation (after Wilson, 1985).

sites on the injected protein. In laboratory assays, heterologous antigen (a related protein from species B) also is tested for its capacity to elicit a reaction from this antibody. Cross-reactivity with the heterologous antigen can be quantified and compared to results of the control reaction involving the original ("homologous") antigen. The difference in antigen–antibody reactivities in tests involving homologous versus heterologous antigens provides a measure of the genetic relationship (usually expressed as immunological distance or ID units) between these proteins from species A and B. A fundamental assumption underlying all immunological approaches is that level of antibody–antigen "recognition" provides a quantitative guide to the genetic similarity of the antigenic proteins involved.

The first immunological assays performed in a phylogenetic context (Nuttall, 1904) involved precipitin tests, whereby antibodies against unpurified proteins were mixed with antigens in solution, and the resultant antibody–antigen precipitate was measured. More recently, these and other crude immunological tests have been supplanted by microcomplement fixation (MCF) methods, in which antibodies against purified proteins form the basis for the refined estimates of antigen-antibody cross-reactivity. An outline of the procedure is as follows (see Fig. 3.1). A protein such as albumin is purified from a reference species, using standard biochemical protocols. The highly purified protein is injected several times into a rabbit, over a 3-month period. One week after the last injection, rabbit antiserum is collected and titered (standardized to a given level of reactivity under specified MCF conditions). The antiserum is stable and may be stored frozen indefinitely. When a heterologous MCF is to be performed, the antiserum is mixed with varying concentrations of antigen from the diluted plasma (purified protein is not required) of a test species. Included in the reaction mixtures is "complement," a group of proteins normally found in vertebrate serum that becomes trapped in the developing latices of the antigen-antibody reaction. The amount of complement "fixed" in the reaction is a function of the cross-reactivity between antibody and antigen. In spectrophotometric assays, levels of complement remaining unbound are monitored as a function of antigen concentration, and the level of immunological response thereby is quantified.

DATA

The data from MCF or other immunological techniques consist of a quantitative measure of divergence (immunological distance or ID) between related forms of a protein carried by the two or more species. The cross-reactivity registered in immunological tests depends on the affinity and specificity of the antibodies to the antigen, properties which are functions of the immunization protocol as well as the genetic relationships between these molecules. Thus, standardization in techniques is essential. High-affinity antibodies as monitored by MCF are capable of detecting single amino acid replacements within the antigenic portions of challenging proteins.

The MCF assay in effect counts the number of amino acid replacements responsible for differences between the antigenic sites of related (preferably orthologous) protein forms. The usual protein of choice in vertebrate studies has been albumin for the following reasons: It is monomeric and encoded by a single gene; it is abundant, ubiquitous, and easily purified; it has a large number

(25–50) of major antigenic sites (Benjamin et al., 1984), at which amino acid substitutions translate into antigenic differences detectable in the MCF assay; and, it evolves at a rate appropriate for studies of intermediate taxonomic levels such as subgenera, genera, or families (the usual categories of application for MCF approaches). Maxson and Maxson (1986) present evidence that albumin ID as measured by MCF is a linear estimator of amino acid replacement differences (see also Prager and Wilson, 1993). Other proteins used in MCF assays have included lysozyme, ovalbumin, and transferrin in vertebrates (Leone, 1964; Prager and Wilson, 1976; Wright, 1974), glycerophosphate dehydrogenase, acid phosphatase, and larval proteins in invertebrates (Beverley and Wilson, 1985; Collier and MacIntyre, 1977; MacIntyre et al., 1978), and alkaline phosphatase in bacteria (Cocks and Wilson, 1972; additional references in Maxson and Maxson, 1990).

For a complete assessment of taxa, MCF assays require that antibodies be produced from each species under consideration. Otherwise, only a subset of cross-reactivities between species pairs can be attempted, leading to missing elements in the pairwise ID matrix (the basis for phylogeny reconstruction). Generating and testing antisera for moderate or large numbers of species is time-consuming, but offers the additional advantage of permitting tests of reciprocity (anti-A vs. B and anti-B vs. A). Differences between reciprocal outcomes provide a measure of experimental error in the MCF procedure (Maxson and Wilson, 1975).

Protein Electrophoresis

PRINCIPLES AND PROCEDURES

This method takes advantage of the fact that nondenatured proteins with different net charge migrate at different rates through starch or acrylamide gels (or other supporting media such as cellulose acetate strips) to which an electric current is applied (Fig. 3.2). The charge characteristics stem primarily from the three amino acids with positive side chains (lysine, arginine, and histidine) and the two with negative side chains (aspartic acid and glutamic acid). The net charge of a protein, which varies with the pH of the running condition, determines the protein's movement toward the anode (positive pole) or cathode (negative pole) in the gel. Protein size and shape also can interact with pore size in the electrophoretic matrix to influence migrational properties.

Because of low cost, safety, and ease of use, gels made from hydrolyzed potato starch are employed most widely. The "starch gel electrophoresis" (SGE) procedure begins with the extraction of water-soluble proteins from a particular source (leaves, roots, liver, heart, blood, skeletal muscle, etc.). The extract from each individual is soaked onto a paper wick, and 20 or more such wicks are

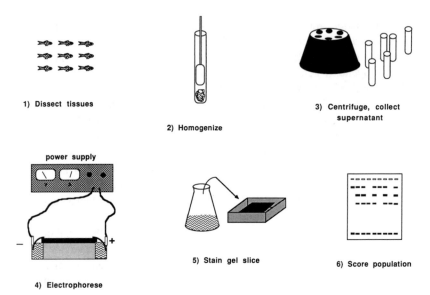

1) Dissect tissues

2) Homogenize

3) Centrifuge, collect
supernatant

power supply

4) Electrophorese

5) Stain gel slice

6) Score population

Figure 3.2. General protocol for protein-electrophoretic surveys (see text).

placed side by side along a slit (the origin) cut into the gel. The gel is placed in
a buffer tray connected to an electrical power supply, and electrophoresis proceeds over several hours. The gel then is removed, sliced horizontally, and the
wafer-thin slices incubated with histochemical stains specific for the enzymes
under assay. Each stain contains a commercially available substrate for the
enzyme, necessary cofactors, and an oxidized salt (usually nitro-blue tetra-
zolium, NBT). For example, the staining solution for lactate dehydrogenase
(LDH) includes lactic acid (the substrate), nicotinamide adenine dinucleotide
(NAD, the cofactor), phenazine methosulfate (PMS, an intermediary catalyst),
and NBT. At the position(s) in the gel to which LDH from the study organism
has migrated, a reaction is catalyzed whereby lactic acid is oxidized to pyruvic
acid and the salt is reduced to a blue precipitate visible to the naked eye as a
discrete band. The band profile is the "zymogram" pattern for the enzyme, and
usually can be interpreted in simple genetic terms.

Histochemical stains single out the products of particular genes from among
the thousands of other undetected proteins also migrating through a gel. Their
development (Hunter and Markert, 1957), coupled with improvements in electrophoretic procedures and media, eliminated need for laborious protein purification procedures that continue to preclude direct amino acid sequencing from
most population applications. Recipes and ingredients for more than 100 histochemical stains are available widely. Not all enzymes resolve well for a given
taxon, however, and a typical multilocus SGE survey involves successful assay

of about 10–30 enzymes, perhaps encoded by 15–50 genes (some enzymes are encoded by multiple loci). Often, hundreds or even thousands of individuals are analyzed.

For example, one starch gel carrying extracts from 25 individuals can be sectioned into about 5 replicate slices and each slice incubated with a different stain. Twenty such gels per day might be run in an active laboratory. Thus, in a single day, a total of 2500 genotypes (25 individuals × 20 gels × 5 enzyme stains) could reasonably be scored. This is a conservative estimate because many stains reveal genotypes for two, three, or four gene copies whose products catalyze essentially the same reaction. Such masses of genetic data are incredible by the standards of premolecular genetics, where the elucidation of even a small handful of Mendelian loci and genotypes in a few individuals required breeding studies conducted over multiple generations. Indeed, within a short time of the onset of the allozyme revolution in the mid-1960s, vastly more genotypic data from natural populations were gathered than in all the 100 years since Gregor Mendel.

DATA

Zymogram patterns normally are interpretable in terms of Mendelian genotypes at particular loci. The Mendelian bases of observed polymorphisms may be verified in several ways. First, experimental crosses may be conducted to provide relatively direct appraisals of the genetics underlying zymogram phenotypes. For example, a cross between two presumptive homozygotes exhibiting different bands on a gel should produce uniformly heterozygous progeny with predictable zymogram appearance (see following text and Fig. 3.3), and backcrosses of such progeny to either parent should produce heterozygotes and appropriate homozygotes in approximately equal frequency. Similar "family tests," often conducted with plants, circumvent the need to conduct crosses by addressing whether an array of progeny exhibits banding patterns consistent with Mendelian expectations given the genotype of the known mother. In early protein electrophoretic surveys, such direct experimental validation of zymogram variation was commonplace, but as experience with the simple genetic bases of the banding patterns for commonly used enzyme systems accumulated, further corroborations through breeding or family studies became less critical. Another approach to verifying the Mendelian basis of zymogram variation involves population genetic considerations, whereby frequencies of electrophoretic types are compared against Hardy-Weinberg expectations for neutral alleles. For populations that are outcrossed, and free from pronounced microspatial subdivision, frequencies of suspected genotypes almost invariably have proved to be close to those predicted from observed electromorph frequencies, thus providing strong support for the Mendelian nature of the variants.

Third, molecular considerations can help document the Mendelian basis of zymogram variations (Fig. 3.3). Most enzymes surveyed electrophoretically have a known quaternary structure. For example, phosphoglucomutase (PGM) is a monomer, composed of a single polypeptide subunit with catalytic activity. Thus, homozygotes at a PGM gene show one band on gels, and heterozygotes show a simple two-band pattern, one band produced by each of the two PGM alleles. An example of a dimer is glucosephosphate isomerase (GPI), whose catalytic activity requires the joining of two polypeptide subunits. The zymogram for GPI heterozygotes—a three-band gel profile in which the middle band is approximately twice the intensity of the flanking bands—reflects random dimeric associations between the polypeptides produced by the two alleles. Purine-nucleoside phosphorylase (PNP) is an example of a trimer, and LDH is a tetramer, such that heterozygotes for these loci normally exhibit four-band and five-band zymograms, respectively, with characteristic band intensities (Fig. 3.3).

Most observed allozyme variants probably are attributable to nucleotide substitutions causing replacements of the charged amino acids, although direct molecular evidence for this conclusion seldom is available. In any event, because

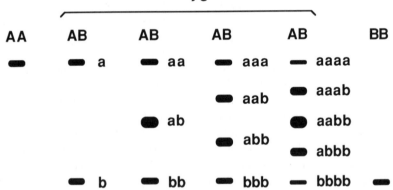

Figure 3.3. Examples of single-locus zymogram patterns. Lowercase letters indicate polypeptide subunits produced by alleles A or B; uppercase letters indicate diploid genotypes. Shown in the four central lanes (columns) are the expected zymogram patterns for a heterozygous individual when the enzyme in question is (from left to right) monomeric, dimeric, trimeric, or tetrameric, respectively. For example, with a tetrameric protein system, a heterozygote typically would exhibit five bands on the gel with the following subunit compositions: aaaa, aaab, aabb, abbb, and bbbb. If the two alleles produce similar polypeptide concentrations and subunit assembly is random, then intensities of the five respective bands should appear in the ratio 1 : 4 : 6 : 4 : 1.

particular DNA mutational profiles underlying electromorph differences normally remain unknown, the multiple allozyme alleles at any locus must be viewed as qualitative multistate traits, the phylogenetic order of which cannot be safely inferred from the observable property, electrophoretic mobility. Allozyme data normally consist, then, of the accumulation of such genotypic information from a number of loci, typically unlinked and scattered about the nuclear genome (Pasdar et al., 1984; Shows, 1983; Wheat et al., 1973). Apart from their many uses as single-locus Mendelian markers in such areas as parentage assessment and gene flow estimation, allozyme frequencies observed at multiple loci also can be employed to compute quantitative measures of genetic distance (Chapter 4) between the populations, species, or other taxa under comparison. Such genetic distances represent a composite numerical summary of all allozyme data included in the survey.

Protein electrophoretic assays also reveal various aspects of gene expression that may be under genetic control. Because gene expression patterns themselves are products of the evolutionary process, they too can be informative as phylogenetic markers. For example, among 26 orders and more than 40 assayed families of living birds, only the woodpeckers (Picidae), honeyguides (Indicatoridae), barbets (Capitonidae) and toucans (Rhamphastidae) consistently exhibit a unique three-band zymogram for malate dehydrogenase (MDH), perhaps attributable to a gene duplication event (Avise and Aquadro, 1987). The taxonomic distribution of this fortuitously discovered MDH pattern helped settle a long-standing debate about whether these birds, which superficially look so different, are indeed allied phylogenetically as their traditional placement within Piciformes would suggest. However, apart from identifying this single clade, the MDH marker was of no further utility for avian phylogenetic assessment [quite in contrast to the conventional class of information on allozymic allele frequencies, where a more comprehensive picture of piciform relationships has emerged (Lanyon and Zink, 1987)]. Thus, the rarity of idiosyncratic allozyme characters is both a phylogenetic blessing and a curse—rarity suggests monophyly and thereby implies that special weight be given to eccentric molecular features, but it also means that the applicability of such markers is limited.

On the other hand, not all idiosyncratic patterns of gene expression have proved to be informative phylogenetically. For example, Mindell and Sites (1987) assayed tissue expression patterns for 30 presumptive isozyme loci in representatives of two avian orders (Charadriiformes and Passeriformes) and observed numerous inconsistencies with the accepted taxonomy. For these systems, they concluded that "widespread homoplasy, opposite polarities and limited predictive capability for the isozyme tissue expression patterns suggest that most may be more useful in studies of gene regulation than in higher level taxonomy." Clearly, caution must be exercized in the utilization of gene expression patterns as phylogenetic markers.

Many enzymes are encoded by two or more loci that arose through gene duplications via polyploidy, aneuploidy, or regional intrachromosomal duplication (Buth, 1983; MacIntyre, 1976; Ohno, 1970). Their zymogram patterns usually are predictable (and interpretable) from rules governing polypeptide assembly into functional enzymes with known quaternary structure. From such evidence, Gottlieb (1988) and Soltis et al. (1987) have identified several duplications of isozyme-encoding nuclear genes among various species of diploid *Clarkia* plants. These duplications individually appeared to be rare and, hence, of potential phylogenetic relevance. On the other hand, three complications dealing ultimately with establishment of gene homologies served to temper enthusiasm somewhat (Sytsma and Smith, 1992). First, the convergent origin of a duplicate gene in independent lineages remains a possibility. Second, postduplication silencing of either member of a duplicate pair may occur, perhaps on separate occasions in different lineages. Third, the presence of duplicate loci might represent a plesiomorphic condition at the hierarchical level of the taxonomy examined. The latter two possibilities indeed have been documented in the *Clarkia* studies (Gottlieb, 1988).

The protein products of duplicated genes frequently have diverged in structure and regulatory control since the duplication event and may show striking ontogenetic changes or tissue specificities of potential relevance to phylogenetic assessment. Thus, whereas most vertebrates express a single GPI gene, bony fish express two unlinked GPI loci, one predominantly in skeletal muscle and the other in liver (Avise and Kitto, 1973; Whitt et al., 1976). All vertebrates (with the exception of lamprey fishes) have muscle- and heart-specific LDH expression involving the "A" and "B" genes (Markert et al., 1975). As gauged by zymogram patterns, the assembly of heterotetramers between the LDH loci sometimes is nonrandom, presumably due to taxon-specific genetic regulatory influences (Murphy, 1988; Sites et al., 1986). Some birds (doves) and mammals have an additional LDH enzyme (produced by the "C" locus) expressed only in the primary spermatocytes (Blanco and Zinkham, 1963; Matson, 1989; Zinkham et al., 1969); also, bony fish carry a third gene for LDH, expressed in a variety of tissues in primitive species and predominantly in eye or liver of advanced teleosts (Horowitz and Whitt, 1972; Markert and Faulhaber, 1965; Shaklee et al., 1973; Whitt et al., 1975). It is doubtful that this third LDH locus in fish is orthologous to those in either the doves or mammals (Fisher et al., 1980; Quattro et al., 1993), but in any event, these LDH loci clearly have evolved distinctive regulatory profiles in different taxonomic groups. Other multilocus protein systems studied extensively with regard to patterns of gene expression and phylogeny include malate dehydrogenase, glycerol-3-phosphate dehydrogenase, creatine kinase (Buth et al., 1985; Fisher and Whitt, 1978; Fisher et al., 1980; Philipp et al., 1983), and the globin superfamily of oxygen-carrying molecules (Dayhoff, 1972; Doolittle, 1987).

Duplicate genes also are subject to evolutionary silencing or loss, and these patterns can be informative phylogenetically. For example, Ferris and Whitt (1978, 1979) utilized patterns of enzyme loss and change of gene expression to reconstruct the phylogeny of the catostomid suckers, a group of freshwater fishes that underwent a polyploidization event some 50 million years ago and subsequently became "diploidized" at approximately 50% of assayed structural genes. The rationale for these reconstructions was as follows. Immediately following the polyploidization event, all loci in the sucker genome must have been duplicated, and the "primitive" condition from that point forward became presence of each duplicate gene. As mutations accumulated, some genes lost expression (and become pseudogenes), whereas other duplicate copies diverged in structure and function (Ferris and Whitt, 1977). These processes presumably were nearly irreversible (however, see Buth, 1979, 1982), such that taxa sharing possession of the derived states (loss or alteration of gene expression) likely belonged to Hennigian clades (assuming that the same losses of gene expression had not occurred independently in separate evolutionary lineages).

DNA ASSAYS

DNA–DNA Hybridization

PRINCIPLES AND PROCEDURES

This method (see Fig. 3.4) relies on the double-stranded nature of duplex DNA, and the fact that paired nucleotides on the two complementary strands are held together by hydrogen bonds (two coupling each adenine–thymine base pair, and three coupling each guanine–cytosine). These hydrogen bonds are the weakest links in DNA, so that when native DNA is boiled in solution, the duplexes dissociate or "melt" into single strands but otherwise remain structurally intact. As the melted sample is cooled, strands collide by chance, and those with complementary nucleotide sequence reassociate into double-stranded molecules as their respective bases pair and reform hydrogen bonds. A rapidly reassociating component represents repetitive DNA (because homologous strands are in high number and collide most frequently). This fraction is removed. The remaining fraction represents single- or low-copy sequences in the genome. Such single-copy DNA strands are then mixed under conditions where duplex formation occurs. The mixtures may involve DNA strands from a single sample or species (in which case homoduplexes are produced) or the mixtures may involve strands from two species (forming heteroduplexes). The culminating step in the hybridization protocol involves characterizing the thermal stabilities of these homoduplexes and heteroduplexes by gradually raising temperatures and monitoring the course of molecular dissociation to single strands. The thermal stability exhibited by any duplex depends largely on the similarity of nucleotide sequences in its two strands, because only properly paired bases are hydrogen-bonded. The measured difference in thermal stability between homoduplexes and heteroduplexes provides a quantitative estimate of the genetic divergence between the two species.

Further details of the DNA hybridization process are as follows. First, DNA

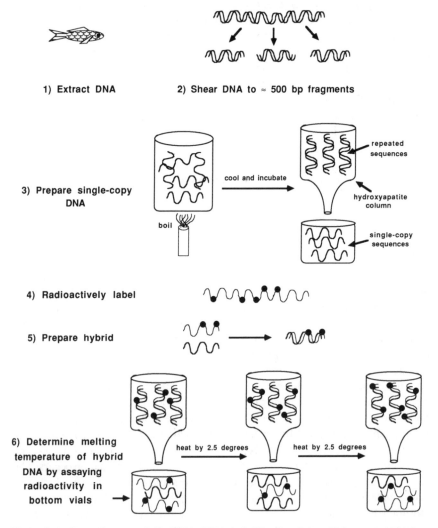

Figure 3.4. General protocol for DNA–DNA hybridization (after Sibley and Ahlquist, 1986).

is extracted from the nucleus of cells, separated from RNA and proteins, and physically sheared into fragments averaging 500 nucleotides in length (to reduce viscosity and to permit subsequent fractionation of repetitive from single-copy DNA). The sheared fragments are boiled, cooled, and their reassociation kinetics employed [as first described by Britten et al. (1974)] to remove most of the repetitive fraction. This is accomplished by incubating the DNA in solution at about 50°C for a short time, such that the repetitive sequences preferentially anneal and most single-copy sequences remain unpaired. This solution is passed through a hydroxyapatite column, which binds double-stranded DNA only. The single-stranded DNA which passes through the column is labeled with radioactive iodine (and becomes known as the tracer) and is mixed with a much larger amount of unlabeled DNA (the driver) from the same or a different species. This mixture is incubated at 60°C for several days to form hybrid DNA molecules that have one labeled and one unlabeled strand. The sample is then placed on a hydroxyapatite column, and gradually heated in a water bath at 2.5°C increments over a 60–95°C range. At each temperature increment, additional duplexes that have melted (a function of degree of base-pair mismatching) are washed from the column into a vial. Counts of radioactivity in the vials record the amount of duplex DNA that melted at the various temperatures.

DATA

The raw data from DNA–DNA hybridization consist of "thermal elution profiles" (Fig. 3.5) that summarize the observed percentages of dissociated, single-stranded DNA as a function of melting temperature. From such cumulative melting curves, various statistics can be generated that provide quantitative estimates of the degree of base-pair mismatch between the DNAs under comparison (e.g., Britten, 1986; Sarich et al., 1989; Sheldon and Bledsoe, 1989). Considerable debate (beyond the scope of current discussion) has centered on which of these distance measures is to be preferred, but in practice all tend to be highly correlated (Kirsch et al., 1990). One such distance measure is based on T_m, defined as the interpolated temperature at which 50% of the hybrid molecules that were formed remain in duplex condition. The differences in T_m values between homoduplex and heteroduplex melting profiles (ΔT_m) constitute the "raw" distance data used in phylogenetic reconstructions based on the DNA hybridization approach.

The relationship between ΔT_m and percent base-pair mismatch is thought to be linear (Britten et al., 1974; Caccone et al., 1988b; Kohne, 1970), but the exact conversion between the two is uncertain. Several studies have examined the relationship by studying the thermal stability properties of synthetic oligonucleotides or other sequences of known base composition (Bautz and Bautz, 1964; Hutton and Wetmur, 1973; Laird et al., 1969; Springer et al., 1992) and have

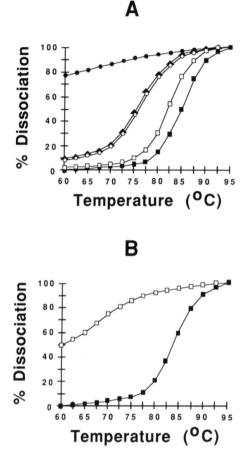

Figure 3.5. Thermal elution profiles from DNA–DNA hybridization. Shown are cumulative melting curves for the single-copy fraction of nDNA in some flightless ratite birds (after Sibley and Ahlquist, 1990). (*A*) Homoduplex DNA of the emu (*Dromaius novaehollandiae*) (closed squares) and heteroduplexes between that species and the southern cassowary (*Casuarius casuarius*) (open squares), greater rhea (*Rhea americana*) (open circles), ostrich (*Struthio camelus*) (closed triangles), and domestic fowl (*Gallus gallus*—a nonratite outgroup) (closed circles). In these comparisons, the cassowary appears genetically closest to the emu, followed in order by the rhea, ostrich, and domestic fowl. (*B*) Homoduplex DNA of the ostrich (closed squares) and the heteroduplex with the greater rhea (open squares). Note that although the melting curves involving the rhea and ostrich are nearly identical when compared against the emu (panel *A*), this does not necessarily imply that these latter species are genetically close to one another—they could differ from the emu in different phylogenetic directions. Indeed, differences between melting curves in panel B indicate a large genetic distance between the rhea and ostrich.

reported that incremental ΔT_m values of 1°C correspond roughly to 0.7–1.7% base-pair mismatch. As a working rule of thumb, a ΔT_m of 1°C often is equated to 1.0% base mispairing (see Britten, 1986; Koop et al., 1986). However, Powell et al. (1986) and Caccone et al. (1988b) obtained somewhat higher conversion ratios (ΔT_m of 1°C equaling 1.5–2.0% base-pair mismatch) in their studies of cloned DNAs of known sequence.

Because DNA hybridization data involve an averaging of genetic differences across a large fraction of the genome (primarily the single-copy portion), they sometimes have been promoted as the strongest available source of phylogenetic information (Sibley and Ahlquist, 1990). Thus, it is somewhat surprising that these methods have not been employed more widely by molecular systematists. DNA–DNA hybridization approaches *have* had tremendous impact in molecular

genetics by revealing important aspects of genomic structure—amounts of repetitive DNA, lengths of repeated sequences, and interspersion patterns among
repetitive and low-copy sequences (e.g., Britten and Kohne, 1968). With regard
to phylogenetic applications, reservations expressed about the DNA hybridization approach include the fact that the raw data consist solely of distance values
(rather than directly observed molecular character states) and that the influences
of factors affecting the kinetics of hybridization (such as differences in base
composition, DNA fragment size, and genome size) are incompletely understood. Some of these factors nonetheless are partially controlled or standardized
in most DNA hybridization studies. For example, effects of base compositional
differences (numbers of A–T versus C–G pairs) among sequences can be ameliorated by use of chaotropic solvents (Werman et al., 1990) and comparisons
can be confined to particular organismal groups (such as birds) where complications arising from confounding variables such as pronounced differences in
genome structure or organization should be minimized.

The development of automated thermal elution devices (such as the
"DNAnalyzer" of Sibley and Ahlquist, 1981) greatly expedited the process of
gathering DNA hybridization data. Indeed, the honor for the largest number of
species yet included in any molecular systematic survey no doubt belongs to
Sibley and Ahlquist (1990), who have conducted some 27,000 DNA–DNA
hybridizations involving about 1700 avian species (only about 350 of these
species provided DNA used as tracer, however, so that only a small subset of all
possible pairwise comparisons was accomplished).

Restriction Analyses

The discovery of restriction endonucleases (Linn and Arber, 1968; Meselson
and Yuan, 1968) revolutionized molecular biology. Type II restriction enzymes
(Kessler, 1987) cleave duplex DNA at particular oligonucleotide sequences,
usually either four, five, or six base pairs in length. For example, *Eco*RI (named
after the bacterium *Escherichia coli* from which it was isolated) acts like a
precise scalpel to cut double-stranded DNA wherever the nonmethylated 5'-
GAATTC-3' sequence occurs. Several hundred such enzymes, most with different recognition sequences, have been isolated and characterized from various
bacterial strains (Roberts, 1984; examples in Box 3.1). In a bacterium, these
enzymes protect against invasion by foreign DNA (host DNA is protected by
bacterial-specific methylation systems). In a molecular genetics laboratory, restriction enzymes find wide application in assays of DNA restriction fragment-
length polymorphisms (RFLPs).

All RFLP analyses involve cutting (restricting) DNA with one or more endonucleases, separating the resulting fragments according to molecular weight by
gel electrophoresis, and visualizing the size-sorted fragments. Differences

Box 3.1. Recognition Sequences and Precise Cutting Sites (*) in Double-Stranded DNA for Some of the Endonucleases Commonly Employed in Restriction Site Surveys. Py and Pu indicate pyrimidines (T or C) and purines (A or G), respectively.

Enzyme	Recognition Sequence	Enzyme	Recognition Sequence
*Ava*I	5' . . . G*Py C G Pu G . . . 3'	*Eco*RV	5' . . . G A T*A T C . . . 3'
	3' . . . G Pu G C Py*C . . . 5'		3' . . . C T A*T A G . . . 5'
*Bam*HI	5' . . . G*G A T C C . . . 3'	*Mbo*I	5' . . . *G A T C . . . 3'
	3' . . . C C T A G*G . . . 5'		3' . . . C T A G* . . . 5'
*Bcl*I	5' . . . T*G A T C A . . . 3'	*Pst*I	5' . . . C T G C A*G . . . 3'
	3' . . . A C T A G*T . . . 5'		3' . . . G*A C G T C . . . 5'
*Bgl*II	5' . . . A*G A T C T . . . 3'	*Pvu*II	5' . . . C A G*C T G . . . 3'
	3' . . . T C T A G*A . . . 5'		3' . . . G T C*G A C . . . 5'
*Cla*I	5' . . . A T*C G A T . . . 3'	*Spe*I	5' . . . A*C T A G T . . . 3'
	3' . . . T A G C*T A . . . 5'		3' . . . T G A T C*A . . . 5'
*Eco*RI	5' . . . G*A A T T C . . . 3'	*Xba*I	5' . . . T*C T A G A . . . 3'
	3' . . . C T T A A*G . . . 5'		3' . . . A G A T C*T . . . 5'

among individuals in these "digestion profiles" may result from base substitutions within cleavage sites, additions or deletions of DNA, or sequence rearrangements, with each source of variation producing characteristic banding changes. Three important and partially interrelated variables in the assays include electrophoretic media employed, means of fragment visualization, and choice of DNA to be analyzed. These general considerations will be discussed first, and methodological details for particular applications will be added later.

The usual electrophoretic media are agarose or acrylamide gels. These form dense matrices through which larger DNA fragments migrate more slowly than smaller fragments under the influence of an electric current. At neutral pH, DNA is negatively charged (due to properties of the sugar-phosphate backbone), and therefore moves toward the anode of electrophoretic gels at rates determined by molecular size. Agarose gels (0.6–2.0% agarose) are most useful for separating DNA fragments in the size range 300–20,000 base pairs (bp) and acrylamide gels (3.5–20.0%) in the range 10–1000 bp. To facilitate estimation of restriction fragment lengths in the sample DNAs, molecular size standards (commercially available) typically are included in each gel.

Visualization of DNA fragments can be accomplished by several means. Some electrophoretic assays begin with highly purified DNA isolated from particular sources (such as mitochondria), in which case DNA fragments in the gel are revealed by chemical or radioactivity "stains." When DNA amounts are high

(> 50 ng per gel band), ethidium-bromide provides a convenient chemical agent for fragment detection. Ethidium-bromide binds to DNA in such a way that staining intensities are proportional to fragment sizes and digestion profiles appear stoichiometric. Silver-staining is similar and reportedly provides greater sensitivity in detecting small DNA quantities [< 100 pg (Guillemette and Lewis, 1983)]. In highly sensitive "end-labeling" procedures, DNA digestion fragments are labeled radioactively with ^{32}P- or ^{35}S-tagged nucleotides prior to electrophoretic separation. After the gel has been run, it is vacuum-dried and overlaid by X-ray film whose development as an autoradiograph reveals positions to which the DNA fragments migrated. With end-labeling, band intensities are independent of fragment size (because all fragments have two labeled ends), and the method, therefore, is useful in revealing smaller fragments when DNA amounts are limited.

Other RFLP assays begin with DNA of heterogeneous classes (e.g., total nuclear DNA preparations), with elucidation of DNA fragments from particular genes accomplished after electrophoresis by the technique of "Southern hybridization" (Southern, 1975). In this method, all DNA fragments in the gel are denatured in a basic solution and then transferred as single strands (by capillary action or electrophoresis) to a nylon or nitrocellulose membrane. The membrane is incubated with a single-stranded "probe"—DNA previously isolated, purified, and radioactively labeled—under conditions where any strands in the membrane that are complementary to those of the probe hybridize with it to form radioactive duplexes. When high-stringency conditions are employed, hybridization with distantly related or nonhomologous DNA is avoided. Thus, the probe in effect picks complementary and (ideally) homologous sequences to itself from among the thousands or millions of undetected fragments that also have migrated through the gel. These fragments with sequence similarity to the probe then are visualized by autoradiography of the "Southern blot."

The probe in Southern hybridizations thus identifies the DNA under assay in a particular study. This probe may constitute, for example, a single gene from a nuclear or cytoplasmic genome, a noncoding stretch of DNA sequence, or an entire animal mtDNA. If the probe contains DNA present in multiple copies in the genome, the Southern blot reveals fragments from all members of the family to which the probe has hybridized. In some such cases, Southern blots may reveal highly complex digestion profiles wherein nearly all individuals are distinguished by their "DNA fingerprints." The probes in Southern hybridizations may come from DNA highly purified by physical means (e.g., mtDNA isolated via CsCl gradient centrifugation), or more normally via cloning of particular genes through biological vectors (Sambrook et al., 1989). For rapidly evolving sequences, utility of the probe may be confined to assays within or among closely related species, whereas for slowly evolving sequences the probes may retain cross-hybridizing utility across broader taxonomic assemblages. The most common limiting factor in Southern hybridization studies has been the availability of suitable probe DNA.

Because different classes of DNA differ dramatically in terms of the nature of genetic information provided by restriction analysis, as well as in additional details of isolation procedure, they will be discussed separately in the sections that follow.

ANIMAL MITOCHONDRIAL DNA

Procedures   A crucial initial step in isolation of animal mtDNA is the efficient separation of cytoplasm (where mitochondria are housed) from cell nuclei (see Fig. 3.6). Soft tissue such as heart, liver, or ovary is minced, gently homogenized, and centrifuged at low speed (700 × g) to remove some of the nuclei and cellular debris. Subsequent centrifugation at higher speed (20,000 × g) pellets the mitochondria which then are washed and lysed. The next step in the purification involves CsCl-EtBr gradient centrifugation (160,000 × g) for 48 h [or less, depending on the speed and type of rotor employed (Carr and Griffith, 1987)]. The mtDNA, which appears as a discreet band in the gradient, is removed by hypodermic needle and separated from remaining contaminants by dialysis. Purified mtDNA samples can be stored frozen indefinitely. This purified mtDNA then can be used either as a probe in Southern blots to reveal mtDNA bands in heterogeneous DNA samples or the purification process can be repeated for each individual in the genetic survey and the RFLPs elucidated directly by chemical staining or radioactive end-labeling. A rate-limiting step in mtDNA analysis by the above protocol involves the lengthy gradient centrifugations. Various shortcuts for mtDNA isolation are available when large quantities of mitochondrial-rich cells (such as oocytes) are available, or when only small amounts of mtDNA are required (Chapman and Powers, 1984; Jones et al., 1988; Palva and Palva, 1985; Powell and Zuninga, 1983).

Many laboratories working with animal mtDNA employ end-labeling procedures. As typically accomplished, mtDNA samples purified from each individual are digested with restriction endonucleases, and the resulting fragments are radioactively tagged and electrophoretically separated. Development of an autoradiograph reveals the mtDNA digestion profile for each enzyme. For plant mitochondrial and chloroplast genomes, which are much larger, the normal laboratory procedures involve Southern blotting using cloned genes or subsets of the genome as probe. [One recently discovered complication that could apply to the Southern blotting approach, but not to end-labeling of purified mtDNA, occurs when homologues of particular cytoplasmic genes occur also in a cell's nucleus as a result of intergenomic transfer (M.F. Smith et al., 1992).]

Data   The structure and genetic basis of variation in metazoan animal mtDNA is probably better understood than that of any comparably sized region

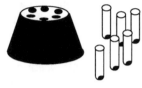

1) Dissect and homogenize
tissues

2) Centrifuge to pellet
mitochondria

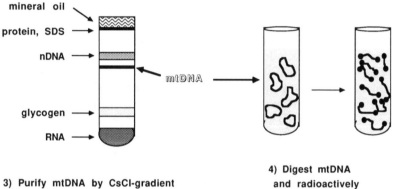

mineral oil

protein, SDS

nDNA

mtDNA

glycogen

RNA

3) Purify mtDNA by CsCl-gradient
centrifugation; remove mtDNA band

4) Digest mtDNA
and radioactively
end-label

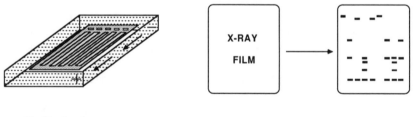

X-RAY
FILM

5) Electrophorese

6) Develop autoradiograph

Figure 3.6. General protocol for mtDNA restriction site analysis by radioactive end-labeling.

of the nuclear genome (see the reviews in Attardi, 1985; Brown, 1985; Cantatore and Saccone, 1987; Gray, 1989; Wallace, 1982; Wolstenholme, 1992). With few exceptions, animal mtDNA is a closed circular molecule, typically 15–20 kilobases (kb) in length, and composed of about 37 genes coding for 22 tRNAs, 2 rRNAs, and 13 mRNAs specifying proteins involved in electron transport and oxidative phosphorylation (Wallace, 1986; Fig. 3.7a). A "control region" of about 1-kb initiates replication and transcription. Gene arrangement in mtDNA appears generally stable, although differences in gene order do distinguish some higher animal taxa (Desjardins and Morais, 1990; Okimoto et al., 1992; Pääbo et al., 1991; Chapter 8). Nearly the entire mtDNA genome is involved in the coding function: introns, large families of repetitive DNA, pseudogenes, and even sizable spacer sequences between genes are rare or lacking.

With regard to the general mode of animal mtDNA evolution, much also is understood (see the reviews in Avise and Lansman, 1983; Avise et al., 1987a; Birley and Croft, 1986; Harrison, 1989; Moritz et al., 1987; Wilson et al., 1985). Individuals commonly are homoplasmic or nearly so, with a single mtDNA sequence predominating in all tissues (however, see Bermingham et al., 1986; Hale and Singh, 1986; Moritz and Brown, 1987, and references therein for exceptions). The reason for this relative sequence homogeneity within individuals probably involves bottlenecks in mtDNA numbers in intermediate germ cell generations (Birky et al., 1989; Chapman et al., 1982; Clark, 1988; Laipis et al., 1988; Rand and Harrison, 1986; Solignac et al., 1984, 1987; Takahata, 1985). Mitochondrial DNA normally evolves rapidly at the sequence level, no doubt due in part to a lack of known repair mechanisms for mutations that arise during replication (Wilson et al., 1985). Some sequences within the control region evolve with exceptional rapidity and have proved to be of utility in high-resolution analyses of population structure (Stoneking et al., 1991). Although addition/deletion changes in mtDNA are not rare, most differences between sequences reflect point mutations, with a strong initial bias for transitions over transversions (Aquadro and Greenberg, 1983; Brown and Simpson, 1982; Brown et al., 1982; Greenberg et al., 1983).

Finally, and most importantly for genetic marker purposes, mtDNA is transmitted predominantly through maternal lines in most species (Avise and Vrijenhoek, 1987; Dawid and Blackler, 1972; Giles et al., 1980; Gyllensten et al., 1985a; Hutchison et al., 1974). Several exceptions to strict maternal inheritance are known (see Avise, 1991b; Gyllensten et al., 1991; Kondo et al., 1990), but even in the few species such as marine mussels (*Mytilus*) where "paternal leakage" is relatively common (Zouros et al., 1992), the paternally- and maternally-derived molecules are not known to recombine genetically in progeny (Hoeh et al., 1991). Genotypes for mtDNA thus represent nonrecombining characters, asexually transmitted usually via females through the pedigrees of what otherwise may be sexually reproducing species. Thus, for simplicity, mtDNA genotypes are referred to as clones or haplotypes, and their inferred evolutionary interrelationships interpreted as estimates of "matriarchal phylogeny" (Avise et al., 1979b). From a functional perspective, mtDNA consists of about 37 genes, but from a phylogenetic perspective the entire mtDNA molecule represents one nonrecombining genealogical unit with multiple alleles.

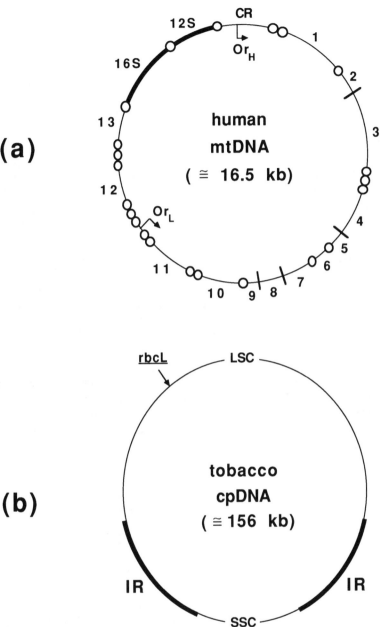

Figure 3.7. Major structural features of animal mitochondrial DNA and plant chloroplast DNA (the two molecules are not drawn to the same scale). (*a*) Human mtDNA, composed of a control region (CR), and genes encoding 2 rRNAs (12S and 16S), 22 tRNAs (open circles), and 13 polypeptides. Also shown are the sites (Or$_H$ and Or$_L$) at which replication is initiated along the complementary DNA strands. (*b*) Tobacco (*Nicotiana tabacum*) cpDNA, composed of large and small single-copy regions (LSC, SSC), and a large inverted repeat (IR). Also shown is the position of the *rbcL* gene, which has figured prominently in nucleotide sequencing studies (Chapter 8).

The raw data in mtDNA restriction surveys consist of fragment length profiles produced by individual enzymes (Fig. 3.8). Because mtDNA is a closed circle, the number of linear fragments is equal to the number of restriction sites recognized by the endonuclease. A useful check on gel scoring is provided by the mtDNA genome size within a given species, to which observed fragment sizes should sum. This feature also facilitates direct comparisons of digestion profiles across studies or among different laboratories (a desirable property not shared fully by allozyme methods, where meaningful comparisons of allelic products require that known electromorph standards be run in all gels). A typical mtDNA population survey may involve use of 10–20 different restriction enzymes and reveal perhaps 50–100 restriction fragments per individual. Because the enzymes employed are commonly five- and six-base cutters, this is equivalent to assaying 250–600 bp of recognition sequence per specimen. The larger mtDNA population surveys sometimes include many hundreds of individuals.

Differences among mtDNA digestion profiles normally arise from point mutations that create or destroy enzyme recognition sequences. Commonly, the pattern of interconversion among digestion profiles resulting from restriction site mutations can be deduced directly from the single-enzyme gel patterns them-

---

Box 3.2. Presence (+) Versus Absence (0) Restriction Site Matrix. This example involves 96 restriction sites representing 20 different mtDNA haplotypes (a–t) observed among 107 sharp-tailed sparrows, *Ammodramus caudacutus* (data from Rising and Avise, 1993).

```
a   ...........00.0......0....................0000...................0....0......0.0.0..........0.....
b   ...........00.0......0...................0000..............00....0......0.0.0..........0.....
c   ...........00.0......0....................000...........0....0......0.0.0..........0.....
d   ...........00.0......0..................0.00.............0....0......0.0.0..........0.....
e   .........00.0......0....................0000..............00....0......0.0.0..................
f   ...........00.0......0....................0000............0....0......0.0.0..........0.....
g   ...........00.0......0....................0000............0....0......00.0.0..........0.....
h   ...........00.0......0...................0.........0......0............00.0.0.....00...0.....
i   ...........00.0......0...................0.0......0.......0............00.0.0.....00...0.....
j   ..........0...0......0...................0.0......0.......0............00.0.0.....00...0.....
k   ...........00.0......0...................0.0......0.......0............00.0.0.....00...0.....
l   ...........00.0......0...................0.0......0......0....00...00...00...0.....
m   ..........0.00......0...................0.0......0.......0............00.0.0.....00...0.....
n   ...........00.00......0...................0.0......0......0............00.0.0.....00...0.....
o   ...........00.0......0...................0.00......0......0............00.0.0.......00.0.0.....
p   ...........00.0......0...................0.00......0......0............00.0.0.....00...0.....
q   ...........00.0......0...................0.00......0......0............00......00.0.0.....
r   ...........00...0......0...................0.0......0......0............00.0.0.....00...0.....
s   ...........00.0......0....0...............0.0......0......0............00.0.0.....00...0.....
t   ...........00.0......0....0...............0.0......0......0............00...0......00...0.....
```

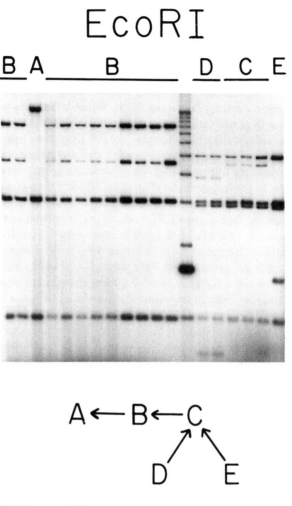

Figure 3.8. Interpretation of mtDNA digestion profiles (from Avise, 1987). Shown is an autoradiograph of *Eco*RI digests of mtDNA from 18 eels (genus *Anguilla*). The seventh lane from the right is a molecular size standard, in which the darkest band is 1.6 kb in size, successive bands above it are approximately 2, 3, 4, 5, 6, 7, 8, . . . kb, and the band below it is 1.0 kb. Five *Eco*RI patterns (A–E) are evident; their interrelationships are summarized in a parsimony network shown under the radiograph (the arrows indicate the direction of restriction site loss, and not necessarily the direction of evolution). For example, pattern A differs from B by loss of an *Eco*RI restriction site, which converts the two larger fragments in the B profile (4.6 and 8.0 kb) to the largest fragment in A (12.6 kb). In turn, C differs from B by gain of an *Eco*RI site which converts B's 8.0-kb fragment to C's fragments of sizes 5.1 kb and 2.9 kb. Pattern E apparently has a "doublet" (two fragments of indistinguishable molecular weight) at 3.1 kb.

selves, using the information from mtDNA fragment sizes (Fig. 3.8). Such information can be accumulated across restriction enzymes and used to generate composite mtDNA clonal descriptions. Data also may be recorded as binary characters and summarized in a presence–absence matrix of restriction sites across individuals or mtDNA clones (Box 3.2). Studies may go further by mapping the positions of sites relative to one another or to landmarks on the mtDNA genome, using double-digestion or partial-digestion procedures (Fig. 3.9). Normally, however, sites can be mapped only to within a few tens or hundreds of base pairs of their true locations.

Some mtDNA gel profiles are sufficiently different or complex that pattern interconversions cannot readily be deduced. One may nonetheless count percentages of shared (and presumably homologous) fragments, but caution is indicated because, unlike site changes, not all fragment changes are independent. For example, a point mutation creating a restriction site will result in the correlated appearance of two smaller fragments with the same total molecular weight as the one larger fragment lost (Fig. 3.9). If many such changes distinguish digestion profiles, the evolutionary interconversions are difficult to deduce. (For similar reasons, digestion profiles produced by restriction enzymes with four-base recognition sites can be difficult to score—they typically involve 10–20 or more mtDNA fragments each—whereas five- and six-base cutters normally produce simple gel patterns with only about 2–10 bands.) Statistical methods that take into account such correlated fragment changes are available for converting percentages of shared fragments to estimates of sequence divergence (e.g., Nei and Li, 1979; Upholt, 1977). However, because of uncertainties that the molecular assumptions underlying these conversion formulas are met in any given study, analyses based on "site" differences are desirable wherever possible.

Additional sources of mtDNA variation stem from occasional differences in mtDNA size, most often due to variations in copy number of localized tandem repeats usually in or near the control region of the molecule (Bermingham et al., 1986; Harrison et al., 1985; Moritz and Brown, 1986, 1987). The larger of these localized repeat regions (they can range in size from a few base pairs to more than 1 kb), are readily distinguished from restriction-site changes because they alter concordantly the sizes of particular restriction fragments across all digestion profiles (smaller fragment-size differences might be overlooked, however, particularly when they reside in high-molecular-weight gel bands). It is important to properly characterize genomic size changes so that they are not misinterpreted as independent restriction-site changes involving multiple enzymes.

Restriction analyses of animal mtDNA normally are applied at the level of conspecific populations and closely related species, but other features of the molecule have proved useful in higher-level phylogenetics. From sequencing studies, it is known that transitions greatly outnumber transversions in the early stages of mtDNA divergence, but because transversions tend to accumulate in a ratchet fashion through time, the observed *ratio* of transitions to transversions may itself provide a useful dating device (Moritz et al., 1987). Other potentially useful phylogenetic features of mtDNA include gene arrangement and gene content [which are known to differ among some higher animal taxa (Hoffmann et al., 1992; Hyman et al., 1988; Moritz et al., 1987; Wolstenholme et al., 1985)], structures of the tRNA and rRNA genes and their products (Cantatore et

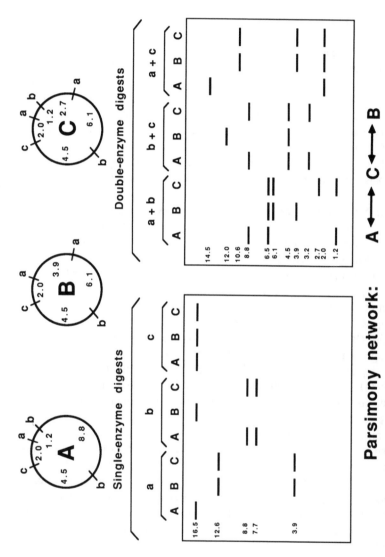

Parsimony network: A ←→ C ←→ B

Figure 3.9. Scoring and mapping of mtDNA restriction sites. Shown are three different mtDNA haplotypes (A, B, and C) as evidenced by digestion patterns produced by three restriction endonucleases (a, b, and c). The restriction maps at the top are unknown at the outset, but have been deduced from observed gel profiles produced in the single- and double-enzyme digestions. Fragment sizes (in kb) are indicated. A parsimony network at the bottom summarizes the most parsimonious pathway of evolutionary interconversion between haplotypes with respect to the sites assayed. Note that, in this case, restriction site changes and the parsimony network (but not the full restriction-site map) also could have been deduced directly from the single-enzyme digestion profiles.

al., 1987; Wolstenholme et al., 1987), the genetic codes utilized, modes of mtDNA replication and transcription, and, as described later, various mtDNA sequences themselves (Chapter 8).

PLANT MITOCHONDRIAL AND CHLOROPLAST DNA

Several proven features in the evolution of animal mtDNA were completely unanticipated (Avise, 1991a; Avise and Lansman, 1983). These include the usual pattern of within-individual homoplasmy (predominance of a single mtDNA sequence), despite between-individual sequence differences, and the rapid pace of nucleotide substitution in the face of what would seem to be severe evolutionary constraints on function as judged by the "genetic economy" of the molecule (Attardi, 1985). However, once the major evolutionary patterns of animal mtDNA were revealed, it might be supposed that they would apply to other cytoplasmic genomes as well. Surprisingly, such has not proved to be the case.

Plant mtDNA is highly variable in size [200–2400 kb (Palmer, 1985; Pring and Lonsdale, 1985; Ward et al., 1981)] and typically exists as a collection of different-sized circles arising from extensive recombinational processes within individuals that interconvert between a "master" molecule and subgenomic circles (Hanson and Folkerts, 1992; Palmer and Herbon, 1986; Palmer and Shields, 1984). Inheritance is usually but not invariably maternal (Birky, 1978; Forsthoefel et al., 1992). Although plant and animal mtDNAs are similar with regard to gene content and general function, their evolutionary patterns differ diametrically (Birky, 1988; Palmer, 1992). For species assayed to date, plant mtDNA appears to evolve rapidly with respect to gene order, but slowly in nucleotide sequence [perhaps 100-fold slower than animal mtDNA (Palmer and Herbon, 1988)]. Reasons for the slow accumulation of point mutations are not understood, but one ad hoc hypothesis is that plant mitochondria may possess relatively error-free DNA replication systems, or perhaps highly efficient enzymes for repair of DNA damage (Palmer and Herbon, 1988). In any event, the complicated recombinational features and technical difficulties in laboratory assay have conspired to limit the utility of plant mtDNA in molecular systematics.

Chloroplast DNA provides yet another story. cpDNA is transmitted maternally in most plants (Birky, 1978; Gillham, 1978; Hachtel, 1980), biparentally in some (e.g., Metzlaff et al., 1981) and *paternally* in several assayed gymnosperms (e.g., Dong et al., 1992; Szmidt et al., 1987; Wagner et al., 1987; see also Schnabel and Asmussen, 1989). cpDNA sequences are known to move occasionally to the nucleus (Baldauf and Palmer, 1990; Baldauf et al., 1990; Gantt et al., 1991), and indeed there is considerable indirect support for past exchanges among the mtDNA, cpDNA, and nDNA genomes (Fox, 1983; Gellissen et al., 1983; Gray, 1989; Nugent and Palmer, 1991; Stern and Palmer,

1984). The molecule varies in size from about 120 to 217 kb in photosynthetic land plants, due largely to extent of reiteration of a large inverted repeat that includes genes for the rRNA subunits (Zurawski and Clegg, 1987) (Fig. 3.7b). With some possible exceptions (Milligan et al., 1989; Wagner et al., 1987), the rate of cpDNA evolution generally appears slow both in terms of primary nucleotide sequence [silent substitution rates have been estimated at only three to four times greater than those of plant mtDNA (Wolfe et al., 1987)] and in terms of gene rearrangement (Curtis and Clegg, 1984; dePamphilis and Palmer, 1989; Palmer, 1990; Ritland and Clegg, 1987). Because of the large size of cpDNA and its leisurely pace of evolution, most systematic treatments have involved restriction site or sequence determinations for particular genes (Clegg et al., 1986; Palmer, 1987; Palmer et al., 1988a; Zurawski and Clegg, 1987) or have monitored the taxonomic distributions of unique cpDNA structural features across higher-level plant taxa (e.g., Downie and Palmer, 1992; Jansen and Palmer, 1987; see Chapter 8). Nonetheless, some studies have uncovered considerable intraspecific cpDNA variation as well (D.E. Soltis et al., 1992).

SINGLE-COPY NUCLEAR DNA

Procedures Restriction analyses of single-copy nuclear (scn) DNA traditionally have relied on Southern blotting procedures, with the probes representing DNA sequences cloned into a biological vector such as lambda phage or a bacterial plasmid (Kochert, 1989; Fig. 3.10). Once suitable probes are available, the general Southern blotting methodology proceeds as summarized in Figure 3.11.

Clones used as probes may represent particular genes of known function or anonymous gene regions drawn at random from single-copy sequences in a genomic library (a library is a collection of cloned DNA fragments). Construction of suitable probes is the most challenging aspect of scnRFLP analysis. Probes for genes of known function sometimes are derived from "c" (complementary) DNA—sequences produced by reverse transcription of a particular messenger RNA. Such probes have been developed for a wide variety of genes and species, and are exchanged routinely among research laboratories. An alternative approach involves generation of anonymous single-copy probes, as follows. Total cell DNA is extracted and digested to completion with a restriction enzyme such as *Hind*III. Fragments of size 500–5000 bp are isolated by gel electrophoresis and cloned into a suitable vector, thereby generating a DNA library. The library then is screened for single-copy sequences by "dot blot" hybridization (Fig. 3.12), whereby the DNA contents of each clone are hybridized under controlled conditions with radioactively labeled total cell DNA. The radioactive signal intensity of the clone dot is a reflection of the genomic copy number of a particular sequence, such that strong signals identify clones carrying

1) Extract DNA

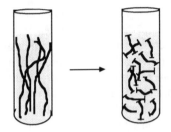

2) Digest DNA with
restriction enzyme

3) Open bacterial plasmids by
digestion with same
restriction enzyme

4) Ligate DNA into plasmids

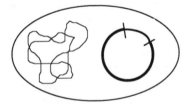

5) Insert recombinant plasmids
into bacteria

6) Culture transformed bacteria
to multiply DNA

Figure 3.10. General protocol for DNA cloning and genomic library construction.

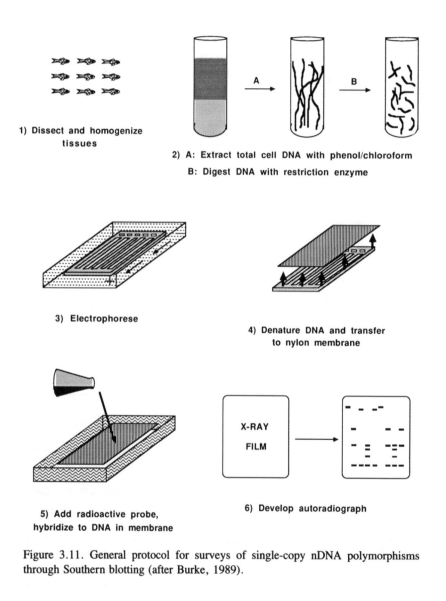

1) Dissect and homogenize tissues

2) A: Extract total cell DNA with phenol/chloroform
   B: Digest DNA with restriction enzyme

3) Electrophorese

4) Denature DNA and transfer to nylon membrane

5) Add radioactive probe, hybridize to DNA in membrane

6) Develop autoradiograph

Figure 3.11. General protocol for surveys of single-copy nDNA polymorphisms through Southern blotting (after Burke, 1989).

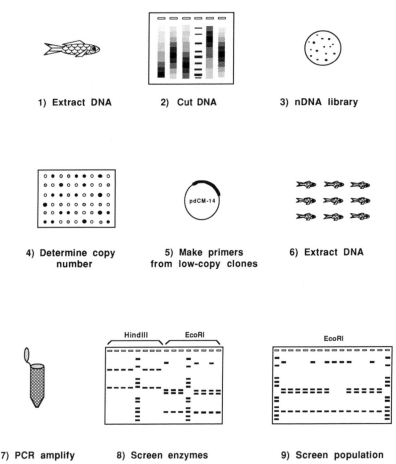

Figure 3.12. General protocol for surveys of scnDNA polymorphisms through the use of the polymerase chain reaction.

repetitive DNA and absent or weak signals identify clones containing single- or low-copy sequences. One important distinction between the use of cDNA and genomic clones as probes is that the former represent processed coding sequences for the transcribed gene product, whereas the latter also may include regions flanking the gene, and introns (noncoding stretches carried within most genes that are interspersed with the sequences specifying amino acids). Some of these noncoding sequences may evolve especially rapidly and provide an additional class of genetic markers.

Recently, alternative methods for generating scnRFLPs have been developed that take advantage of the polymerase chain reaction (PCR; Saperstein and Nickerson, 1991). For single-copy sequences identified in a nuclear genomic library, primers for the PCR reaction (described later) are generated and employed to amplify homologous DNA from each sampled individual. These amplified DNAs then are digested by restriction enzymes, electrophoresed, and stained directly (e.g., with EtBr). The process is summarized in Figure 3.12. This PCR-based method offers several advantages over the Southern blotting approach (Karl and Avise, 1993): Because only a small amount of DNA is required for the PCR reaction, assays can be conducted even when tissue sources are limited; the DNA analyzed is of defined length (i.e., bounded by primers), so size differences underlying digestion profiles readily can be distinguished from restriction-site differences, and scoring generally is facilitated; because DNA is amplified in vitro with the PCR, DNA methylation can be eliminated as a potential source of fragment variation (amplified DNA is completely unmethylated); and finally, because large amounts of DNA can be produced by amplification, the PCR-based method eliminates the need for radionucleotides and the long exposure times (up to 7 days) required to produce a scorable signal by autoradiography of Southern blots. On the negative side, considerable effort and expense are entailed in library construction, screening for single-copy sequences, and generation of primers for the PCR reactions.

Data Single-copy sequences are those that occur with a frequency of one per haploid genome (as opposed to various classes of repetitive DNA that occur in multiple copies). All methods for restriction analysis of single-copy DNA are intended to reveal polymorphisms (nucleotide substitutions, addition/deletions, inversions, etc.) within such particular stretches of sequence. The raw data are in many respects analogous to those provided by protein electrophoresis: For each gene region assayed, diploid individuals can be described as homozygous or heterozygous for various restriction-site polymorphisms (Fig. 3.13); the Mendelian nature of the polymorphisms can be verified by breeding studies or by agreement of allele frequencies with Hardy-Weinberg expectations for populations suspected of random mating; genotypic descriptions can be accumulated across many enzyme/gene combinations; a population may exhibit multiple alleles at each "locus"; and the alleles thus identified usually remain unordered phylogenetically (unless multisite haplotype characterizations can be accomplished in regions of low recombination—see the following paragraphs and also the discussion of gene trees in Chapter 4). The major advantage over protein electrophoresis is that, in principle, a nearly unlimited pool of genetic variants may be tapped. Thousands of single- or low-copy regions, to each of which many restriction enzymes can be applied, are present in most organisms (single-copy DNA constitutes 20–80% of most plant and animal genomes—see the review in Sibley and Ahlquist, 1990). Also, the assays reveal polymorphisms at both silent and replacement nucleotide positions (as well as that due to other sources such as duplications/deletions and insertions of transposable elements).

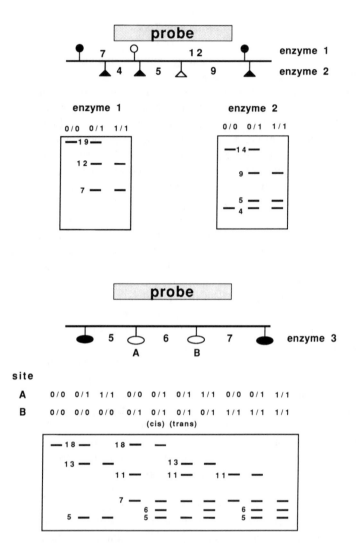

Figure 3.13. Interpretation of restriction digestion profiles for Mendelian polymorphisms in diploid organisms, as assayed by Southern blotting using scnDNA probes. Shown are gel patterns produced by each of three enzymes whose restriction site positions are indicated by circles, triangles, and ovals. Closed symbols indicate nonvariable restriction sites; open symbols indicate variable sites that are either present (1) or absent (0) along a given chromosome. Heterozygous individuals are designated as "0/1" and respective homozygotes by "1/1" and "0/0." In the bottom figure, note that each lane (individual) in the gel is characterized by two diploid genotypes, reflecting various combinations of presence versus absence at each of two adjacent but variable restriction-site positions. Other numbers in the figure indicate the sizes (in kb) of various restriction fragments.

The major disadvantage in comparison to protein electrophoresis is that the methods are expensive, both in effort and materials. Perhaps for this reason, rather few applications of scnRFLP approaches in evolutionary biology have as yet appeared, despite the fact that these methods are widespread in related research endeavors including the mapping of disease genes and quantitative trait loci (Botstein et al., 1980; B. Martin et al., 1989; Paterson et al., 1988; Weller et al., 1988), and in breeding studies and strain verification in domestic species (Apuya et al., 1988; Beckmann et al., 1986).

Even when DNA sequences assayed by the above procedures are short (i.e., 0.5–2 kb), particular segments sometimes exhibit two or more linked restriction polymorphisms (Fig. 3.13). When multiple RFLPs are revealed within a defined gene region, their phases become of additional interest. Phase refers to the pattern of association (cis versus trans configuration) among multiple variants along a chromosome. Suppose, for example, that two linked polymorphisms in a diploid organism each involve presence (1) versus absence (0) of a restriction site, such that the four possible states for a nuclear chromosome are 1,1 and 0,0 (cis configurations) and 1,0 and 0,1 (trans configurations). These states are evident directly from the assayed diploid genotypes of all individuals except double heterozygotes. In populations where the unambiguous diploid genotypes reveal a consistent phase for the variable sites (complete disequilibrium), it is probably safe to assume that the double heterozygotes carry two chromosomes with these same linkage phases (Karl et al., 1992). But when disequilibrium is incomplete, the phases of double heterozygotes cannot always be recovered from the gel digestion patterns (Fig. 3.14). The difficulties of phase determination in diploids become even greater when more than two linked polymorphisms are involved. For these reasons (and because of the likelihood of intragenic recombination in the history of the assayed region), complete haplotype determinations are seldom accomplished from scnRFLP approaches. Thus, unlike mitochondrial RFLP data, where haplotype phase is obvious and assessment of phylogeny among alleles routine, evolutionary relationships among scnRFLP alleles of a particular locus are seldom determined.

RIBOSOMAL RNA GENES AND OTHER MIDDLE-REPETITIVE GENE FAMILIES

When Southern blotting involves the use of a DNA probe that contains homologies to repetitive genomic sequences, the probe hybridizes to all such sequences and thereby simultaneously reveals restriction fragment changes at multiple members of the gene family. Ribosomal RNA genes in the nuclei of eukaryotic cells usually exist as tandemly repeated elements, each repeat unit composed of a highly conserved coding sequence of total length about 6 kb, plus shorter and more variable noncoding spacer regions (Fig. 3.15). The repeated rDNA modules may occur at one or several chromosomal sites. The copy number of rDNA repeats per genome varies from several hundred in some mammals and insects to many thousands in plants (Long and Dawid, 1980).

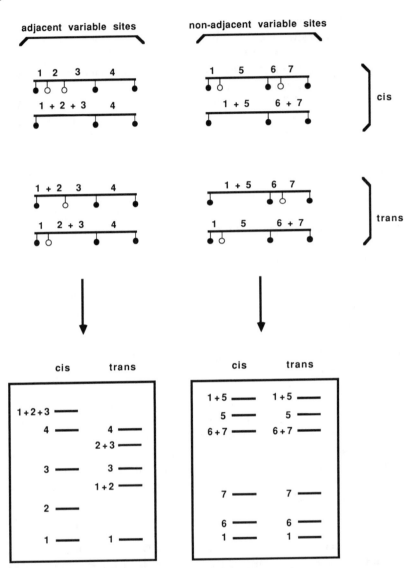

Figure 3.14. Effects of the position of variant restriction sites on haplotype phase determination in diploid individuals that are double-heterozygotes (after Quinn and White, 1987). When the variant sites (open circles) for a restriction enzyme are adjacent (left-hand column), RFLP digestion profiles differ between the cis and trans phases. When one or more invariant sites (closed circles) intervene between the variant sites (as in the right column), cis and trans heterozygotes produce identical gel profiles and, hence, cannot be distinguished.

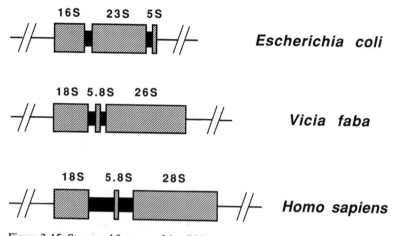

Figure 3.15. Structural features of the rDNA repeat module (drawn to approximate scale) in *E. coli,* and in a representative plant and animal (after Appels and Honeycutt, 1986). Hatched regions indicate loci encoding the "small" (16S and 18S) and "large" (23S, 26S, 28S) subunits of ribosomal RNA, as well as the "5S" rRNA elements. Black regions indicate internal transcribed spacers, which often differ in length.

The ready availability of probes for ribosomal RNA genes has prompted a number of studies of population variation and differentiation in these genetic regions based on Southern blotting methods (Appels and Dvorak, 1982; Arnold et al., 1991; Davis et al., 1990; Rieseberg et al., 1990a,b; Rogers et al., 1986; Saghai-Maroof et al., 1984; Schaal et al., 1987; Williams et al., 1985). These studies have revealed RFLP markers that often distinguish closely related species, and sometimes conspecific populations. The genetic differences normally involve varying lengths of the repeat unit due to heterogeneity in size of the spacer regions, with additional variation occasionally stemming from restriction site changes in both the coding and spacer segments (Schaal, 1985).

The occurrence of multiple rDNA copies per genome opens the possibility of intraindividual polymorphism, which indeed has been observed within both plants and animals. A potential difficulty in interpreting genetic markers provided by any multigene family involves understanding the degree to which mechanisms of concerted evolution may have homogenized the repeated DNA sequences (Arnheim, 1983; Ohta, 1980; Ohta and Dover, 1983), and hence the extent to which different family members can be viewed as providing independent bits of phylogenetic information. Various complications ultimately related to the problem of distinguishing orthology from paralogy (see Chapter 1) can arise. For example, Williams et al. (1987) showed that variants for rRNA genes have a nonrandom distribution among the X and Y chromosomes in *Drosophila,* suggesting that within-chromosome homogenizing mechanisms are more effi-

cient than those working among chromosomes; Arnold et al. (1988) showed that
biased gene conversion has influenced the distributions of sequences for rRNA
genes in a hybrid zone between grasshopper subspecies. Thus, the use of nuclear
rRNA genes (and other repetitive DNA families) as genetic markers in microevolutionary studies could be compromised by the heterogeneity of molecular
processes governing the evolutionary turnover and, hence, genetic relationships
among the repeat elements themselves (Schaal et al., 1991). On the other hand,
Hamby and Zimmer (1992) conclude that ''the most remarkable feature of rDNA
is the overall sequence homogeneity among members of the gene family.'' For
use as simple genetic markers, an ideal situation would be where concerted
evolution was so pronounced that all copies of a repetitive sequence within an
individual were quickly homogenized, and each specimen could thus be characterized by an unambiguous, specifiable genotype. However, such extreme
concerted evolution also would carry the consequence of confining the information content of a multigene family to that of a single ''locus.''

Despite these potential complications, RFLP markers from rDNAs have contributed to studies of geographic population structure and patterns of introgression
in hybrid zones (Arnold et al., 1987; Baker et al., 1989; Cutler et al., 1991; Learn
and Schaal, 1987). Far more prominent, however, have been nucleotide sequence
analyses of both nuclear and cytoplasmic ribosomal RNA genes (Hillis and Dixon,
1991; Mindell and Honeycutt, 1990), which have provided much of the molecular
data for phylogenetic reconstructions among deeper branches in the tree of life
(Chapter 8). Naturally, these studies have focused on slowly evolving coding
regions of the rRNA genes. At these macro-evolutionary scales, the fraction of
overall rDNA sequence heterogeneity attributable to intraindividual or intraspecific polymorphism becomes negligible, and safely can be neglected.

Southern blotting procedures have been employed to assess levels of genetic
variability in other kinds of multigene families as well. For example, the major
histocompatibility complex (MHC) is a family of tightly linked homologous loci
that encodes cell surface antigens involved in the immunological response. In
humans, mice, and other mammals, particular MHC genes are known to be
extremely polymorphic, some with scores of alleles (Hedrick et al., 1991;
Hughes and Nei, 1988, 1989; Klein, 1986). Feline probes homologous to one
class of MHC loci were employed by Yuhki and O'Brien (1990; see also Winkler
et al., 1989) to assess molecular variation in African cheetahs and Asiatic lions,
species suspected by other criteria to possess low genome-wide variability due to
historical bottlenecks in population size (Chapter 9). Mammalian MHC probes
have also revealed DNA polymorphisms in birds (Gibbs et al., 1991).

MINISATELLITE SEQUENCES AND DNA FINGERPRINTING

The genetic complexity inherent in repetitive DNA families sometimes can be
turned to advantage in terms of providing individual-specific genetic markers.

The term "DNA fingerprinting" usually is associated with a molecular approach introduced by Jeffreys et al. (1985a), in which Southern blot assays of hypervariable regions of DNA reveal gel banding profiles that distinguish most or all individuals (barring monozygotic twins) within a sexually reproducing species. The DNA probes originally employed by Jeffreys (1987) hybridize to conserved core sequences (10–15 bp long) scattered in numerous arrays about the human genome as part of a system of "dispersed tandem repeats" (Fig. 3.16), also referred to as minisatellite DNA sequences or as VNTR (variable number of tandem repeat) loci. Each repetitive unit within an array is about 16–64 bp long. Increases and decreases in the lengths of particular arrays result from changes in the repeat copy number arising from high rates of unequal crossing-over during meiosis [indeed, there is speculation that the minisatellite sequences themselves provide hotspots for recombination (Jarman and Wells, 1989)].

The original Jeffreys' probes were isolated from a myoglobin intron in humans and applied to problems in human forensics (Dodd, 1985; Gill et al., 1985; Jeffreys et al., 1985b, 1985c), but these probes also cross-hybridize to reveal DNA banding profiles in many species including other mammals (Hill, 1987; Jeffreys and Morton, 1987; Jeffreys et al., 1987), birds (Brock and White, 1991; Burke and Bruford, 1987; Hanotte et al., 1992a; Meng et al., 1990), fishes (Baker et al., 1992), and even some invertebrates such as corals and snails (Coffroth et al., 1992; Jarne et al., 1990, 1992). Other probes for hypervariable minisatellites have been identified (such as one from the M13 phage) that behave similarly in providing DNA fingerprints in additional vertebrate taxa (Georges et al., 1987; Longmire et al., 1990, 1992; Vassart et al., 1987), as well as in some invertebrate animals (Zeh et al., 1992), plants (Rogstad et al., 1988), and microbes (Ryskov et al., 1988).

The complex gel profiles characteristic of DNA fingerprints appear when an enzyme is employed that cleaves outside the repeat arrays (Fig. 3.16). From each homologous chromosome position in an individual, either one or two bands will be revealed, depending on whether the specimen is homozygous or heterozygous with respect to number of tandem repeats in the array. The presence of several such arrays scattered about the genome results in multilocus digestion profiles typically consisting of some 20 or more scorable bands per individual in the 4–23-kb size range. In an animal population, dozens of alleles characterized by differing lengths may be segregating at each chromosomal position. The multilocus nature of the data and the sheer complexity in single gel profiles provide a convenient means for establishing genetic identity versus nonidentity, as well as for assessing parentage [because barring spontaneous de novo mutation (Jeffreys et al., 1988b), each band in a individual's DNA fingerprint must derive from either its biological mother or father].

In multilocus DNA fingerprints, the advantage of genetic complexity for individual diagnosis becomes a liability in other contexts. Generally, it remains

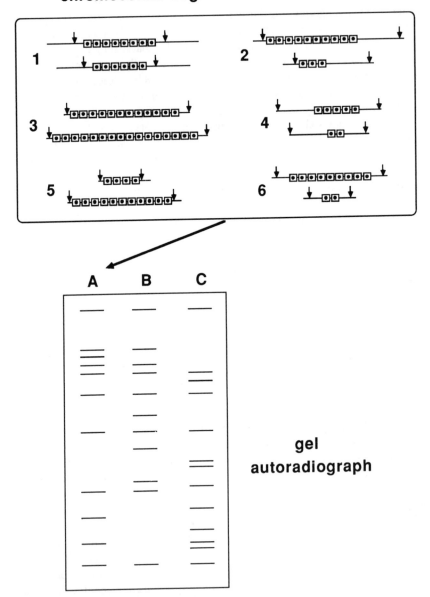

Figure 3.16. DNA fingerprinting from VNTR loci. Shown are six dispersed chromo-somal segments (on the same or different chromosomes), each of which may harbor variable numbers of the tandem repeat elements. Solid circles within the repeats indicate the conserved core sequence to which the probe hybridizes in a Southern blot. A restriction enzyme that cuts (arrows) outside the repeat regions thus reveals a complex digestion profile on a gel autoradiograph.

unknown which bands in a fingerprint belong to which loci (array), so allelism is not established. Thus, whether individuals are homozygous or heterozygous at particular "genes" seldom is ascertained, nor can the allele or genotype frequencies characterizing populations be determined. These problems seriously compromise attempts to estimate genetic relatedness from the measurable attribute in DNA fingerprints—the fraction of shared marker bands (Lynch, 1988; but see also Kuhnlein et al., 1990 and Packer et al., 1991)—or to estimate gene flow or other population parameters employing the usual statistical algorithms that require straightforward Mendelian markers.

These shortcomings of multilocus DNA fingerprinting methods for applications other than genetic identity and parentage assessment have prompted development of methods for analyzing particular minisatellite regions, one at a time. These include use of refined DNA probes and more stringent hybridization conditions in Southern blotting, or PCR-based methods (Horn et al., 1989; see next section) to reveal genetic variation tied to individual hypervariable loci (Jeffreys et al., 1988a; Nakamura et al., 1987; Wong et al., 1986, 1987). For example, under stringent assays, a "3'HVR" probe consisting of tandem 17-bp repeats hybridizes to a single region on human chromosome 16 downstream from the α-globin complex (Higgs et al., 1986; Jarman et al., 1986). Each such assayed VNTR locus produces only one or two bands in a particular DNA digestion, depending on whether the individual in question is homozygous or heterozygous for the number of repeats (or in other flanking regions) between restriction sites. In human populations, several such hypervariable loci exhibit dozens of alleles, some of which may be difficult to distinguish unambiguously by gel mobility alone (Devlin et al., 1991). Another recent variant of the "single-locus" fingerprinting approach circumvents these ambiguities in allelic identification by characterizing patterns of base substitution (in addition to repeat copy number) within particular minisatellite loci (Jeffreys et al., 1990, 1991).

The accumulation of genetic information from several hypervariable single-locus regions combines the advantages of relatively straightforward allelic interpretation (characteristic of the allozyme and scnDNA approaches described earlier) with the individual-diagnostic power of multilocus DNA fingerprinting. However, as with allozyme and scnDNA methods (but unlike the situation for mtDNA), detailed phylogenetic relationships among the various alleles of a VNTR locus normally are not assessed. Use of single-locus VNTR assays is the current method of choice by companies and agencies involved in human forensic practice (Balazs et al., 1989; Budowle et al., 1991; Chapter 5). Single-locus VNTR probes also are being developed for other taxa such as birds (Burke et al., 1991; Hanotte et al., 1991). A current limitation of this approach is that the probes often prove highly taxon-specific (unlike the multilocus Jeffreys' probes employed under low-stringency conditions). However, some minisatellite probes do cross-hybridize under high stringency conditions to highly variable loci in numerous related species (Hanotte et al., 1992b).

Recently, another class of tandem repetitive elements referred to as "micro-satellites" has been discovered in various animals (Fries et al., 1990; Litt and Luty, 1989; Stallings et al., 1991; Tautz, 1989; Weber and May, 1989) and plants (Nybom et al., 1992). Each microsatellite locus consists of reiterated short sequences (particular di-, tri- or tetra-nucleotides) tandemly arrayed (Hamada et al., 1984), with variations in repeat copy number accounting for a profusion of distinguishable alleles. In humans, for example, hundreds of microsatellite loci have been characterized, each typically with several alleles (5–10 or more) that may be revealed as RFLPs following PCR amplifications using conserved flanking regions (see Valdés et al., 1993). These highly-allelic Mendelian polymorphisms [known also as "simple sequence length polymorphisms" or SSLPs (Schlötterer et al. (1991)] should find widespread application in various areas of population biology such as gene flow estimation and parentage assessment (Ellegren, 1991). Although microsatellite approaches are still in their infancy with regard to non-human population applications (at the time of this writing), they evidence the broader point that additional sources for molecular markers likely will be developed and employed widely in the near future.

DNA Sequencing and the Polymerase Chain Reaction

PRINCIPLES AND PROCEDURES

Two methods for direct sequence determination of purified DNA have been available for nearly two decades. One approach, introduced by Maxam and Gilbert (1977, 1980) (Fig. 3.17), relies on chemical cleavage reactions specific to individual bases (A, T, C, or G). The ends of a targeted stretch of DNA are radioactively labeled and the DNA is divided into four subsamples which then are treated with different chemical reagents that cleave at base-specific positions. For example, one subsample is treated with dimethyl sulfate and piperidine, which results in DNA cleavage only at the G positions. Reactions are carried out under conditions such that only a small, random fraction of sites actually is cleaved in any molecule, so that the composite digestion contains a collection of fragments of varying length, each terminated at a G position. Then fragments are separated electrophoretically in a polyacrylamide gel and visualized by autoradiography. Parallel reactions specific for the other three bases are carried out and the fragments separated on adjacent lanes of the gel. Thus, the DNA sequence can be read directly from the ladderlike banding pattern appearing on an autoradiograph (Fig. 3.17).

An alternative sequencing procedure, introduced at the same time by Sanger et al. (1977), is the usual technique of choice today. This method relies on the controlled interruption of in vitro DNA replication (Fig. 3.17). Double-stranded DNA is denatured to single strands, and a short DNA segment (the primer) known

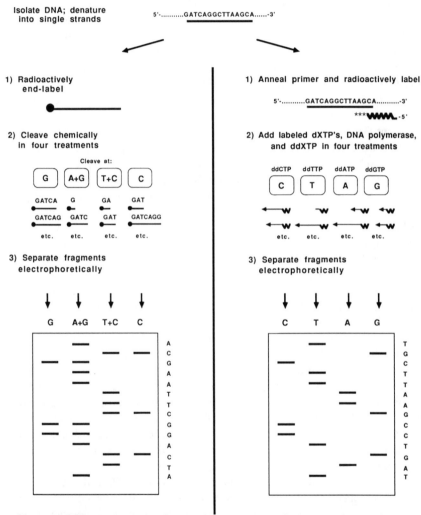

Maxam-Gilbert approach   **Sanger approach**

Figure 3.17. General protocols for DNA sequencing via the Maxam–Gilbert (left panel) and Sanger (right panel) approaches (after Hillis et al., 1990).

to be complementary to a sequence on the target DNA is annealed to the target sample. This primer/template mixture is divided into four subsamples, each of which is subjected to a primer extension reaction catalyzed by DNA polymerase. All four reactions contain the four deoxynucleotides (dA, dC, dG, and dT), plus a single dideoxynucleotide (ddN, a nucleotide that lacks the 3'-OH group present in deoxynucleotides). The synthesized DNA strand is made radioactive, either by labeling the end of the primer or by the incorporation of a labeled deoxynucleotide during synthesis. DNA sequence extension occurs by attachment of nucleotides to a free 3'-OH, so that wherever ddNTP has been incorporated into the growing strand, further strand extension is arrested. The polymerase reaction is carried out under conditions such that the incorporation of ddNTPs is rare and random. Thus, different DNA molecules in a subsample achieve varying lengths before termination at a particular base. As under the Maxam-Gilbert approach, fragments from the four subsamples are separated electrophoretically in a polyacrylamide gel, visualized by autoradiography, and the DNA sequence read directly.

In the past, sequencing applications were limited by the availability of purified, homologous DNA sequences from different organisms. Such sequences had to be isolated and amplified in vivo by laborious procedures of cloning into microbial vectors (Fig. 3.10). The recent development of the polymerase chain reaction (Mullis and Faloona, 1987; Saika et al., 1985, 1988) has changed this situation dramatically by permitting rapid in vitro DNA amplifications. As a result, DNA sequencing has grown explosively in the last decade and has become among the most popular of methods for phylogenetic reconstruction (Miyamoto and Cracraft, 1991).

The PCR technique involves three steps (Fig. 3.18): (a) denaturation of double-stranded DNA by heating; (b) annealing of extension primers to sites flanking the region to be amplified; and (c) primer extension, in which strands complementary to the region between the flanking primers are synthesized under the influence of a DNA polymerase (*Taq*) which is thermostable. The double-stranded products are cycled repeatedly through steps (a)–(c). In each round of denaturation-annealing-extension, the target sequence is roughly doubled in the reaction mixture, so that after 20 or more rounds the product assumes overwhelming preponderance and can be sequenced directly as purified DNA spanning the primers. Indeed, the amplification primers for PCR also can be employed as sequencing primers in the Sanger reactions, allowing a direct coupling of these approaches. The PCR and sequencing procedures are becoming increasingly automated and can be carried out with commercially-available temperature cyclers coupled to a sequencing apparatus.

The primers employed to initiate the PCR process are short sequences (about 20–30 nucleotides long) that exhibit high sequence similarity (particularly in the 3' end) to regions flanking the target sequence. For example, several such primers have been identified in animal mtDNA (Box 3.3), which for reasons dis-

1) **Isolate DNA**

2) **Denature and anneal primers**

3) **Primer extension**

4) **Denature and anneal primers**

5) **Primer extension**

6) **Denature and anneal primers**

7) **Primer extension**

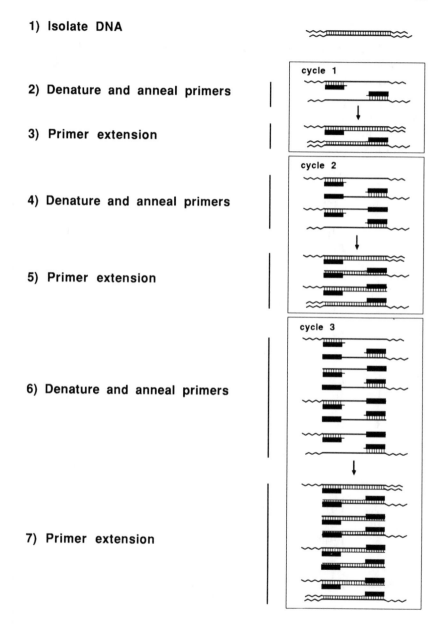

8) **Repeat cycles**

Figure 3.18. General protocol of the polymerase chain reaction for amplifying DNA (after Oste, 1988).

Box 3.3. Examples of Conserved Primer Sequences Within Various Mitochondrial Genes That Have Proved Useful for PCR Amplifications.

Sequences are given in the 5′ to 3′ direction; Y indicates either a cytosine (C) or thymine (T); R either an adenine (A) or guanine (G); and W indicates an A or T. Several of these primers were developed for fishes, but in many cases they successfully amplify DNA from other animals as well.

Gene	Primer	Primer Sequence	Reference
Control region	L15926	TCAAAGCTTACACAGTCTTG-TAAACC	Kocher et al., 1989
	H16498	CCTGAAGTAGGAACCAGATG	Meyer et al., 1990
	L16518	CATCTGGTTCTTTCTTCAGGG-CCAT	Meyer, 1993
12S rRNA	H1109	GTGGGGTATCTAATCCCAGTT	Meyer, 1993
	L1091	AAAAAGCTTCAAACTGGGATT-AGATACCCCACTAT	Kocher et al., 1989
	H1478	TGACTGCAGAGGGTGAC-GGGGCGGTGTGT	Kocher et al., 1989
Cytochrome oxidase I	L5950	ACAATCACAAAGAYATYGG	Normark et al., 1991
	H7196	AGAAAATGTTGWGGGAARAA	Normark et al., 1991
ATPase 6	H8517	GGGRACTTTGACTGGTACT	Meyer, 1993
	L8580	AGCCCCACATACCTAGGTA-TCCC	Meyer, 1993
	H8907	GGGGTTCCTTCAGGCAAT-AAATG	Meyer, 1993
Cytochrome oxidase III	L9225	CACCAAGCACACGCATACCA-CAT	Meyer, 1993
	H9407	AAAGTTCCTGTGGTGTG-CGGGGG	Meyer, 1993
Cytochrome b	L14841	AAAAAGCTTCCATCCAA-CATCTCAGCATGATGAAA	Kocher et al., 1989
	L15020	GCYAAYGGCGCATCCT-TYTTYTT	Meyer, 1993
	H15149	AAACTGCAGCCCCTCAGAA-TGATATTTGTCCTCA	Kocher et al., 1989

S$_1$ S$_2$ F$_1$ Examples of F$_2$ or backcross genotypes

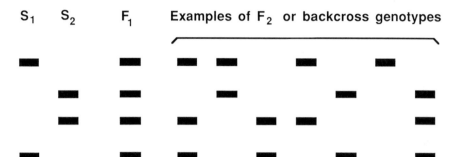

Figure 3.19. Nature of RFLP data generated by the "RAPD" procedure. In this approach, short random primers are employed in PCR reactions to amplify anonymous DNA segments (in this case, from four unlinked gene regions). The presence of a band indicates successful amplification; absence indicates unsuccessful amplification, perhaps resulting from naturally occurring mutations in the primer recognition site. Thus, RAPDs in diploid organisms usually behave in a dominant/recessive fashion (although some cases of codominant inheritance are known—e.g., Fritsch and Rieseberg, 1992). As exemplified in this figure, the RAPD approach may be especially useful in characterizing first-generation or later-generation hybrids between genetically distinct species, S$_1$ and S$_2$.

cussed later is the molecule toward which the majority of PCR efforts in population biology thus far has been directed. The PCR also can be coupled to DNA assays other than sequencing, for example in the generation of nuclear RFLPs from the amplified sequences, as in the microsatellite and various other approaches described earlier. Another PCR-based RFLP approach is the "RAPD" method (pronounced "rapid," for random amplified polymorphic DNA) (Williams et al., 1990). This technique involves screening DNA for interpretable polymorphisms using short ($\cong$ 10 bp) primers of arbitrary sequence to amplify at random from a few anonymous genomic sequences (Welsh et al., 1991). The method typically yields polymorphisms with dominance-recessive characteristics (Fig. 3.19). At the time of this writing, the RAPD approach has not yet been employed widely in population biology, but it has been promoted strongly by some authors (see Hadrys et al., 1992; Hedrick, 1992).

DATA

In a sense, all assay methods short of DNA sequencing can be thought of as indirect attempts at sequence acquisition. Indeed, the availability of nucleotide sequence data in principle allows recovery of genetic information at less-detailed levels (e.g., amino acid sequences, RFLP maps), whereas the converse is not true (Box 3.4). However, for logistic reasons, DNA sequences (typically 500 bp or more in length per sample) usually are gathered from only one or a few genes

Box 3.4. Nucleotide Sequences in the mtDNA Cytochrome b Gene of Marine Turtle Species (Taxa a–n) (after Bowen et al., 1993a).

The entire data set involved more than 500 nucleotide positions per sample (only 63 of which are shown here), which is typical for many sequencing studies.

1. Nucleotide sequences. Each dash indicates a base identical to that of the reference species, "a."

Taxon	Nucleotide Sequence																				
	ACC	GGA	ATC	TTC	TTG	GCA	ATA	CAC	TAT	TCA	CCA	GAT	ACT	TCC	CTG	GCA	TTC	TCA	TCA	ATC	ATC
a	ACC	GGA	ATC	TTC	TTG	GCA	ATA	CAC	TAT	TCA	CCA	GAT	ACT	TCC	CTG	GCA	TTC	TCA	TCA	ATC	ATC
b	—	—	—	—	C-A	—	—	-T	-C	—	—	-C	-TC	—	A-A	-C	-T	—	—	-T	-C
c	—	—	—	—	C-A	—	—	-T	-C	—	—	-C	-TC	—	A-A	-C	-T	—	—	-T	-C
d	-T	—	—	—	C-A	—	—	-T	-C	—	—	—	-TC	—	A-A	-C	-T	—	—	—	TC-
e	-T	—	—	—	C-A	—	—	-T	-C	—	—	—	-TC	—	A-A	-C	-T	—	—	—	TC-
f	-T	—	G-	—	C-A	—	—	-T	-C	—	—	—	-TC	-T	A-A	-C	-T	—	—	—	TC-
g	-T	—	—	—	C-A	—	—	-T	-C	—	—	—	-TC	—	A-A	-C	-T	—	—	—	-C
h	—	—	—	—	C-A	—	—	—	-C	—	—	-C	-TC	—	A-A	-T	-T	—	—	-T	GCT
i	—	—	—	-T	C-A	—	—	—	-C	—	—	-C	-TC	—	A-A	-T	-T	—	—	-T	GCT
j	—	—	—	—	C-A	—	—	—	-C	—	—	-C	-TC	—	A-A	-T	-T	—	—	-T	GCT
k	—	—	—	-T	C-A	—	—	—	-C	—	—	-C	-TC	—	A-A	-T	-T	—	—	-T	GCT
l	—	—	—	—	C-A	—	—	—	-C	—	—	—	-TC	—	A-A	-T	—	—	—	—	GC-
m	—	—	—	—	-A	—	—	—	—	—	—	-C	-T-	—	-C	-C	—	—	—	G-T	GCT
n	—	—	—	—	-A	—	—	—	—	—	—	-C	-T-	—	-C	-C	—	—	—	G-T	GCT

2. Same data coded as purines ("0") versus pyrimidines ("1").

Taxon	ACC	GGA	ATC	TTC	TTG	GCA	ATA	CAC	TAT	TCA	CCA	GAT	ACT	TCC	CTG	GCA	TTC	TCA	TCA	ATC	ATC
a	011	000	011	111	110	010	010	101	101	110	110	001	011	111	110	010	111	110	110	011	011
b	—	—	—	—	—	—	—	—	—	—	—	—	—	—	0-	-1	—	—	—	—	—
c	—	—	—	—	—	—	—	—	—	—	—	—	—	—	0-	-1	—	—	—	—	—
d	—	—	—	—	—	—	—	—	—	—	—	—	—	—	0-	-1	—	—	—	—	-1-
e	—	—	—	—	—	—	—	—	—	—	—	—	—	—	0-	-1	—	—	—	—	-1-
f	—	—	—	—	—	—	—	—	—	—	—	—	—	—	0-	-1	—	—	—	—	-1-
g	—	—	—	—	—	—	—	—	—	—	—	—	—	—	0-	-1	—	—	—	—	—

h	–	–	–	–	–	–	–	–	–	–	–	–	–	–	0–	–	–	–	–	–	–
i	–	–	–	–	–	–	–	–	–	–	–	–	–	–	0–	–	–	–	–	–	–
j	–	–	–	–	–	–	–	–	–	–	–	–	–	–	0–	–	–	–	–	–	–
k	–	–	–	–	–	–	–	–	–	–	–	–	–	–	0–	–	–	–	–	–	–
l	–	–	–	–	–	–	–	–	–	–	–	–	–	–	0–	–1	–	–	–	–	–
m	–	–	–	–	–	–	–	–	–	–	–	–	–	–	–1	–1	–	–	–	–	–
n	–	–	–	–	–	–	–	–	–	–	–	–	–	–	–1	–1	–	–	–	–	–

3. Same data translated into amino acid sequences (by reference to the mitochondrial genetic code).

	thr	gly	ile	phe	leu	ala	met	his	tyr	ser	pro	asp	thr	ser	leu	ala	phe	ser	ser	ile	ile
a	thr	gly	ile	phe	leu	ala	met	his	tyr	ser	pro	asp	thr	ser	leu	ala	phe	ser	ser	ile	ile
b	–	–	–	–	–	–	–	–	–	–	–	–	ile	–	met	–	–	–	–	–	thr
c	–	–	–	–	–	–	–	–	–	–	–	–	ile	–	met	–	–	–	–	–	thr
d	–	–	–	–	–	–	–	–	–	–	–	–	ile	–	met	–	–	–	–	–	ser
e	–	–	–	–	–	–	–	–	–	–	–	–	ile	–	met	–	–	–	–	–	ser
f	–	–	val	–	–	–	–	–	–	–	–	–	ile	–	met	–	–	–	–	–	ser
g	–	–	–	–	–	–	–	–	–	–	–	–	ile	–	met	–	–	–	–	–	thr
h	–	–	–	–	–	–	–	–	–	–	–	–	ile	–	met	–	–	–	–	–	ala
i	–	–	–	–	–	–	–	–	–	–	–	–	ile	–	met	–	–	–	–	–	ala
j	–	–	–	–	–	–	–	–	–	–	–	–	ile	–	met	–	–	–	–	–	ala
k	–	–	–	–	–	–	–	–	–	–	–	–	ile	–	met	–	–	–	–	–	ala
l	–	–	–	–	–	–	–	–	–	–	–	–	ile	–	met	–	–	–	–	–	ala
m	–	–	–	–	–	–	–	–	–	–	–	–	ile	–	–	–	–	–	–	val	ala
n	–	–	–	–	–	–	–	–	–	–	–	–	ile	–	–	–	–	–	–	val	ala

Any of these coded data sets could be analyzed phylogenetically by appropriate computer programs. Note that when the data are coded as purines versus pyrimidines, only transversions would be counted in the resulting phylogenetic estimates; and when the data are coded as amino acids, only replacement substitutions would be counted. These data codings exemplify a trade-off common to most sequencing studies: Although these latter two treatments weight heavily for mutational events that are rare and thus less likely to be homoplasious over short evolutionary time scales (see Chapter 4), much information of potential phylogenetic significance (particularly at lower sequence divergence levels) can be lost by neglecting the silent transitions. For example, in the data sets above, 20 of the 63 nucleotide positions shown were variable overall, whereas only four positions displayed transversions and only five represented replacement substitutions.

in a given study and from relatively small numbers of individuals. Thus, these data provide high resolution for molecular aspects of sequence differences (e.g., changes in coding versus noncoding regions; transitions versus transversions; silent versus replacement substitutions in coding regions; base substitutions versus addition/deletion changes or nucleotide rearrangements, etc.), but at the expense of sacrificing genetic information from a broad base of loci and individuals. Contemplated biological applications for DNA sequencing must weigh such considerations.

There are several other clear and unique advantages to sequencing methods, particularly when coupled to PCR. The PCR requires as a template only minute quantities of undegraded DNA, as has been obtained from such unlikely sources as single feathers (Taberlet and Bouvet, 1991), hairs (Morin et al., 1992; Taberlet and Bouvet, 1992; Vigilant et al., 1989), animal excrement (Höss et al., 1992), museum-preserved material (Higuchi et al., 1984), and even fossils (Chapter 8). Amplification of DNA by the PCR is much faster and easier than by cloning. Finally, sequences that evolve at different rates may be chosen for analysis (Chapter 4), so that sequencing studies can be tailored to phylogenetic analyses at any desired level of evolutionary separation.

Potential concerns about PCR-based assays include the following. Because of the extreme sensitivity of the PCR, sample contamination (e.g., by microbes, physical handling, or any source of contact with nontarget DNA) poses a serious difficulty. Degree of fidelity of the PCR amplification has been of some concern (Dunning et al., 1988; Ennis et al., 1990; Pääbo and Wilson, 1988; Saiki et al., 1988), because any misincorporation of nucleotides in the early rounds of amplification will result in an amplified sequence differing at least slightly from the original template. Low-frequency errors in amplification have been observed, but because these are normally far less than the incidence of natural nucleotide site polymorphisms within populations, their effects in most phylogenetic applications (particularly at higher taxonomic levels) probably are negligible (Kwiatowski et al., 1991). Finally, most PCR applications (other than RAPDs) require prior knowledge of sequences flanking the target DNA. This latter obstacle has diminished considerably as sequence data have accumulated from numerous genes and species.

REFERENCES TO LABORATORY PROTOCOLS

For the interested reader, the following references provide far more detailed descriptions of the various laboratory methods.

Protein immunology: Champion et al., 1974; Maxson and Maxson, 1990.

Protein electrophoresis: Harris and Hopkinson, 1976; Murphy et al., 1990; Selander et al., 1971; Shaw and Prasad, 1970.

DNA–DNA hybridization: Sibley and Ahlquist, 1990; Werman et al., 1990.

DNA restriction analysis: Dowling et al., 1990; Hames and Higgins, 1985; Hoelzel, 1992; Karl and Avise, 1993; Lansman et al., 1981; Quinn and White, 1987; Sambrook et al., 1989; Watson et al., 1992.

DNA fingerprinting: (single-locus and multilocus): Bruford et al., 1992.

DNA sequencing and the PCR: Hillis et al., 1990; Innis et al., 1990; Palumbi et al., 1991; Sanger et al., 1977.

SUMMARY

1. A rich potpourri of laboratory methods exists for revealing genetic markers. Protein assays with the greatest impact in the fields of population biology and evolution have been the immunological approach of microcomplement fixation (MCF) and multilocus starch-gel electrophoresis (SGE). The most influential of the DNA assay methods have been DNA–DNA hybridization, analyses of restriction fragment-length polymorphisms (RFLPs), and nucleotide sequencing.

2. Each of these assay methods is applied most fruitfully to particular classes of genes or their products: MCF to abundant, purifiable proteins encoded by single genes, such as albumin; SGE to enzymatic proteins for which specific histochemical stains are available; DNA–DNA hybridization to the single-copy fraction of the nuclear genome; RFLP analyses to a variety of genetic systems, including mitochondrial DNA, chloroplast DNA, single-copy nuclear DNA, ribosomal RNA genes, and the various categories of repetitive elements underlying DNA fingerprints; and nucleotide sequencing to virtually any segments of DNA, particularly those that can be amplified via the polymerase chain reaction.

3. The multivarious laboratory methods available differ widely in the nature of genetic information provided. The next chapter will attempt to organize thoughts about how these diverse molecular approaches can be cataloged with regard to phylogenetic information content.

4

Interpretive Tools

*Thus the hereditary properties of any given organism
could be characterized by a long number written in a
four-digital system.*

G. Gamow, 1954

Once molecular data have been gathered, how are they analyzed and interpreted in a phylogenetic context? The answer, of course, depends on the particular biological problem addressed and the nature of the molecular information. The biological settings are nearly as diverse as one's imagination allows. The interpretive tools are those of population genetics and phylogenetics. Some approaches to molecular data analysis are highly idiosyncratic to a particular biological problem or data base (examples appear throughout Part II), whereas others are standard and generalizable. This chapter will introduce some of the fundamentally important principles and procedures in this latter class of analytical tools.

CATEGORICAL SUBDIVISIONS OF MOLECULAR GENETIC DATA

One initial way to organize thought about analytical approaches is to subdivide the diverse types of molecular data into categories for which shared philosophical approaches to analysis and interpretation may pertain. The following are some of these partitions (Fig. 4.1).

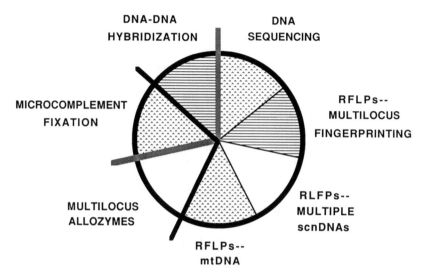

Figure 4.1. Alternative ways to "slice the pie" of molecular genetic data classes. The heavy black line separates protein assays from those dealing directly with DNA. The heavy gray line divides the methods according to whether the raw data consist of qualitative character states or distance values only. The stippled slices of the pie indicate those techniques that normally supply information from only one gene (or linkage group) at a time, as opposed to the remaining methods that usually access genetic data from multiple loci either simultaneously (lined slices) or cumulatively (open slices).

Protein Versus DNA Information

An obvious consideration is whether a molecular technique reveals variation at the level of proteins or, alternatively, at the level of DNA. Methods in the former category unmask primarily the genetic changes in coding regions that have altered amino acid sequence. On the other hand, some nucleic acid techniques provide access to a far greater panoply of architectural changes in both coding and noncoding regions of the genome. Thus, informative genetic markers may be derived from synonymous as well as nonsynonymous (amino acid altering) nucleotide substitutions in protein-coding sequences, from genetic changes in introns and gene-flanking regions, and from additions and deletions of genetic material, sequence rearrangements, or other DNA-level features.

Qualitative Versus Distance Data

Some molecular techniques, notably DNA–DNA hybridization and microcomplement fixation, provide information solely in the form of numerical or quantitative distance values between taxa (see Springer and Krajewski, 1989). Most other molecular approaches provide raw data in the form of qualitative character states—electromorphs, restriction sites or fragments, or nucleotide sequences. The latter can be converted to quantitative estimates of genetic distance, if so desired, but the converse is not true; that is, qualitative character states cannot be recovered from distance data alone. This distinction between qualitative and distance data is important for two fundamental reasons. First, many biological applications (e.g., forensic identification, parentage determination, kinship assessment, gene flow estimation, and the characterization of hybrids) require qualitative genetic markers, whereas other applications such as phylogenetic estimation can employ either qualitative or distance data. Second, several phylogenetic or tree-building algorithms (e.g., parsimony and cladistic analyses) require qualitative data, whereas other algorithms utilize matrices of genetic distances among taxa.

The concept of genetic distance is fundamental to molecular systematics. A genetic distance between two sequences, individuals, or taxa is a quantitative estimate of how divergent they are genetically. The units of distance depend on the nature of the molecular information summarized. For example, values of Nei's (1972) D for protein electrophoretic data are interpreted as the net numbers of codon substitutions per locus that have accumulated since separation of any two populations and can be either corrected or uncorrected for presumed multiple substitutions ("hits") at the same amino acid site. An analogous measure for DNA restriction or sequence data is p, the estimated number of base substitutions per nucleotide (or percent sequence divergence if uncorrected for multiple hits). Genetic distance (ΔT_m) values from DNA–DNA hybridization have no immediate molecular interpretation beyond the measured difference in thermal stability *per se*, although additional independent information may permit a calibration of ΔT_m to magnitude of sequence divergence. Similarly, the immunological distance (*ID*) values from microcomplement fixation are, at face value, merely descriptors of antigen-antibody cross-reactivity (although again, with additional information they may be calibrated to numbers of amino acid substitutions). Definitions of some of the standard distance statistics for various types of molecular genetic data are summarized in Box 4.1.

The converse of genetic distance is "genetic similarity." Thus, when genetic distance is low, genetic similarity is high, and vice versa. In the early literature of molecular evolution, it was customary to refer to "percent homology" between DNA sequences or other molecular characters under comparison. This practice now is discouraged. The word homology properly refers to organismal features (such as genes) that trace to a shared ancestral condition, so that sequences are either homologous or they are not. (Complications can arise, however, when a complex gene or genome contains sequence elements of both homologous and nonhomologous origin). In any event, truly homologous

Box 4.1. Examples of Genetic Distance Statistics

Allozymes

For protein-electrophoretic data, the two most commonly employed distance measures are as follows:

(a) Rogers' (1972) distance. For a given locus with m alleles, let x_i and y_i be the frequencies of the ith allele in populations X and Y, respectively. Rogers' D is defined as

$$D = [0.5 \Sigma(x_i - y_i)^2]^{0.5}, \qquad (4.1)$$

where the summation is over all alleles. When data from more than one gene are considered, the arithmetic mean of such values across loci provides the overall genetic distance estimate. Rogers' D can take values between zero and one; Rogers' similarity $S = 1 - D$.

(b) Nei's (1972) standard genetic distance (see also Nei, 1978). At any locus, Nei's "genetic identity" (similarity) is defined as

$$I = \Sigma x_i y_i / (\Sigma x_i^2 \, \Sigma y_i^2)^{0.5}. \qquad (4.2)$$

For multiple loci, the overall similarity or identity is

$$I = J_{xy} / (J_x J_y)^{0.5}, \qquad (4.3)$$

where J_x, J_y, and J_{xy} are the arithmetic means across loci of Σx_i^2, Σy_i^2, and $\Sigma x_i y_i$, respectively. Nei's I can range from zero to one. Nei's standard genetic distance then is calculated as

$$D = -\ln I. \qquad (4.4)$$

This D, which can take values between zero and infinity, is interpreted as the mean number of codon substitutions per locus, corrected for multiple hits.

Many such distance estimates for protein-electrophoretic data have been proposed (see Nei, 1987) and their relative merits in various applications debated at length. However, for particular data sets, most distance measures are correlated highly (although they may differ in magnitude, especially at larger values).

DNA restriction fragments or sites

(a) Restriction fragments. Upholt (1977) was the first to derive the relationship between the proportion of fragments shared in the mtDNA digestion profiles of two taxa, and an estimate of their nucleotide sequence divergence. Let N_x, N_y, and N_{xy} be the number of restriction fragments observed in sequences X and Y and shared by X and Y, respectively. The overall proportion of shared fragments is calculated as

$$F = 2N_{xy} / (N_x + N_y). \qquad (4.5)$$

Then the number of base substitutions per nucleotide (or, approximately, the percentage of nucleotides substituted) is estimated by

$$p = 1 - [0.5(-F + \{F^2 + 8F\}^{0.5})]^{1/r}, \qquad (4.6)$$

where r is the number of base pairs in the enzymes' recognition site. The value of p must be computed separately for enzymes recognizing four-, five-, and six-base sequences, and the final distance value is the average of these estimates (weighted by the total numbers of fragments produced by the respective enzyme classes). Using a slightly different approach, Nei and Li (1979) derived a relationship between F and p essentially identical to that of Upholt.

(b) Restriction sites. For "site" data, let N_x, N_y, and N_{xy} be the number of sites observed in sequences X and Y and shared by X and Y, respectively. The overall proportion of shared sites is calculated as

$$S = 2N_{xy} / (N_x + N_y). \tag{4.7}$$

Then the number of base substitutions per nucleotide is estimated by

$$p = -\ln S / r. \tag{4.8}$$

Values of p again must be calculated separately for enzymes cleaving at four-, five-, and six-base recognition sites and the overall distance computed as the weighted mean of these values.

Nucleotide sequences

Only the simplest case will be considered, in which sequences of the same length can be aligned without ambiguity. Let z_d be the number of nucleotides which differ between two sequences, and z_t be the total number of nucleotides compared. Then the percent sequence difference is simply

$$p = z_d / z_t. \tag{4.9}$$

For sequences exhibiting little divergence, p is a close approximation to the accumulated number of nucleotide substitutions per site because no correction is needed for multiple substitutions at a site. Particularly when sequence divergences are larger, corrections for multiple hits have been suggested. One simple correction was provided by Jukes and Cantor (1969):

$$D = -3/4 \ln(1 - 4p/3). \tag{4.10}$$

This equation was derived under the assumption that at any nucleotide position, substitutions occur with equal probability to any of the remaining nucleotides. Other corrections have been developed by varying these assumptions, for example by allowing different rates for transitional versus transversional substitutions as in the widely employed "two-parameter" model of Kimura (1980).

Note that unlike the allozyme genetic distances that are based on population allele frequencies and, hence, provide distances between populations or species, these latter estimates apply to sequence divergence between particular genes or alleles. If the sequences come from haploid individuals (as is effectively true for uniparentally inherited cytoplasmic genomes), the calculated values also can be interpreted as between-individual distances at that locus. When many such sequences within a population are assayed, mean genetic distance [or nucleotide diversity (Box 2.1)] then is estimated by

$$\text{mean } p = \Sigma f_i f_j \, p_{ij} \tag{4.11}$$

where f_i and f_j are the frequencies of the ith and jth sequences in the sample, and p_{ij} is the estimated sequence divergence between the ith and jth sequences (Nei, 1987). If two or more populations have been assayed, estimates of net sequence divergence (p_{corr}) between populations also can be calculated, based on a correction for within-population polymorphism:

$$p_{\text{corr}} = p_{xy} - 0.5[p_x + p_y] \tag{4.12}$$

where p_{xy} is the mean genetic distance in pairwise comparisons of individuals between populations X and Y, and p_x and p_y are the mean genetic distances among individuals within these respective populations. This correction assumes that sequence diversity within the extant populations is representative of the level of variation present in the common ancestor.

sequences among an array of organisms can exhibit a wide range of genetic similarities, the values depending in part on how long ago the extant taxa separated from common ancestors.

Detached versus Connectible Information

Some types of molecular data can be connected readily across studies, whereas others cannot. Good examples of connectible data are DNA sequences. Once a nucleotide sequence is available for any gene or species, newly obtained sequences can be compared against the original without the need to repeat the earlier assays. Other kinds of information are less connectible, in the sense that data from one study may be impossible to link directly with similar data from others. For example, a ΔT_m value (from DNA–DNA hybridization assays) between the genomes of species A and B is of no immediate service in assessing the relationship of these species to taxa C and D for which another ΔT_m value may be available. In contrast, DNA sequences from these same four species *could* be used to estimate a phylogeny, irrespective of when or where the sequences were determined.

The distinction between connectible versus detached data is not the same as that between qualitative versus distance information. For example, protein electrophoresis provides qualitative genotypic data, but the electromorphs themselves are distinguished by gel mobilities relative to one another. Thus, it is difficult to compare the particular electromorph genotypes reported in one study with those of another, unless shared standards had been employed in both. Another point is that a connectibility for data does not necessarily imply an ease of phylogenetic analysis and, indeed, can create special difficulties due to the combinational properties of such information. Assays for DNA sequencing, for example, have become so proficient that "Methods of data acquisition have vastly outstripped current systems of data management" and "contemporary algorithms for phylogenetic and evolutionary analysis are not equal to the tasks presented by sequence data sets now in hand" (Clegg and Zurawski, 1992). Thus, the phylogenetic opportunities provided by connectible data also pose profound challenges in the development of appropriate analytic methodologies.

Single-Locus versus Multilocus Data

As normally applied, some molecular techniques (such as microcomplement fixation and DNA sequencing) entail the acquisition of data from individual loci, whereas others (such as DNA–DNA hybridization and multilocus fingerprinting) inherently access genetic information from multiple gene regions (Fig. 4.1). This distinction is important because the amount of genetic information assessed influences the interpretations to be drawn from data. A major advantage of

DNA-DNA hybridization approaches is that the estimated ΔT_m's reflect an average across the thermal stability properties of multitudinous sequence regions; and the individual-diagnostic power of multilocus DNA fingerprinting stems in part from independent assortment among the multiple dispersed arrays of repetitive elements.

For most applications involving genetic markers, the number of functional genes assayed is less important than the number of linkage groups represented, which influences how many independent bits of phylogenetic information are revealed. For example, animal mtDNA is composed of some 37 functional genes, all transmitted as a nonrecombining unit primarily through female lines. Thus, from a phylogenetic perspective, the entire 16-kb mtDNA molecule is only a single "gene."

Multilocus assay methods can be categorized further into those which assess information from multiple loci simultaneously (e.g., DNA–DNA hybridization and multilocus DNA fingerprinting) versus sequentially (multilocus protein electrophoresis and assays of scnDNA RFLPs) (Fig. 4.1). Only the latter normally provide information interpretable in simple genetic terms (i.e., as Mendelian genotypes at particular loci). Although the number of genes included in such allozyme and scnDNA assays typically is small or moderate, even a handful of interpretable genetic polymorphisms considered in aggregate can provide remarkable power in applications such as forensics, parentage assessment, gene flow estimation, and the characterization of hybrids.

Utility of Data Along the Phylogenetic Hierarchy

Another way to slice the molecular-techniques pie is with regard to the level of evolutionary separation at which the various methods can be applied (Box 4.2). Most assay methods provide an empirical window of opportunity that is fairly narrow relative to the broad field of potential phylogenetic applications. For example, microcomplement fixation and DNA–DNA hybridization methods are suitable for phylogenetic studies at intermediate taxonomic levels, when species' separations date to approximately 2–100 million years ago (mya). At shorter divergence times, genetic distances estimated by these methods tend to be small and not significantly different from zero. At the microevolutionary end of the continuum, multilocus DNA fingerprinting is highly appropriate for questions of genetic identity versus nonidentity and parentage. Studies of RFLPs from mtDNA and scnDNA have been most fruitful at levels of conspecific populations and closely related species, as have allozyme surveys. Among the available molecular methods, only DNA sequencing can find application at virtually any taxonomic level. This flexibility stems from the fact that different DNA sequences evolve at highly different rates (see the next section), such that studies can be tailored around the choice of appropriate sequence elements.

Box 4.2. Levels of Evolutionary Divergence at Which Various Molecular Genetic Methods Normally Provide Informative Phyogenetic Markers (Modified from Hillis and Moritz, 1990).

Hierarchical Level	Protein Immunology	Protein Electrophoresis	DNA–DNA Hybridization	RFLP Analyses of			DNA Sequencing
				mtDNA	scnDNA	VNTR Loci	
Genetic identity/ nonidentity	—	*	—	*	*	**	*
Parentage	—	*	—	*	**	**	*
Conspecific populations	—	**	—	**	**	*	*
Closely related species	*	**	*	*	*	—	*
Intermediate taxonomic levels	**	*	**	—	—	—	**
Deep separations (>50 mya)	*	—	*	—	—	—	**

(**)—highly informative; (*)—marginally informative, but not an ideal approach for reasons of cost-ineffectiveness or other difficulties; (—)—inappropriate use of method. Not all categorizations are absolute. For example, some isozyme characters such as presence/absence of duplicate gene products can be useful at higher taxonomic levels.

Nonetheless, because of labor and expense, obtaining DNA sequences from large numbers of individuals and genes in a population context is not particularly cost-effective, and sequencing studies usually are conducted at intermediate or higher levels of the phylogenetic hierarchy.

Of course, not only the molecular method but also the amount of genetic information obtained influence the resolution obtainable in a given application. For example, restriction-site or sequencing studies of animal mtDNA often involve assay of about 500 nucleotide pairs per individual. At the conventional sequence divergence rate of 2% per million years (see next section), roughly 1 of 500 base pairs is expected to change after 100,000 years of maternal lineage separation, thus establishing 100 millenia as an approximate lower limit of resolving power for de novo mutations with this level of effort. If the entire 16,000-bp mtDNA genome could be assayed, the ability to detect a single nucleotide change would be increased by more than 30-fold, suggesting a lower limit on resolving power of just a few thousand years. However, these statements refer only to the accumulation of novel mutations and not to significant shifts in allelic frequencies of ancestral polymorphisms (which could take place in as little as a single generation by genetic drift, natural selection, or migration).

MOLECULAR CLOCKS

General Concepts

Zuckerkandl and Pauling (1965) were the first to propose that various proteins and DNA sequences might evolve at constant rates over time, and thereby provide internal biological timepieces for dating past evolutionary events. The concept of a molecular clock fits well with neutrality theory because, as discussed in Chapter 2, the rate of neutral evolution in genetic sequences is equal simply to the mutation rate to neutral alleles. However, clock concepts are not necessarily incompatible with selectionist scenarios: If a large number of assayed genes was acted on by multivarious selection processes over long periods of time, short-term fluctuations in selection intensities might tend to average out and the magnitudes of overall genetic distance between isolated taxa could well be correlated strongly with the time elapsed since common ancestry.

Few concepts in molecular evolution have been more debated (or abused) than those concerning molecular clocks. At the outset, three general points must be understood. First, the debate is not whether molecular clocks behave metronomically, like a working timepiece—they do not. If molecular clocks exist, both neutralists and selectionists predict at best a "stochastically constant" behavior, like radioactive decay (Ayala, 1982c; Fitch, 1976). Second, not all molecular phylogenetic applications hinge critically on the reliability of molecular clocks. For example, genetic characters likely to be of monophyletic origin (such as gene duplications or differences in gene arrangement) remain powerful as phylogenetic markers, regardless of the rate of evolution at the level of DNA sequence; and, appraisals of genetic identity and parentage (e.g., by DNA fingerprinting) hardly depend on a steady pace for DNA sequence evolution. Many tree-building algorithms based on genetic distance matrices relax assumptions of rate homogeneity among lineages (see the next section), and indeed branching orders, in principle, can be inferred directly from distributions of qualitative character states using techniques such as cladistic or parsimony analyses that remain valid irrespective of whether molecules evolve in strictly time-dependent fashion. Nevertheless, data and concepts pertaining to molecular clocks *are* of fundamental importance in many phylogenetic contexts.

A third point about molecular clocks is that, from an empirical standpoint, different DNA sequences unquestionably evolve at markedly different rates (see the reviews in Li and Graur, 1991; Nei, 1987; A.C. Wilson et al., 1987). These rate heterogeneities are apparent at several levels: (a) across nucleotide positions within a codon [mean rates for synonymous substitutions normally are several times higher than those involving nonsynonymous changes in protein-coding regions (Figs. 4.2a,b)]; (b) among nonhomologous genes within a lineage [nonsynonymous rates can vary by orders of magnitude, as between the slowly evolving histones and rapidly evolving relaxins (Fig. 4.2a)]; (c) among classes of DNA within a genome [e.g., introns and pseudogenes evolve more rapidly than do nondegenerate sites in protein-coding genes (Fig. 4.2c)]; and (d) among genomes within an organismal lineage (e.g., synonymous substitution rates in cpDNA are several-fold lower than those of plant nuclear genomes, and vertebrate mtDNA evolves on average about 5–10 times faster than does most single-copy nuclear DNA). Under neutrality theory (Kimura, 1983), such rate heterogeneities across nucleotide sites, genes, and genomes within a phylogenetic lineage are interpreted to reflect varying intensities of purifying selection associated with differing levels of functional constraint on DNA sequences, perhaps in conjunction with a variation in the underlying rate of mutation to strictly neutral alleles (Britten, 1986). Ironically, these kinds of extreme rate heterogeneities can be highly beneficial for phylogenetic studies, by permitting choice of appropriate DNA regions geared to the time-scale requirements of a particular phylogenetic problem (Box 4.2). For example, slowly evolving sequences for ribosomal RNA genes have been extremely informative in reconstructing deep branches in the tree of life (Chapter 8), whereas rapidly evolving mtDNA sequences have revolutionized phylogenetic studies of animals at the intraspecific level (Chapter 6). Within a given molecule such as mtDNA, the more slowly accumulating replacement substitutions and transversions may be most informative at the levels of species, genera, or families, whereas rapidly accumulating silent substitutions and transitions often provide numerous markers for the study of local populations (Bowen et al., 1993a).

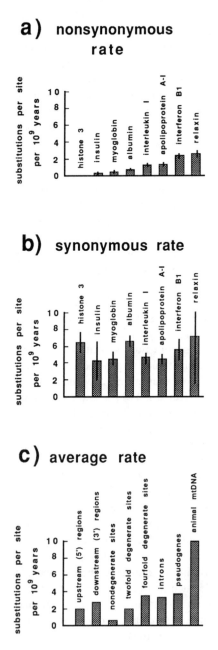

Figure 4.2. Observed rates of nucleotide substitution in various genes and gene regions (generated from the information in Li and Graur, 1991). (**a**) and (**b**) Nonsynonymous and synonymous substitution rates, respectively, in eight protein-coding genes sequenced from both humans and rodents. The rate calculations are based on the assumption that observed sequence differences accumulated over 80 million years. (**c**) Average rates of substitution in different parts of these and other genes. Nondegenerate nucleotide sites are those at which all possible substitutions are nonsynonymous; at fourfold degenerate sites, all possible substitutions are synonymous, and at twofold degenerate sites, one of three possible nucleotide changes is synonymous and the other two are nonsynonymous.

Clock Calibrations and Controversies

In discussions of phylogeny reconstruction, a more controversial form of rate heterogeneity involves possible differences in the evolutionary tempo of homologous DNA sequences across organismal lineages. For dating cladogenetic events, approximate uniformity in such evolutionary rates clearly would be desirable (A.C. Wilson et. al., 1987) and, indeed, numerous "universal" rate calibrations have been suggested for a variety of genes and assay methods.

For example, a conventional calibration for evolutionary rate of animal mtDNA [derived from the slope of the linear portion of the divergence curve (Fig. 4.3A)] is about 2% sequence divergence per million years between pairs of lineages separated for less than 10 million years, or 1×10^{-8} substitutions per site per year per evolutionary line (Brown et al., 1979). In referring to mtDNA in higher animal taxa, Wilson et al. (1985; see also Shields and Wilson, 1987a) concluded that "no major departures from this rate are known for the molecule as a whole." However, this conclusion is now controversial for the following empirical reasons. First, the *ratios* of mean mtDNA/scnDNA divergence rates have been shown to differ significantly between animal groups (Caccone et al., 1988a; DeSalle et al., 1987; Powell et al., 1986; Vawter and Brown, 1986), and whether the rate heterogeneity is attributable to variation in mtDNA, scnDNA, or both is unclear. Second, different nucleotide positions and genes within mtDNA evolve at varying rates within a lineage (Brown et al., 1982; Gillespie, 1986; Moritz et al., 1987) and particular mtDNA genes (such as cytochrome oxidase) reportedly show rate differences as high as fivefold across taxa (Brown and Simpson, 1982; Crozier et al., 1989). Third, recent studies have reported significant mean mtDNA rate heterogeneities across lineages of Hawaiian *Drosophila* (DeSalle and Templeton, 1988), mammalian orders (Hasegawa and Kishino, 1989), vertebrate classes (Avise et al., 1992a; Bowen et al., 1993a), and homeothermic versus heterothermic vertebrates (Kocher et al., 1989). The argument has been raised that the pace of mtDNA evolution among animal groups may be linked causally to differences in metabolic rate and/or to the generation-length and body-size differences with which metabolic rate is negatively correlated (Avise et al., 1992a; A.P. Martin et al., 1992; Thomas and Beckenbach, 1989). A related suggestion involves the concept of "nucleotide generation time"—the time elapsed between episodes of DNA replication or repair. The idea is that nucleotide positions with higher turnover (shorter replication intervals) are subjected to more mutational opportunities per unit of sidereal time; species with higher metabolic rates and briefer organismal life spans tend to have shorter nucleotide generation times and, therefore, may have higher absolute rates of molecular evolution (Martin and Palumbi, 1993).

Examples of reported calibrations for other putative molecular clocks are summarized in Fig. 4.3. Again, controversies surround the validity and univer-

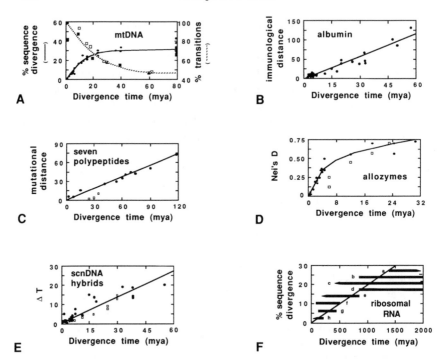

Figure 4.3. Examples of "clock calibrations" reported for various types of molecular genetic data. All dates along the abscissa came from fossil or biogeographic evidence. (**A**) Solid line: mitochondrial DNA sequence divergence for various mammals (after Brown, 1983). The slope of the linear portion of this curve gives the conventional mtDNA clock calibration of 2% sequence divergence per million years between recently separated lineages (note that beyond about 15–20 million years, mtDNA sequence divergence begins to plateau, presumably as the genome becomes saturated with substitutions at the variable sites). Hatched line: percentage observed transitions for various mammals (open squares, after Moritz et al., 1987) and *Drosophila* (closed squares, after DeSalle et al., 1987). (**B**) Albumin immunological distances (as estimated by microcomplement fixation) for various carnivorous mammals and ungulates (after Wilson et al., 1977). (**C**) Accumulated codon substitutions in seven proteins (cytochrome c, myoglobin, α- and β-hemoglobin, fibrinopeptides A and B, and insulin) for various mammalian species [after Langley and Fitch (1974) and Nei (1975)]. The three squares below the solid line involve primate comparisons. (**D**) Codon substitutions per locus (Nei's *D*) based on allozyme comparisons for carnivores (dots) and primates (squares) (after Wayne et al., 1991a). (**E**) Δ*T* values from DNA–DNA hybridizations involving carnivores (dots) and primates (squares) (after Wayne et al., 1991a). (**F**) Percent sequence divergence in the 16S ribosomal RNA gene for various eubacterial forms: (a) cyanobacteria; (b) chloroplasts; (c) microaerophiles; (d) mitochondria; (e) obligate aerobes; (f) *Photobacterium;* (g) *Rhizobium* and *Bradyrhizobium;* and (h) *Escherichia* (after Ochman and Wilson, 1987). The wide horizontal bars indicate considerable uncertainty about the actual divergence times from nonmolecular evidence. The slope of the line is drawn arbitrarily to represent an evolutionary rate of 1% sequence divergence per 50 million years.

sality of such results, particularly as applied across taxonomic groups. The inherent appeal of molecular clocks appears to have led to some egregious claims. For example, the early protein-electrophoretic literature conveyed the strong impression that allozyme distances could be used reliably and accurately to date speciations; many papers concluded that observed genetic distances were consistent with the suspected separation times for particular species as gauged by nonmolecular evidence. But different authors were employing (perhaps unwittingly) allozyme rate calibrations that differed by more than 22-fold from one another! (see the review in Avise and Aquadro, 1982.) Given such a huge range of potential "clock" calibrations in the literature, it is difficult to imagine any observed allozyme distance that could not be accommodated with a given fossil-based scenario! Ironically, if these researchers individually were correct that a molecular clock was ticking *within* each taxonomic group, then collectively no single allozyme clock could apply *across* these same taxa.

Ayala (1986) summarized evidence for the erratic behavior of particular protein clocks across lineages, as did Britten (1986) and Brunk and Olsen (1990) for the overall rate of scnDNA sequence divergence as gauged by DNA-DNA hybridization. Some investigators have concluded that although mean molecular rates in the nuclear genome may vary among taxa, they do so in a predictable or, at least, a consistent fashion. For example, molecular evolution arguably appears slower in primates than in rodents and appears especially slow in the hominoids (Goodman et al., 1971; Koop et al., 1989; Li and Tanimura, 1987; Li et al., 1987; Maeda et al., 1988; but see also Caccone and Powell, 1989; Easteal, 1991; Kawamura et al., 1991; Sibley and Ahlquist, 1987). Several of the published interpretations are similar to those presented earlier for mtDNA. Thus, for avian and mammalian species, Sibley and colleagues [long-term advocates of the concept of a "uniform rate of DNA evolution" (Sibley and Ahlquist, 1984)] presented recent evidence that short generation time or increased number of germ-line cell divisions may be associated with a higher mean rate of scnDNA sequence evolution (Catzeflis et al., 1987; Sibley et al., 1988). More generally, apparent variation in nuclear substitution rates among taxa has been attributed to differences in generation lengths (Gaut et al., 1992; Kohne, 1970; Laird et al., 1969; Li et al., 1987), numbers of DNA replications in germ-line cells (Wu and Li, 1985), repair efficiencies during DNA replication (Britten, 1986), or degree of exposure to mutagens [including the free radicals whose DNA-damaging effects appear correlated with metabolic rates among species (Adelman et al., 1988)].

On the other hand, some researchers consistently have maintained that sidereal time is the best predictor of genetic divergence and that molecular clocks can be calibrated universally across organismal groups (Wilson, 1985). As stated by Wilson et al. (1987): "Molecular evolutionary clocks have ticked at much the same rate per year in many eubacterial genes as in the nuclear genes of animals and plants.'' They also propose an intriguing hypothesis for this conclusion, based on the following assumptions: (a) most nucleotide substitutions involve neutral mutations; (b) the mutation rate per year is higher in short-generation organisms; (c) the fraction of mutations that is effectively neutral is lower in larger populations (because of the greater impact there of deterministic forces including natural selection); and (d) species with shorter generations tend to have larger populations. If these speculations hold, a greater mutation rate in short-generation species might be counterbalanced by a lower fraction of effectively neutral mutations, such that overall evolutionary rates remain relatively constant among diverse taxa.

Absolute and Relative Rate Comparisons

In any event, how are molecular rates assessed, and how is it that the heated debates about rate heterogeneity have continued for so long without final resolution? One difficulty is that molecular-distance measures often become nonlinear with time at various evolutionary depths (e.g., Figs. 4.3A,D), and some arguments against molecular clocks have stemmed from appraisals involving inappropriate regions of divergence curves. Another problem involves the difficulty of determining confidence limits for genetic-distance estimates (Nei, 1987). Thus, any distance in Fig. 4.3 is merely a point value with estimation errors of various types. Another aspect of statistical concern is whether, for a given molecule, the mean of substitution rates among lineages is equal to the variance, as predicted under neutrality theory for a Poisson-like process. Several studies have concluded that the empirical variance in substitution rates slightly exceeds the mean (Langley and Fitch, 1974; Ohta and Kimura, 1971), and this has been interpreted as evidence against a uniform clock (Gillespie, 1986, 1988; Takahata, 1988).

Perhaps the most serious difficulty in the calibration of molecular distance against sidereal time is that firm independent knowledge from fossil or biogeographic evidence is required also. Unfortunately, such information is insecure or lacking for most taxa (if this were not true, there would be little motivation for molecular phylogenetic appraisals!). For example, all separation dates in Fig. 4.3 came from the fossil record and the range in estimates of divergence time is, in some cases, extremely wide (e.g., Fig. 4.3F). Fossils seldom provide the solid anchor of separation times that molecular biologists sometimes suppose—preserved remains often are scanty and confined to a few phenotypic attributes whose phylogenetic relevance is suspect. These problems are especially acute for morphologically simple creatures like bacteria.

Biogeographic evidence also can be difficult to interpret, even in the cleanest of instances. To cite one example, it is well documented that the Isthmus of Panama rose about three million years ago and must, therefore, have curtailed any former gene flow between tropical marine faunas in the eastern Pacific and western Atlantic oceans. Today, the green turtle (*Chelonia mydas*) is distributed circumtropically and shows a distinct break in mtDNA phylogeny that distinguishes all assayed populations from the Atlantic versus Pacific (Bowen et al., 1992; Chapter 6). However, the magnitude of mtDNA sequence divergence between these clades is only $p_{corr} = 0.006$, a value tenfold lower than expected under the postulated three million years of separation, provided the mammalian mtDNA evolutionary rate of 2% sequence divergence per million years applies (Fig. 4.3A). Perhaps mtDNA evolution in green turtles is slower than in mammals by an order of magnitude (Avise et al., 1992a). Alternatively, turtle mtDNA might evolve at a more usual pace, but Atlantic and Pacific populations were in genetic contact more recently (perhaps through dispersal of green turtles around South America or South Africa during Pleistocene interglacial periods when climates may have been warmer than now). Similar genetic studies (of allozymes and mtDNA) have been conducted on several Atlantic-Pacific geminate species pairs of fishes and sea urchins separated by the Panamanian Isthmus, with conflicting interpretations drawn regarding possible heterogeneity in evolutionary rates of homologous genes (Bermingham and Lessios, 1993; Grant, 1987; Lessios, 1979, 1981; Vawter et al., 1980).

To circumvent the uncertainties of the species' divergence dates from fossil or biogeographic evidence, molecular evolutionists also employ "relative-rate" tests that do not depend on knowledge of absolute divergence times (Sarich and Wilson, 1973). Each test requires at least two related species (say A and B) and an outside reference species (C) known to have branched off prior to the separation of A and B. The rationale is illustrated in Fig. 4.4a. By definition, the true evolutionary distance between A and B (d_{AB}) is equal to the sum of their branch lengths from a common ancestor at point O (i.e., $d_{OA} + d_{OB}$). Similarly,

$$d_{AC} = d_{OA} + d_{OC}, \quad \text{or rearranged,} \quad d_{OA} = d_{AC} - d_{OC} \quad (4.13)$$

and

$$d_{BC} = d_{OB} + d_{OC}, \quad \text{or rearranged,} \quad d_{OB} = d_{BC} - d_{OC} \quad (4.14)$$

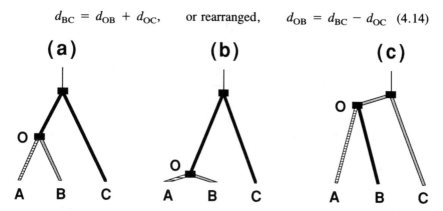

Figure 4.4. Rationale (**a**) and potential difficulties (**b,c**) for the relative rate test (see text).

Subtracting Eq. (4.14) from Eq. (4.13) yields

$$d_{OA} - d_{OB} = d_{AC} - d_{BC} \qquad (4.15)$$

According to the molecular clock, d_{OA} and d_{OB} should be equal ($d_{OA} - d_{OB}$ = O) and, hence, $d_{AC} = d_{BC}$. Genetic distances d_{OA} and d_{OB} cannot be measured empirically, but d_{AC} and d_{BC} can. The relative-rate test asks whether d_{AC} and d_{BC} as estimated by a particular molecular approach are statistically different and, hence, incompatible with the molecular clock hypothesis.

Many relative rate tests for molecular data have been conducted (see the review in Li and Graur, 1991). For example, from their DNA–DNA hybridization studies, Sibley and Ahlquist (1986) report that "genetic distances between the outlier and each of the other species . . . are always equal, within the limits of experimental error" and "thousands of such trios of species . . . yield the same result and attest to the uniform average rate of the DNA clock in birds." Rats, mice, and hamsters also have passed the relative-rate test (Li et al., 1987; O'hUigin and Li, 1992). On the other hand, failures to pass the test contributed to the conclusion that nDNA sequences evolve more rapidly in rodents than in primates (Li et al., 1987; Wu and Li, 1985) and that certain cpDNA sequences evolve more rapidly in annual angiosperms than in other plants (Bousquet et al., 1992).

The relative-rate test is not without difficulties, as illustrated in Fig. 4.4. If A and B separated very recently compared to C (Fig. 4.4b), d_{AC} and d_{BC} may appear equal in an empirical test even under a highly erratic clock (because the vast majority of evolution has taken place in the long and shared OC branch). Conversely, if the separation of A and B was close to that of the common ancestor with C (Fig. 4.4c), an incorrect assumption about the hypothesized branching order of the three species might exist, such that d_{AC} and d_{BC} could appear different even under a perfectly constant clock. Thus, errors both of false acceptance and false rejection of evolutionary clocks can be envisioned in particular tests of relative rate.

Closing Thoughts About Molecular Clocks

What can be concluded from the immense effort expended on assessment of molecular clocks? It is now undeniable that some and probably most molecular systems evolve at heterogeneous rates across at least some taxa. Thus, if precise clocks exist, they are local rather than universal. Yet the time elapsed since common ancestry (whether calibrated in absolute time units or generation lengths is uncertain) remains perhaps the single best predictor of molecular divergence, especially when genetic distance is measured across large numbers of loci. What justifies this bold statement? The evidence is mostly indirect, inconclusive in individual instances, but compelling cumulatively: (a) the nodes in numerous molecular phylogenetic trees generally have proved consistent with independent time estimates, insecure as these may be (e.g., Fig. 4.3); (b) evolutionary rates for molecules proceed mostly independently of those for morphology and other phenotypic attributes, where rates can vary wildly under the influence of differing selection regimes; and (c) given current understanding of the mechanistic basis of molecular evolution (particularly at the level of DNA sequence), it would be most surprising if mean genome-wide genetic distances did not increase with time.

In many instances, molecular timepieces with less than full precision nonetheless can provide significant improvement over phylogenetic understanding gained from nonmolecular data. Consider for example the single-celled microbes, whose phylogeny was totally unknown prior to the application of molecular information. Phylogenetic patterns in ribosomal RNA genes and other loci have revealed stunning genetic relationships and subdivisions among microbial taxa, including those that early in evolution entered into endosymbiotic relationships with protoeukaryotic cells (Chapter 8). Molecular clocks keep far from perfect time, but to dismiss the time-dependent properties of molecular evolution out of hand would be to deny access to an invaluable and sometimes sole source of temporal information.

PROCEDURES FOR PHYLOGENY RECONSTRUCTION

Phylogenetic trees are graphical representations consisting of nodes (taxonomic units) and branches (pathways connecting nodes) that summarize the evolutionary relationships among organisms (Fig. 4.5). In most studies, the

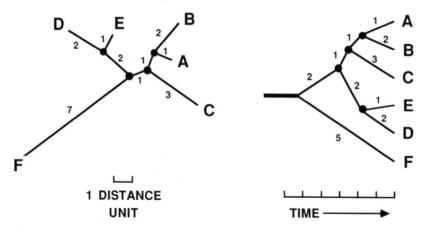

Figure 4.5. Alternative representations of a phylogenetic tree for six extant OTUs. **Left:** Unrooted network with scaled branches. **Right:** Rooted tree with unscaled branches (the root is the heavy line and the tree is oriented along a temporal axis). Internal nodes are indicated by black dots. Note that branch angles have no meaning— branches may be rotated freely about any internal node without materially affecting tree topology.

operational taxonomic units (OTUs) are species or higher taxa, such that the tree topology can be characterized properly as nonanastomotic. In some cases, the OTUs also may be well-isolated conspecific populations, individuals, or nonrecombined alleles of a gene. The external nodes in a phylogenetic tree represent extant OTUs, and internal nodes are the ancestral units. Peripheral branches lead to external nodes, and interior branches connect internal nodes. Branch lengths reflect the number of evolutionary changes along each ancestral-descendant pathway. If the genetic distance between any two OTUs is equal to the sum of all branch lengths connecting them, the tree is said to be strictly additive. Departures from additivity provide one measure of the degree of distortion in phylogeny estimation due to homoplasy in the data (or perhaps to improper behavior of the distance measures or phylogenetic algorithms employed).

Phylogenetic trees may be graphed in several ways (Fig. 4.5). A tree is scaled when branch lengths are drawn proportional to the numbers of genetic changes; otherwise, it is unscaled (although branch lengths may be indicated numerically along the diagram). A tree is rooted when an internal node is specified that represents the common ancestor of all OTUs under study; otherwise, it is unrooted and commonly referred to as a network. A tree is bifurcating when two immediate descendant lineages come from each node, and multifurcating when three or more lineages do so.

One important class of applications for molecular data is estimation of phylogenetic trees. The process is challenging for several reasons. First, even small numbers of OTUs can be connected into astronomical numbers of different trees, only one of which actually is correct. For nOTUs, the number of different bifurcating rooted trees (N_{Tr}) possible is given by

$$N_{Tr} = [(2n - 3)!]/[2^{n-2}(n - 2)!] \qquad (4.16)$$

and the number of bifurcating unrooted trees (N_{Tu}) for $n \geq 3$ is

$$N_{Tu} = [(2n - 5)!]/[2^{n-3}(n - 3)!] \qquad (4.17)$$

(Felsenstein, 1978). Thus, the number of possible trees increases rapidly as the number of taxa increases, and even the small value of $n = 10$ yields $N_{Tr} = 34{,}459{,}425$ and $N_{Tu} = 2{,}027{,}025$. Many phylogenetic algorithms work by searching among trees for those that exhibit desirable properties according to some specified optimality criterion (e.g., shortest total branch length under parsimony). Unfortunately, when the number of OTUs is more than about a dozen, it is usually impractical for even the fastest computers to examine all possible trees, and various truncated search procedures must be implemented. A second difficulty is that any molecular reconstruction is a function both of the genetic data themselves and of the distance measures and phylogenetic algorithms applied to them, and these influences can be difficult to tease apart. Finally, true phylogeny seldom is known with certainty from independent evidence (however,

Table 4.1. Hypothetical distance matrix for five OTUs (see text)

	A	B	C	D	E
A	—	0.08	0.19	0.70	0.65
B		—	0.17	0.75	0.70
C			—	0.80	0.60
D				—	0.12
E					—

see Atchley and Fitch, 1991), so appraisals of phylogenetic methods normally rest on indirect evidence. The net result of these difficulties in evaluating alternative tree-building approaches is that much scope has existed for argumentation over which method of phylogenetic reconstruction is "best."

What follows are brief descriptions of some of the algorithms most commonly employed in molecular phylogenetics. These may be divided usefully into quantitative (distance) methods versus qualitative (character-state) approaches (Avise, 1983b). Extended treatments of these and other phylogenetic procedures (including "maximum likelihood," which is beyond the current scope) can be found in Felsenstein (1982, 1988), Sneath and Sokal (1973), Swofford and Olsen (1990), and references therein.

Distance Approaches

All distance-based approaches begin with an OTU × OTU matrix, the body of which consists of estimated pairwise genetic distances between taxa. For n OTUs, there are $n(n - 1)/2$ pairwise distances (excluding "self" comparisons along the matrix diagonal). Clearly, because of phylogenetic connections among taxa, such estimates cannot be treated as independent values from a statistical perspective. Indeed, the historical interconnections that the distances reflect *are* the primary focus. In Table 4.1 is presented a hypothetical distance matrix for five OTUs (10 pairwise comparisons) that will provide a basis for illustrating various distance algorithms for tree construction.

UPGMA CLUSTER ANALYSIS (SNEATH AND SOKAL, 1973)

Cluster analyses group OTUs according to overall similarity or distance and are the simplest methods computationally. The clustering procedure most commonly employed is the "unweighted pair group method with arithmetic averages" (UPGMA), which operates as follows. A distance matrix (such as in Table 4.1) is scanned for the smallest distance element, and the OTUs involved are joined at an internal node drawn in an appropriate position along a distance axis (Fig. 4.6a). In our example, OTUs A and B are joined first, at distance level $d = 0.04$ (because the sum of lengths of branches connecting A and B is the observed $d = 0.08$). This distance element in the matrix is then discarded. The matrix is scanned

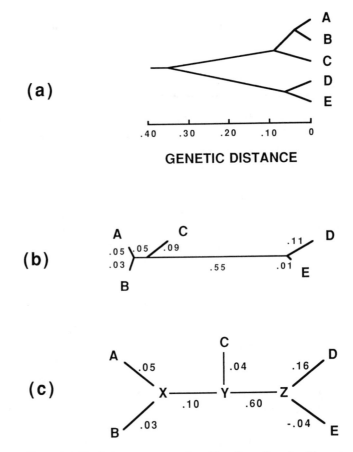

Figure 4.6. Evolutionary trees produced by alternative algorithms applied to the genetic distance matrix in Table 4.1. (a) UPGMA dendrogram. (b) Scaled neighbor-joining (N–J) network [a Fitch–Margoliash (F–M) network proved nearly identical to this N–J tree for these data]. (c) Distance Wagner network, drawn as unscaled but with distances indicated along branches. The latter network also exemplifies a common graphical presentation in which internal nodes are labeled. In this instance, note how rooting along the Y–Z branch and appropriate rotation of branches can cause the Wagner, N–J, and F–M networks to resemble closely the UPGMA dendrogram.

again for the smallest remaining distance, which in this case is $d = 0.12$ joining D and E. These OTUs are clustered at level $d = 0.06$. The next smallest distance in the matrix is $d = 0.17$ between C and A. However, A is already part of a previously formed cluster with B, so C cannot be joined directly to A but rather must be connected through the A–B internal node. This clustering level is determined by the arithmetic mean of the distances between C and the OTUs in the previous cluster [$d = (0.19 + 0.17)/2 = 0.18$]. Thus, C joins the A–B group at $d = 0.09$. All that remains is to join the A–B–C cluster with the D–E cluster. The level of joining is determined from the mean of all pairwise distances between the OTUs in these clusters [$d = (0.70 + 0.65 + 0.75 + 0.70 + 0.80 + 0.60)/6 = 0.70$]. Note that the final cycle of clustering in UPGMA exhausts all remaining distance values in the matrix. The final "tree" (UPGMA figures usually are referred to as phenograms or dendrograms) is shown in Fig. 4.6a.

Several points about UPGMA should be clarified. In each cycle of the clustering procedure, OTUs or previously formed clusters are grouped according to the smallest *mean* distance between the taxa involved (rather than smallest single distance element in the original matrix). Each OTU contributes equally to the calculation of these mean distances (hence the term "unweighted"). All extant OTUs are depicted as "right justified" along the genetic distance axis (Fig. 4.6a), such that the branches from a node are in this respect unscaled. The dendrogram is rooted implicitly, at the point where the last (i.e., deepest) clusters join. Finally, in some UPGMA presentations, the numerical scale of the distance axis is "doubled" (i.e., the scale in Fig. 4.6a could be drawn equivalently with respective distances 0, 0.20, 0.40, 0.60, 0.80), such that the distance values refer to the joint (rather than single) branch lengths. The particular convention on distance axes that has been followed in a given study can be checked easily by reference to the original distance matrix.

The major assumption of UPGMA clustering is the equal rate of evolution along all dendrogram branches (thus, any true rate heterogeneity would remain undetected in a UPGMA reconstruction). Despite this strong assumption, UPGMA performs unexpectedly well in recovering tree topologies, particularly branch lengths, in empirical tests using computer simulations (Nei et al., 1983; Sourdis and Krimbas, 1987; Tateno et al., 1982). This seems to be due to the fact that genetic-distance estimates are subject to large stochastic error and the distance-averaging aspects of UPGMA tend to reduce the effects of this error (Nei, 1987). In any event, because UPGMA groups taxa that differ least genetically (without finer points of consideration), UPGMA is a simple and intuitively appealing example of the phenetic approach to data summary (Chapter 2).

FITCH–MARGOLIASH METHOD (1967)

The "F–M" procedure was one of the first distance algorithms to relax the assumption of uniform evolutionary rate, and is still widely employed. The general approach is illustrated most easily by reference to three taxa. Consider OTUs A–C in Table 4.1, for which we wish to construct the branch lengths x, y, and z in the following network.

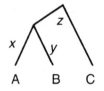

By definition, $d_{AB} = x + y$, $d_{AC} = x + z$, and $d_{BC} = y + z$. Although x, y, and z cannot be observed directly, they may be estimated from the observable distances among extant OTUs. From Table 4.1, $x + y = 0.08$; $x + z = 0.19$; and $y + z = 0.17$. Simultaneous solution of these three equations with three unknowns yields the desired branch lengths, as follows:

$$
\begin{array}{ll}
& (x + y = 0.08) \\
\text{plus} & \underline{(x + z = 0.19)} \\
\text{yields} & (2x + y + z = 0.27) \\
\text{minus} & (y + z = 0.17) \\
\text{yields} & \quad 2x = 0.10 \quad \text{or} \quad x = 0.05.
\end{array}
$$

The unique solutions are $x = 0.05$, $y = 0.03$, and $z = 0.14$. Note that these branch lengths produce a perfectly additive tree, but this seldom is true when $n > 3$ OTUs are considered. Note also that the branches x and y connecting A and B to their immediate common ancestor can differ in length (hence the capacity under F–M to infer rate heterogeneity). As applied to more than three OTUs, the F–M procedure cycles through this tree-fitting process until all branch lengths are obtained [see Fitch–Margoliash (1967) or Li and Graur (1991) for details].

NEIGHBOR-JOINING METHOD (SAITOU AND NEI, 1987)

This method (N–J) is related conceptually to cluster analysis, but also allows for unequal rates of molecular change among branches. It does so by constructing, at each step of the analysis, a transformed distance matrix that has the net effect of adjusting branch lengths between each pair of nodes on the basis of mean divergence from all other nodes. For example, below the diagonal in the top matrix of Box 4.3 is a transformed distance matrix for OTUs A–E, generated from the original matrix (Table 4.1) as follows. The modified distance between OTUs A and B ($d^* = -1.03$) equals the observed distance between A and B ($d = 0.08$) minus the sum of the distances between each of these taxa and all others (i.e., $0.08 + 0.19 + 0.70 + 0.65 + 0.08 + 0.17 + 0.75 + 0.70 = 3.32$), the latter quantity divided in this case by 3 (or generally by 2 less than the number of OTUs in the matrix). Similar calculations fill out the modified distance matrix, which then is searched for the minimum (most negative) value. In our example, the smallest transformed distance is between D and E ($d^* = -1.36$), so these are joined first, through the internal "node 1." The branch lengths joining D and E to node 1 then are calculated as shown in Box 4.3. Note that unlike UPGMA, but as under F–M, these lengths can differ. The process continues through $n - 1$ cycles, with joined extant OTUs replaced by internal nodes as the distance matrices shrink in size. The final tree for our example is shown in Fig. 4.6b. Those interested in additional details and formal steps of the operation should consult Studier and Keppler (1988) or Swofford and Olsen (1990).

DISTANCE WAGNER METHOD (FARRIS, 1972)

The F–M and N–J algorithms attempt to fit the distance matrix to an additive tree, with the view that empirical distances in the matrix may be either under-estimates or overestimates of their true values. The net effect is that some pathways connecting OTUs in the tree are longer, and others shorter, than their empirical distances in the matrix. For example, in the N–J network (Fig. 4.6b), the sum of branch lengths connecting C to E (0.65) is greater than the estimated distance in Table 4.1 ($d = 0.60$), whereas the sum of branch lengths connecting C to D (0.75) is smaller than its corresponding empirical value ($d = 0.80$).

The distance Wagner procedure is similar, but assumes that the observed distances are lower bounds on true values (as perhaps would be the case if genetic distances were uncorrected for superimposed evolutionary changes). Path lengths in Wagner trees equal or exceed corresponding observed distances. A tree that minimizes the total of all branch lengths under this stipulation satisfies the optimality criterion under the distance Wagner method, and various algorithms for approaching this task have been proposed (Farris, 1972; Swofford, 1981; Tateno et al., 1982; Waterman et al., 1977). Thus, searches for trees with minimum total length under distance Wagner and related procedures bear some analogy to qualitative parsimony approaches described later.

Under the Farris (1972) algorithm, a distance Wagner tree for our five OTUs is generated as follows (Box 4.4). First, A and B are joined because they exhibit the smallest genetic distance in the original matrix (Table 4.1). OTU C is added next, to an internal node X along the A–B path, and the three branch lengths [distances $d(A, X)$, $d(B, X)$, and $d(C, X)$] are calculated as shown in Box 4.4b. The inferred branch lengths $d(D, X)$ and $d(E, X)$ also are calculated and stored for later use. Then, a decision is made about which remaining OTU (D or E) next joins the network, and to which existing pathway it should be added (A–X, B–X, or C–X). In our case, E is joined to the C–X branch (through a newly created internal node Y) because this inferred genetic distance is the smallest among the possibilities (Box 4.4d). The new branch lengths [$d(E, Y)$, $d(C, Y)$, and $d(X, Y)$] are calculated as before. The final step in our case is to join D to the network,

Box 4.3. Cycling Operation of the Neighbor-Joining Algorithm

	A	B	C	D	E	r	r/3
A	—	0.08	0.19	0.70	0.65	1.62	0.54
B	−1.03	—	0.17	0.75	0.70	1.70	0.57
C	−0.94	−0.99	—	0.80	0.60	1.76	0.59
D	−0.63	−0.61	−0.58	—	0.12	2.37	0.79
E	−0.58	−0.56	−0.68	−1.36	—	2.07	0.69

Distance D to node 1 = $0.12/2 + (0.79 - 0.69)/2 = 0.11$
Distance E to node 1 = $0.12 - 0.11 = 0.01$

	A	B	C	Node 1	r	r/2
A	—	0.08	0.19	0.62	0.89	0.44
B	−0.82	—	0.17	0.66	0.91	0.46
C	−0.75	−0.79	—	0.64	1.00	0.50
Node 1	−0.78	−0.76	−0.82	—	1.92	0.96

Distance C to node 2 = $0.64/2 + (0.50 - 0.96)/2 = 0.09$
Distance node 1 to node 2 = $0.64 - 0.09 = 0.55$

	A	B	Node 2	r	r/1
A	—	0.08	0.08	0.16	0.16
B	−0.26	—	0.10	0.18	0.18
Node 2	−0.26	−0.26	—	0.18	0.18

Distance A to node 3 = $0.08/2 + (0.16 - 0.18)/2 = 0.03$
Distance node 2 to node 3 = $0.08 - 0.03 = 0.05$

	B	Node 3
B	—	0.05
Node 3		—

r is the sum of the observed distances between the OTU of that row and other extant OTUs or nodes. All values were rounded to two decimal points.

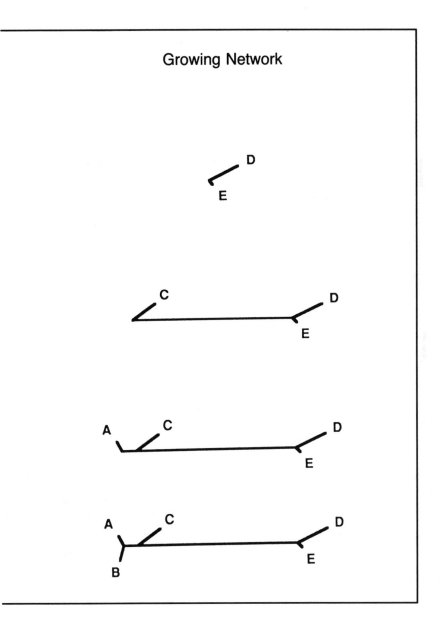

Growing Network

Box 4.4. Stepwise Operation of the Farris Algorithm for Identifying a Distance Wagner Network from the Data in Table 4.1.

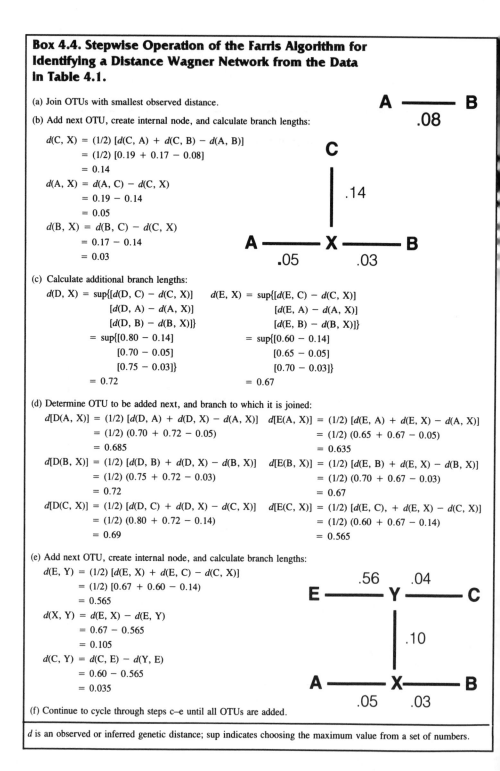

(a) Join OTUs with smallest observed distance.

A —— B
.08

(b) Add next OTU, create internal node, and calculate branch lengths:

$d(C, X) = (1/2) [d(C, A) + d(C, B) - d(A, B)]$
$= (1/2) [0.19 + 0.17 - 0.08]$
$= 0.14$

$d(A, X) = d(A, C) - d(C, X)$
$= 0.19 - 0.14$
$= 0.05$

$d(B, X) = d(B, C) - d(C, X)$
$= 0.17 - 0.14$
$= 0.03$

(c) Calculate additional branch lengths:

$d(D, X) = \sup\{[d(D, C) - d(C, X)]$
$[d(D, A) - d(A, X)]$
$[d(D, B) - d(B, X)]\}$
$= \sup\{[0.80 - 0.14]$
$[0.70 - 0.05]$
$[0.75 - 0.03]\}$
$= 0.72$

$d(E, X) = \sup\{[d(E, C) - d(C, X)]$
$[d(E, A) - d(A, X)]$
$[d(E, B) - d(B, X)]\}$
$= \sup\{[0.60 - 0.14]$
$[0.65 - 0.05]$
$[0.70 - 0.03]\}$
$= 0.67$

(d) Determine OTU to be added next, and branch to which it is joined:

$d[D(A, X)] = (1/2) [d(D, A) + d(D, X) - d(A, X)]$
$= (1/2) (0.70 + 0.72 - 0.05)$
$= 0.685$

$d[D(B, X)] = (1/2) [d(D, B) + d(D, X) - d(B, X)]$
$= (1/2) (0.75 + 0.72 - 0.03)$
$= 0.72$

$d[D(C, X)] = (1/2) [d(D, C) + d(D, X) - d(C, X)]$
$= (1/2) (0.80 + 0.72 - 0.14)$
$= 0.69$

$d[E(A, X)] = (1/2) [d(E, A) + d(E, X) - d(A, X)]$
$= (1/2) (0.65 + 0.67 - 0.05)$
$= 0.635$

$d[E(B, X)] = (1/2) [d(E, B) + d(E, X) - d(B, X)]$
$= (1/2) (0.70 + 0.67 - 0.03)$
$= 0.67$

$d[E(C, X)] = (1/2) [d(E, C), + d(E, X) - d(C, X)]$
$= (1/2) (0.60 + 0.67 - 0.14)$
$= 0.565$

(e) Add next OTU, create internal node, and calculate branch lengths:

$d(E, Y) = (1/2) [d(E, X) + d(E, C) - d(C, X)]$
$= (1/2) [0.67 + 0.60 - 0.14)$
$= 0.565$

$d(X, Y) = d(E, X) - d(E, Y)$
$= 0.67 - 0.565$
$= 0.105$

$d(C, Y) = d(C, E) - d(Y, E)$
$= 0.60 - 0.565$
$= 0.035$

(f) Continue to cycle through steps c–e until all OTUs are added.

d is an observed or inferred genetic distance; sup indicates choosing the maximum value from a set of numbers.

at a position determined by its smallest inferred distance to existing branches. In general, the Farris algorithm repeats these tree-construction rules, keeping track of internal nodes and branch lengths at each stepwise addition of an extant OTU. The final distance Wagner network for OTUs A–E is presented in Fig. 4.6c.

COMPARISON OF DISTANCE MATRIX METHODS

Much debate has concerned which of these (or other) distance algorithms produces the "best" tree. One basis for choice is "goodness-of-fit," a measure of how well the inferred distances in the tree match the empirical distance values in the original matrix. Branch lengths between OTUs read from the tree (output distances) can be compared to input distances in the matrix by any of several suggested statistics, including a "cophenetic correlation" (Sneath and Sokal, 1973), "percent standard deviation" (Fitch and Margoliash, 1972), or an "*F*" statistic (Prager and Wilson, 1978). As an example, the latter is defined as

$$F = 100 \sum_{i=1}^{s} | I_i - O_i | \Big/ \sum_{i=1}^{s} I_i \qquad (4.18)$$

where for the s pairwise comparisons among OTUs, I and O are the input and output distances, respectively. Smaller values of F indicate better fit. As might be expected, procedures such as F–M or N–J that explicitly adjust tree branches to achieve a fit to an additive tree generally outperform methods such as UPGMA that do not (Avise et al., 1980; Berlocher, 1981; Prager and Wilson, 1978). On the other hand, the nature of the distance measure itself (e.g., whether it could, in principle, yield an additive tree) also can affect the outcome, as can to some extent the definition of the statistic (such as F) used to assess the agreement between the tree and data.

A second basis for choice among distance algorithms involves degree of congruence among trees derived from different sets of data. Because a given array of species has a single phylogenetic topology along which all characters have evolved, methods of data summary producing more highly congruent trees might be judged superior (Farris, 1971). Several measures of congruence among trees have been suggested (e.g., Farris, 1973; Mickevich, 1978). Unfortunately, such comparisons seldom have been applied to molecular information because of the usual absence of multiple independent data sets for a particular taxonomic group.

A third approach for comparing the performance of phylogenetic algorithms involves computer simulation of evolutionary change along specified model trees. From observed genetic distances among extant computer OTUs, phylogenies are estimated and an algorithm's performance is evaluated by how well it recovers the known tree (Fiala and Sokal, 1985; Jin and Nei, 1991; Saitou and Nei, 1987). Potential shortcomings of this approach are (a) the difficulty of assessing the biological plausibility of assumptions underlying the simulated evolution of genetic characters through the computer tree and (b) the danger of circular reasoning when the best phylogenetic algorithm proves to be the one whose assumptions most closely match those underlying the simulation. All phylogenetic algorithms involve assumptions (explicit or implicit). A continuing challenge is to identify and understand the assumptions and to assess their appropriateness for the evolutionary mode of the characters under assay. Note also that facility with implementing a phylogenetic algorithm, even by hand, is no guarantee that the underlying assumptions will become apparent (if you do not believe this, try working through the F–M, distance Wagner, or N–J examples provided earlier).

A fourth and powerful method for evaluating algorithm performance was introduced by Hillis et al. (1992). They serially propagated bacteriophage T7 in the presence of a mutagen, experimentally dividing the culture at various time intervals such that a known phylogeny was produced. Then the terminal lineages were assayed for restriction-site maps and the empirical data were used to infer the evolutionary history by various phylogenetic methods. All five algorithms employed, which included F–M, N–J, and UPGMA (as well as a qualitative parsimony approach—see next section), produced the correct branching order of the known topology, but differed slightly in ability to recover correct branch lengths. Of course, such direct appraisals of phylogenetic methods for living organisms will be possible only in a few systems such as T7 where mutation rates are high and thousands of generations occur each year (see Chapter 6).

Among the distance methods considered, only UPGMA automatically produces a rooted tree. To root any of the other networks, two procedures may be followed. First, outgroup OTUs (those known from independent evidence to have branched off earlier from other taxa under study) can be included in the empirical analysis, in which case the root is placed between an outgroup and the node leading to ingroup members. Alternatively, if an approximate uniform rate of evolution is assumed for long time periods, the network may be rooted at the midpoint of the longest pathway between any extant OTUs (note that this assumption is less stringent than that of rate homogeneity in all tree branches). For example, the longest pathway in the N–J network in Fig. 4.6b is of length 0.76 (between B and D). Placement of the root midway along this path produces a rooted tree that in this case resembles closely the UPGMA dendrogram (Fig. 4.6a).

Character-State Approaches

Some molecular techniques provide discrete character information, such that a data matrix can be developed that assigns a character state (x_{ij}) to each OTU (i)

for each character (*j*) (e.g., Boxes 3.2 and 3.4). The types of molecular characters and their states vary with circumstance and must be specified carefully in each study. For example, in an allozyme survey, the gene for LDH could be considered one character, with different states being the observed electromorphs; or the character might be a particular nucleotide position in a DNA sequence, with possible states A, T, C, or G. Such characters which can assume three or more states are called multistate traits. Binary characters are those that can assume only two states. For example, the presence versus absence of a RFLP restriction site at a particular map location exhausts the two possible states for that character. In some cases, characters might be defined at a more inclusive level. For example, the collection of different restriction-site maps produced by a given endonuclease, or the collection of DNA sequences for a given gene, might justifiably be considered the multiple states of their respective characters.

Multistate characters may be unordered or ordered. Electromorphs of an allozyme locus are examples of unordered character states because their evolutionary interrelationships cannot be deduced directly from the observed electrophoretic mobilities. Similarly, the alternative states at a given nucleotide position normally are considered unordered because there is no a priori reason to assume a particular evolutionary pathway for the interconversion among A, T, C, and G. On the other hand, restriction-site maps for a given enzyme (or DNA sequences for a given gene) may be treated as ordered multistate characters when their probable evolutionary transformations have been deduced from reasonable criteria such as parsimony (Figs. 3.8 and 3.9). However, such character-state phylogenies are themselves evolutionary inferences (hypotheses) derived from the application of phylogenetic procedures to information accumulated from lower-level character-state descriptions (individual restriction sites or nucleotide positions, in these cases). Thus, character-state matrices for most computer-based phylogenetic algorithms consist of data coded at these more fundamental levels.

The polarity of character states refers to the *direction* of evolution and, hence, is distinct from the concept of character order. Polarized characters are those for which ancestral and descendant states have been determined. Thus binary characters, and ordered or unordered multistate characters, may be either polarized or nonpolarized. Use of inferred character polarity is essential in Hennigian cladistic analysis (Fig. 2.5) but is not necessary in all forms of parsimony. What follows are brief descriptions of the qualitative character-state algorithms most commonly employed in molecular phylogenetics.

Hennigian Cladistics

The philosophy underlying Hennigian cladistics was described in Chapter 2. The critical feature is that synapomorphic character states must be identified and that they alone provide the basis for clade identification. If there were no disagreements in clade delineation among putative synapomorphs and each character state was monophyletic, a Hennigian reconstruction should capture the true evolutionary tree. Molecular features that are likely of monophyletic origin (e.g., unique gene arrangements or other idiosyncratic genetic attributes) are especially well suited for Hennigian cladistic interpretations. Unfortunately, most large molecular (and nonmolecular) data sets include some conflicts among characters, no doubt due in part to the possibility of polyphyletic mutational origins of individual character states (such as the electromorphs of a locus, nucleotides at a sequence position, or sites in a restriction map). To resolve the apparent dilemmas in character-state distributions across taxa, parsimony approaches are employed widely.

MAXIMUM PARSIMONY

A most-parsimonious tree is one that requires the smallest number of evolutionary changes to explain the observed differences among OTUs. Consider Figure 4.7, which presents a hand-generated estimate of a parsimony network for 10 extant OTUs based on 9 variable characters. (Hypothetical OTUs HYP1 and HYP2, not directly observed, were added arbitrarily to make all branches of unit

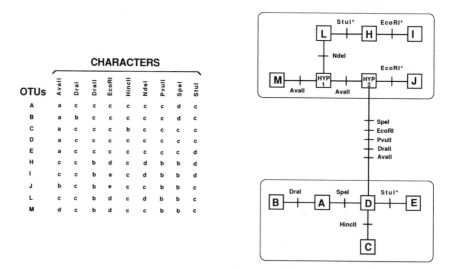

CHARACTERS

OTUs	AvaI	DraI	DraII	EcoRI	HincII	NdeI	PvuII	SpeI	StuI
A	a	c	c	c	c	c	c	d	c
B	a	b	c	c	c	c	c	d	c
C	a	c	c	c	b	c	c	c	c
D	a	c	c	c	c	c	c	c	c
E	a	c	c	c	c	c	c	c	d
H	c	c	b	d	c	d	b	b	d
I	c	c	b	e	c	d	b	b	d
J	b	c	b	e	c	c	b	b	c
L	c	c	b	d	c	d	b	b	c
M	d	c	b	d	c	c	b	b	c

Figure 4.7. Estimate of an unrooted maximum parsimony network (**right**) based on an OTU X character matrix (**left**) for a subset of mtDNA clones observed in green turtles (after Bowen et al., 1992). The lowercase letters are mtDNA digestion profiles produced by nine restriction enzymes (letters adjacent in the alphabet denote profiles that differed by a single restriction site). Inferred restriction site changes along branches of the network are indicated.

length in the upper portion of the tree.) Construction of the network could be initiated by connecting any OTU to its nearest genetic neighbors via reference to the character-state matrix. For example, haplotype D is one mutation step removed from each of three other haplotypes (A, C, and E) that are two steps removed from one another. Haplotype B, in turn, is one step from A, two steps from D, and three steps each from C and E. Thus, the distribution of character states for OTUs A–E yields the singular most-parsimonious network shown at the bottom right of Figure 4.7. Note that this portion of the network is strictly additive.

Generation of the complete network for all 10 extant OTUs illustrates complications that may arise. First, there is a large genetic gap distinguishing OTUs A–E from the assemblage H, I, J, L, M, so where the branch connecting these groups should be placed is not obvious initially. Here, D and HYP2 are joined because they differ by five steps, whereas any other intergroup branches would involve six steps or more. Second, some genetic character states appear in different (presumably distantly related) portions of the network. For example, the character state *Stu*I-*d* appears in some representatives of both the upper and lower OTU groups, probably due to polyphyletic origins from *Stu*I-*c*. Similarly, the state *Eco*RI-*e* appears in OTUs I and J that are not adjacent genetically as judged by the other assayed characters. These character states contribute to homoplasy by introducing additional steps along network branches beyond those differentiating OTUs in the original character-state matrix. Nonetheless, in this example, the sum of all pairwise output distances in the network (256) is only slightly greater that the sum of all input distances (250), indicating strong goodness-of-fit between tree and data.

In usual practice, computer-based parsimony algorithms operate by searching alternative trees for minimum total length. Sometimes, many trees of different topology prove equally parsimonious or require similar numbers of steps. Nonetheless, such networks constitute only a small fraction of the vast universe of potential trees, most of which, therefore, can be eliminated from further consideration.

Actually, parsimony approaches comprise a family of related methods with varying assumptions about how character-state transformations occur. The most commonly employed parsimony algorithms and their assumptions are as follows:

a.   Wagner parsimony (Farris, 1970; Kluge and Farris, 1969). This approach (not to be confused with the distance Wagner method) allows free reversibility of character states in the tree, with changes in either direction equally likely (thus alternative rootings produce no change in tree length). Characters may be binary or ordered multistate, although transformations among multistate characters must occur through intervening states only. The network in Figure 4.7 is based on Wagner assumptions and includes both binary and ordered multistate characters.

Fitch's (1971) modification of the Wagner approach allows direct transformations among any unordered multistate characters.

b. Dollo parsimony (Farris, 1977). This parsimony approach assumes that an ancestral condition for each binary or ordered multistate character can be specified and that each nonancestral character is uniquely derived (multiple reversions to the ancestral condition are allowed). The analysis seeks to minimize total tree length under these conditions. In principle, Dollo parsimony should be appropriate when probabilities of change among character states are highly asymmetric. For example, because a mutation in any nucleotide position of a restriction site eliminates enzyme recognition, loss of a particular restriction site might be more likely than its gain (DeBry and Slade, 1985; Templeton, 1983, 1987). However, if derived states are polyphyletic even occasionally, Dollo parsimony can exhibit pathologic behavior (Swofford and Olsen, 1990).

c. Camin–Sokal (1965) parsimony. This form of parsimony carries the stringent assumption that all evolutionary change is irreversible. Thus, the approach goes beyond Dollo parsimony by disallowing reversions to the ancestral condition. The method is not employed widely with genetic data because most molecular characters probably violate this assumption.

d. Generalized parsimony (Swofford and Olsen, 1990). In effect, all parsimony methods (including Wagner parsimony) make assumptions about the "costs" of transformation among character states. A generalized parsimony approach allows flexibility in assignment of these costs among characters, ideally utilizing independent evidence about the relative frequencies of different kinds of molecular changes. Understandably, such approaches are expensive computationally, and usable algorithms still are in their infancy.

Conclusions About Phylogenetic Procedures

For the practicing empiricist, what recommendations can be made concerning this plethora of phylogenetic algorithms? Of course, it is important to attempt a match between assumptions of the phylogenetic procedure(s) and the nature of evolution for the molecular characters under assay. Second, it generally is advisable to attempt multiple methods of data analysis, particularly if these entail philosophically distinct approaches. For example, both UPGMA clustering (a distance-based method epitomizing the phenetic philosophy) and Wagner parsimony (a character-state approach more akin to cladistic thought) reasonably can be applied to many molecular data sets and the results compared. Generally, most of the significant features in a given data set (e.g., major genetic breaks reflected as long internodal distances) are revealed irrespective of the phylogenetic algorithm employed, whereas problematic differences among trees tend to involve topology shifts where OTUs or internal nodes are bunched closely.

Third, at least one character-state approach should be included when the data permit. Because tree topologies derived from qualitative approaches provide explicit inferences about character-state distributions along branches, they are extremely rich in empirical content (*sensu* Popper, 1968); in other words, they offer testability and accountability (Avise, 1983b; Baverstock et al., 1979; Patton and Avise, 1983). Consider again Figure 4.7, where character-state changes are specified explicitly along branches. Such detailed evolutionary understanding is not possible with distance analyses alone. Furthermore, if portions of the tree topology are suspect by external evidence, the problem areas and the characters responsible can be identified. Perhaps the offensive characters were scored incorrectly or mistransferred during data analysis. Perhaps they arose polyphyletically (an important observation in its own right), in which case the character states might be subjected to further examination (sequencing in this case) to assess the molecular basis of the apparent convergence. The point is that further hypothesis testing flows readily from explicit accounts of character-state distributions.

Fourth, it is highly desirable to include confidence statements about putative clades revealed in a phylogenetic reconstruction. One common approach is bootstrapping (Felsenstein, 1985a; Hedges, 1992), which involves resampling (with replacement) from the existing data sets and assessing the frequency with which particular groups or clades appear in trees generated from the resampled data. Thus, bootstrapping indicates how well various groupings of OTUs are supported by existing data (though not necessarily how well the available data represent unsampled genetic characters). Finally, use of outgroups is to be encouraged because this allows rooting of the tree and helps to establish character state polarities.

Because of the large size of most molecular data sets, computer programs are required to calculate population genetic parameters and distance matrices and to implement phylogenetic algorithms. The software packages employed most commonly are PHYLIP (available from Joseph Felsenstein, University of Washington, Seattle), PAUP (written by David L. Swofford and distributed by the Illinois Natural History Survey, Champaign), BIOSYS-1 (Swofford and Selander, 1981, also available from the Illinois Natural History Survey), FREQPARS (Swofford and Berlocher, 1987, available from Swofford at the Smithsonian Institution, Washington, D.C.), Hennig86 (James S. Farris, Port Jefferson Station, New York), and MacClade (written by Wayne P. Maddison and David R. Maddison, and available from Sinauer, Sunderland, MA). Brief descriptions of these programs can be found in Swofford and Olsen (1990).

Before closing this section, it should be emphasized that all phylogenetic algorithms discussed earlier are based on two fundamental assumptions: that the characters are (a) homologous and (b) independent. The concept of independence warrants elaboration. Characters are independent in a mechanistic sense if changes in one character occur independently of those in another, such that the character states do not co-vary because of pleiotropic effects in the underlying mutational process. For example, restriction fragment gains and losses tend to co-vary across digestion profiles, whereas the responsible site changes do not, and for this reason it is preferable (when possible) to code RFLP data as presence/absence of restriction sites rather than fragments (or at least to accommodate the covariance of fragments in the phylogenetic analysis). The assumption of independence, critical to most computational algorithms, is probably valid for most molecular characters (unlike the situation for many morphological traits) and indeed is a major strength of multicharacter molecular approaches.

However, there is another sense in which some molecular characters may be partially nonindependent evolutionarily. When molecular characters are tightly linked, as is usually the case for restriction sites or sequence data at a particular locus, molecular states tend to co-vary *in transmission* across organismal generations (barring intragenic or interallelic recombination). Some molecular systems such as animal mtDNA presumably are free of recombination almost entirely. Thus, although such character states are independent in the mechanistic sense of mutational origins, they may not elucidate independent transmission pathways through an organismal pedigree. Recognition of this fact has contributed to the fundamental distinction, now recognized widely in molecular phylogenetics, between a "gene tree" and an organismal phylogeny (Avise, 1989a; Doyle, 1992; Neigel and Avise, 1986; Pamilo and Nei, 1988; Tajima, 1983; Tateno et al., 1982; Wilson et al., 1985).

GENE TREES VERSUS SPECIES TREES

When OTUs are the alleles of a locus (e.g., haplotype sequences or restriction-site maps), a reconstructed phylogeny represents a gene tree. Any group of organisms has a single pedigree that extends back through time as an unbroken chain of parent-offspring genetic transmission, but due to the nondeterministic nature of biparental Mendelian heredity in sexually-reproducing species, not all genes will have trickled through this organismal pedigree in identical fashion. Thus, gene trees are likely to differ somewhat from one (unlinked) locus to the next (Ball et al., 1990). Furthermore, a gene tree may differ in topology from that of the population tree or species tree through which it has been transmitted [even in the absence of introgressive hybridization (Chapter 7)], due to an inevitable process called "stochastic lineage sorting" that may be exemplified as follows.

Consider a single population pedigree through which haplotypes have descended. A simple case conceptually involves mtDNA inherited through female lines, but the principles apply to haplotypes of particular nuclear genes as well. As shown in Figure 4.8, some females by chance leave no daughters (those mtDNA lineages terminate), whereas others produce one or more daughters that may contribute mtDNA to successive generations. Thus, as an inevitable con-

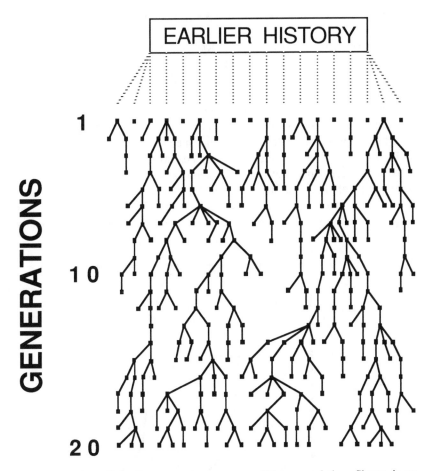

Figure 4.8. The allelic lineage sorting process within a population. Shown is an mtDNA gene tree through 20 generations, where each node represents an individual female, and branches lead to daughters. The tree was generated by assuming a Poisson distribution of progeny numbers with a mean of one daughter per female (after Avise, 1987).

sequence of differential organismal reproduction, the mtDNA gene tree is continually "self-pruning"—certain branches are lost as others proliferate. At equilibrium, the expected frequency distribution of times to common ancestry can be approximated (Box 2.3). In general, it is unlikely that two or more founding lineages by chance will survive beyond $4N$ generations, where N is the population size (Fig. 4.9).

Now consider the lineage sorting process extended to two daughter taxa (A and B) stemming from a common ancestral population. With regard to a gene tree within these sister populations or species, three phylogenetic categories are possible (Fig. 4.10): (I) *reciprocal monophyly,* in which case all alleles within each sister taxon are genealogically closer to one another than to any heterospecific alleles; (II) *polyphyly,* where some alleles in each taxon are genealogically closer to heterospecific alleles than to homospecific alleles; and (III) *paraphyly,* in which case all alleles within one daughter taxon are one another's closest relatives, whereas some alleles in the second taxon are genealogically closer to heterospecific alleles. The first category (Fig. 4.10, case I) illustrates how the depths of gene trees can vary even when their branching topologies agree with the species tree (e.g., the alleles in B trace to an ancient node b whereas the

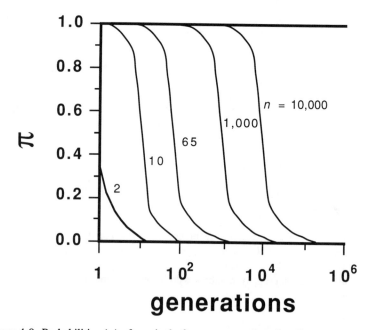

Figure 4.9. Probabilities (π) of survival of two or more founding lineages through time. Shown are probability curves for populations of various size in which females produce daughters according to a Poisson distribution with mean 1.0 (after Avise et al., 1984a).

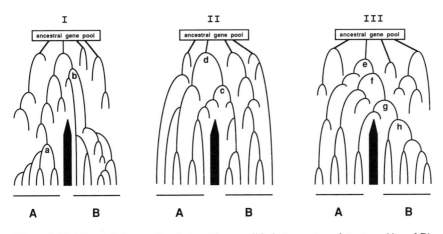

Figure 4.10. Three phylogenetic relationships possible between two sister taxa (A and B) with respect to an allelic genealogy (after Avise et al., 1983). Lowercase letters point out important ancestral nodes to which extant alleles or haplotypes trace. The solid dark bars indicate barriers to reproduction (extrinsic or intrinsic). The phylogenetic categories in the gene tree are as follows: I, reciprocal monophyly; II, polyphyly; III, paraphyly.

alleles in A trace to a recent node a). The latter two categories of relationship (Fig. 4.10, cases II and III) illustrate how a gene tree also can differ in fundamental branching order from a species tree. These discordances arise because many allelic separations inevitably predate the species split (unless the ancestral form went through an extreme population bottleneck just prior to speciation). Figure 4.11 shows diagrammatically how these three categories of phylogenetic relationship may characterize one-and-the-same pair of sister taxa at different times following their speciation.

If genetic distances among haplotypes are measured in units of time since common ancestry, these categories of phylogenetic relationship between sister taxa (with respect to a gene tree) may be defined formally by the inequalities in Table 4.2. Neigel and Avise (1986) employed computer simulations to monitor the phylogenetic status of sister taxa with respect to mtDNA lineages (Fig. 4.12). Shortly after speciation, the probability is high that sister taxa exhibit a polyphyletic gene-tree status. At intermediate times since speciation (typically N–$3N$ generations, where N is the population size of each sister taxon), probabilities of poly-, para-, and monophyly are intermediate as well. Only after about $4N$ generations do sister taxa appear reciprocally monophyletic with high probability. Similar results apply to nuclear genes (Nei, 1987), although the times to monophyly are extended accordingly because of the expected fourfold larger effective population sizes for nuclear loci (Box 2.3).

When balancing selection maintains polymorphisms within the species, the expected times to reciprocal monophyly in a gene tree may be extended consid-

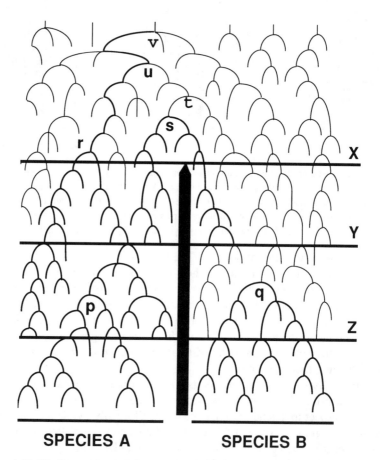

Figure 4.11. The lineage sorting process extended to two sister taxa (after Avise and Ball, 1990). Shown are distributions of allelic lineages at a single gene through an ancestral population subdivided at time X into two daughter populations or species. With respect to this gene tree, the sister species are polyphyletic between time levels X and Y (follow tracings from nodes s, t, u, and v), show paraphyly between levels Y and Z, and are reciprocally monophyletic beyond level Z.

Table 4.2. Definitions of the phylogenetic status of two sister taxa with respect to a gene tree they contain[a]

Phylogenetic Category	Phylogenetic Status	Distance Relationship
I	A and B monophyletic	max d_{AA} < min d_{AB} and max d_{BB} < min d_{AB}
II	A and B polyphyletic	max d_{AA} > min d_{AB} and max d_{BB} > min d_{AB}
IIIa	A paraphyletic with respect to B	max d_{AA} > min d_{AB} and max d_{BB} < min d_{AB}
IIIb	B paraphyletic with respect to A	max d_{AA} < min d_{AB} and max d_{BB} > min d_{AB}

[a]After Neigel and Avise, 1986. Maximum distances within either taxon (max d_{AA} or max d_{BB}) versus minimum distance between sister taxa (min d_{AB}) are the deciding criteria (see Fig. 4.10 and text).

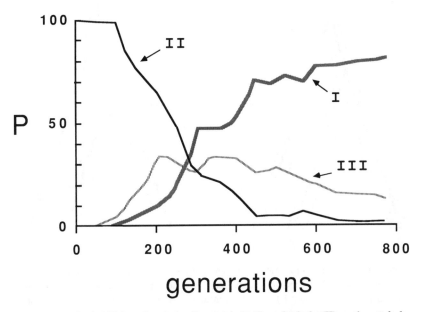

Figure 4.12. Probabilities of reciprocal monophyly (I), polyphyly (II), and paraphyly (III) for two sister taxa G generations following a simulated speciation (after Neigel and Avise, 1986). In each of 100 replicate computer runs, the daughter species were founded by 20 and 30 individuals respectively and allowed to grow rapidly to carrying capacity $N = 200$.

erably beyond those expected under the neutral model. Much recent interest has attended the discovery of two such apparently balanced polymorphisms that have persisted for millions of years and across several speciation events. These involve the major histocompatibility loci in rodents and primates (Figueroa et al., 1988; Lawlor et al., 1988; McConnell et al., 1988; Takahata and Nei, 1990), and the self-incompatibility locus in Solanaceae plants (Ioerger et al., 1990).

Discordances between species-splitting patterns and the topologies of gene trees also can characterize taxa that separated anciently but whose speciations occurred close together in time (Fig. 4.13). The same kinds of lineage sorting processes are responsible; in this case, lineages from the polymorphic ancestral gene pool that happen to have reached fixation in the descendent taxa may by chance be those that produce a gene-tree/species-tree discordance (Takahata,

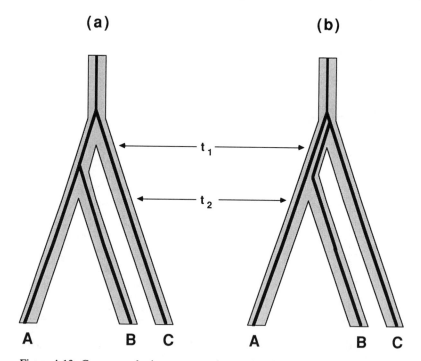

Figure 4.13. Gene genealogies across ancient nodes close in time. Shown are two topologically distinct gene trees (dark lines) possible within a species tree consisting of two sister taxa (A and B) and an outgroup (C). In diagram (a), the gene tree and species tree have the same branching pattern, whereas in (b) the branching topologies differ. For neutral alleles, the probability of the discordance exemplified by diagram (b) is given by $\frac{2}{3}e^{-T/2N_e}$ (Nei, 1987), where $T = t_1 - t_2$ (t_1 is the time of the first speciation and t_2 is the time of the second speciation), and N_e is the effective population size.

1989; Wu, 1991). In the primate literature, intense debates have concerned which "true" tree characterizes the phylogeny for human, chimpanzee, and gorilla, a related triad of species whose branching pattern appears to consist of two closely spaced nodes like those in Figure 4.13. Many molecular assays have been applied, but not all agree in outcome (Chapter 8). Perhaps with respect to gene trees, no single outcome should be expected.

These perspectives stemming from molecular research reveal several points of qualitative importance to phylogenetics, beyond the immediate fact that gene trees and species trees can differ in branching topology. First, with regard to relationships in particular molecular (or other) characters, the phylogenetic status of a given pair of species is itself an evolutionarily dynamic characteristic, with a usual time course subsequent to speciation being polyphyly→paraphyly→monophyly (Fig. 4.12). Second, the phylogenetic status of species is a function both of the pattern of population splitting and of the historical demography within the populations involved (Avise et al., 1984a). Thus, for example, species with a larger effective population size (Box 2.2) will tend to retain polyphyletic or paraphyletic status for longer times than will species with small N_e, all else being equal. Third, in accounting for the appearance of "heterospecific" alleles within a species, it now is apparent that possibilities involving lineage sorting from an ancestral gene pool must be considered in addition to the usual scenarios of interspecies transfer mediated via introgressive hybridization (Chapter 7).

Because of its rapid pace of evolution, "haploid" packaging within most organisms, and nonrecombining mode of transmission, mtDNA has provided the vast majority of empirical data suitable for estimating gene trees over microevolutionary time scales. In principle, data from nuclear loci can be exploited similarly, but two major difficulties arise. First is the technical problem of isolating individual haplotypes of a locus from diploid organisms. Box 4.5 describes several genetic systems or experimental approaches where this difficulty might be circumvented. A second potential complication, particularly at the intraspecific level, is intragenic recombination (or gene conversion). Such shuffling of genetic material among alleles, if frequent over the time scales relevant to a genealogical reconstruction, will obscure the otherwise linear evolutionary histories of particular haplotypes within a species (Hudson, 1990). Patterns of nonrandom association (disequilibrium) among tightly linked polymorphic markers can help to reveal how frequently recombination may have occurred in the history of a gene region (Clark, 1990; Stephens, 1985). Nonetheless, for secure recovery of a gene genealogy, attention normally must be confined to DNA segments with little or no recombination (Box 4.6).

Gene trees and species trees are equally "real" phenomena, merely reflecting different aspects of the same phylogenetic process. Thus, occasional discrepancies between the two need not be viewed with consternation as sources of "error" in phylogeny estimation. When a species tree is of primary interest,

Box 4.5. Some Special Genetic Systems and Potential Approaches for the Isolation of DNA Haplotypes.

Approach	Rationale	Comments	Example Reference
mtDNA	Most individuals effectively homoplasmic and haploid for mtDNA; nonrecombining genetic transmission	By far, most widespread source for gene-tree data	Avise, 1989a
Sex chromosomes (e.g., X or Y in mammals, Z or W in birds)	Heterogametic sex is haploid; limited recombination, particularly in Y or W chromosomes	Few genes as yet identified or surveyed	Bishop et al. 1985; Vulliamy et al. 1991
Species with haplo-diploid sex determination	Males are haploid in many hymenopteran insects	No known attempts as yet for full gene trees	Hall, 1990
Species with prominent haploid phase of life cycle	For example, gametophyte stage of mosses is haploid	No known attempts as yet for full gene trees	McDermott et al., 1989
Species with prominent haploid tissue	Endosperm in seeds of gymnosperms is a haploid product (gametophyte) from the female parent	Does not apply to angiosperms, where endosperm is triploid	
Haploid species	Haploid microorganisms should be suitable, provided sexual reproduction and recombination are limited	Relatively few attempts	Nelson et al., 1991
DNA amplification in vitro (PCR) from single gametes, or single DNA molecules	Gametes are haploid; each molecule represents one haplotype	In principle, single gametes or molecules can be isolated by serial dilution	H. Li et al., 1988; Boehnke et al., 1989; Ruano et al., 1990
DNA amplification in biological vectors	Cloning passes DNA through a bottleneck of one molecule	Relatively laborious	Scharf et al., 1986
Extraction of individual chromosomes	Use of inbred strains, or of controlled crosses producing individuals with chromosomes identical by descent; especially powerful when applied to genes within chromosome inversion systems where recombination is limited or absent	Methods most readily available in *Drosophila*	Aquadro et al., 1986, 1991

When such genetic systems also exhibit limited recombination, they could provide ideal opportunities for construction of gene trees.

Box 4.6. Intraspecific Gene Trees for Nuclear Loci

Most early attempts to study nuclear gene trees at the intraspecific level involved *Drosophila* species because experimental crosses can be conducted readily to "extract" individual chromosomes from diploid fruit flies taken from nature, thereby generating strains in which particular chromosomal homologues are "identical by descent" (Box 4.5). Figure 4.14 shows one such genealogy estimated for ADH haplotypes extracted from *D. melanogaster*.

Construction of a gene genealogy requires a history of limited or no recombination among homologous haplotypes over the time scales of the phylogeny. It has proven difficult to predict which gene regions are likely to be sufficiently free of such recombination. For example, similar attempts to construct an intraspecific allelic phylogeny for ADH in a related fruit fly *D. pseudoobscura* have been thwarted by an absence of strong non-random associations among linked restriction sites, apparently due to a history of higher interallelic exchange (Schaeffer and Miller, 1992; Schaeffer et al., 1987). The xanthine dehydrogenase region in *D. pseudoobscura* also shows few nonrandom associations among restriction sites (Riley et al., 1989), as do several assayed loci other than ADH in *D. melanogaster* [e.g., the *notch, white, zeste-tko,* and perhaps *amylase* gene regions (Aguadé et al., 1989a; Langley and Aquadro, 1987; Langley et al., 1988; Schaeffer et al., 1988)]. Many demographic as well as molecular factors can influence the history of effective recombination within a gene region. For example, the allelic contents of species that are highly subdivided spatially should tend to exhibit greater disequilibrium than those of "high gene flow" species in which the alleles are routinely brought together such that recombination among them at least is possible. Different chromosomal regions also are known to differ inherently in recombination rates, with likely consequences extending to several aspects of genome structure including patterns of intragenic marker associations (Aguadé et al., 1989b; Begun and Aquadro, 1992).

When stretches of DNA with limited recombination can be identified, the gene genealogies that they imply might be used to map phenotypic mutations (Templeton et al., 1992). The rationale for this endeavor is that any such mutations linked with the nonrecombining marker region under study would be embedded within the same evolutionary history that is represented by the allelic cladogram.

Aquadro et al. (1991) have capitalized on the recombination-suppression properties of chromosomal inversions in *D. pseudoobscura* to generate a phylogeny for an amylase gene (*Amy*) that is contained within the inverted region of the third chromosome. The reduction in effective recombination in inversion heterozygotes is dramatic and occurs because crossing over inside the inverted region normally produces dysfunctional duplication and deficiency products that are shunted to polar bodies where they fail to participate in zygote formation (and *Drosophila* males lack recombination). The significance of recombination suppression in this system is that the maintenance of linked and presumably adaptive complexes of genes within the inverted region is facilitated. A gene tree based on restriction site maps for 28 *Amy* haplotypes is presented in Figure 4.15, from which the following major conclusions emerged (Aquadro et al., 1991): (a) restriction-site differences are considerably greater among the various gene arrangements than among haplotypes within the same gene arrangement; (b) the gene phylogeny based on *Amy* appears concordant with the inversion phylogeny generated independently from cytologic considerations; and (c) from application of a molecular clock, the inversion polymorphism may be about two million years old.

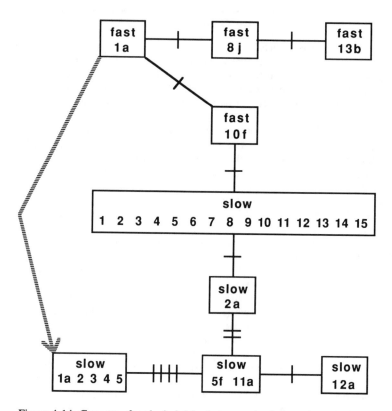

Figure 4.14. Gene tree for alcohol dehydrogenase haplotypes in *Drosophila melanogaster* (after Aquadro et al., 1986 as redrawn by Avise, 1989a). The haplotype labeled "slow" in the largest box serves as an arbitrary reference in which the numbers refer to the standard states for 15 restriction sites and other characters arrayed in the 5′ to 3′ direction of the ADH gene. Each genetic variant from the reference is represented by a lowercase letter. "Slow" and "fast" refer to a replacement substitution (character 10) that underlies two common protein electromorphs at the ADH locus. The presumed transformations among character states can be read cumulatively through the network. For example, haplotype "slow 5f 11a" differs from the reference in genetic states for characters 2, 5, and 11. Lines crossing branches of the network indicate numbers of character-state changes inferred along a path. The haplotype at the lower left represents a probable interallelic recombinant between the two haplotypes to which it is connected by paths. From this phylogeny and additional genetic evidence, it was surmised that the "fast" allozyme genotype represents a derived condition having evolved recently from the ancestral "slow" allozyme allele (Aquadro et al., 1986; Ashburner et al., 1979; Stephens and Nei, 1985).

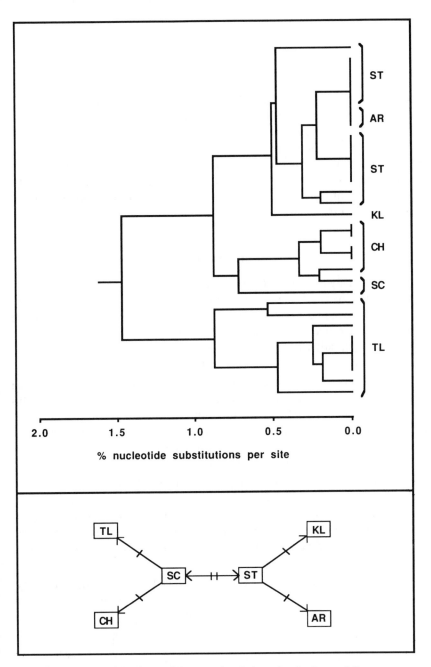

Figure 4.15. Molecular and karyotypic phylogenies in *Drosophila pseu-doobscura* (see Box 4.6). **Above:** Gene tree for 28 molecular haplotypes observed at the amylase (*Amy*) locus (after Aquadro et al., 1991). Also indicated are the chromosomal inversion types from which these *Amy* haplotypes were extracted: ST (Standard); AR (Arrowhead); KL (Klamath—from *D. persimilis*); CH (Chiricahua); SC (Santa Cruz); and TL (Treeline). **Below:** Cytogenetic phylogeny of these same gene arrangements.

gene trees can assist in understanding the population demographies underlying the speciation process, as well as the species-splitting patterns themselves (Chapters 6 and 7). Of course, for such purposes, it would be desirable to include information from multiple gene genealogies. Each gene tree also is of inherent interest because it describes the evolutionary history of genetic changes within a localized bit of the genome. In studies of the ages and origins of particular adaptations (or of genetic disorders), such single-locus reconstructions become the primary foci of attention (Figs. 4.14 and 4.15).

SUMMARY

1. The analytical methods of population genetics and phylogenetics are applicable to molecular data, but the particular algorithms employed are multivarious and depend both on the nature of the biological problem and the nature of the molecular data. Molecular information can be classified into several alternative but partially overlapping categories to which distinctive general methods of data analysis and interpretation pertain.

2. The concept of molecular clocks has played a major role in molecular phylogenetics. Nonhomologous sequences are known to evolve at greatly differing rates, but considerable controversy still exists about the magnitude of the heterogeneity in rates of homologous sequence evolution among organismal lineages. Both absolute and relative rate tests have been employed widely in the assessment of molecular evolutionary tempos.

3. Phylogenetic algorithms may be classified into distance-based versus character-state approaches. Important examples of the former class include cluster analyses, the Fitch–Margoliash method, neighbor-joining, distance Wagner, and other algorithms that utilize a matrix of genetic distances between taxa. Among the character-based methods are Hennigian cladistics and several forms of parsimony analysis.

4. Important distinctions exist between a gene tree and an organismal phylogeny. These two aspects of phylogeny provide different but mutually informative perspectives on the evolutionary process.

Part II

Applications

5

Individuality and Parentage

*With the recognition . . . that the . . . genome is replete
with DNA sequence polymorphisms such as RFLP's, it
was only a small leap to imagine that DNA could, in
principle, provide the ultimate identifier.*

E.S. Lander, 1991

Most species of sexually reproducing organisms harbor sufficient variation that
appropriate molecular genetic assays can distinguish one individual from another
with high probability. Furthermore, the transmission of genetic markers across a
single generation, as interpreted against the established rules of Mendelian inheritance, provides a powerful means for determining parent-offspring links.
Issues of genetic identity versus nonidentity, and of parentage (maternity and
paternity) fall at the extreme micro-evolutionary end of the phylogenetic continuum. The classes of genetic markers that have proved most suitable are those
that provide highly variable qualitative character states with known transmission
properties, e.g., allozymes and hypervariable single-locus and multilocus
RFLPs.

GENETIC IDENTITY VERSUS NONIDENTITY

Human Forensics

One of the first legal cases in the United States to admit DNA as evidence,
Pennsylvania v. Pestinikas in 1986, involved use of PCR-based assays to analyze

tissue samples from an exhumed corpse (Moody, 1989). The first criminal conviction in the United States based in part on DNA evidence came in a 1987 rape trial (*State v. Andrews, Orange County, Florida*) and established a legal precedent for the use of "DNA typing" to link a suspect to biological material (e.g., blood, semen, or hair follicles) left at a crime scene (Kirby, 1990; Roberts, 1991). By the year 1990, more than 2000 court cases in 49 states had used DNA evidence in either civil litigations or criminal proceedings (Chakraborty and Kidd, 1991). As with conventional fingerprinting, DNA typing merely provides physical evidence that potentially associates victims and suspects with one another or to a crime location and, therefore, must be used in conjunction with other lines of evidence to establish guilt or innocence. Indeed, the parallels with traditional fingerprinting appear so strong that at least one legal expert has predicted that "DNA analysis will be to the end of the 20th century what fingerprinting was to the 19th" (Melson, 1990, p. 189).

One illustrative example of DNA typing in a homicide case involved a mortuary worker accused of murder and incineration of his estranged wife at a crematorium in Wichita, Kansas (account in the *New York Times,* Nov. 21, 1988, as related by Kirby, 1990). Circumstantial independent evidence had implicated the worker in his wife's death, but he staunchly maintained that she had not been at the mortuary near the time of her disappearance. However, bloodstains discovered on the side of the crematorium proved by DNA typing to match other remaining tissue from the deceased woman. The mortuary worker was convicted of first-degree homicide and aggravated kidnap. In another unusual but illustrative example of the power of DNA typing methods, Hagelberg et al. (1991) used the PCR to amplify DNA sequences from the 8-year-old skeletal remains of a murder victim. By comparing microsatellite DNA markers in the remains with those of the presumptive parents, the victim's identity was established. Not all forensic applications of DNA typing involve crimes this macabre, but molecular genetic methods similarly have provided useful physical evidence in numerous cases of homicide, rape, burglary, assault, hit-and-run accidents, missing persons, and others.

Early forensic work involved typing various blood groups and serum proteins that exhibited circumscribed genetic variability and hence provided only limited evidence on individual identity and uniqueness. The approach now employed most widely in human forensics utilizes RFLP data accumulated from several hypervariable VNTR loci, assayed one at a time. Within human populations, each such locus reveals large numbers of DNA fragments on gels [on the order of scores or hundreds (Fig. 5.1)] that differ in length due to variations in the numbers of the small tandem repeat units. Because any DNA fragment size is measured with some error, the determination of distinct allelic classes from the quasi-continuous distribution of fragment lengths is not entirely straightforward (Devlin et al., 1992), and, in practice, various grouping procedures are em-

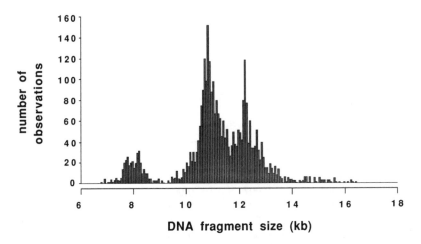

Figure 5.1. Frequency distribution of restriction fragments (sizes in kb) from the D2S44 locus in Caucasian samples (after Devlin et al., 1992). Data from Lifecodes, Inc.

ployed to pool fragments of similar length into allelic classes or "bins," the widths of which are functions of the magnitude of experimental error in fragment migration across replicates (Budowle et al., 1991). Even so, each VNTR locus employed by forensics laboratories exhibits many distinguishable alleles (usually 10–30), with most being uncommon or rare (Table 5.1). One consequence is that the great majority of individuals appears heterozygous (Box 5.1), exhibiting two DNA gel bands. Another consequence is that the probability of a single-locus genetic match between randomly chosen individuals is low. As illustrated in Box 5.1, genotypic frequencies from several unlinked VNTR loci then are combined to calculate the multilocus probabilities of observing a given DNA profile in a random draw from the population.

Several laboratories operated by private companies or governmental units (notably Cellmark, Lifecodes, and the Federal Bureau of Investigation) routinely conduct DNA typing using VNTR loci. Provided the relevant tissues samples left at the crime scene have yielded assayable DNA, two evidential outcomes are possible: (a) the samples do not match the suspect by DNA typing, in which case the evidence may be declared exculpatory; or (b) the tissues are declared a match. In the western judicial tradition where a suspect is considered "innocent until proven guilty," the latter situation clearly focuses the following question: What is the likelihood that such a DNA match occurs by chance? At face value, such probabilities calculated from available VNTR data are infinitesimally small (3×10^{-9} in the example in Box 5.1), such that matches usually are interpreted as establishing genetic identity "beyond reasonable doubt."

Table 5.1. Frequencies of alleles (bins) at four hypervariable VNTR loci in Caucasians[a]

| | VNTR Locus D1S7 | | Binned allele frequencies in Caucasian samples at VNTR loci | | |
| | Frequencies in | | | | |
Binned Allele	Caucasians ($n = 605$)	Blacks ($n = 372$)	D2S44 ($n = 802$)	D17S79 ($n = 563$)	D4S139 ($n = 460$)
1	0.004	0.007	0.005	0.010	0.004
2	0.006	0.009	0.003	0.003	0.010
3	0.009	0.011	0.016	0.007	0.006
4	0.012	0.007	0.024	0.004	0.014
5	0.011	0.016	0.046	0.015	0.033
6	0.014	0.020	0.034	0.223	0.024
7	0.010	0.011	0.123	0.199	0.040
8	0.029	0.035	0.106	0.263	0.047
9	0.021	0.023	0.084	0.200	0.054
10	0.014	0.030	0.049	0.029	0.071
11	0.028	0.030	0.083	0.032	0.108
12	0.031	0.026	0.039	0.010	0.190
13	0.046	0.044	0.041	0.006	0.129
14	0.067	0.069	0.039		0.095
15	0.057	0.065	0.087		0.036
16	0.061	0.073	0.089		0.036
17	0.069	0.054	0.075		0.103
18	0.055	0.051	0.022		
19	0.060	0.047	0.018		
20	0.063	0.063	0.008		
21	0.079	0.062	0.008		
22	0.077	0.060			
23	0.077	0.074			
24	0.032	0.017			
25	0.019	0.027			
26	0.050	0.071			

[a]Also shown for comparison are allele frequencies at the D1S7 locus in a sample from the Black population. These data were introduced by the Federal Bureau of Investigation to a criminal case in Athens, Georgia, May, 1991, and are part of a larger data base that included frequencies from additional VNTR loci in Caucasians, Blacks, and Hispanics.

Box 5.1. Probabilities of Single-Locus and Multilocus DNA Profiles Based on the VNTR Data for Caucasians in Table 5.1.

(a) Probability that an individual is heterozygous at the D1S7 locus:

$$h = 1 - \Sigma[(0.004)^2 + (0.006)^2 + \ldots + (0.050)^2] = 0.945.$$

(b) Examples of probabilities of particular allelic combinations at individual loci:

5/22	heterozygote at D1S7:	$0.011 \times 0.077 \times 2.0 = 0.001694$
6/8	heterozygote at D2S44:	$0.034 \times 0.106 \times 2.0 = 0.007208$
5/9	heterozygote at D17S79:	$0.015 \times 0.200 \times 2.0 = 0.006000$
12/12	homozygote at D4S139:	$0.190 \times 0.190 \qquad = 0.036100$

(c) Probability of the multilocus DNA profile in (b):

$$0.001694 \times 0.007208 \times 0.006000 \times 0.036100 = 3 \times 10^{-9}.$$

(d) Relevance: Suppose a crime suspect exhibited the multilocus genotype shown in (b). Then if assumptions of the model are met, the probability of a match with a randomly drawn genotype from the available Caucasian sample is about 1 in 333 million.

These calculations assume random associations of alleles within and among loci (Hardy-Weinberg equilibrium and gametic phase equilibrium, respectively).

However, these conclusions are based on particular assumptions whose validity has been questioned. Indeed, recent criticisms raised within the scientific community have placed DNA forensic procedures themselves on trial (Lander, 1989; Lewontin and Hartl, 1991). One class of problems involves technical matters relating to allelic discrimination, handling of samples, and the repeatability of laboratory tests. Such difficulties can be alleviated by the adoption of tighter standards for quality control, as recommended recently by a panel of the NRC (National Research Council, 1992). Another contentious issue involves the probability calculations for genotypic matches. The calculations exemplified in Box 5.1 involve the potentially erroneous assumption that genotype frequencies across loci are independent. One likely source of nonindependence is population subdivision. Human populations are not entirely homogeneous, but rather exhibit genetic substructure that is expected to produce allelic correlations due to nonrandom mating and historical separations. (For example, alleles for blond hair and blue eyes, though present at loci that are independent in a physiological and transmission sense, nonetheless are highly correlated in humans due to racial and populational histories.) On the other hand, in statistical analyses of the empirical data bases for the VNTR loci employed in human forensics (taken from the FBI and Lifecodes), few significant genetic correlations within or among loci have been found (Risch and Devlin, 1992; Weir, 1992).

The effect of population substructure on forensic conclusions is a matter of degree (Nichols and Balding, 1991). Consider an extreme example in which a suspect belongs to a small, inbred community that differs dramatically in allele frequency from the North American Caucasian population. Use of the Caucasian data base (Table 5.1) as a reference for the calculation of genotypic probabilities clearly would be inappropriate, and the direction of error could work against the defendant. Thus, the likelihood of a genotypic match between the suspect and another member of the local community who actually may have committed the crime is much greater than the probability of a match within a broader population. To circumvent this problem, each relevant human "subgroup" might be specified separately and the appropriate probability calculations based accordingly for each case. Unfortunately, such extensive genetic characterization is infeasible logistically, even if the appropriate subgroups somehow could be identified. Such concerns led Lewontin and Hartl (1991) to conclude that current applications of VNTR typing have serious flaws as forensic evidence. Morton (1992) responded that the Lewontin-Hartl (1991) objections to genotypic probability calculations are themselves "absurdly" conservative in favor of the defense, and Chakraborty and Kidd (1991) conclude that "If DNA evidence is excluded from courtroom applications, the prospect of convicting true criminals, as well as exonerating the falsely accused, will be substantially diminished."

The degree of human population substructure should not be overstated. In terms of allozyme and blood-group polymorphisms, Lewontin (1972) earlier had argued that more than 90% of overall genetic diversity in humans occurred within (rather that between) races and concluded that "our perception of relatively large differences between human races and subgroups . . . is indeed a biased perception, and that, based on randomly chosen genetic differences, human races and populations are remarkably similar to each other. . . ." Such conclusions apply to at least some VNTR loci as well (Balazs et al., 1989), as illustrated by the similar spectra of allele frequencies at D1S7 in the Caucasian versus Black populations (Table 5.1). On the other hand, whereas frequencies of most allelic classes at VNTR loci have proved similar across ethnic groups, allelic frequencies at some such loci do differentiate among certain human populations (Balazs et al., 1992; Krane et al., 1992).

A variety of conservative calculation procedures can be followed that tend to diminish bias against the defense. These include the use of wider bins for grouping DNA fragments and employment of observed rather than expected frequencies of single-locus genotypes in the reference population (to circumvent assumptions of Hardy–Weinberg equilibrium). In one case in Columbus, Georgia (*Georgia* v. *Caldwell*), calculations based on such modifications increased the probability of a genetic match by 100-fold: from 4×10^{-8} to 4×10^{-6} (the suspect nonetheless was convicted). The NRC report (1992) proposed another ingenious solution, highly conservative in favor of the defense, known as the "ceiling principle." They recommend that allele frequencies be estimated at all marker loci in 15–20 human populations representing a diversity of ethnic groups. For each allele, its highest frequency in any population or 0.05, whichever is higher, should then be employed for estimating the expected genotypic frequencies against which to evaluate the genotypic profile of the suspect. This method overestimates the expected frequency of genotypes in the reference data base, but the effect is such that any errors introduced are in the direction of decreasing the chances that an innocent suspect is convicted.

Fortunately, the current controversy over DNA forensic practice likely will become moot in the near future with the adoption of tighter standards for quality control, development of refined laboratory techniques for distinguishing alleles (e.g., Jeffreys et al., 1991), inclusion of data from additional hypervariable loci and more populations, and adoption of the conservative NRC guidelines for calculating probabilities of a genotypic match.

Ramets and Genets

BACKGROUND AND CONCEPTS

Many and perhaps most species of plants and invertebrate animals reproduce facultatively by either sexual or asexual (clonal) means (Jackson et al., 1985). For example, the quaking aspen (*Populus tremuloides*) can produce sexual seeds but also proliferates vegetatively via buds that sprout from the roots of a mature tree. Death of the mother stem then may result in the physical disconnection of clonemates. Clonal proliferation in other plants may involve runners, stolons, rhizomes, bulbs, root or stem suckers, plant fragments, or even asexual (apomictic) seeds that combine the advantages of cloning with wide dispersal potential (Cook, 1980). The latter may arise when a nonmeiotic cell in the ovarian wall initiates seed formation or when failure of a reduction division in a germ cell lineage produces eggs with a full complement of chromosomes from the maternal parent. In many corals such as the staghorn (*Acropora cervicornis*), a colony consists of numerous asexually-derived polyps that are genetically identical to one another and to the sexually-produced planula larvae from which they arose. The polyps are housed jointly in a secreted calcareous skeleton that occasionally breaks, thereby producing "daughter" colonies that are genetically identical but physically disjunct. Some coral species also produce dispersive asexual larvae (Stoddart, 1983a). Various mechanisms of asexual proliferation, including clonal production of larval-like propagules, somatic fragmentation, parthenogenesis, or production of multiple individuals by division of an early embryo or zygote (polyembryony), are a normal part of the natural histories of many invertebrate and some vertebrate animals (Blackwelder and Shepherd, 1981; Jackson, 1986).

As phrased by Harper (1985), "It is the nature of many plant and animal growth forms that the organism dies in bits and continues growth as separated parts." In such species with clonal reproduction, challenging questions arise, such as: What constitutes an individual? What are the units of selection (Buss, 1983, 1985)? Harper (1977) defined the genetic individual or "genet" to include all entities (however physically organized) that have descended from a single sexually-produced zygote and, hence, that are genotypically identical to one another (barring mutation). By contrast, a "ramet" is an individual in a physical or functional sense—a physiologically or morphologically coherent module having arisen through clonal replication. Thus, a genet may consist of many modular ramets, asexually derived. Many evolutionary interpretations of field data hinge critically on the correct distinction of clonemates from nonclonemates. For example, secure genetic knowledge of which ramets ultimately derive from the same zygote is necessary for drawing proper inferences about (a) sex ratios within sexual-asexual populations, (b) magnitudes and patterns of effective gene flow, (c) degrees of outcrossing and the mating system, (d) extents of interclonal competition, and (e) the evolutionary ages of clones (Cook, 1983, 1985).

In many cases, the breeding system of a species is unknown but asexual reproduction is suspected. Molecular genetic markers can help settle the issue. For example, some populations of the mustard plant (*Arabis holboellii*) have characteristics suggestive of apomictic capabilities (reproduction without fertilization), including the appearance of pollen and embryos with unreduced chromosome number. Roy and Rieseberg (1989) confirmed the occurrence of apomixis in *Arabis* by showing that assayed siblings were genetically identical to their respective parent plant at several polymorphic allozyme loci. Many marine benthic algae release spores into the water column, but whether these are sexual or asexual propagules remained uncertain. For one such species (*Enteromorpha linza*), Innes and Yarish (1984) employed allozyme markers to document clonal spore production. Using DNA fingerprint assays, Nybom and Schaal (1990) detected a large number of DNA fingerprint genotypes in a population of the predominantly sexually-reproducing black raspberry (*Rubus occidentalis*), whereas a related species suspected of frequent asexual reproduction (the blackberry, *R. pensylvanicus*) exhibited far fewer genotypes. On the other hand, use of allozyme markers documented sexual reproduction in the free living amoeba *Naegleria lovaniensis* (Pernin et al., 1992) and in the fungal pathogen *Crumenulopsis sororia* (Ennos and Swales, 1987).

Many marine invertebrates brood their young, and it is of interest to know whether these larvae are the products of sexual or asexual reproduction. Using allozyme assays, Black and Johnson (1979) showed that brooded young of the intertidal anemone *Actinia tenebrosa* were genetically identical to their parents, indicating asexual reproduction. Similarly, Ayre and Resing (1986) documented asexual reproduction for two coral species (*Tubastraea diaphana* and *T. coc-*

cinea). On the other hand, in two other coral species assayed allozymically (*Acropora palifera* and *Seriatopora hystrix*), nonparental genotypes were detected in the majority of the larval broods, thus indicating reproduction by sexual means. Clearly, brooded larvae can be produced both sexually and asexually in various invertebrate species (Ayre and Resing, 1986).

In many animal populations, the first suggestion of parthenogenesis [whereby progeny develop directly from an unfertilized female gamete (Soumalainen et al., 1976; see Fig. 5.3)] often comes from the indirect evidence of a strongly female-biased sex ratio in nature. Clonal reproduction then may be confirmed with genetic markers—true ameiotic parthenogens derived from a single female are genetically uniform, barring postformational mutations (Hebert and Ward, 1972), and clonal parthenogenetic populations that arose through recent hybridization exhibit "fixed heterozygosity" at loci distinguishing the parental species (Dessauer and Cole, 1986). For example, Echelle and Mosier (1981) used allozyme evidence to confirm that a population of silverside fishes (in the *Menidia clarkhubbsi* complex) reproduces by clonal means, as did Dawley (1992) for two killifish populations that proved to have arisen through crosses between the sexual species *Fundulus heteroclitus* and *F. diaphanus*. In the parthenogenetic aphids *Myzus persicae* and *Sitobion avenae*, DNA fingerprint assays revealed characteristic genetic signatures confirming suspected modes of clonal reproduction (Carvalho et al., 1991). As discussed later, molecular markers similarly have substantiated clonal reproduction in many other invertebrates, fishes, amphibians, and reptiles.

The hallmark of clonal reproduction is the stable transmission of genotypes across generations, without the shuffling effects of genetic recombination (the only source of variation, therefore, being mutation). Thus, the term "clonal reproduction" sometimes is used also to describe genetic transmission in strictly self-fertilizing hermaphroditic organisms, where the intense inbreeding that characterizes this reproductive mode could have resulted in near homozygosity at most loci. Although such organisms may retain meiosis and syngamy (union of gametes, in this case from a single parent), genetic segregation and recombination in effect are suppressed once homozygosity through inbreeding is achieved. For example, the Floridian population of the cyprinodontid fish *Rivulus marmoratus* [the only known example of a hermaphroditic vertebrate animal with internal self-fertilization (Harrington, 1961)] exists in nature as highly homozygous "clones," as gauged by intraclonal acceptance of fin grafts [indicating near identity at histocompatibility loci (Harrington and Kallman, 1968; Kallman and Harrington, 1964)] and by complete homozygosity at 31 genes whose protein products were assayed electrophoretically (Vrijenhoek, 1985). Nonetheless, recent DNA fingerprinting studies (B.J. Turner et al., 1990, 1992) revealed considerable genetic variation that had remained undetected in the earlier assays. This last result highlights a cautionary note that applies to studies of "clonal diversity'' in any organism—the absolute number of clones recognized can depend on the discriminatory power of the assay as well as the reproductive biology and evolutionary history of a species. Thus, it is not sufficient to be concerned merely with the number of identifiable clones; the evolutionary relationships among the distinguishable genotypes and the biological processes that generated these genetic differences also must be considered.

SPATIAL DISTRIBUTIONS OF CLONES

Clearly, the geographic distribution of genotypes in sexual-asexual species will be influenced by the relative frequencies of sexual versus clonal reproduction and the dispersive characteristics of propagules produced by these respective reproductive modes. Early attempts to illuminate these population genetic structures utilized indirect, phenotypic criteria for clonal identifications, e.g., distributions of morphological attributes or phenologies in plants (Barnes, 1966), agonistic behaviors among sea anemones (Sebens, 1984), and histocompatibility responses (acceptance versus rejection of tissue grafts) in marine invertebrates and unisexual vertebrates (Cuellar, 1984; Neigel and Avise, 1983a; Schultz, 1969). Recent attempts to assign ramets to genets and to map the spatial distributions of clones have involved more direct molecular genetic assays.

Plants and Fungi   Maddox et al. (1989) used allozyme genotypes at four polymorphic loci to map the microspatial distributions of goldenrod (*Solidago altissima*) clones in fields of various ages. Different allozyme genotypes almost certainly reflect sexual recruitment (via seeds) into the population, whereas the multiple ramets of a clone reflect asexual proliferation via rhizomes. Patterns of dispersion of goldenrod genotypes differed among the fields: Clones were localized in the youngest plots (e.g., Fig. 5.2), whereas older fields exhibited higher spatial intermixtures of clones and fewer remaining rhizomic connections among ramets. Apparently, colonization of a field by sexually produced seeds is followed by ramet proliferation and eventual spatial mixing of clones over microgeographic scales.

Other plant species may have wider clonal distributions (Ellstrand and Roose, 1987; Silander, 1985). Clonal growth in the bromeliad *Aechmea magdalenae* results in the spread of ramets over distances of several meters, as judged by distributions of multilocus allozyme genotypes (Murawski and Hamrick, 1990). Using similar allozyme approaches, Hermanutz et al. (1989) mapped genets in the arctic dwarf birch (*Betula glandulosa*), a sexual species that at its northern limit also reproduces by a process called ''vegetative layering'' whereby prostrate branches beneath the moss layer produce new ramets vegetatively. Single clones were dispersed over areas of at least 50 m^2 (the approximate sizes of the assayed plots). In the columnar cactus *Lophocereus schottii*, dispersal of asexual propagules can take place via detached stem pieces that on occasion may be

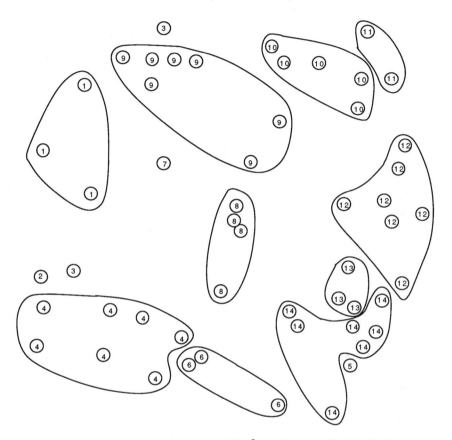

Figure 5.2. Microspatial map (total area 0.75 m²) of allozymically identified clones in *Solidago altissima* (after Maddox et al., 1989). Circles denote current living ramets and numbers indicate different electrophoretic genotypes. Larger ellipses encompass the probable ramets belonging to each genet.

washed downstream by floodwaters. Parker and Hamrick (1992) found that although most genetically identified clonemates in this species were spatially aggregated at scales of a few meters, some widely separated individuals (> 70 m apart) had identical allozyme genotypes that might reflect such instances of longer-distance clonal dispersal. Dispersal of asexual genotypes over large distances also has been suggested for other riparian species such as willows (Douglas, 1989).

A form of apomictic reproduction in plants that could result in unusually widespread dispersion of clones is agamospermy, the formation of unreduced spores, seeds, or embryos by asexual processes. In the marine alga *Enteromorpha linza,* which can produce asexual spores that disperse in the water column, particular clones (allozymically identified) were distributed over the entire survey transect of more than 150 km of shoreline (Innes, 1987). Within each of two obligate agamospermous populations of dandelions (*Taraxacum* sp.), all individuals proved genetically identical at 15 allozyme loci, whereas related sexual populations were highly diverse genetically (Hughes and Richards, 1988). However, other apomictic dandelion populations showed considerable genetic variation and coexistence of multiple clones (Ford, 1985; Ford and Richards, 1985). Only limited genetic information exists on the broader geographic distribution of genets within these or other agamospermous taxa (Bayer, 1989; Hughes and Richards, 1989). One potential complication in interpreting the clonal structures of ancient and widespread "agamospecies" is in distinguishing sexually-derived genetic variation from that which may have arisen via postformational mutations (Brookfield, 1992; see beyond).

According to Cook (1980), perhaps the record for size and age of a plant clone involves the quaking aspen. Based on a distinctive morphological appearance and spatial arrangement, one suspected genet appeared to be represented by more than 47,000 ramets (covering 107 acres) that may trace to a single seed deposited several thousand years ago at the close of the Wisconsin glaciation (Kemperman and Barnes, 1976). On the other hand, looks may be deceiving. In allozyme surveys of other quaking aspen populations, Cheliak and Patel (1984) found that several "clones" provisionally identified by morphology actually were composed of several distinct electrophoretic genotypes that probably had arisen through recombination (and hence sexual reproduction). They concluded that environmental influences on phenotypic appearance in aspens invalidate morphological appraisals as a reliable guide to clone identification. The recent documentation of extensive genetic variation in quaking aspens as revealed by DNA fingerprinting raises the hope that clonal identifications and distributions in this species soon may become definitive (Rogstad et al., 1991).

Genetic documentation *is* available for the honey mushroom, *Armillaria bulbosa,* in which one genetic clone identified by mtDNA and nuclear RAPD markers has been claimed as being among the largest and oldest of organisms on Earth (M.L. Smith et al., 1990, 1992). This pathogenic fungus of tree roots in mixed hardwood forests can spread vegetatively by cordlike aggregations of hyphae that weave across the forest floor. Molecular genetic markers revealed that one presumably interconnected clone of *A. bulbosa* in northern Michigan had spread across 37 acres, may weigh in aggregate more than 90,000 kg (about the size of an adult blue whale) and is perhaps some 1500 years old. A second and smaller clone nearby covered a mere 5 acres! Recently, a report has appeared of an even larger fungal clone (in *Armillaria ostoyae*) covering 1500 acres in Washington State (Anonymous, 1992), although genetic confirmation in this case appears to be lacking.

Invertebrate Animals   Most seastars (Asteroidea, Echinodermata) can reproduce asexually by fission, whereby detached arms regenerate new bodies. To determine the relative importance of fissiparity versus sexual reproduction in the seastar *Coscinasterias calamaria,* Johnson and Threlfall (1987) employed allozyme markers in conjunction with laboratory and field observations. Results indicated that on local scales, clonal reproduction predominated such that many individuals within 50 m of one another were clonemates. The importance of sexual recruitment in this species was documented as well, through the observation of distinct clonal genotypes characterizing different regions of the study area in western Australia. Similarly, colonies of the hard coral *Pavona cactus* can proliferate by fragmentation, and an allozyme survey revealed the significance of this asexual process in distributing clonemates over distances of up to nearly 100 m along reefs in eastern Australia (Ayre and Willis, 1988). In the coral *Pocillopora damicornis,* which produces dispersive planulae larvae by asexual means, most clonal genotypes were clustered spatially, but nonetheless particular clones as revealed in protein assays reportedly occurred over distances of up to several kilometers (Stoddart, 1984a, 1984b). On the other hand, allozyme surveys of the intertidal sea anemone *Oulactis muscosa* (a species that can reproduce by fission) revealed a population genetic structure consistent with recruitment almost exclusively by sexual reproduction (Hunt and Ayre, 1989). In the sea anemone *Actinia tenebrosa* (a species that can produce brooded young asexually), particular clonal genotypes as identified by allozyme assays were distributed over hundreds of meters of shoreline in Australia (Ayre, 1984). In yet another species of sea anemone (*Metridium senile*), allozyme surveys of various populations in northeastern North America indicated marked differences among sites in the frequencies with which sexual versus clonal recruitment had taken place (Hoffman, 1986).

Much discussion has concerned the reliability of histocompatibility bioassays as a guide to clonal distinctions in marine invertebrates. Within many coral and sponge species, artificial grafts between colony branches exhibit either an "acceptance" or "rejection" reaction. Considerable indirect evidence suggests that these respective responses signal clonal identity versus nonidentity, at least in some species (e.g., Hildemann et al., 1977; Neigel and Avise, 1983b; review in Avise and Neigel, 1984). However, protein electrophoretic studies conducted to date have not fully corroborated this possibility (Curtis et al., 1982; Neigel and Avise, 1985; Resing and Ayre, 1985; however, see Hunter, 1985). Thus, occasional instances of graft acceptance between colonies differing in allozyme genotype have been reported, as have instances of graft rejection between colonies identical in genotype at a small number of moderately polymorphic allozyme loci (Table 5.2). Nonetheless, because of the limited genetic data as well as technical difficulties with both the field and laboratory procedures (Grosberg, 1988; Neigel and Avise, 1985), much more attention should be devoted to the issue of how reliably histocompatibility responses reveal clonal makeup in various marine

Table 5.2. Tissue graft response and multilocus allozyme genotype (three polymorphic loci) in 65 assayed pairs of sponges within a local population of *Niphates erecta* (after Neigel and Avise, 1985).[a]

Allozyme Genotype	Tissue Graft Response	
	Acceptance	Rejection
Same	23	5
	(26)	(12)
Different	5	32
	(21)	(35)

[a]Shown in parentheses are similar results obtained for 94 assayed pairs within the coral species *Montipora dilatata* and *M. verrucosa*, where a single polymorphic allozyme locus was monitored (after Heyward and Stoddart, 1985). For both the corals and sponges, the associations between tissue graft response and clonal identity as revealed by allozyme genotype were highly significant statistically.

invertebrate species. Other traditional guides to presumed clonal identity, such as patterns of aggression among sea anemonies, and morphotypic appearances among colonies of corals and sponges also have been called into question from reappraisals based on molecular genetic markers (e.g., Ayre, 1982; Ayre and Willis, 1988; Solé-Cava and Thorpe, 1986).

Because of their power and sensitivity, various DNA fingerprinting approaches should find wide application in studies of clonal population structure in marine invertebrates [provided that the frequency of sexual reproduction is considerably greater than the minisatellite mutation rate (Brookfield, 1992)]. However, few empirical examples thus far are available. Coffroth et al. (1992) utilized Jeffrey's and related minisatellite probes to study DNA fingerprints and clonal structure in a gorgonian coral (*Plexaura* sp. A) that reproduces by fragmentation as well as by sexual production of dispersive larvae. Among 73 assayed colonies on 7 reefs in Panama, 29 different genotypes were identified by these molecular methods. The only cases of identical DNA fingerprints involved colonies located on the same reef, usually in close proximity. On two reefs, both DNA fingerprinting and histocompatibility assays were conducted, and these approaches revealed similar numbers of clones—17 and 13, respectively. Also worthy of mention were experimental controls demonstrating that (a) multiple samples from a single colony produced identical DNA fragment profiles and (b) zooxanthellae symbiotic to the corals were not the source of the DNA bands scored. Based on this experience with *Plexaura*, DNA fingerprinting would appear to hold much promise for further genetic analyses of corals and other invertebrates capable of clonal reproduction.

Many invertebrates can proliferate clonally by parthenogenesis. For example, 17 of 33 North American species in the earthworm family Lumbricidae exhibit a parthenogenetic reproductive mode that apparently evolved from an ancestral hermaphroditic condition (Jaenike and Selander, 1979). In one such species (*Octolasion tyrtaeum*), Jaenike et al. (1980) used allozyme markers to identify eight distinct clones, two of which were widespread and common in diverse soil types over the several thousand square kilometers of study area in the eastern United States. Thus, some asexual lineages apparently can occupy broad niches and achieve tremendous ecological success, at least over the short term. Many freshwater gastropods (snails) likewise reproduce parthenogenetically (Jarne and Delay, 1991). Allozyme studies of the polyploid parthenogen *Thiara balonnensis* in Australia revealed that local populations generally consist of one clone only, with genetic distance among clones correlated with geographic distance (Stoddart, 1983b). This pattern of variation was postulated to result from the gradual evolution of new clones by mutational processes (as opposed to occasional sexual reproduction) in conjunction with geographic separations.

A converse example of high clonal diversity over microgeographic scales was encountered in another species that exhibits obligate parthenogenesis throughout much of its range, the cladoceran *Daphnia pulex* (Hebert and Crease, 1980). Initial surveys revealed a total of 22 allozymic clones in 11 populations, with up to 7 genotypes coexisting in a single lake. Subsequent studies of *D. pulex* using mtDNA and allozyme markers revealed many additional clones and also demonstrated that obligate parthenogenesis had a polyphyletic origin from facultative parthenogenesis within this species (Crease et al., 1989; Hebert et al., 1989). In a related species *D. magna,* tens to hundreds of clones sometimes coexist within a pond (Hebert, 1974a, 1974b; Hebert and Ward, 1976). Where *D. magna* occurs in temporary habitats, it is a cyclic parthenogen: Drought-resistant sexual eggs (requiring fertilization) are produced each year, and these zygotes reestablish populations that then are maintained by two or three generations of clonal parthenogenesis until the pond again dries up. In this case, clonal coexistence might only be short term because new recombinationally-derived clones enter the population each year from sexual eggs. Interestingly, *D. magna* in permanent habitats reproduces by continued parthenogenesis and tends to exhibit fewer clonal types (Hebert, 1974c).

AGES OF CLONES

A common belief is that populations of clonally reproducing organisms must have short evolutionary life spans, due to an accumulation of deleterious mutations and gene combinations that cannot be purged in the absence of recombination [a phenomenon referred to as "Muller's ratchet" (Muller, 1964)], or to a presumed lack of sufficient recombinationally-derived genotypic diversity to allow adaptive responses to environmental challenges (Darlington, 1939; Felsenstein, 1974; Maynard Smith, 1978; Williams, 1975). However, the evidence cited above suggests that some clonal lineages in invertebrates and plants can achieve wide distributions and enjoy at least moderate-term ecological success.

Remarkably, about 70 vertebrate "species" also are known to reproduce by clonal or quasi-clonal means (Dawley and Bogart, 1989) (the term "biotype" is preferred for such forms because traditional species concepts hardly apply). These biotypes typically consist solely of females that propagate by parthenogenesis or related reproductive modes (Fig. 5.3). Essentially all unisexual vertebrates arose through hybridization between related sexual species, and this aspect of their evolutionary histories will be deferred to Chapter 7. Here we consider the ages of particular vertebrate unisexual lineages, as inferred from recent molecular assays primarily involving mtDNA.

Two conceptual approaches to assessment of vertebrate clonal ages have been attempted from molecular data. The first involves estimation of the genetic distance between a unisexual and its closest sexual relative. In a review of 24

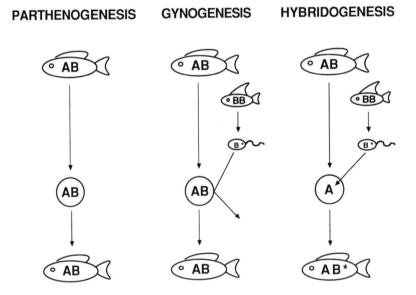

Figure 5.3. Three known modes of unisexual reproduction in vertebrates (after Avise et al., 1992c). In parthenogenesis, the female's nuclear genome is transmitted intact to the egg, which then develops into an offspring genetically identical to the mother; in gynogenesis, the process is the same except that sperm from a related bisexual species is required to stimulate egg development; and in hybridogenesis, an ancestral genome from the maternal line is transmitted to the egg without recombination, whereas paternally-derived chromosomes are discarded premeiotically, only to be replaced each generation through fertilization by sperm from a related sexual species.

unisexual vertebrate lineages that have been so compared (Avise et al., 1992c), 13 proved indistinguishable in mtDNA assays from an extant genotype in the related sexual taxon, indicating a very recent evolutionary separation; 5 additional lineages differed from nearest assayed bisexual taxon at sequence divergence estimates of less than 1%, suggesting times of origin within the last 500,000 years (Fig. 5.4). A few unisexual haplotypes did show greater sequence differences from related sexual forms, and these translate into literal estimates of evolutionary durations of perhaps a few million years. However, a serious reservation about such estimates is that closer relatives within the sexual progenitor may have become extinct after unisexual separation, or otherwise remained unsampled in the collections, such that unisexual ages could be grossly overestimated by this approach. Indeed, because of the low mtDNA lineage diversity observed within most unisexual taxa relative to their sexual cognates (Fig. 5.4), most authors have concluded that unisexuals arose very recently, even when genetically close mtDNA lineages were not observed among the sexual relatives sampled (e.g., Vyas et al., 1990).

One recent example in which an ancient clonal age *was* promulgated involves gynogenetic mole salamanders in the genus *Ambystoma*. From comparisons of mtDNA sequences in the unisexuals versus extant sexual relatives, Hedges et al. (1992a) and Spolsky et al. (1992) estimated evolutionary durations for the gynogens of about four to five million years. However, one reservation about the relevance of this conclusion to clonal persistence arguments is that these salamanders are unusual among the vertebrate unisexuals in that their evolution may not be strictly clonal—other molecular data suggest that they continually acquire nuclear DNA from sexual species, presumably via occasional incorporation of sperm into the egg. If so, the antiquity of these "clonal" salamanders applies strictly only to the mtDNA lineages that they contain.

A second approach to the estimation of clonal ages involves assessing the scope of genetic variability within unisexual clades that by independent evidence are of monophyletic (single-hybridization) origin. This method avoids confounding postformational processes indicative of an old lineage with genetic diversity that may have arisen from multiple hybrid origins. Quattro et al. (1992a) examined mtDNA and allozyme variability within a hybridogenetic clade of fishes in northwestern Mexico (*Poeciliopsis monacha-occidentalis*) that by independent zoogeographic evidence and tissue-graft analyses was of monophyletic origin. The genetic data confirmed the monophyly of the clade and also documented considerable genetic diversity within it, including the accumulation of several mitochondrial and allozymic mutations in the *monacha* portion of the genome (which comes from the female parent). From the magnitude of this diversity, the authors estimated that the unisexual clade was more than 100,000 generations old. One reservation about the relevance of this conclusion to clonal persistence arguments is that because of the hybridogenetic reproductive mode of these

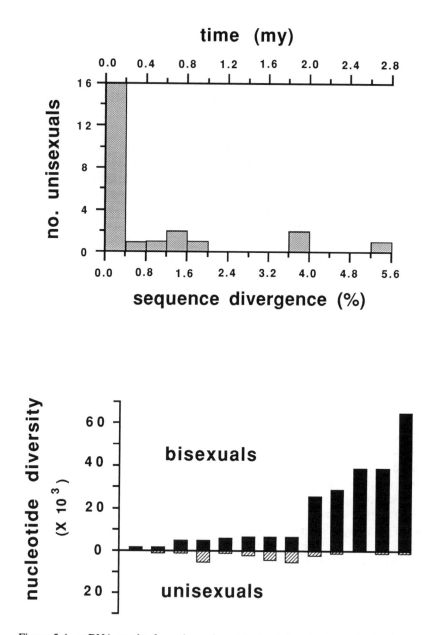

Figure 5.4. mtDNA results for unisexual vertebrates (after Avise et al., 1992c).
Top: Frequency distribution of the smallest genetic distances observed between
mtDNA genotypes in unisexual clades and those in the closest assayed sexual
relatives (also shown are the associated evolutionary ages of the unisexuals based on
the conventional mtDNA clock calibration of 2% sequence divergence per million
years between lineages). **Bottom:** Nucleotide diversities in 13 sexual species and
their respective unisexual derivatives, arranged in rank order from left to right by the
magnitude of variation within the bisexuals.

Poeciliopsis fishes (Fig. 5.3), inheritance only of the maternal component of the lineage is strictly clonal [the somatic complement of each generation includes new genetic input from the paternal side (*P. occidentalis*), such that the overall genetic system of the hybridogenetic unisexuals is referred to as "hemiclonal"].

In any event, Maynard Smith (1992) argues that in evolutionary terms, 100,000 years "is but an evening gone" and for this reason results for *Poeciliopsis* do not contradict the conventional wisdom that organismal clones are short-lived. Regardless of one's perspectives about whether such time scales are "long" or "short" in the context of clonal persistence debates, molecular data have provided the first critical information regarding the evolutionary durations of vertebrate lineages lacking recombination.

Genetic Mosaics

Occasionally, what appears to be a single ramet actually represents two or more genets that have fused into one morphologically (and perhaps physiologically) integrated module. A case in point involves strangler figs (*Ficus* spp.). These trees often begin growth when bird- or mammal-deposited seeds germinate in the humus-filled crotches of host trees. Shoots grow upward and roots downward around the host bole. The roots eventually cross and fuse to form a unified woody sheath surrounding the host, which then may die such that only the fig tree remains. Thompson et al. (1991) showed by allozyme analyses that individual fig trees often were genetic mosaics consisting of multiple genotypes. Thirteen of the 14 sampled trees showed detectable genetic differences among branches, such that at least 45 genetic individuals altogether were represented. Presumably, the mosaicism is attributable to postgermination fusions among roots that trace to multiple seeds deposited in the host tree.

Fusions among ramets also are common in many marine invertebrates including corals, sponges, bryozoans, and hydroids (Grosberg, 1988; Jackson, 1985). These somatic fusions involve postlarvae more often than mature colonies, and in some cases the larvae may be asexual products of a single genet. However, in other situations the fusing larvae are known or suspected to be sexually-produced siblings, thus generating genetic mosaics. Aggregation and fusion among kin may be facilitated by cosettlement of larvae with limited dispersal, but evidence suggests that cellular recognition systems also are involved, and these may reflect the same highly polymorphic histocompatibility loci that influence graft acceptance versus rejection responses in mature colonies (e.g., Grosberg and Quinn, 1986). Unfortunately, molecular genetic markers as yet have not been widely capitalized on to address the frequency or pattern of mosaicism in colonial marine invertebrates.

The correct identification of genetic mosaics is important in several evolutionary regards. For example, outcross pollinations or fertilizations could occur within a mosaic "individual," thus influencing the genotypes of progeny and the perceived mating system of a plant or animal species. If mosaics are common in nature, the number of genets in a population could be seriously underestimated by a mere census of the number of ramets, with consequences extending to any evolutionary parameters that are influenced by effective population size (such as the expected magnitude of genetic drift). The occurrence of genetic mosaics also raises important issues regarding the degree of physiological and functional integration of composite individuals.

In a quite different sense, the cells of all eukaryotic organisms can be viewed as genetic mosaics containing nuclear and organellar genomes with formerly independent evolutionary histories. Mitochondrial and plastid genomes almost certainly represent the descendents of bacteria that early in evolution entered into endosymbiotic relationships with protoeukaryotic host cells bearing precursors of the nuclear genome (Margulis, 1970). As elaborated in Chapter 8, the most compelling evidence for the mosaic nature of the eukaryotic cell comes from phylogenetic assessments of genetic markers provided by slowly evolving genes.

Clonal Reproduction in Microorganisms

Eukaryotic protozoans such as the agents of malaria, sleeping sickness, Chagas' disease, and leishmaniases infect more than 10% of the world's human population and account for tens of millions of deaths every year. Sexual reproduction often is assumed for these parasites (most of which are diploid) because recombination among strains has been observed in the laboratory. Nonetheless, recent data from molecular markers have demonstrated that several parasitic protozoans in nature can and often do reproduce by clonal means (Tibayrenc et al., 1990, 1991a). For example, analyses of scnRFLPs revealed that globally distributed virulent strains of the protozoan *Toxoplasma gondii* are nearly homogeneous genetically, likely because of clonal reproduction and a single evolutionary origin from nonvirulent strains that exhibit moderate polymorphism and are capable of sexual reproduction (Sibley and Boothroyd, 1992). Such findings are of medical importance because they can influence the strategies for diagnosis of the disease agents, as well as for developing vaccines and curative drugs (Tibayrenc et al., 1991b).

Among other protozoans, the case for clonal reproduction in nature is particularly strong for *Trypanosoma cruzi*, the agent of Chagas' disease. Genetic studies of *T. cruzi* isolated from humans, insect vectors, and mammals sampled throughout the geographic distribution of the disease in South and Central America revealed several population genetic signatures characteristic of predominant clonal reproduction (Tibayrenc et al., 1986; Tibayrenc and Ayala, 1987, 1988; Zhang et al., 1988). These include fixed heterozygosity and other evidence for an absence of segregation genotypes at individual loci, an overrepresentation of identical multilocus allozyme genotypes that also were geographically wide-spread, and a correlation in clonal identification between independent sets of genetic markers (Boxes 5.2 and 5.3). Some of the clones proved remarkably divergent from one another, a finding that now should be taken into account by the medical community, which traditionally has tended to view this and other such protozoan species as undifferentiated quasi-panmictic entities. Similar lines of molecular genetic evidence for clonality are available also for several other parasitic protozoans (Table 5.3).

In the cereal rust *Puccinia striiformis*, Newton et al. (1985) used nuclear-encoded allozymes and cytoplasmically-transmitted RNAs (from mycoviruses inside the rust cells) to assay numerous accessions of this fungus from around the world. Representatives of one geographically widespread group of wheat-attack-ing forms (*P.s. tritici*) proved completely uniform by these assays, whereas some other related fungal species showed much higher levels of genetic variability. *P. striiformis* has no known sexual stage, so the presumed clonal transmission

Table 5.3. Population genetic evidence for clonality in various parasitic protozoans that are agents of human disease (after Tibayrenc et al., 1991b).[a]

| Organism | Criterion | | | | Evidence for Clonality[b] |
	(a)	(b)	(c)	(d)	
Entamoeba histolytica	0	0	+	ND	Moderate
Giardia sp.	0	0	+	+	Moderate
Leishmania major	ND	+	+	ND	Strong
L. donovani/*infantum*	ND	+	+	ND	Strong
L. tropica	+	+	+	ND	Strong
L. Old World	ND	+	+	ND	Strong
Naegleria sp.	+	ND	+	ND	Weak
Plasmodium falciparum	0	0	+	ND	Weak
Trichomonad sp.	0	0	+	ND	Weak
Trypanosoma brucei s.l.	+	+	+	+	Strong
T. congolense	+	+	+	ND	Moderate
T. cruzi	+	+	+	+	Strong

[a]The listed criteria for clonality are described in Box 5.2. +, the criterion is satisfied; ND, not done; 0, data not available because the test is ploidy-level dependent (most taxa appear diploid, although *Plasmodium* is haploid and the ploidy levels of *Entamoeba*, *Giardia*, and *Trichomonad* remain uncertain).

[b]Overall weight of available evidence, which includes also sample sizes of assayed strains and numbers of scored polymorphic loci.

probably contributes to the genetic results. However, further comparative studies of pathogenic fungi with different life cycles will be required before final conclusions are drawn.

Bacteria too have been the subject of molecular genetic analyses to assess reproductive mode in nature. Following the discovery in the 1940s of genetic recombination mediated by conjugative transfer, transformation, and transduction in laboratory strains of *Escherichia coli,* the notion developed that extensive genetic exchange probably characterized most bacterial taxa (Hedges, 1972). However, until recently most research on the systematics, epidemiology, and pathogenicity of bacteria involved physiologic, serologic, or other phenotypic characteristics that seldom could be tied to specifiable alleles or loci that would permit reliable elucidation of population genetic structure and reproductive mode (Selander et al., 1987a). This situation changed beginning with allozyme studies first conducted by Milkman (1973, 1975), and since then a wealth of protein and DNA evidence has accumulated, suggesting that clonal reproduction predominates in many bacterial forms in nature. Because bacterial prokaryotes normally are haploid, the lines of evidence for clonality consist primarily of criteria (c) and (d) in Box 5.2.

For example, early work with *E. coli* revealed that despite high allozymic variation ($H \cong 0.50$, an order of magnitude greater than values for most higher eukaryotes), the number of distinctive multilocus genotypes was limited severely. Furthermore, identical or closely similar electrophoretic types were observed among some samples from geographically widespread, unassociated hosts and persisted over periods as long as 100 years (Ochman and Selander, 1984; Selander and Levin, 1980; Whittam et al., 1983a, 1983b). This and subsequent molecular research demonstrating strong nonrandom allelic associations across loci [gametic phase disequilibrium (Box 5.3)] has led to the view that chromosomal genetic transmission in *E. coli* is fundamentally clonal in nature, with rather infrequent exchange of genes by means of recombination. [Two qualifications to this conclusion should be stressed: (a) plasmid or other extrachromosomal elements of adaptive relevance *can* be exchanged commonly among *E. coli* (and other bacterial) strains (Hartl and Dykhuizen, 1984; Valdés and Piñero, 1992) and (b) occasional exchanges involving chromosome segments apparently do occur also, producing some limited degree of reticulation among otherwise isolated chromosomal lineages and leading to the appearance of mosaic chromosomes whose segments trace to different *E. coli* ancestors (Milkman and Bridges, 1990; Milkman and Stoltzfus, 1988)].

In any event, many ramifications stem from the new paradigm regarding the prevalence of clonal inheritance in *E. coli.* For example, the multiple clones occurring within an individual host probably represent successive invasions of multiple founding genotypes (Caugant et al., 1981) rather than products of recombination among a small number of pioneering strains as formerly was

Box 5.2. Population Genetic Criteria Strongly Suggestive of Clonal Reproduction in Diploid Populations (modified from Tibayrenc et al., 1991b)

Inferences (Observations)	Comments and Caveats (with Regard to Eliminating the Possibility of Frequent Sexual Reproduction)
Absence of meiotic segregation at single marker loci	
(a) Fixed heterozygosity (most or all individuals appear heterozygous)	Observation also incompatible with self-fertilization; must consider possibility of mis-scoring due to gene duplication or polyploidy
(b) Significant deficit in frequencies of some expected diploid genotypes; other deviations from Hardy-Weinberg equilibrium	Missing heterozygotes also consistent with self-fertilization; must exclude effects of population subdivision, assortative mating, selection, etc.
Absence of recombination among multiple marker loci	
(c) Overrepresented, widespread identical genotypes; significant deficit of expected recombinant genotypes; nonrandom associations of alleles among loci [gametic phase disequilibrium (Box 5.3)]	Should consider possible effects of selection or population subdivision; must take into account low expected frequencies of multilocus genotypes when allelic variation is high
(d) Correlation between independent sets of genetic markers	Should consider possible effects of population subdivision or correlated selection pressures

Box 5.3. Gametic Phase Disequilibrium

Gametic phase disequilibrium is the nonrandom association between alleles of different loci. By "nonrandom" is meant that the multilocus combinations of alleles depart significantly from expectations based on products of the single-locus allelic frequencies. Consider the simplest possible case, involving two loci (A and B) each with two alleles (A_1, A_2 and B_1, B_2) whose frequencies are p_1, p_2, q_1, and q_2, respectively. Four di-locus gametic genotypes (or haplotypes) are possible:

	Alleles at Locus A	
	A_1 (p_1)	A_2 (p_2)
B_1 (q_1)	A_1B_1 (p_1q_1)	A_2B_1 (p_2q_1)
B_2 (q_2)	A_1B_2 (p_1q_2)	A_2B_2 (p_2q_2)

(Alleles at Locus B)

If alleles are associated at random in haplotypes (gametic phase equilibrium), the expected frequencies of these di-locus genotypes are p_1q_1, p_1q_2, p_2q_1, and p_2q_2. One convenient quantitative measure of a departure from this expectation is given by the gametic phase disequilibrium parameter, defined as

$$D = P_{11}P_{22} - P_{12}P_{21},$$

where P_{11} and P_{22} are the observed frequencies of haplotypes in the "coupling" phase, and P_{12} and P_{21} are observed frequencies in the "repulsion" phase. It can be shown that in a large random mating population, any initial disequilibrium [$D(0)$] among neutral alleles will tend to decay toward zero (provided that $c \neq 0$) according to the equation

$$D(G) = (1-c)^G D(0),$$

where $D(G)$ is the disequilibrium remaining at generation G, and c is the probability of a recombination event between the two loci each generation (or the "recombination fraction"). Thus, for unlinked loci ($c = 0.5$), disequilibrium decays by one-half each generation. Rates of disequilibrium decay slow as the recombination fraction decreases. When the loci examined involve one nuclear gene and one cytoplasmic gene, the analogous nonrandom gametic associations are referred to as "cytonuclear disequilibria" (Asmussen et al., 1987). Because nuclear and cytoplasmic genes are unlinked, any initial disequilibrium between such loci is expected to decay monotonically to zero by one-half per generation, in a random mating population.

Gametic phase disequilibrium can arise from any historical or contemporary process that has restricted recombination among loci, such as physical linkage of genes on a chromosome, or other factors that can generate nonrandom associations among the alleles of unlinked loci as well, including population subdivision, founder effects, a self-fertilization mating system (Chapter 6), or selection favoring particular multilocus allelic combinations (Lewontin, 1988). Genomes or portions thereof characterized by a rarity or absence of recombination may be viewed as linked "supergenes" with regard to evolutionary dynamics. One likely consequence involves genetic hitchhiking, whereby alleles that are neutral mechanistically nonetheless may spread through a population because of chance association with an allele favored by natural selection at another locus. Such hitchhiking on rare favorable mutations could lead to *periodic selection* that has the net effect of purging population genetic variability in nonrecombining systems (Levin, 1981). The *selective sweeps* involved in periodic selection may account in part for the observation that genetic variability in mtDNA (Chapter 2) and in bacterial species (Milkman, 1973) is vastly lower than might have been expected given suspected mutation rates to neutral alleles and apparent population sizes. Such selective sweeps may account also for the observation of greatly reduced nucleotide diversity in chromosomal regions of eukaryotic nuclear genomes that are characterized by low recombination rates (Begun and Aquadro, 1992).

supposed; widespread dispersal of strains (even on a worldwide scale) apparently can take place; prospects for identifying the phylogenetic origins of phenotypically characterized strains of *E. coli* and its relatives [such as *Shigella* (Whittam et al., 1983b)] are enhanced when the confounding effects of frequent recombination can be neglected; and, in general, the dynamics of bacterial lineages may prove to be governed to a considerable degree by evolutionary factors including "periodic selection" (sequential replacement by clones of higher fitness, with net effect of purging population genetic variability) that, in principle, apply with special force to any low-recombination genetic system (Box 5.3).

Similar evidence for the prevalence of clonal chromosomal inheritance in other bacterial species including pathogenic forms is leading to a fundamental reassessment of bacterial taxonomy and epidemiology (Selander and Musser, 1990). The following are a few examples (Selander et al., 1987b). Based on the distinctiveness of sets of clones as identified by protein electrophoresis, *Legionella* consists of two undescribed species masquerading as *L. pneumophila* (Selander et al., 1985). Some of the more than 50 electrophoretic types (ETs) are distributed worldwide, and the same clone may cause both Legionnaires' disease and Pontiac fever. In *Bordetella,* a pathogen responsible for human whooping cough and a variety of respiratory diseases in animals, numerous clones and several genetically distinct species proved to exhibit strong host specificities (Musser et al., 1987). For example, clone ET-1 of *B. bronchiseptica* is a pig specialist, ET-6 a dog specialist, and the named species *B. parapertussis* and *B. pertussis* are other clonal forms of *B. bronchiseptica* that have become highly specialized as human pathogens. In *Haemophilus influenzae,* certain clones are distributed worldwide, and one distinctive clonal group (ET-91–94) causes meningitis and septicemia in human neonates. Another clone (ET-1) is known to have increased greatly in frequency in the United States between 1939 and 1954 and now causes about 30–40% of the invasive disease (Musser et al., 1985). Other *Haemophilus* clonal groups exhibit no close associations with particular disease conditions (Musser et al., 1985, 1986). Unusually high clonal variation characterizes *Neisseria meningitidis,* but only a few among the hundreds of multilocus genotypes appear to have been responsible for most major epidemics worldwide over the past 60 years. For example, one epidemic disease that started in Norway in the mid-1970s and spread through Europe is caused by a group of clones in the ET-5 complex that bears little genetic relationship to other electrophoretic types (Caugant et al., 1986). Another severe epidemic that appeared in Cuba in the late 1970s also is caused by ET-5 complex clones. These same strains apparently were brought to Miami via Cuban refugees and initiated an outbreak in Florida in 1980–1981. For the bacterium *Staphylococcus aureus,* among the dozens of electrophoretic types isolated from two continents, a single clone accounted for 88% of the cases of urogenital toxic shock syndrome in women (Musser et al., 1990). All of these population genetic studies of clonal structure exemplify the power of molecular markers in addressing problems in bacterial evolution that are of diagnostic and epidemiologic relevance.

On the other hand, not all bacterial taxa have proved to be predominantly clonal. Based on a great diversity of allozyme and RFLP genotypes observed in a wild population of *Bacillus subtilis,* Istock et al. (1992) concluded that recombination must be frequent relative to binary fission. This species is known to have a proclivity for spontaneous transformation, whereby DNA is exchanged by direct cell to cell contact. Such transformations presumably underlie the recombinational events that generated the high genotypic diversity reported.

Results for *B. subtilis* notwithstanding, available population genetic data indicate that recombination events within many bacterial species are far too rare to produce random allelic associations. Nonetheless, even for these species occasional chromosomal recombinations are by no means ruled out, and these could have important consequences over evolutionary time scales. Direct genetic evidence for chromosomal recombination does exist for *E. coli* and a growing list of bacterial taxa (DuBose et al., 1988; Selander et al., 1991). Thus, a particular *E. coli* strain, for example, could have its chromosomes derived in bits and pieces from multiple ancestors, the degree of mosaicism depending on the ancestors' remoteness, and the historical number of recombinational events involved (also an inverse function of closeness of physical linkage of the genes (Hartl and Dykhuizen, 1984). If recombination does generate, even occasionally, selectively advantageous genotypes which then are propagated by clonal means, such events could have an important impact on bacterial adaptation and evolution over the long term.

Gender Determination

Ascertainment of an individual's sex can prove difficult in many situations, for example, in early life history stages, in species with little dimorphism in secondary sexual characters, or in species with internal gonads (such as birds). Yet, knowledge of sexual identity is critical in ethological studies, sex ratio estimation, management of matings among captive animals, and other areas of population biology. In some groups such as many turtles and reptiles, sex is influenced by the temperature at which eggs are incubated (Bull, 1980), but in most taxa gender has a clear genetic basis. For these latter species, assay of molecular markers linked to sex-determining genes or chromosomes can, in principle, provide a powerful means of gender identification.

Quinn et al. (1990) isolated a piece of DNA homologous to the W (female-specific) chromosome of the snow goose (*Chen caerulescens*) and used this molecular probe to determine the sex of more than 150 birds from which blood samples had been taken. Similarly, Griffiths and Holland (1990) isolated a W-specific DNA probe for the herring gull (*Larus argentatus*), as did Sinclair et al. (1990) for a Y (male-specific) chromosome in humans. Using DNA fingerprinting assays, Millar et al. (1992) found W-specific bands in the brown skua (*Catharacta lonnbergi*) and used these gender markers to document significantly different sex ratios in adult birds versus chicks. It remains to be seen how widely these or similar probes can be employed in gender examination across broader taxonomic arrays. One potential complication already has been documented— occasional cross-homology between the Z and W chromosomes of birds (and the X and Y chromosomes of mammals), due perhaps to occasional transposition or recombination events between these sex chromosomes or to sequence conservation from an ancestral pair of homologues (Burgoyne, 1986; Ohno, 1967; Page et al., 1982, 1984). Nonetheless, Sinclair et al. (1990) discovered a Y-specific probe that could be used in Southern blot analyses to determine sex in a wide range of mammals.

A remarkable application of sex-specific probes to gender identification in nature involved humpback whales (*Megaptera novaeangliae*), which like other baleen whales lack any obvious secondary sexual characteristics that distinguish males from females. A human Y-chromosome clone was employed successfully as a hybridization probe in RFLP analyses to determine the gender of 72 free-ranging whales from which skin biopsies had been collected by a special dart (C.S. Baker et al., 1991). The sex of yet another individual was identified by assays of DNA extracted from sloughed skin collected from the whale's swimming path!

PARENTAGE

Procedures for genetic assessment of parentage are similar in principle to those used to assess genetic identity/nonidentity, with the added complication that the rules of Mendelian transmission genetics must be taken into account when comparing the genotypes of sexually-produced progeny against those of putative parents. Parentage analyses typically address some version of the following question: Are the adults who are associated behaviorally or spatially with particular young the true biological parents of the offspring in question? If the answer proves to be no, a genetic "exclusion" has been achieved. Whether the actual mothers and fathers also can be specified depends on the size and genetic composition of the pool of candidate parents and on the level of genetic variability monitored. Sometimes one biological parent is known from independent evidence and the problem simplifies to one of paternity (or maternity) exclusion or inclusion. In other cases, neither parent is known with certainty prior to the molecular study.

Knowledge of biological parentage often is important in behavioral and evolutionary studies. For example, matings are difficult to observe directly in nature for many species, but reproductive behaviors and patterns of gene flow (Chapter 6) nonetheless can be deduced from molecular information on maternity and paternity. Proper interpretations of behavioral interactions between presumed family members depend on knowledge of genetic ties, including parentage. Even when matings can readily be observed, questions of genetic parentage remain of interest. Thus, in many birds and mammals, copulations are known to occur outside the socially bonded pair, but the extent to which these result in illegitimate young has been uncertain and constitutes a major deficiency in the understanding of sexual selection and the evolution of mating systems (Gyllensten et al., 1990; Mock, 1983; Trivers, 1972). By revealing genetic parentage, molecular data provide more direct assessments of realized reproductive success and, thus, at least partially circumvent the danger of equating mating prowess or other components of the reproductive process with actual gene transfer across generations. Finally, knowledge of biological parentage is critical for correct interpretations of the transmission genetics or heritabilities of any morphological characters as deduced from field data on presumed parent-offspring associations (Alatalo et al., 1984).

Parentage analyses utilize the cumulative information from multiple polymorphic loci assayed either individually (e.g., allozymes, scnRFLPs) or jointly (multilocus DNA fingerprinting). General interpretive procedures may be introduced by the following examples:

(a) Maternity and paternity both uncertain, exclusions attempted. T.W. Quinn et al. (1987, 1989) compared goslings within each of several broods of the snow goose (*Chen caerulescens*) against their adult male and female nest attendants (putative parents) using genetic markers from multiple single-copy nuclear

Box 5.4. Diploid Genotypes (at 14 Nuclear RFLP Loci) of Nest Attendants and Goslings in Three Families of the Snow Goose (after T.W. Quinn et al., 1987).

As elaborated in the text, families 1 and 3 show strong genetic evidence of nonattendant parentage for some goslings.

	A	B	C	D	E	F	G	H	I	J	K	L	M	N
Family 1														
male attendant	2,2	2,2	2,3	1,2	1,1	1,1	1,4	2,2	1,2	1,2	1,2	1,2	2,2	2,2
female attendant	2,2	2,2	2,2	1,1	1,1	1,1	1,3	1,2	2,2	1,1	1,1	1,2	1,2	1,2
gosling 1	2,2	2,2	2,2	1,2	1,1	1,1	1,1	1,2	1,2	1,1	1,2	1,1	1,2	1,2
gosling 2	2,2	2,2	2,2	1,1	1,1	1,1	3,4	2,2	2,2	1,2	1,2	1,2	2,2	2,2
gosling 3	2,2	2,2	2,2	1,2	1,1	1,1	1,3	2,2	1,2	1,1	1,1	2,2	2,2	2,2
gosling 4	2,3^a	2,2	1,1^d	1,1	1,1	1,1	1,2^a	1,2	1,2	1,1	1,1	1,1	1,1^c	1,1^c
Family 3														
male attendant	1,2	2,2	2,4	1,1	1,1	1,1	1,1	1,2	1,1	1,1	1,1	1,2	1,1	1,2
female attendant	2,2	2,2	1,2	1,1	2,2	1,1	1,2	1,2	2,2	1,1	1,1	1,1	2,2	1,2
gosling 9	2,2	2,2	1,2	1,1	2,2	1,1	1,1	2,2	1,2	1,2^a	1,1	1,1	1,2	1,2
gosling 10	2,2	2,2	2,4	1,1	1,2	1,1	1,2	1,1	1,2	1,1	1,1	1,1	2,2^c	1,1
gosling 11	1,2	2,2	2,4	1,1	1,1^b	1,1	1,1	1,2	2,2^c	1,1	1,1	1,1	1,1^b	1,1
gosling 12	2,3^a	2,2	2,2	1,1	1,1^b	1,1	2,2^c	1,1	1,2	1,1	1,1	1,1	1,1^b	2,2
gosling 13	1,2	2,2	2,2	1,1	1,2	1,1	1,2	2,2	1,2	1,1	1,1	1,2	1,2	1,2
Family 4														
male attendant	2,2	2,2	3,3	1,2	1,1	1,1	1,1	1,2	1,2	1,1	1,1	1,1	1,1	1,2
female attendant	2,2	1,2	1,1	1,1	1,1	1,1	1,1	2,2	1,1	1,1	1,2	1,2	1,2	1,2
gosling 14	2,2	1,2	1,3	1,2	1,1	1,1	1,1	1,2	1,2	1,1	1,2	1,1	1,2	1,2
gosling 15	2,2	1,2	1,3	1,2	1,1	1,1	1,1	1,2	1,2	1,1	1,1	1,2	1,1	1,2
gosling 16	2,2	2,2	1,3	1,2	1,1	1,1	1,1	2,2	1,2	1,1	1,2	1,1	1,2	1,2
gosling 17	2,2	2,2	1,3	1,2	1,1	1,1	1,1	2,2	1,2	1,1	1,1	1,2	1,1	2,2

^{a}Excludes one unspecified parent; ^{b}excludes putative mother; ^{c}excludes putative father; ^{d}excludes both putative parents

RFLP loci. From their data (Box 5.4), the following observations and deductions were made. Two goslings in family 3 (numbers 11 and 12) proved to be homozygous at some loci (E and M) for alleles not present in the female attendant. Such cases excluded the putative mother and were interpreted to reveal instances of intraspecific brood parasitism (IBP), whereby other females (not assayed) must have contributed eggs to the nest. Other goslings (e.g., number 10 in family 3) proved to be homozygous (locus M) for alleles not present in the male attendant. Such cases excluded the putative father and were interpreted to reveal likely instances of extra-pair fertilization (EPF) by other males in the population.

Some heterozygous loci (e.g., J in gosling 9, family 3) exhibited one allele not observed in either nest attendant and a second allele present in both attendants. Such loci exclude one of the putative parents, but do not alone determine which attendant is disallowed. Finally, some loci (e.g., C in gosling 4, family 1) were homozygous for alleles not observed in either putative parent, thus excluding both. Overall, the genetic data confirmed suspicions from field observations that IBP (and probably EPF) are relatively common in snow geese populations (see also Lank et al., 1989).

(b) Maternity known, paternity to be decided among a few candidate males. Burke et al. (1989) applied multilocus DNA fingerprinting assays to the dunnock sparrow (*Prunella modularis*), a species with a variable mating system tending toward polyandry in which two males sometimes mate with and defend the territory of a single female. In the DNA fingerprints, paternally-derived bands in progeny were identified as those that could not have been inherited from the known mother. Then, the true father was determined by comparing bands from the fingerprints of candidate sires against these paternal alleles in progeny. For example, Figure 5.5 shows DNA fingerprints from one known mother (M), her four offspring (D–G), and two candidate sires (P_α and P_β). In this family, the genetic data demonstrate that progeny G was sired by P_α, whereas D, E, and F were fathered by P_β. Thus, individual broods of polyandrous dunnock females indeed can be multiply-sired.

(c) One parent or two? Many plants and invertebrate animals are hermaphroditic, i.e., an individual produces both male and female gametes. Such individuals might self-fertilize (in which case offspring have a single parent) or matings may be facultative or compulsory with other individuals (thus producing two-parent progeny). From wild-caught females whose mating habits are in question, genetic examination of progeny often can reveal whether some progeny carry alleles not present in the mother and, hence, derive from outcross fertilizations. Furthermore, comparisons of genotypic frequencies in a natural population against Hardy–Weinberg expectations can aid in deciding whether cross-fertilization or self-fertilization predominates (because the latter is a most intense form of inbreeding, whose continuance leads to pronounced deficits in heterozygote frequency and the eventual appearance of completely homozygous strains). In examples of these approaches, allozyme data were employed to show that cross-fertilization is the prevailing mode of reproduction in several species of hermaphroditic freshwater snails in the genera *Bulinus* and *Biomphalaria* (Rollinson, 1986; Vrijenhoek and Graven, 1992; Woodruff et al., 1985), that intermediate levels of self-fertilization characterize the Florida tree snail *Liguus fasciatus* (Hillis, 1989) and the coral *Goniastrea favulus* (Stoddart et al., 1988), and that self-fertilization predominates in populations of the sea anemone *Epiactis prolifera* (Bucklin et al., 1984). In similar allozymic studies of 19 species of

offspring

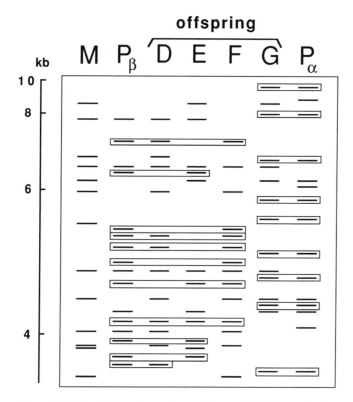

Figure 5.5. Parentage analyses by multi-locus DNA fingerprinting. Shown is a gel designed to assess whether offspring D–G with known mother M were sired by P_α, P_β, or both (after Burke et al., 1989). Boxes encompass paternally derived bands in progeny that permitted choice between the two candidate sires. See text for further explanation.

terrestrial slugs in the families Limacidae and Arionidae, the majority of taxa were shown to be predominant outcrossers (Foltz et al., 1982, 1984).

[As mentioned earlier, another form of uniparental reproduction is parthenogenesis, which experimentally can be distinguished from self-fertilization by examination of the diploid genotypes among the offspring of a heterozygous parent. Fixed heterozygosity among progeny is inconsistent with expectations of Mendelian segregation under self-fertilization but is a hallmark of ameiotic parthenogenetic reproduction. Using such allozymic evidence, Hoffman (1983) documented that a laboratory population of one slug species formerly suspected of self-fertilization (*Deroceras laeve*) actually reproduced by parthenogenesis.

Further discussion of parentage in the context of parthenogenetic reproduction will be deferred to Chapter 7.]

Paternity Analysis

In species with internal fertilization, or where close physical or behavioral associations connect mothers with offspring, maternity may be evident, whereas paternity is uncertain. Such is often the case in humans, where indeed the bulk of effort by molecular forensic laboratories is devoted to paternity assessment (Lewis and Cruse, 1992). In many species, establishment of paternity is complicated further by the fact that females may store viable sperm for considerable periods of time. Many insects possess specialized sperm storage organs (spermathecae). In lizards, female reproductive tracts may retain viable sperm for as long as two months, and in some snakes for perhaps five to seven years (Devine, 1984). Some avian and mammalian females also may store sperm for periods of days to weeks (Birkhead and Møller, 1992; Parker, 1984), although in general both the life spans of sperm and the durations of sperm storage by females are considerably longer in birds than in mammals (Gomendio and Roldan, 1993).

MULTIPLE PATERNITY IN ANIMALS

It might be supposed that male-specific DNA sequences (e.g., on the Y chromosome in mammals and insects) would be of special use in paternity assessment. Indeed, Williams and Strobeck (1986) reported multiple paternity for the male progeny of some individual *Drosophila melanogaster* females, as gauged by the observed diversity in RFLP patterns for ribosomal RNA gene markers that were known to be confined to the Y chromosome. However, few such case histories utilizing male-specific chromosomes as yet are available, no doubt for two reasons: only a limited number of polymorphisms thus far has been uncovered on these sex chromosomes and, in any event, paternity assignments for female progeny could not be accomplished by this approach.

However, many paternity studies *have* utilized protein-electrophoretic or autosomal DNA markers. By applying these molecular assays to single-female litters, broods, or clutches, concurrent multiple paternity has been documented for a wide variety of species in nature. Notable examples have come from mice (Birdsall and Nash, 1973; Xia and Millar, 1991), shrews (Tegelström et al., 1991), prairie dogs (Hoogland and Foltz, 1982), snakes (Gibson and Falls, 1975), marine turtles (Harry and Briscoe, 1988), salamanders (Tilley and Hausman, 1976), fishes (Chesser et al., 1984; Simonsen and Frydenberg, 1972; Travis et al., 1990), insects (Milkman and Zeitler, 1974; Parker, 1970), spiders (Martyniuk and Jaenike, 1982), isopods (Heath et al., 1990; Sassaman, 1978), limpets and other snails (Gaffney and McGee, 1992; Mulvey and Vrijenhoek, 1981), and lobsters (Nelson and Hedgecock, 1977).

Several of these studies involved socially polygynous species and thus confirmed suspicions that multiple copulations or inseminations could indeed result in multiple successful fertilizations of a progeny cohort. For example, female Belding's ground squirrels are known to mate with several different males, and allozymic results established that an estimated 78% of litters were multiply-sired, usually by two or three males (Hanken and Sherman, 1981). In a similar study of the willow leaf beetle (*Plagiodera versicolora*), more than 50% of wild caught females produced egg clutches expressing multiple sires (McCauley and O'Donnell, 1984) and in *Drosophila pseudoobscura*, single-female broods were estimated to have been fathered by 1.7 different males on average (Cobbs, 1977). On the other hand, not all molecular genetic analyses have uncovered evidence for multiple paternity. For example, using allozyme assays, Foltz (1981) demonstrated a high degree of monogamy in the oldfield mouse *Peromyscus poliontus;* and in a DNA fingerprinting study, all 99 assayed offspring from more than 25 families of the California mouse (*P. californicus*) resulted from matings within individual male-female pairs [a finding that according to Ribble (1991) provided "the first convincing demonstration of exclusive monogamy in a mammal" (but see also Kleiman, 1977)].

Of greater surprise were genetic observations that single broods in some supposedly monogamous species frequently contain illegitimate progeny resulting from copulations involving one or more parents other than the care-givers. Perhaps nowhere has such evidence for cuckoldry been discussed more widely than in the Passeriformes (perching birds), which as a group traditionally had been considered among the most monogamous of organisms (Gill, 1990). Gowaty and Karlin (1984) were among the first to report genetic evidence for multiple paternity in a purportedly monogamous bird (the eastern bluebird, *Sialia sialis*), a species now estimated from genetic data to produce between 8% and 35% illegitimate young (Gowaty and Bridges, 1991a). Many additional examples of multiple concurrent parentage in birds since have come to light (Table 5.4), with illegitimate young often in high frequency. For example, an allozyme survey of the indigo bunting (*Passerina cyanea*) established that at least 37 of 257 offspring (14%) carried genotypes incompatible with the putative father (Westneat, 1987). Statistical corrections based on detection probabilities (because only a few polymorphic protein markers were employed) raised the estimated frequency of EPFs in indigo buntings to between 27% and 42%. These latter estimates agree with a DNA fingerprinting survey of another bunting population, where 22 of 63 nestlings (35%) were shown to have resulted from extra-pair fertilizations (Westneat, 1990). Not all genetic reappraisals of passeriform species have produced evidence for cuckoldry; among 176 young from 32 families of warblers in the genus *Phylloscopus*, no illegitimate young were found using sensitive single-

Table 5.4. Examples of concurrent multiple paternity or maternity in avian species as revealed in studies that employed molecular genetic markers.

Species	No. Broods Assayed	No. Progeny Assayed	Observed % Broods with Non-Kin Progeny[a]	Observed % Non-Kin Progeny[a]	Corrected % Non-Kin Progeny[b]	Assay	EPF, IBP, or Both[c]	Reference[d]
Indigo bunting (Passerina cyanea)	98	257	24	14	27–42	Allozymes	EPF	Westneat, 1987
Indigo bunting (Passerina cyanea)	25	63	48	35	35	DNA fingerprints	EPF	Westneat, 1990
White-fronted bee-eater (Merops bullockoides)	65	97	8	7	9–12	Allozymes	Both	Wrege and Emlen, 1987
Cliff swallow (Hirundo pyrrhonota)	105	349	21	10	24	Allozymes	Both?	Brown and Brown, 1988
Eastern kingbird (Tyrannus tyrannus)	19	60	47	30	39–53	Allozymes	IBP, Both?	McKitrick, 1990
House wren (Troglodytes aedon)	18	97	22	6	31	Allozymes	IBP, Both?	Price et al., 1989
Red-winged blackbird (Agelaius phoeniceus)	36	111	45	28	28	DNA fingerprints	EPF	Gibbs et al., 1990
Field sparrow (Spizella pusilla)	17	52	41	19	—	Allozymes	Both	Petter et al., 1990

[a]Defined here as progeny for which at least one of the putative parents is excluded by genetic evidence.

[b]Correction based on empirical detection probabilities for the polymorphic loci assayed [by procedures such as those detailed in Westneat et al. (1987)].

[c]EPF: extra-pair fertilization; IBP: intraspecific brood parasitism.

[d]Additional examples may be found in Birkhead et al. (1990), Bollinger and Gavin (1991), Evarts and Williams (1987), Gavin and Bollinger (1985), Gelter and Tegelström (1992), Joste et al. (1985), Lifjeld et al. (1991), Morton et al. (1990), Mumme et al. (1985), Seutin et al. (1991), Sherman and Morton (1988), and Wetton et al. (1987, 1992).

locus and multilocus VNTR assays (Gyllensten et al., 1990). Nonetheless, the unexpectedly high frequency of extra-pair fertilizations reported for many birds is prompting a fundamental reassessment of avian mating systems (Westneat et al., 1990).

Studies of multiple parentage are most informative when combined with behavioral or life history observations. For example, in eastern bluebirds the frequency of "nondirectly descendant nestlings" (NDDN) proved to be significantly greater for (a) males in their first breeding season rather than older males, (b) males paired with females who more frequently were off territory during their fertile periods than for males with sedentary mates, and (c) males that stayed closer to females and thus exhibited what might be interpreted as mate-guarding behavior (a counter intuitive result, unless it is supposed that males can sense a propensity for cheating by certain females and attempt to monitor them accordingly). Gowaty and Bridges (1991b) interpreted some of these observed trends as consistent with the postulate that female bluebirds actively pursue extra-pair fertilizations (rather than acting as passive or coerced on-territory recipients of EPF-seeking males, as might be assumed under traditional mating system theory). Whether any fitness advantages accrue to EPF-seeking females remains unknown, although two conventional lines of speculation are that greater genetic diversity among progeny might be selectively advantageous, and/or that extra-pair copulations provide "fertilization insurance."

In other avian genetic studies integrated with demographic or behavioral observations, Morton et al. (1990) discovered through DNA fingerprinting that older males in the colonial-nesting purple martin (*Progne subis*) achieve far higher fecundities than do younger males, a result attributed to forced copulations by older males that result in age-biased extra-pair fertilizations. In a similar DNA fingerprinting study of polygynous red-winged blackbirds (*Agelaius phoeniceus*), Gibbs et al. (1990) found that the proportion of illegitimate chicks is significantly greater in marshes with higher male densities and that the cuckolding males often are territorial neighbors. In a study of barn swallows (*Hirundo rustica*), DNA fingerprints revealed that notwithstanding a pairing preference of females for males with longer and more symmetrical tail streamers (Møller, 1992), such males did not receive the expected fitness advantage because of a significantly increased susceptibility to cuckoldry (H.G. Smith et al., 1991). The authors interpreted this result as indicative that longer-tailed males are hampered in ability to guard mates and that this provides an evolutionary counterbalance to the sexual selection for longer tails via female choice. In a detailed behavioral and DNA-fingerprinting analysis of a population of blue tits (*Parus caeruleus*) in Belgium, Kempenaers et al. (1992) found that mate-guarding by males was relatively ineffective in protecting paternity: "Attractive" males (those with many visits from neighboring females) actually suffered less loss in paternity (fewer extra-pair young in their own nests) than did "unattractive" males. These males also were larger and survived better. Results were interpreted as support-ive of a "genetic quality hypothesis," whereby females somehow assess male quality and mate preferentially with superior males, regardless of other social or populational considerations.

In some organisms such as the striped-backed wren (*Camphylorhynchus nuchalis*), young remain in natal groups and appear to assist adult kin in rearing new broods. Under sociobiological theory, postponement of dispersal and breed-ing to assist in the rearing of others' progeny may be favored by natural selection if the helpers' contributions to production of close kin exceeds expected repro-ductive success had they dispersed (Brown, 1987; Hamilton, 1964). Studies based on DNA fingerprinting by Rabenold et al. (1990) demonstrated another, more direct avenue by which the fitness of helpers was enhanced. Some of the behaviorally dominant males in wren social groups were found to share paternity with auxiliary males previously thought to be nonreproductive. Thus, such re-production by subordinate males may further help to explain their long tenure as "helpers at the nest." In contrast, among females the only reproductives proved to be dominant individuals, a result interpreted as consistent with the observed aggressiveness of young female wrens in competition for breeding sites outside the natal group.

Troops of rhesus macaque monkeys (*Macaca mulatta*) are characterized by strong dominance hierarchies (both genders) whose behavioral underpinnings are postulated to have evolved in response to selective pressures favoring high-ranking individuals. Do males of higher social rank truly exhibit higher fitness through greater access to receptive females (Dewsbury, 1982)? In a landmark genetic study of primates, Duvall et al. (1976) applied allozyme analyses in conjunction with behavioral observations to a captive group of rhesus macaques and discovered that only 7 of 29 offspring (24%) produced over the 2 years of the study actually were fathered by the α (top-ranking) male. Even low- and mid-ranking adolescent monkeys sired several offspring and, therefore, clearly had access to ovulating females. In a similar study of another rhesus troop, Curie-Cohen et al. (1983) found that over an 8-year period the dominant male sired only 13–32% of the offspring even though he participated in 67% of the ob-served copulations, and that the second ranking male sired 30–48% of the offspring despite participation in only 14% of the observed matings. Working with six groups of the same species, Smith (1981) found that reproductive success as determined by genetic markers *was* significantly correlated with rank, although the pattern was such that changes in rank position appeared to follow rather than precede changes in reproductive success. Examples of further genetic work on rhesus monkeys include observations that a reproductive advantage is enjoyed by the sons of high-ranking mothers, at least prior to their dispersal from their natal groups (Smith and Smith, 1988), and that males and females of similar social rank tend not to produce disproportionate numbers of progeny (Small and

Smith, 1982). The overall picture to emerge from these and other studies is that social dominance or even copulation frequency within rhesus troops are at best imprecise predictors of breeding success of males [although the possibility remains that lifetime reproductive fitness of socially dominant males might be higher due to increased probabilities of survival for their offspring, rather than increased reproductive activity per se (Bernstein, 1976)].

Rhesus macaques typically have large troops and are seasonal breeders, with females exhibiting more or less synchronized fertile cycles during a well-defined period. Stern and Smith (1984) speculated that in such cases it is more difficult for males to monopolize females than in social systems where troop size is smaller and estrus is dispersed temporally. A related macaque species (*M. fascicularis*) breeds nonseasonally and was employed as a test of this hypothesis. However, genetic paternity analyses of 44 *M. fascicularis* offspring born over a 28-month period revealed no evidence to support a positive association between male social rank and number of offspring sired (Shively and Smith, 1985).

A different outcome was uncovered in studies of the red howler monkey (*Alouatta seniculus*), where in none of nine surveyed troops could genetic evidence exclude the dominant male as the father of offspring conceived during his tenure (Pope, 1990). In both single-male and multimale harems, only the top-ranking male was observed to mount females, and genetic evidence supported the idea that dominant-male social status confers reproductive success. Red howler troops are small and spatially cohesive, and this may facilitate behavioral monitoring by the α male. Pope (1990) further suggested that females may avoid mating with subordinate males to avoid infanticide, because infants conceived during successful and attempted status changes by males frequently are killed.

Preliminary studies of molecular paternity also have been conducted on other primates including baboons, marmosets, lemurs, guenons, mandrills, gorillas, chimpanzees, and several monkeys and other macaque species (deRuiter, 1992; R.D. Martin et al., 1992). In comparing results, two complications have been as follows: DNA fingerprinting methods have worked better for some species than others (for technical reasons) and levels of genetic variability have proved to differ among populations and species (T.R. Turner et al., 1992). Nonetheless, results for the red howler monkey notwithstanding, an emerging generality is that social status and observed copulation frequency often are poor guides to male reproductive success. Thus, for many primate species, these traditional methods of fitness estimation appear to be grossly inadequate predictors of successful progeny production.

PATERNITY IN PLANTS

In plants, fatherhood results from the spread of pollen, as mediated for example by insect pollinators or wind. The same types of molecular genetic analyses as described above can be applied to questions of pollen source (Adams et al., 1992; Devlin and Ellstrand, 1990). The task again is simplified when the mother is known (e.g., as the bearer of the seeds in question), but can remain difficult when there is a large pool of potential pollen donors. Paternity may be addressed with regard to the multiple seeds within a fruit and/or with regard to the entire seed set of a maternal parent.

Many plant species are monoecious (hermaphroditic). Not all such plants can self-fertilize, however, for several reasons: (a) male and female flowers of an individual may mature at different times or be spatially separated on the plant, (b) the stamens and stigma within a perfect flower (a flower possessing both male and female parts) may be positioned such that mechanical pollen transfer is unlikely, or (c) self-incompatibility genes may be present. These sterility genes are known to carry multiple alleles that appear to have been selected to prevent possible deleterious effects of the intense inbreeding that self-fertilization entails. (Operationally, when the maternal parent and a pollen grain share the same allele at a self-sterility locus, sporophytic tissue discriminates against gametophytic tissue, for example by inhibiting growth of the pollen tube down the style.) Nevertheless, self-fertilization apparently has evolved independently from the outcrossing mode on numerous occasions (Stebbins, 1970; Wyatt, 1988), probably at least 150 times in the Onagraceae alone (Raven, 1979).

Thus, one of the first genetic questions regarding a monoecious species concerns the frequency with which self-fertilization (as opposed to outcrossing) takes place. When the female parent is known, the problem simplifies to one of paternity assessment, with the issue in this case being how often the individual plant that mothered an array of progeny or seeds can be excluded as the father of those genotypes. For example, any offspring that exhibits an allele not present in its known mother must have arisen through an outcross event (barring mutation). Of course, the power to detect outcrossing depends on the number of polymorphic marker loci monitored and the number and frequencies of alleles. Several statistical models have been developed to quantitatively estimate rates of selfing versus outcrossing (s and t, respectively, where $s + t = 1$) from genotypic information at one or more loci (Brown and Allard, 1970; Ennos and Clegg, 1982; Ritland and Jain, 1981; Schoen, 1988; Shaw et al., 1981). For example, the widely employed "mixed-mating" model assumes that the mating process can be divided into two distinct components: self-fertilization and random mating (i.e., random independent draws of pollen from the total population of plants) (Brown, 1989; Clegg, 1980). This model may be especially appropriate for wind-pollinated species. A variant of this model that is probably more applicable to insect-pollinated species assumes that outcross events within a family are correlated because they may involve successive pollen draws from a single male parent (Schoen and Clegg, 1984). In reality, mean genetic relatedness within a progeny array often may fall somewhere between full sibs and half sibs, and furthermore the sibship composition no doubt varies from array to array.

From allozymes and other genetic markers, these "mating system parameters" (*s* and *t*) have been estimated empirically for numerous populations and species of monoecious plants. In an early summary of the literature by Lande and Schemske (1985), the overall frequency distribution of outcrossing rates proved to be bimodal, with most species either predominantly selfing or outcrossing (Fig. 5.6). These authors interpreted the bimodality in mating system as consistent with a scenario in which outcrossing is selected for in historically large species with substantial inbreeding depression, whereas selfing is favored in species where pollinator failure or population bottlenecks have greatly reduced the level of inbreeding depression (via prior purging of deleterious recessive alleles). Empirical evidence does exist for a high variance among plant species in degree of inbreeding depression, with outcrossers typically exhibiting the greatest reductions in fitness under inbreeding (Schemske and Lande, 1985). Nevertheless, few monoecious plant species are "fixed" for either pure outcrossing or pure selfing, and different populations within some species show tremendous variation along the selfing-outcrossing continuum (Fig. 5.6). Furthermore, the bimodality of mating systems noted by Schemske and Lande (1985) may reflect a bias due to the disproportionate representation of selfing grasses and outcrossing trees in the early literature (Aide, 1986).

In monoecious species where outcrossing has been established, or in any dioecious species, the next genetic question likely to arise is: "Which plants were the pollen donors for particular outcrossed offspring?" As illustrated in Box 5.5, molecular genetic markers again can supply the answer. The approach involves comparing the diploid genotype of each seed or progeny with that of its known mother, and thereby deducing (by subtraction) the haploid genotype of the fertilizing pollen. Then candidate fathers are screened for diploid genotype, and paternity excluded for those males whose genotypes could not match the deduced pollen contribution to the progeny. Sometimes all males except the true father can be excluded. When multiple candidates remain, statistical procedures exist for assigning "fractional paternity" based on the probabilities of being the father. In the first large-scale application of these approaches, Ellstrand (1984) employed six highly polymorphic allozyme loci to establish paternity for 246 seeds from 9 maternal plants within a closed population of the wild radish, *Raphanus sativus*. Multiple paternity was found for all assayed progeny arrays from a maternal plant and for at least 85% of all fruits, with the minimum paternal donor number averaging 2.27. The wild radish is a self-incompatible, insect-pollinated species. Subsequent work established that most multiply-sired fruits resulted from the simultaneous deposition by a single insect vector of pollen from several plants ["pollen carryover" (Marshall and Ellstrand, 1985)] and that a considerable fraction of the seed paternity for some plants (up to 44%)

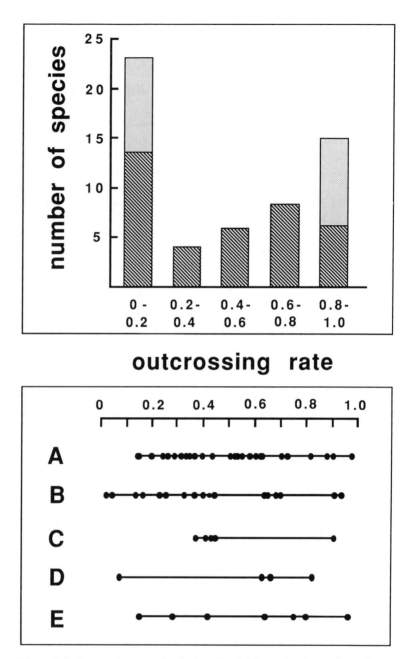

Figure 5.6. Outcrossing rates in plants estimated by molecular markers. **Top:** Frequency distribution of mean outcrossing rate as estimated from allozyme markers for 55 monoecious plant species (after Schemske and Lande, 1985). Hatched bars are animal-pollinated species, shaded bars are wind-pollinated species (Aide, 1986). **Bottom:** Interpopulational variation in outcrossing rate in each of five plant species: (A) *Lupinus succulentus,* (B) *L. nanus,* (C) *Clarkia exilis,* (D) *C. tembloriensis,* and (E) *Gilia achilleifolia* (after Schemske and Lande, 1985). Solid circles are population means and horizontal lines represent observed ranges within each species.

Box 5.5. Illustration of Paternity Assignment for Five Progeny from a Known Mother [taken from a much larger allozyme data set in Ellstrand (1984)].

The body of the table consists of observed diploid genotypes at each of six loci in the wild radish.

	LAP	PGI	PGM1	PGM2	6PGD	IDH		
				Allozyme Locus				
Potential Fathers								
A	2,2	1,2	2,3	2,2	3,3	1,1		
B	2,2	2,3	1,3	1,1	1,3	1,1		
C	1,2	1,2	1,3	1,1	1,2	1,2		
D	1,5	1,1	1,2	1,3	1,3	1,1		
E	2,3	2,2	1,2	1,2	1,1	1,3		
F	2,2	1,3	2,2	1,2	1,3	1,1		
G	1,1	1,2	1,2	1,2	3,3	1,1		
H	1,1	1,2	1,2	1,2	1,3	2,2		
I	1,2	1,1	1,1	1,2	1,3	1,1		
J	1,2	2,3	1,2	1,2	3,3	1,2		
K	2,2	1,2	1,3	2,2	3,3	1,1		
L	1,2	1,1	1,1	2,3	1,3	1,1		
M	2,5	1,1	1,2	2,3	1,2	1,3		
N	1,1	1,1	1,2	1,1	1,1	1,1		
O	1,3	1,2	1,2	1,2	3,3	1,1		
Known Mother								
Z	1,2	1,1	1,1	1,2	1,3	1,1		
							Deduced Paternal	
							Gamete	Assignment
Offspring								
P	2,2	1,2	1,3	1,2	1,1	1,2	223-12	C
Q	2,2	1,2	1,3	1,2	2,3	1,1	223-21	C
R	1,2	1,2	1,3	1,1	1,2	1,1	-23121	C
S	1,2	1,1	1,2	2,3	1,1	1,3	-12313	M
T	2,2	1,1	1,1	1,3	2,3	1,3	211323	M

derived from immigrant pollen from populations at least 100 m away (Ellstrand and Marshall, 1985).

In a similar allozymic study of a small forest herb *Chamaelirium luteum*, Meagher (1986) established paternity likelihoods for 575 seeds with known mothers. The distribution of intermate (pollen-flow) distances indicated that more nearby fertilizations had taken place than expected on the basis of random mating, but nonetheless some mating pairs were separated from one another by more than 30 m. A follow-up study of established seedlings (where maternity was unknown) confirmed this pollen-dispersal profile and also demonstrated that pollen and seed dispersal distances were quite similar (Meagher and Thompson, 1987). Surprisingly, no obvious relationship was found in this species between the size of the male plant (indicative of reproductive effort) and the paternal success (Meagher, 1991). Hamrick and Murawski (1990) conducted similar genetic paternity analyses on several tropical woody species and showed that (a) a significant proportion of the pollen received by individuals came from relatively few pollen donors, (b) many matings (30–50%) appeared to take place between nearest-neighbors, and (c) some matings (10–25%) nonetheless did involve long-distance pollen flow (greater than 1 km). The overall breeding structure to emerge from the latter study appeared to have two components: a leptokurtic pattern of pollen dispersal within populations, superimposed on a more even distribution of "background" pollen originating from outside the population.

Results from genetic determinations of pollen source sometimes can assume important economic and management ramifications for crop species. For example, commercial seed orchards provide a significant fraction of the zygotes used to establish pine plantations in the southeastern United States. One such seed orchard for the loblolly pine (*Pinus taeda*) in South Carolina was composed of grafted ramets of 50 loblolly clones that had been chosen and maintained for phenotypically desirable traits. Using multilocus allozyme markers, Friedman and Adams (1985) discovered that at least 36% of the seeds from this orchard had been fertilized by outside pollen, despite a surrounding 100-m-wide buffer zone positioned explicitly to prevent such genetic contamination by nonselected males. In an experimental population of cultivated cucumbers (*Cucurbita pepo*), Kirkpatrick and Wilson (1988) showed that approximately 5% of progeny had been fathered by a wild cucumber relative (*C. texana*). The authors suggested that such weed-crop genetic exchange may have occurred over thousands of years and provided an important source of genetic variability for selected cultivars.

SPERM AND POLLEN COMPETITION, PRECEDENCE

The widespread occurrence of polygyny and the multiple paternity of progeny arrays means that sperm from two or more males commonly must be placed in direct competition for fertilization of eggs within a female's reproductive cycle.

Several morphological characteristics and reproductive behaviors of males have been interpreted as adaptations to meet the genetic challenges resulting from this supposed competition with another male's sperm (Parker, 1970). For example, in many worms, insects, spiders, snakes, and mammals, the male secretes a plug that serves temporarily as a "chastity belt" to block the female's reproductive tract from subsequent inseminations. Other widespread male behaviors that have been interpreted as providing paternity assurance in the face of potential sperm competition include prolonged copulation (up to one week in some butterflies), multiple copulations with the same female, and postcopulatory guarding (Parker, 1984). From a female's perspective, mechanisms to prevent sperm competition are not necessarily desirable, and this may lead to intersexual conflicts of interest. Among the suggested advantages accruing to females from sperm competition and multiple paternity are an enhanced opportunity to acquire the "best" available male genotypes, fertility insurance, and an increase in the genetic diversity among progeny (Knowlton and Greenwell, 1984). Recent evidence also suggests that the reproductive tracts of some females may play a more active role than previously supposed in post-copulatory choice among sperm for fertilizations (Birkhead and Møller, 1993).

From the progeny arrays of multiply-inseminated females, molecular markers can be employed to determine which among the competing males' sperm have achieved the fertilizations. Is the first-mating male at a reproductive advantage, or does the last-mating male achieve the highest fertilization success? Or is there no mating-order effect, the probability of fertilization instead merely being proportional to the numbers of sperm inseminated by multiple males (the "raffle" scenario)? These questions have been addressed using genetic markers for numerous animal species (see the reviews in Birkhead and Møller, 1992; Smith, 1984). In insects, it is frequently the case [but not always (Laidlaw and Page, 1984)] that the last male to mate with a female sires most of the offspring (Fig. 5.7). For example, in the bushcricket *Poecilimon veluchianus,* Achmann et al. (1992) showed by DNA fingerprinting that the last-mated male achieves more

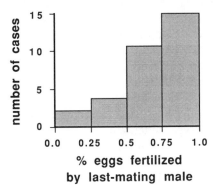

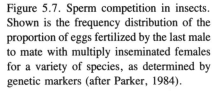

Figure 5.7. Sperm competition in insects. Shown is the frequency distribution of the proportion of eggs fertilized by the last male to mate with multiply inseminated females for a variety of species, as determined by genetic markers (after Parker, 1984).

than 90% of the fertilizations. Mating in this species involves transfer of a large spermatophore to a female, who often copulates with several males and then may eat some spermatophores after copulation. The genetic findings appeared to eliminate the possibility that nourishment gained by a female from the spermatophore "gift" of an early-mating male reflects a paternal investment strategy enhancing that male's fitness.

The term "sperm displacement" conventionally has been employed to describe the enhanced reproductive success exhibited by last-mating males. In some insects such as the locust (*Locusta migratoria*), an active "sperm flushing" process indeed has been observed that probably contributes to the phenomenon (Parker, 1984); in the dunnock sparrow, males before copulating peck at the cloaca of a female, apparently causing her to eject sperm from previous matings (Birkhead and Møller, 1992). However, in most cases the mechanisms of sperm displacement either remain unknown, or appear less active. In *Drosophila,* for example, there is no known evidence for any substance in the male's ejaculate that incapacitates the sperm of competitor males. In chickens and ducks, semen from different inseminations is stored in separate layers in the female, with the most recent contribution remaining on top and, therefore, perhaps most likely to fertilize the next available egg (McKinney et al., 1984). For such cases, more neutral terms such as "sperm predominance" (Gromko et al., 1984) or "sperm precedence" are to be preferred.

In some birds, both raffle competition and sperm precedence are known to operate, but over different time scales. If inseminations occur more than about 4 hours apart, then last-male sperm precedence tends to operate, but a sperm raffle characterizes the process when two males inseminate a female in rapid succession (Birkhead and Møller, 1992).

In a few insects (Fig. 5.7) and other animal species, *first*-mating males appear to have the fertilization advantage. For example, in the intertidal copepod *Tigriopus californicus,* allozyme studies showed that virtually all of a female's progeny are fathered by her first mate (Burton, 1985). In this species, a male often clasps a female for a period of several days before her sexual maturation. In light of the genetic observations, Burton (1985) interpreted this prolonged clasping behavior by males as a precopulatory mate-guarding strategy to assure that a potential mate has not been inseminated previously.

In the relatively asocial ground squirrel *Spermophilus tridecemlineatus,* synchronously breeding females are scattered spatially at low densities. As a consequence of this natural history, the mating system probably conforms to what has been labeled "scramble-competition polygyny," and indeed, behavioral observations suggest that the strongest phenotypic correlate of male mating success is male mobility during the breeding season, with the attendant increased likelihood of encountering females in estrus (Schwagmeyer, 1988). Through use of allozyme markers, Foltz and Schwagmeyer (1989) discovered that in wild populations of this species, the first male to copulate with a multiply-mated female sires the majority (on average 75%) of the resulting progeny. Results were interpreted to indicate that the mating advantage for first males during precopulatory scramble competition also translates into a genetic advantage during the ensuing postcopulatory sperm competition.

A remarkable example of first-male fertilization advantage recently was reported for the spotted sandpiper, *Actitis macularia.* In this polyandrous avian species with strong tendencies for behavioral sex-role reversal, territorial females pair with, defend, and lay clutches for several males. Molecular studies based on DNA fingerprinting showed that males pairing early in the mating season cuckold their females' later mates by means of sperm storage in the females' reproductive tracts (Oring et al., 1992). Thus, not only does an early-pairing male have a greater confidence of paternity, but he thereby also appropriates the reproductive efforts of subsequent males toward enhancement of his own fitness!

An intriguing hypothesis of ''sperm sharing'' has been advanced for some species of hermaphroditic freshwater snails (Monteiro et al., 1984). Under this suggestion, a snail might pass on the sperm from a previous mate to another partner, such that the transmitting individual acts mechanically as a male but achieves no genetic contribution to progeny. However, an empirical test of this hypothesis based on allozyme markers failed to support the sperm-sharing hypothesis (Rollinson et al., 1989). Instead, hermaphroditic snails proved capable of passing on their own sperm while still producing eggs fertilized by sperm received from an earlier mating.

In plants, opportunities also exist for competition among male gametes from different donors, as, for example, via differing rates of pollen tube growth through stigmatic tissue toward the egg (Snow, 1990). Using allozyme markers to establish paternity, Marshall and Ellstrand (1985) demonstrated that most of the seeds in multiply-sired fruits of the wild radish (*Raphanus sativus*) resulted from the first in a series of sequential pollen donors. Further study revealed that microgametophyte competition among several pollen donors appeared more pronounced than among male gametophytes produced from a single pollen source (Marshall and Ellstrand, 1986). In the morning glory *Ipomoea purpurea,* similar allozymic analyses also revealed a strong fertilization advantage for first-pollinating males, even when pollen donations from a second source occurred immediately after the first (Epperson and Clegg, 1987). In paternity studies of yet another herbaceous plant *Hibiscus moscheutos,* allozyme markers revealed that individuals with fast-growing pollen tubes sired a disproportionate number of seeds following mixed experimental pollinations (Snow and Spria, 1991).

Maternity Analysis

Maternity also is sometimes in question. Indeed, one of the first applications of DNA fingerprinting involved human maternity assessment in an immigration dispute (Jeffreys et al., 1985c). The case concerned a Ghanaian boy born in the United Kingdom who left to join his father and subsequently returned alone to be reunited with his mother. However, there was suspicion that a substitution had occurred, either for an unrelated boy or for a son of one of the mother's several sisters living in Ghana. At request of the family's solicitor, a DNA-fingerprint analysis was conducted of the boy, his putative mother, and several of the child's undisputed siblings (the task was complicated because the boy's father was uncertain). The first step involved deducing paternal-specific bands in the boy's DNA fingerprint. These were DNA fragments present in at least one of the siblings but absent from the mother. The second step involved subtracting these paternal-specific bands from the boy's DNA fingerprint. All 40 remaining fragments matched those present in the woman, indicating that she indeed was the child's biological mother. These genetic data were provided to the immigration authorities who then dropped objections and granted the boy residence in the United Kingdom.

Tamarin et al. (1983) introduced an ingenious method for maternity assignment in small mammals that, when used in conjunction with genetic markers, offers special promise for assessing parentage in natural populations. The approach involves injection of pregnant or lactating females with unique combinations of gamma-emitting radionuclides (e.g., ^{58}Co, ^{85}Sr, ^{65}Zn) that transfer to progeny via the placenta or the mother's milk. The isotopic profiles of the young are determined spectrophotometrically and are matched against those of prospective mothers to establish maternity (a caveat could apply to the analysis if mothers sometimes nurse offspring other than their own). Sheridan and Tamarin (1986) combined this method of maternity assignment with protein-electrophoretic analyses to assess parentage in 40 offspring from a natural population of meadow voles (*Microtus pennsylvanicus*). Knowledge of maternity facilitated the paternity analyses and led to the conclusion that about 38% of the adult males in the population bred successfully in the surveyed time period, fathering at most two litters each.

It might be supposed that rapidly-evolving mtDNA molecules would provide ideal genetic markers of maternity, and indeed they do with regard to extended female lineages within a species (Chapter 6). However, although mtDNA has proved sufficiently variable within some species to provide "fingerprints" that distinguish most individuals under conventional RFLP assays (Avise et al., 1989), no explicit applications of mtDNA to questions of single-generation maternity seem to have been published (however, see Kessler and Avise, 1985). In practice, the multilocus genotypic variability generated by recombination among nuclear genes far surpasses that anticipated even for the rapidly mutating but nonrecombining mtDNA. Thus, genetic approaches that employ multilocus allozyme or nuclear DNA-fingerprinting assays have provided most of the opportunities for maternity assessment, as illustrated by the following examples.

Each spring, pregnant females of the Mexican free-tailed bat (*Tadarida brasiliensis*) migrate to caves in the American Southwest and form colonies, often of several million individuals. Most females produce single pups that within hours of birth are deposited on the cave ceilings or walls in dense creches. Lactating females return to the creches and nurse pups twice each day. Traditional thought was that nursing must be indiscriminate such that the mothers act "as one large dairy herd delivering milk passively to the first aggressive customers" (Davis et al., 1962). However, McCracken (1984) challenged this view with protein-electrophoretic evidence indicating that nursing was selective along genetic lines. This conclusion stemmed from comparing observed allozyme genotypes in female-pup nursing pairs with the expected frequencies of such genotypic combinations if nursing were random. A highly significant deficit of maternal genetic exclusions (relative to expectations from population genotype frequencies) indicated selective nursing by females of their own pups (or at least of pups with similar genotypes). McCracken estimated that only 17% of the assayed females were nursing pups that could not be their offspring.

VandeBerg et al. (1990) employed protein-electrophoretic markers to validate pedigrees in captive squirrel monkeys (genus *Saimiri*). Among 89 progeny for which parentage had been inferred from behavioral observations, assignments for 7 individuals proved incorrect, and retrospective examination of colony records in conjunction with further genetic typing permitted a correction of the pedigree records. Five of the errors had involved cases of mistaken paternity, but, interestingly, two involved mistaken maternity. These latter cases apparently were the consequence of infant swapping between dams shortly after birth, an "allomaternal" behavior that previously had gone unrecognized.

More commonly, questions about maternity arise in oviparous animals such as birds, fishes, and insects, where prolonged care of eggs outside the female's body opens possibilities for intraspecific brood parasitism or other means of egg or progeny mixing. In birds, traditional methods for inferring IBP include monitoring nests for supernormal clutch sizes, noticing the appearance of eggs deposited outside the normal laying sequence of the resident female, or detecting intraclutch differences in the physical appearance of eggs in those species where interclutch differences in egg patterning are pronounced. Recent approaches have involved more direct maternity assessment via molecular-genetic markers.

The finding of frequent intraspecific brood parasitism in snow geese (Box 5.4) already has been mentioned, and additional examples of IBP in avian species are presented in Table 5.4. For example, in wild zebra finches (*Taeniopygia guttata*), a DNA-fingerprint examination of 92 offspring from 25 families revealed that about 11% of offspring and 36% of broods resulted from IBP and that the average number of parasitic eggs per clutch was greater than 1.0 (Birkhead et al., 1990). In house wrens (*Troglodytes aedon*), a similar genetic study based on allozymes led to the conclusion that as many as 30% of chicks were produced by females other than the nest attendant (Price et al., 1989). Such unexpectedly high frequencies of IBP in avian species as revealed by molecular markers (and by more traditional approaches) have rekindled great interest in this remarkable "egg-dumping" behavior by females (Petrie and Møller, 1991).

SUMMARY

1. Qualitative markers from highly polymorphic molecular systems provide powerful tools for assessing genetic identity versus nonidentity, and parentage (maternity and paternity).

2. In human forensics, genetic typing by RFLP assays of VNTR loci provides the late-20th century analogue of traditional fingerprinting and has found wide application in civil litigations and criminal cases. Conservative procedures for calculating probabilities of a genotypic match can serve to ameliorate any potential biases against the defense.

3. For plant and animal species that are known or suspected of reproducing by asexual (clonal) as well as sexual means, a variety of qualitative molecular markers has been used to assess the reproductive system, to describe the spatial distributions of particular genets (all clonal descendants from a single zygote), and to estimate the ages of clonal lineages. Some clones have proved to be unexpectedly old.

4. In several microorganisms including various bacteria and eukaryotic protozoans, molecular markers have revealed unexpectedly strong proclivities for clonal reproduction. These findings are of medical as well as academic interest because they may influence strategies for diagnosis of disease agents and for development of vaccines and curative drugs.

5. Molecular markers have found application in identifying genetic mosaics in nature and in determining gender, where these features may not be obvious from external phenotypes alone.

6. Molecular assessments of genetic parentage involve linking an offspring to its biological mother and/or father when the parents are uncertain from other evidence. Methods of analysis are influenced by whether paternity, maternity, or both are uncertain and by the size of the pool of candidate parents.

7. Single-female clutches or broods in many species, including those formerly thought to be monogamous, often have proved through molecular analysis to consist of multiply-sired young. The widespread occurrence of multiple pa-

ternity raises the possibility of sperm or pollen competition, a phenomenon studied extensively through molecular markers.

    8.   Multiple maternity of broods or clutches also has been documented by molecular markers in several species. In birds, for example, this results from the previously under-appreciated phenomenon of intraspecific brood parasitism, or "egg-dumping."

    9.   Molecular studies of maternity and paternity are most informative and exciting when coupled with observations on behavior or natural history and interpreted within the context of behavioral–ecologic theory.

6

Kinship and
Intraspecific Phylogeny

*. . . community of descent is the hidden bond which natu-
ralists have been unconsciously seeking.*

C. Darwin, 1859

Clonal identity and parentage (Chapter 5) are extreme examples of close kinship,
but now we shall be concerned with genetic relatedness within and among
broader groups of extended kin. Questions of genetic relatedness arise in virtually all discussions of social species where particular morphologies and behaviors might have evolved as predicted under theories of kin selection and inclusive
fitness (Box 6.1). Interest in kinship also arises for any species whose populations are structured spatially, perhaps along family lines. At increasingly greater
depths in time, *all* conspecific individuals are related genealogically through an
extended pedigree that constitutes the intraspecific phylogeny of a species.

CLOSE KINSHIP AND FAMILY STRUCTURE

Procedures for the molecular-genetic assessment of close kinship also require
qualitative genetic markers with known transmission properties, such as allozymes or scnRFLPs. However, unlike the straightforward situation in paternity
and maternity assessment, where the single genetic pathway connecting an offspring with each of its biological parents permits explicit predictions about a

Box 6.1. Genetic Relatedness Within Groups, Inclusive Fitness, and Kin Selection

Genetic relatedness. In discussions of close kinship, it is useful to have available a quantitative measure of genetic relatedness (r) between individuals. An intuitive interpretation of a coefficient of relatedness is provided by an answer to the following question: What is the probability that an allele carried by the focal individual also is possessed by the relative in question? Or in other words, what is the expected proportion of alleles shared between the genomes of these individuals? In normal diploid species, $r = 1/2$ for full siblings and parent-offspring pairs (Fig. 6.1), $r = 1/4$ for half-sibs or for an individual with his uncles, aunts, grandparents, and grandchildren, $r = 1/8$ for first cousins, $r = 0$ for nonrelatives, etc. Generally, for any known pedigree, values of r can be determined directly by such pathway analyses (Cannings and Thompson, 1981; Michod and Anderson, 1979).

In the more usual situations in nature where pedigrees are unknown, several statistical methods have been developed and refined for estimating mean coefficients of relatedness among group members from polymorphic genetic markers such as those provided by allozymes (Crozier et al., 1984; Pamilo and Crozier, 1982; Queller and Goodnight, 1989). For example, Pamilo (1984a) derived an estimate for r that can be expressed in terms of heterozygosities observed at a locus ($h_{obs,m}$) and those expected under Hardy-Weinberg equilibrium ($h_{exp,m}$) within a colony m with N_m individuals, in comparison to the heterozygosities observed (H_{obs}) and expected (H_{exp}) within the broader population composed of c colonies:

$$r = \frac{H_{exp} - (1/c)\Sigma\, h_{exp,m} - (1/c)\Sigma[(1/N_m - 1)]\,[h_{exp,m} - (1/2)\,h_{obs,m}]}{H_{exp} - (1/2)\,H_{obs}} \tag{6.1}$$

This coefficient of relatedness may be interpreted as a genotypic correlation among group members in a subdivided population [see Pamilo (1984a) for derivations and discussion].

Inclusive fitness and kin selection. Classic genetic fitness usually is defined as the average direct reproductive success of an individual possessing a specified genotype in comparison to that of other individuals in the population. Inclusive fitness, which entails a broader view of the transmission of genotypes across generations, incorporates both the individual's personal or classic fitness as well as the probability that the individual's genotype in question may be passed on through relatives. These latter transmission probabilities are influenced by the coefficients of relatedness involved. Concepts of inclusive fitness have been advanced as an explanation for the evolution of "self-sacrificial" behaviors, wherein alleles influencing such altruism may have spread in populations under the influence of kin selection. According to Hamilton's (1964) rule, a behavior is favored by selection whenever

$$\Delta w_x + \Sigma r_{yx}\Delta w_y > 0, \tag{6.2}$$

where Δw_x is the change the behavior causes in the individual's fitness, Δw_y is the change the behavior causes in the relative's fitness, and r_{yx} is the genetic relatedness of the individuals involved. In general, under Hamilton's rule, an allele will tend to increase in frequency if the ratio of the cost C that it entails (loss in expected personal reproduction through self-sacrificial behavior) to the benefit B that it receives (through increased reproduction of relatives) is less than r:

$$C / B < r. \tag{6.3}$$

Under the proverbial example of altruistic behavior, an individual's alleles would, thus, tend to increase in frequency if personal fitness was sacrificed for a comparable gain in personal fitness by more than two full sibs, four half-sibs, or eight first cousins!

circumscribed set of expected genotypes in these individuals, most extended kinship assessments are complicated by the fact that numerous alternative transmission pathways potentially link more distant relatives. Thus, in genetic studies of broader kinship, particularly those based on small numbers of loci, the focus typically shifts from attempts to enumerate relationships among particular individuals (but see later section on noneusocial colonies and groups and also Queller and Goodnight, 1989) to a concern with probability patterns of mean genetic relatedness within groups.

The general types of concepts and reasoning that are involved in kinship assessment may be introduced qualitatively by the following example (from Avise and Shapiro, 1986). Juveniles of the serranid reef fish *Anthias squamipinnis* occur in social aggregations ranging in size from a few individuals to more than 100. Although eggs and larvae of this species are pelagic, drifting in the open ocean, Shapiro (1983) raised the intriguing hypothesis that juvenile cohorts might consist of close genetic relatives (predominantly siblings from a single spawn) that had stayed together through the pelagic phase and settled jointly. If true, kin selection would have to be considered as a factor potentially influencing behaviors within the social groups and, furthermore, marine biologists would have to reevaluate the conventional wisdom that the products of separate spawns are mixed thoroughly during the pelagic phase. To critically test the Shapiro hypothesis, genotypes were surveyed at each of three polymorphic allozyme loci in eight discrete social cohorts of juvenile *A. squamipinnis* from a single reef in the Red Sea. Allelic frequencies are presented in Box 6.2.

The hypothesis of close genetic affinity within *A. squamipinnis* cohorts yields several genetic predictions that were not borne out by the allozyme data. For example, if progeny within a cohort were full sibs, they should (a) exhibit at most four alleles at a locus (but instead several cohorts displayed more than four alleles at the *Pgm* and *Sod* loci), (b) exhibit alleles in frequencies consistent with transmission primarily from only two parents (but instead allele frequencies departed dramatically from these expectations, and the best goodnesses of fit to the empirical data required a *minimum* of 11–16 equally contributing parents per cohort), (c) share any rare alleles that were observed in the study (but instead all rare alleles present in more than one individual were distributed across cohorts), (d) often exhibit strong heterozygote excess relative to expectations from within-cohort allele frequencies, as, for example, in a mating between alternate homozygotes or between a homozygote and a heterozygote (but instead no significant excesses were revealed), and (e) exhibit genotypic relatedness values [r (Box 6.1)] close to 0.5 (but instead estimates of r were invariably near zero). In addition, if *A. squamipinnis* cohorts existed primarily as full-sib assemblages, there should be (f) large between-cohort variances in allele frequency (but observed variances were close to zero), and (g) large differences among cohorts in

Box 6.2. Observed Allele Frequencies at Three Allozyme Loci in Eight Local Cohorts of the Reef Fish *Anthias squamipinnis* (after Avise and Shapiro, 1986).

		Cohort								
Gene	Allele	A	B	C	D	E	F	G	H	Total
Ldh	2n =	124	140	128	158	152	140	134	106	1,082
	a	0.613	0.679	0.680	0.620	0.605	0.643	0.649	0.613	0.638
	b	0.379	0.314	0.320	0.380	0.395	0.350	0.351	0.377	0.358
	c	0.008	—	—	—	—	0.007	—	0.010	0.003
	d	—	0.007	—	—	—	—	—	—	0.001
Pgm	2n =	124	140	128	162	156	140	162	106	1,118
	a	0.944	0.936	0.937	0.938	0.904	0.957	0.926	0.943	0.935
	b	0.048	0.043	0.016	0.050	0.070	0.022	0.049	0.028	0.042
	c	0.088	0.021	0.031	0.012	0.026	0.021	0.025	0.019	0.020
	d	—	—	0.008	—	—	—	—	—	0.001
	e	—	—	0.008	—	—	—	—	0.010	0.002
Sod	2n =	124	140	128	162	156	140	162	106	1,118
	a	0.758	0.707	0.734	0.729	0.679	0.722	0.698	0.792	0.723
	b	0.113	0.150	0.180	0.148	0.211	0.164	0.191	0.142	0.164
	c	0.089	0.100	0.070	0.080	0.071	0.057	0.087	0.047	0.076
	d	0.040	0.036	0.008	0.037	0.026	0.036	0.012	0.019	0.027
	e	—	—	0.008	—	0.013	0.021	0.012	—	0.007
	f	—	0.007	—	—	—	—	—	—	0.001
	g	—	—	—	0.006	—	—	—	—	0.001

Within-cohort genetic relatedness based on the application of Eq. (6.1) to the three loci yields estimates of $r = -0.01, 0.01$, and 0.02, hence providing strong evidence against the proposition that cohorts consist of close kin (see text).

single-locus and multilocus genotypic frequencies (but no such differences were observed). Overall, these several lines of reasoning applied to the genetic data demonstrated that the assayed cohorts of *A. squamipinnis* do not consist exclusively or predominantly of close genetic relatives, but rather represent nearly random samples of progeny from many matings.

In general, from such molecular data, estimates of the average genetic relatedness within groups can be quantified by approaches such as outlined in Box 6.1. Estimated values of r then may be compared against theoretical expectations for organisms of known pedigree relationship, and thereby used as a genealogical backdrop against which to evaluate hypotheses regarding social behaviors and the possible influence of kin selection.

Eusocial Colonies

Species considered eusocial consist of groups of individuals possessing the following traits (Wilson, 1971): cooperation in the caring for young; reproductive division of labor, with more-or-less sterile individuals working on behalf of the reproductives; and overlapping generations of colony workers. Eusociality long has intrigued biologists because its evolution entails a remarkable transition to sterility of worker individuals—a most extreme form of reproductive altruism. The epitome of eusociality has been reached in colonial hymenopteran insects, including nearly all ants and the highly social bees and wasps, where sterile workers (who are all females) provide exaggerated examples of "helpers at the nest."

Hamilton (1964) first proposed that the evolution of such extreme reproductive altruism might have been facilitated by the altered and asymmetric genetic relationship among parents and siblings stemming from haplo-diploid sex determination. In these hymenopteran species, males develop from unfertilized eggs and hence are haploid for a single set of chromosomes inherited from their mother. Females develop from fertilized eggs and are diploid. Thus, daughters of a monogamous female share three-quarters of their alleles [all alleles from their father and half from their mother (Fig. 6.1B)], a situation that contrasts with expected relationships in normal diploid species where full sibs share only one-half of their alleles. Thus, according to Hamilton's insight, an ancestral hymenopteran with any behavioral predisposition toward rearing sisters, even at the risk of producing fewer young, might have increased her inclusive fitness by adopting this behavioral strategy. In other words, if the genetic ties within a generation are closer that those between generations, an individual might profit under the currency of inclusive fitness by investing in a parent's reproductive success rather than her own.

However, this prediction weakens considerably when the reproductive females (queens) are multiply mated (polyandry, in this context) and/or when multiple queens lay eggs in a nest (polygyny) because mean genetic relatedness among colony workers then is lowered dramatically [by a magnitude depending on the numbers of parents, their genetic relationships to one another, and their contributions to progeny (Fig. 6.1; Wade, 1982; Wilson, 1971)]. Empirical surveys using allozyme markers have been conducted on a variety of social hymenopterans to assess mean within-colony genetic relatedness (Table 6.1). For some species (such as *Rhytidoponera* ants), average relatedness among colony workers, indeed, has proved to be near the value of $r = 0.75$ expected when a singly-inseminated queen founds a monogynous nest. However, genetic relatedness estimates for other species have proved to be significantly lower, often far below the critical threshold of $r = 0.50$ where the original supposed genetic advantage of reproductive altruism by workers would seem to be absent. Such

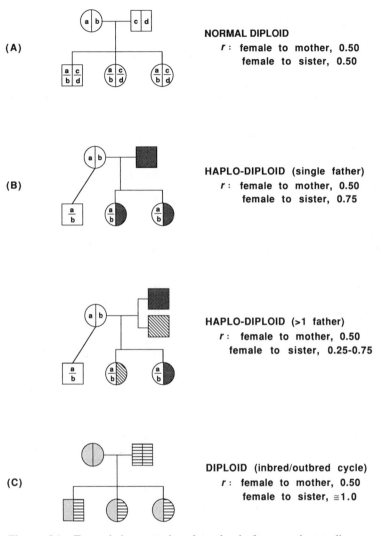

(A)

NORMAL DIPLOID
r: female to mother, 0.50
   female to sister, 0.50

(B)

HAPLO-DIPLOID (single father)
r: female to mother, 0.50
   female to sister, 0.75

HAPLO-DIPLOID (>1 father)
r: female to mother, 0.50
   female to sister, 0.25-0.75

(C)

DIPLOID (inbred/outbred cycle)
r: female to mother, 0.50
   female to sister, ≅1.0

Figure 6.1. Transmission genetics through single-generation pedigrees. Shown are expected patterns under rules of (A) normal diploid inheritance, (B) haplo-diploidy as in eusocial hymenopteran insects (with and without multiple insemination of queens), and (C) diploidy under an extreme inbreeding-outbreeding cycle (see text). Circles (females) and squares (males) bisected by vertical diameter lines represent diploid individuals; males without such lines are haploid. Also shown are mean coefficients of genetic relatedness (r) expected among females under these various reproductive modes. In the case of haplo-diploidy with more than one father (polyandry), values of r depend on the number of fathers and their proportionate contributions to the worker population of a colony. For example, if all of n males contribute equally, then the average coefficient of genetic relatedness becomes $r = 0.5[0.5 + 1/n]$; in general, mean $r = 0.5[0.5 + \Sigma f_i^2]$, where f_i is the proportionate contribution by the ith male. Reproduction by multiple queens within a colony (polygyny) would serve further to lower the mean r.

Table 6.1. Examples of coefficients of genetic relatedness (r) estimated among females within colonies of various eusocial hymenopteran insects. All estimates stem from protein-electrophoretic analyses.[a]

Species	Comparison	r	Comments on colonies and queens[b]	Authority
ANTS				
Rhytidoponera chalybaea (A)	Workers	0.76	Monogynous, monoandrous	Ward, 1983
Rhytidoponera confusa (A)	Workers	0.70	Monogynous, monoandrous	Ward, 1983
Nothomyrmecia macrops	Workers	0.17	Occasionally polygynous	Ward and Taylor, 1981
Myrmecia pilosula	Workers	0.17	Polygynous	Craig and Crozier, 1979
Myrmica rubra	Workers	0.02–0.54	Polygynous	Pearson, 1983
Solenopsis invicta	Workers	0.01–0.08	Polygynous, queens singly-mated	Ross and Fletcher, 1985
Formica aquilonia (Espoo)	Workers	0.09	Polygynous	Pamilo, 1982
Formica polyctena	Workers	0.19–0.30	Polygynous	Pamilo, 1982
Formica sanguinea	Workers	0.31–0.42	Polygnous, queens multiply-mated	Pamilo and Varvio-Aho, 1979
Formica transkaucasica	Workers	0.33	Polygynous, queens singly-mated	Pamilo, 1981, 1982
WASPS				
Agelaia multipicta	Workers	0.27	Polygynous	West-Eberhard, 1990
Parachartergus colobopterus	Females[c]	0.11	Probably polygynous	Queller et al., 1988
Polybia occidentalis	Females[c]	0.34	Polygynous	Queller et al., 1988
Polybia sericea	Females[c]	0.28	Polygynous	Queller et al., 1988
Cerceris antipodes	Females[c]	0.25–0.64	Polygynous	McCorquodale, 1988
Microstigmus comes	Females[c]	0.60–0.70	Monogynous or polygynous, queens often singly-mated	Ross and Matthews, 1989a, 1989b
BEES				
Apis mellifera	Workers	0.25–0.34	Highly polyandrous	Laidlaw and Page, 1984

[a]Additional examples may be found in Hölldobler and Wilson (1990) and Ross and Carpenter (1991).

[b]In most cases, the mating system is not well known.

[c]Reproductive and nonreproductive females not distinguished.

genetic results, coupled with field observations, have revealed that queens often are multiply-inseminated, and furthermore that polygyny is common (see the reviews in Cole, 1983; Hölldobler and Wilson, 1990; Ross and Carpenter, 1991). These findings have raised a conundrum about the evolution of insect eusociality: Because cogenerational individuals within a nest are less closely related to one another than they would be to their own sons or daughters, "if kin selection is a powerful force, what prevents evolution from leading to a more competitive state in which the workers (who have ovaries) try to take over reproduction?" (Hölldobler and Wilson, 1990).

One possibility is that a colony is divided into cliques of closely related individuals who are able to distinguish close kin from more distant kin, and direct altruistic behaviors accordingly. Experiments indicate that kin discrimination is possible for some hymenopteran species but not others (Hölldobler and Wilson, 1990). For example, worker honey bees (*Apis mellifera*) are able to "assess" their relatedness to other individuals and preferentially rear queens from larvae that are related most closely (Visscher, 1986), whereas workers of the ant *Rhytidoponera confusa* appear to lack such abilities (Crosland, 1988).

Another suggestion is that the high levels of polygyny and polyandry observed represent derived behaviors rather than the ancestral conditions under which eusociality evolved. Under this hypothesis, eusociality tends to arise through kin selection when populations are highly structured along family lines, whereas subsequent maintenance and elaboration into advanced eusociality can occur even when within-colony relatedness decreases due to polygyny and polyandry. Eusocial colonies once formed may operate so smoothly and successfully that the inclusive fitness of workers remains higher than if workers became egg-layers, such that reversion to a less eusocial condition simply is infeasible. In ants, it is difficult to test the hypothesis that polygyny and polyandry are derived conditions because most species are strongly eusocial. In the primitively eusocial bee *Lasioglossum zephyrum,* an empirical molecular genetic estimate of $r = 0.7$ is sufficiently high to indicate that kin selection could operate in this species (Crozier et al., 1987). However, in several species of primitively eusocial polistine wasps, r values within colonies sometimes have proved to be moderate or low (Strassmann et al., 1989). Although it is not certain that these species provide valid representations of the ancestral behavioral condition, the findings do demonstrate that low within-colony relatedness is not confined to the advanced hymenopteran societies.

Finally, various ecological-genetic hypotheses have been advanced to explain the conundrum involving low within-colony r. For example, genetic variability among nestmates might diminish susceptibility to infectious parasites (Shykoff and Schmid-Hempel, 1991a, 1991b) or promote broader tolerance to variable environments; caste determination might have a partial genetic basis for which polyandry or polygyny conceivably allow fuller expression (Crozier and Page,

1985); or, perhaps, collaborating queens fare proportionately better than individual queens in competition for limited nest sites (Herbers, 1986). Under this latter hypothesis, inclusive fitness concepts could remain in partial effect if such cofoundresses are genetic relatives, as sometimes (but not always) appears to be the case: In various wasp, bee, and ant species, molecular-genetic appraisals of cofounding queens have revealed mean relatedness values ranging from $r \approx 0.0 - 0.7$ (Metcalf and Whitt, 1977; Ross and Fletcher, 1985; Schwartz, 1987; Stille et al., 1991; Strassmann et al., 1989).

In any event, the altered coefficients of genetic relatedness stemming from haplo-diploidy cannot provide a universal or complete explanation for the evolution of eusociality because (a) other haploid-diploid arthropod species outside the Hymenoptera (e.g. some mites, thrip insects, and beetles) do not exhibit eusociality and (b) some diploid species (notably termites) do. In discussing the evolution of termite eusociality, Syren and Luykx (1977) and Lacy (1980) noted that several termite species possess sex-linked multichromosome translocation complexes that serve to elevate genetic relatedness both between sisters and between brothers. However, the low genetic relatedness between male and female siblings under these translocation systems remains difficult to accommodate with the evolution of termite eusociality (Andersson, 1984; Leinaas, 1983). Another proposed model that could influence inclusive fitness by altering genetic relatedness within and among groups involves cyclic inbreeding-outbreeding (Bartz, 1979; see also Pamilo, 1984b and Williams and Williams, 1957). When male and female mates are unrelated but are each the product of intense inbreeding, their offspring can be nearly identical genetically but only 50% like either parent (Fig. 6.1C). When such conditions hold, any behavioral predispositions of siblings to stay together and to assist parents in rearing the young might be favored for the same reasons of inclusive fitness as set forth above for the haplo-diploid hymenopterans (however, see Crozier and Luykx, 1985). Termites possess several natural history features that favor close social interactions and might set the stage for such a breeding cycle, such as living in protected and contained nests conducive to multigenerational inbreeding and passing symbiotic intestinal flagellates from old to young individuals by anal feeding (an arrangement that necessitates at least some degree of social behavior!) (Wilson, 1971).

A remarkable mammalian analogue of the eusocial system of termites has been discovered in the naked mole-rat, *Heterocephalus glaber* (Jarvis, 1981; Sherman et al., 1991). Brood care and other duties in this colonial underground rodent are performed cooperatively by mostly nonreproductive workers or helpers, who represent the young from previous litters and who assist the queen in rearing progeny that are fathered by a few select males within the burrow system. Are colony mates especially close genetically, such that kin selection might plausibly account for the extreme reproductive selflessness displayed by subordinate individuals? Using DNA-fingerprint assays, Reeve et al. (1990) observed remarkably high probabilities (0.88–0.99) of DNA band-sharing within colonies (similar to estimated levels of band-sharing in DNA fingerprints of highly inbred mice or monozygotic twins in cows and humans). From these molecular data, they estimated a mean coefficient of genetic relatedness significantly greater than 0.50 ($r = 0.81 \pm 0.10$) and concluded that more than 80% of the matings within a colony likely were among siblings or between parents and offspring. Results appear consistent with intense within-colony inbreeding and, hence, leave open the possibility of an influencing role for kinship and inclusive fitness in the evolution of *H. glaber* eusociality. Nonetheless, behavioral and life history characteristics no doubt are important as well, as are phylogenetic constraints, as indicated by the differences among the eight African mole-rat species in the degree to which colonial and eusocial behavior is exhibited (Allard and Honeycutt, 1992; Honeycutt, 1992).

Noneusocial Colonies and Groups

Most group-living species exhibit far less social organization and subdivision of labor than do the eusocial hymenopterans and mole-rats, but genetic relatedness among group members remains of interest. For example, eastern tent caterpillars (*Malacosoma americanum*) are characterized by cooperative nest (tent) building, as well as cooperative foraging along pheromonal trails. The adult moths of this diploid species lay egg masses in trees, from which the first-instar larvae emerge to feed on young leaves at the twig tips. Later, caterpillars move to centralized locations to initiate tent construction. In a temporal genetic study using allozymes, Costa and Ross (1993) found that the mean genetic relatedness within colonies of newly-emerged larvae (from a single egg mass) was $r = 0.49$, not significantly different from the expected value of $r = 0.50$ for full siblings. However, relatedness values declined over the next eight weeks, to $r = 0.38$ (or to $r = 0.25$ if colonies on single-tent trees were disregarded). The temporal reduction in intracolony relatedness represents an erosion of the initial simple family structure, apparently due to the exchange of individuals among colonies of a tree when foragers encounter pheromonal trails of nonsiblings. Results indicate that foreigners (nonsiblings) are not overtly discriminated against by these caterpillars, but rather can be accepted into a colony.

Day-roosting colonies of the bat *Phyllostomus hastatus* in Trinidad are subdivided into compact clusters of adult females that remain highly stable over several years and are attended by a single adult male who from allozyme evidence fathers most of the babies born to females within the harem (McCracken and Bradbury, 1981). The stable groups of adult females are the fundamental units of social structure in this species and it has been hypothesized that they arise from active cooperative interactions among females on shared foraging grounds. Are the harems composed of matrilineally related individuals such that kin selection should be considered as a possible factor influencing social or cooperative behavior? Using allozyme assays, McCracken and Bradbury (1977, 1981) demonstrated that females within a harem are unrelated and represent random samples from the total adult population. Combined with field observations, results indicate that juveniles are not recruited into parental social units and that contemporary kin selection, therefore, cannot explain the maintenance of behavioral cohesiveness in these highly social mammals.

In some other mammalian species, females comprising a group or colony *are* close relatives. For example, black-tailed prairie dogs (*Cynomys ludovicianus*) live in social groups called coteries that typically consist of one or two adult males born in foreign coteries and of several adult females and young that are closely related due to a strong tendency by females to remain within the natal coterie for life (Hoogland and Foltz, 1982). Genetic analyses (based on pedigree and allozyme data) verified that despite the matrilineal population structure, colonies are outbred due to coterie-switching by males and the avoidance of male-daughter matings (Foltz and Hoogland, 1981, 1983; see also Chesser, 1983). Several other ground-dwelling squirrels (in the family Sciuridae) also have varying degrees of social organization built around matrilineal kinship (Michener, 1983). Evidence on kinship in many other group-living animal species is reviewed by Wilson (1975). Traditionally, such genealogical understanding has come from difficult and labor-intensive field observations of mating and dispersal (Fletcher and Michener, 1987), but as in the black-tailed prairie dog, in several cases molecular markers have been employed as additional sources of information.

For example, DNA-fingerprinting assays have been applied to prides of African lions that from prior field observations were thought to exist as matriarchal groups (Gilbert et al., 1991; Packer et al., 1991). A lion pride typically contains two to nine adult females, their dependent young, and a coalition of two to six adult males that have joined the pride from elsewhere. Incoming males collaborate to evict resident males and often kill the dependent young from the prior coalition. Based on observed proportions of bands shared in multilocus DNA fingerprints of nearly 200 animals, it was concluded that female companions within prides always are related closely, male coalition partners either are related closely (particularly within larger coalitions) or are unrelated (in some of the smaller coalitions involving only two to three males), and mating partners usually are unrelated (Fig. 6.2). Furthermore, fingerprinting analyses of parentage revealed that resident males fathered all cubs conceived during their tenure and that the variance in male reproductive success increased greatly as coalition size increased. From these observations, the authors concluded that males only act as nonreproductive "helpers" when coalitions are composed of close relatives.

Several points should be made about these important genetic studies. First, unlike most analyses cited earlier that estimated mean relatedness within groups based on genotype frequencies at particular allozyme or other loci, the multilocus

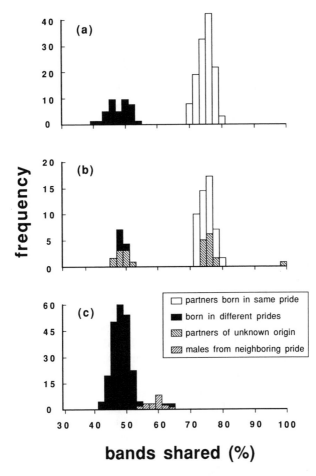

Figure 6.2. Frequency distributions of minisatellite band-sharing in Serengeti lions (after Packer et al., 1991). Band-sharings are indicated between (**a**) females born in the same versus different prides, (**b**) male coalition partners known to have been born in the same versus different (or in some cases unknown) prides, and (**c**) coalition males and resident females.

DNA-fingerprinting approaches allowed assessments of genetic relatedness among particular pairs of individuals. Second, the extent of minisatellite band-sharing initially was calibrated against known or suspected kinship among lions for which long-term field observations and pedigrees were available. Third, such calibrations between band-sharing and *r* proved to be nonlinear and also differed between the two lion populations studied (in the Serengeti National Park and the Ngorongoro Crater). Fourth, the large variances in levels of band-sharing among individuals with a given known *r* meant that for lions of unknown relationship, only the most general of kinship classes (nonrelatives, intermediate relatives, or very close kin) could be inferred. Overall, notwithstanding the considerable genetic insights gained on lion prides, these studies also indicate that the multilocus fingerprinting approach has important limitations for precise estimation of *r* between individuals and normally will require independent background information and population-specific calibration for proper interpretation.

One reason that DNA-fingerprint calibrations may differ among populations, and "unrelated" individuals exhibit varying levels of background band-sharing, could involve earlier populational histories including possible bottlenecks and inbreeding. An extreme example appears to involve the dwarf fox *Urocyon littoralis*, which has colonized several Channel Islands off Southern California within the last 20,000 years. All assayed foxes from one of the smallest and most isolated islands, San Nicolas, exhibited *identical* bands in multilocus DNA fingerprints, and several other island populations also showed greatly enhanced levels of band-sharing (75–95%) relative to foxes from different islands (16–56%) and relative to values typifying outbred populations of many other vertebrate species (10–30%) (Gilbert et al., 1990). Patterns of band-sharing between islands also produced an estimated phylogeny that agreed well with archaeozoological and geological records concerning the history of island colonization for this species. Results indicate that multilocus DNA fingerprinting approaches can in some cases provide useful information on genetic relationships within and among small, isolated, and potentially inbred populations. Nonetheless, attempts to utilize these multilocus fingerprint assays for assessing relationships in larger, outbred populations have met with little success, primarily because the band variability is simply too complex (Baker et al., 1992; Prodöhl et al., 1992). Caution also is indicated because van Pijlen et al. (1991) have found that different DNA-fingerprint probes sometimes can produce grossly different estimates of band-sharing from the same array of populations.

Most cetacean species (whales and dolphins) live in social groups called pods, and these have been the subject of several genetic studies using DNA fingerprinting and other molecular methods (Hoelzel, 1991a). In the long-finned pilot whale (*Globicephala melas*), social groups typically consist of 50–200 animals whose strong herding instincts have been exploited by native peoples to drive entire pods into shallow bays for mass slaughter. Preliminary DNA-fingerprint analyses of tissue samples from several such Faroe Island harvests revealed that adult males are not related closely to adult females within a pod, and furthermore that nearly 90% of fetuses could not have been fathered by a resident male (Amos et al., 1991a, 1991b). From these and additional behavioral observations, the authors conclude that social groups in the pilot whale are built around matrilineal kinship, with considerable interpod genetic exchange mediated by males. Matrilineal relationships within other cetacean species have been examined more directly by mtDNA analyses. For example, among pods of killer whales (*Orcinus orca*) near Vancouver Island, British Columbia, two distinct mtDNA types have been observed that in preliminary assays appear to correspond to a long-recognized behavioral distinction between sympatric groups with fish-hunting versus mammal-hunting social traditions (Hoelzel, 1991b; Hoelzel and Dover, 1991b). On the other hand, in similar mtDNA analyses of groups of spinner dolphins (*Stenella longirostris*), no clustering of matrilines within specific schools or neighboring morphotypes was detected, suggesting significant and recent genetic interchange (Dizon et al., 1991).

Questions concerning kinship within groups also can arise in plants. The whitebark pine (*Pinus albicaulis*) often displays a multistem form, and the stems within a clump have proved to be genetically distinct individuals that are allozymically more similar to one another than to individuals in other clumps (Furnier et al., 1987). This family structure appears to be a direct result of the seed-caching behavior of the primary dispersal agent for the whitebark pine seeds—the Clark's nutcracker bird. Limber pines (*Pinus flexilis*) also frequently exhibit a multiple-trunk growth form that may result either from damage to leader shoots of a single genetic individual or to growth from multiple zygotes (seeds) deposited in caches by birds. From allozymic analyses, nearly 20% of the multitrunk clusters included two to four genetically different individuals and the mean genetic relatedness within these clusters was estimated to be $r = 0.19$, or slightly less than expected for half-sibs (Schuster and Mitton, 1991). The authors note that such grouping of related individuals offers a potential for interactions such as kin selection or sib competition, factors seldom considered for plants.

Kin Recognition

The occurrence in many species of groups of close kin raises additional questions about whether individuals can somehow assess their genetic relatedness to others, and perhaps adjust competitive, cooperative, altruistic, or other behaviors accordingly (Waldman, 1988; Wilson, 1987). In studying such issues, ethologists monitor behavioral interactions among organisms exhibiting varying levels of genetic relatedness. The great majority of such analyses has employed study organisms whose relatedness was known or suspected from direct field observations or from pedigree records in captive settings (Fletcher and Michener, 1987;

Hepper, 1991). However, in a few cases, molecular genetic markers have assisted with the relatedness assignments. For example, in the clusters of limber pines described above, natural fusions or grafts among woody tissues from different trunks commonly are observed. From allozyme data, Schuster and Mitton (1991) discovered that fused trees were related significantly more closely than trees that were unfused. Whether such fusions are adaptively advantageous for the interactants (e.g., though translocation of water and nutrients, or added physical stability) and, hence, may have evolved through kin selection, remains an open issue.

In a free-living population of Belding's ground squirrels (*Spermophilus beldingi*) in California, Holmes and Sherman (1982) employed protein-electrophoretic techniques to distinguish full siblings from maternal half-sibs (resulting from multiple mating). Subsequent behavioral monitoring indicated that full sisters fought significantly less often and aided each other more than did half-sisters. Such nepotism (favoritism shown kin) must require an ability by ground squirrels to judge relatedness. Additional experiments indicated that the proximate cues by which this is accomplished in *S. beldingi* appear to involve physical association during rearing, as well as "phenotypic matching," whereby an individual behaves as if it had compared phenotypic traits (genetically determined) against itself or a nestmate template (Holmes and Sherman, 1982).

Another postulated advantage to kin recognition involves behavioral avoidance of close inbreeding (Hoogland, 1982). Like many amphibians, the American toad (*Bufo americanus*) exhibits site fidelity to natal ponds for breeding and, thus, individuals are likely to encounter siblings as potential mates (Waldman, 1991). Can siblings recognize close kin and avoid incestuous mating? Waldman et al. (1992) monitored mtDNA genotypes in 86 amplexed pairs of toads and found significantly fewer matings between possible siblings (haplotypes shared) than expected from the haplotype frequencies in the local population. From this preliminary genetic evidence, the authors suggest that "siblings recognize and avoid mating with one another." They further suggested that the proximate cues employed might include advertisement vocalizations by males because the resemblance among male calls proved to be positively correlated with genetic relatedness as assessed by band similarities in nuclear DNA fingerprints. Thus, females potentially could employ male vocalizations (and/or other genetically-based clues such as odors) in kinship assessment.

GEOGRAPHIC POPULATION STRUCTURE AND GENE FLOW

Populations of nearly all species, social or otherwise, exhibit at least some degree of genetic differentiation among geographic locales (Ehrlich and Raven, 1969), if for no other reason than because siblings usually tend to begin life spatially near one another and their parents, and mating partners seldom represent random draws from across the geographic range of a species (Turner et al., 1982). In an influential early study of microgeographic population structure based on allozymes, Selander (1970) demonstrated fine-scale spatial clustering of genotypes of house mice (*Mus musculus*) within and among barns on the same farm, apparently due to tribal family structure and genetic drift in small populations. Such genetic structure sometimes exists even in the most improbable of settings. For example, mosquitofish (*Gambusia affinis* and *G. holbrooki*) are abundant and highly dispersive, yet large conspecific samples revealed statistically significant differences in allozyme allele frequencies over a few hundred meters of shoreline in a stream or reservoir (Kennedy et al., 1985, 1986); also, significant temporal variation at a locale was observed over periods of a few weeks to years (McClenaghan et al., 1985). Across broader geographic scales, populations can show additional differentiation due to spatial habitat structure, isolation by distance, or other factors. Thus, further genetic structure in mosquitofish, hierarchically arranged, characterized populations across ponds and streams within a local area, reservoirs within a river drainage, drainages within a region, and regional collections of drainages that house deep genetic differences associated with species-level separations probably dating to the Pleistocene (Scribner and Avise, 1993a; M.H. Smith et al., 1989; Wooten et al., 1988). Molecular analyses of geographic population structure similarly have been conducted on hundreds of animal species at a variety of spatial and temporal scales.

Populations of most plant species also vary in genetic composition, sometimes over microspatial areas of a few kilometers or even meters (Levin, 1979). For example, large populations of the wild wheat *Triticum dicoccoides* show pronounced differences in genetic structure over distances of less than 5 km, due, in part, to limited gene flow and to a self-fertilization reproductive mode (Golenberg, 1989). The grasses *Agrostis tenuis* and *Anthoxanthum odoratum* show sharp clinal variation in several genetically based characters across the meter-wide ecotones between pastures and lead–zinc mines, as a result of strong disruptive selection for heavy-metal tolerance and flowering time (Antonovics and Bradshaw, 1970; McNeilly and Antonovics, 1968). In many species, effective gene flow via pollen and seed dispersal is sufficiently limited that estimates of neighborhood size (the population within which mating is random) commonly include less than a few hundred individuals, occupying areas less than 50 m^2 (Bos et al., 1986; Calahan and Gliddon, 1985; Fenster, 1991; Levin and Kerster, 1971, 1974; Smyth and Hamrick, 1987). As with animal populations, additional genetic structure normally is to be expected over increasing spatial scales.

A continuing challenge is to describe population genetic architectures within species (Box 6.3) and to identify and order the evolutionary forces responsible in particular instances. Broadly speaking, these forces involve migration or gene flow (Box 6.4), random genetic drift, various modes of natural selection, mutational divergence, and the opportunity for genetic recombination mediated by

---

Box 6.3. Descriptive Statistics for Population Structure

Several approaches to the statistical description of population structure are available (Sokal and Oden, 1978a, 1978b; Weir, 1990). For simplicity, we will confine attention to "*F*-statistics" because these have been employed most widely. Wright (1951) introduced a method of describing the genetic population structures of diploid organisms in terms of three *F*-statistics or allelic correlations, F_{IS}, F_{IT}, and F_{ST}, whose theoretical interrelationships can be written as

$$(1 - F_{IT}) = (1 - F_{ST}) (1 - F_{IS}). \qquad (6.4)$$

Wright defined F_{IS} as the correlation between homologous alleles within individuals with reference to the local population, and F_{IT} as the corresponding allelic correlation with reference to the total population. F_{IS} and F_{IT} are often called fixation indices (F_I). Their estimated values also may be interpreted as describing departures from the expected Hardy-Weinberg genotypic frequencies within local populations and within the total population, respectively (Nei, 1973, 1977):

$$F_I = 1 - (h_{obs}/h_{exp}), \qquad (6.5)$$

where h_{obs} and h_{exp} are the observed and expected frequencies of heterozygotes at a locus (Box 2.1). Thus, positive values for fixation indices indicate positive correlations among uniting gametes (heterozygote deficits), likely due to local inbreeding (F_{IS}), or local inbreeding plus population subdivision (F_{IT}).

F_{ST} may also be interpreted as the variance (V_p) of allele frequencies among populations, standardized relative to the maximum value possible given the observed mean allele frequency ($\bar{p}$):

$$F_{ST} = V_p/\bar{p} (1 - \bar{p}). \qquad (6.6)$$

This statistic is used commonly as a measure of population subdivision, and as shown in Box 6.4 provides a convenient approach for estimating interpopulational gene flow in models that assume selective neutrality. The reader should be alerted to the many nuances (beyond the scope of current discussion) involved in calculating *F*-statistics from empirical data, such as how to handle the difficulties of multiple alleles at a locus, multiple loci, and various sources of sampling error within and among subpopulations (Weir and Cockerham, 1984).

---

the mating system. Finer considerations require a partitioning of these general categories into biological factors relevant for each group of organisms. For example, numerous ecological and life history factors are predicted to influence plant population structures (Table 6.2), and in a comparative summary of more than 150 empirical studies of 124 plant taxa based on allozymes, Loveless and Hamrick (1984) found that interpopulational genetic differences indeed generally were consistent with predictions based on a plant's breeding system, floral morphology, life cycle, timing of reproduction, and successional stage (see also

Box 6.4. Genetic Exchange Among Populations

Gene flow is the transfer of genetic material between populations resulting from movements of individuals or their gametes. Usually, gene flow is expressed as a migration rate m, defined as the proportion of alleles in a population each generation that is of migrant origin. Gene flow is notoriously difficult to monitor directly, but commonly is inferred from spatial distributions of genetic markers by several statistical approaches. Most of these approaches are based on equilibrium expectations derived from theoretical models of population structure under neutrality theory, e.g., the "island model" wherein a species is assumed to be subdivided into populations (demes or islands) of equal size N, all of which exchange alleles with equal probabilities, or the "stepping-stone" model wherein only adjacent demes exchange alleles. Allelic frequencies in finite populations are influenced by random genetic drift also, which is a function of effective population size (Box 2.2). The influences of drift and gene flow are difficult to tease apart and, thus, most statistical procedures as applied to spatial genetic information permit estimates only of the product Nm, which can be interpreted as the absolute number of individuals exchanged between populations per generation. Also, Nm is of particular interest because under neutrality theory, at equilibrium the level of divergence among populations is a function of the numbers of migrants rather than the proportions of individuals exchanged. The most common approaches to Nm estimation are as follows:

(a) From F-statistics. Wright (1951) showed that for neutral alleles in an island model, equilibrium expectations are:

$$F_{ST} \cong 1 \: / \: (1 \: + \: 4Nm) \text{ or } Nm \cong (1 \: - \: F_{ST}) \: / \: 4F_{ST}. \tag{6.7}$$

Nei (1973) defined a related measure of between-population heterogeneity (gene diversity or G_{ST}) that bears the same relationship to Nm and also is employed widely. Takahata and Palumbi (1985) suggested modifications of these basic statistics for extranuclear haploid genomes such as mtDNA, and Lynch and Crease (1990) proposed an analogue of the F_{ST} or G_{ST} indices (N_{ST}) that is applicable to any data at the nucleotide level.

(b) From private alleles. Private alleles are those found only in one population. Slatkin (1985a) showed by computer simulations of a variety of artificial populations that the logarithm of the average frequency of private alleles [$p(1)$] is related approximately linearly to the logarithm of Nm:

$$\ln [p(1)] = -0.505 \ln (Nm) - 2.440 \tag{6.8}$$

This result proved insensitive to most changes in parameters of the model, except that a correction for Nm due to differences in the mean number of individuals sampled per population was recommended (see Barton and Slatkin, 1986). The basic rationale underlying Slatkin's (1985a) method is that private alleles are likely to attain high frequency only when Nm is low. In practice, when sufficient genetic information is available, the F_{ST} and private allele methods are expected to yield comparable estimates of gene flow under a wide variety of population conditions (Slatkin and Barton, 1989).

(c) **From allelic phylogenies.** Unlike the two approaches described above that can be applied to phylogenetically unordered alleles (such as those provided by allozymes), this method requires knowledge of the phylogeny of nonrecombining segments of DNA (such as mtDNA haplotypes). Given the correct gene tree and knowledge of the geographic populations in which the allelic clades are found, a parsimony criterion is applied to estimate the minimum number of migration events consistent with the phylogeny. Slatkin and Maddison (1989) show that the distribution of this minimum number is a simple function of *Nm*, which therefore can be estimated from empirical data by comparison to tabulated results from their computer-simulated populations.

None of these procedures should be interpreted as providing a precise estimate of genetic exchange among demes; rather, each offers a qualitative (albeit numerical) guideline as to whether populations likely experience high, moderate, or greatly restricted gene flow. Roughly speaking, the average exchange of one individual per generation ($Nm \cong 1$) between populations, irrespective of deme size, is marginally sufficient in theory to prevent dramatic genetic differentiation by genetic drift alone (Allendorf, 1983). The value $Nm = 1$ corresponds to mean $F_{ST} = 0.20$ [Eq. (6.7)] or $p(1) \cong 0.085$ [Eq. (6.8)]. Thus, as a rule of thumb, "high-gene-flow" species are expected to exhibit lower estimates of F_{ST} and $p(1)$ than these respective values, whereas "low-gene-flow" species should display much higher values. Additional discussion of these procedures for gene flow estimation may be found in Hudson et al. (1992) and Slatkin (1985b, 1987).

Neigel et al. (1991) introduced a philosophically different approach to gene flow analysis that yields an estimate of single-generation dispersal distance (rather than *Nm*). The method is based on expected spatial distributions of lineages of various evolutionary age in a gene tree, assuming an evolutionary clock for the molecule and assuming that dispersal of lineages has occurred via a multigeneration "random-walk" process from specifiable centers of origin for each clade. As applied to an empirical mtDNA gene tree for continent-wide populations of the deer mouse *Peromyscus maniculatus* (Lansman et al., 1983), this method yielded estimates of single-generation dispersal ($\cong 200$ m) that agree well with direct mark-recapture data for this species. This approach should be most suitable for low-dispersal species and for rapidly-mutating, nonrecombining genetic markers such as mtDNA. In such situations, mutations that delineate new descendent lineages may be dispersed at rates sufficiently low to prevent the attainment of an equilibrium between genetic drift and gene flow that many of the earlier models assume.

Table 6.2. Examples of ecological and life history factors and their predicted influences[a] on the population genetic structures of plants (after Loveless and Hamrick, 1984).

Factor	Genetic Heterozygosity Within Populations	Genetic Structure Within Populations	Genetic Structure Among Populations
Breeding System			
Self-fertilizing	Low	High	High
Mixed mating	Moderate	Moderate	Moderate
Outcrossing	High	Low	Low
Floral Morphology			
Monoecious	Depends on % selfing	Depends on % selfing	Depends on % selfing
Dioecious	High	Low	Low
Reproductive Mode			
Apomictic	Depends on other factors	Depends on other factors	Potentially high
Sexual	Potentially high	Depends on other factors	Depends on other factors
Pollination Mechanism			
Sedentary animal	Potentially low	Potentially high	High
Dispersive animal	High	Low	Low
Wind	High	Low	Low
Seed Dispersal			
Limited	??	Potentially high	High
Long range	High	Low	Low
Seed Dormancy			
Absent	Depends on other factors	Depends on other factors	Depends on other factors
Present	Increases potential	Reduces potential	Reduces potential
Phenology			
Asynchronous	No prediction	Increases potential	Increases potential
Synchronous	No prediction	Reduces potential	Reduces potential
Life Form			
Annual	Reduced??	Increases potential	Increases potential
Long-lived	Increased??	Reduces potential	Reduces potential
Timing of Reproduction			
Monocarpic[b]	No prediction	Increases potential	Increases potential
Polycarpic	No prediction	Reduces potential	Reduces potential
Successional Stage			
Early	??	Depends on other factors	Increases potential
Late	??	Depends on other factors	Reduces potential
Geographic Range			
Narrow endemic	Low	Low	High
Widespread	Potentially High	Depends on other factors	Depends on other factors
Population Size, Density			
High	High	Depends on other factors	Depends on other factors
Low	Low	Depends on other factors	Depends on other factors

[a]Predictions remain qualified because categories may be interrelated and other confounding variables may pertain.

[b]Having only one fruiting period during the life cycle.

Hamrick and Godt, 1989). Similarly, a comparative summary of the allozyme literature for more than 300 animal species (Table 6.3) led Ward et al. (1992) to conclude that mobility is an important factor influencing the apparent magnitude of population structure. Thus, many insects and birds, which tend to be vagile organisms, show significantly lower levels of mean population structure than do relatively sedentary creatures such as some amphibians (Table 6.3). In the sections that follow, a few illustrative cases are highlighted that address how particular demographic or other factors may impinge on molecular population genetic structure. Where possible, attempts will be made to draw parallels between results for taxonomically distinct organisms, such as plants and animals.

Autogamous Mating Systems

In a classic series of studies employing allozyme markers, Allard and colleagues documented the dramatic influence that the mating system can assume, in conjunction with natural selection, in shaping the multilocus genetic architecture of plant species. The slender wild oat (*Avena barbata*) is a predominantly self-fertilizing species, first introduced from its native range in the Mediterranean to California during the Spanish period some 400 years ago, and more extensively during the Mission period some 200–250 years ago. Within California, it

Table 6.3. A comparative summary of population structures[a] for 321 animal species surveyed by multilocus protein electrophoresis (after Ward et al., 1992).

Taxonomic Group		Population Differences[a] (with Standard Errors)	No. of Species
Vertebrates			
Mammals		0.242 ± 0.030	57
Birds		0.076 ± 0.020	16
Reptiles		0.258 ± 0.050	22
Amphibians		0.315 ± 0.040	33
Fishes		0.135 ± 0.040	79
	TOTAL	0.202 ± 0.015	207
Invertebrates			
Insects		0.097 ± 0.015	46
Crustaceans		0.169 ± 0.061	19
Mollusks		0.263 ± 0.036	44
Others		0.060 ± 0.021	5
	TOTAL	0.171 ± 0.020	114

[a]Shown are the proportions of total genetic variation within species due to genetic differences between geographic populations, as reflected in the "coefficient of gene differentiation": $(H_T - H_S)/H_T$, where H_S and H_T are the mean heterozygosities estimated within local populations and within the entire species, respectively (Nei, 1973).

has since achieved a remarkable population-genetic structure characterized by a great predominance of two apparently coadapted multilocus gametic types (Allard et al., 1972; Clegg and Allard, 1972). One genotype, "1,2,2,2,1,B,H" (where the numbers and letters refer to alleles at each of five allozyme loci and two morphology genes), is characteristic of the semiarid grasslands and oak savannahs bordering the central Sacramento-San Joaquin Valley, whereas the complementary gametic type ("2,1,1,1,2,b,h") is more common in the strip of coastal ranges and the higher foothills of the Sierra Nevada mountains.

These associations between genotype and the environment (notably xeric versus mesic soils) also are maintained over microgeographic scales in transitional (ecotonal) areas (Hamrick and Allard, 1972), notwithstanding the continued production of recombinant genotypes through occasional outcrossing [the estimated outcrossing rate is $t \cong 0.02$ (Clegg and Allard, 1973)]. The authors conclude that genetic variability in *A. barbata* in California has been internally organized and spatially structured within just 400 years by intense natural selection operating in conjunction with severe constraints on recombination afforded by the mating system (Allard, 1975). These studies added great empirical force to theoretical arguments that any factors such as linkage or inbreeding that tend to restrict intergenic recombination can facilitate the operation of natural selection in moulding coadapted multilocus gene complexes (Box 6.5; Clegg et al., 1972; Jain and Allard, 1966).

"Autogamy" refers to the process of self-fertilization or self-pollination. In a review of the literature on spatial genetic variation in plants, Heywood (1991) concluded that autogamous species often display remarkable levels of local genetic differentiation that far surpass those of predominantly outcrossing species. For example, in the annual plant *Plectritis brachystemon* that is highly autogamous (outcrossing rate 2%), most of the allozyme diversity proved to be *inter*populational; whereas in the sympatric congener *P. congesta* that is predominantly allogamous (outcrossing rate 70%), most of the allozymic diversity is *intra*populational (Layton and Ganders, 1984). Another conclusion was that the spatial genetic architectures observed for predominantly autogamous species frequently are similar to those of taxa that reproduce by a mixture of apomixis (Chapter 5) and sexual outcrossing (Heywood, 1991). Especially when such species occur in low-density situations, the effects of genetic drift, selection on multilocus genotypes, and the restricted recombination engendered by the mating system, all can collaborate to produce striking spatial variation in genotypic distributions.

Analogous population genetic analyses have been conducted on a hermaphroditic land snail that also reproduces by facultative self-fertilization (*Rumina decollata*), and these studies have demonstrated the potential for adaptive convergence in the genetic architectures of animals and plants (Selander and Kaufman, 1975a; Selander et al., 1974). This gastropod, native to the Mediterranean region, was introduced to the eastern United States before 1822 and

---

Box 6.5. Self-Fertilization, Restricted Genetic Recombination, and Multilocus Organization

Self-fertilization is the most extreme form of inbreeding. One consequence of predominant selfing is a severe restriction on inter-genic recombination, and an associated enhanced opportunity for the maintenance of coadapted gene complexes, as can be illustrated qualitatively as follows. Assume that at each of four di-allelic loci, the haploid genotypic combinations 1,1,1,1 and 2,2,2,2 (numbers refer to the allele at each locus) confer high viability on their diploid bearers and that the other 14 multilocus haplotypes (1,2,1,1; 2,1,2,1; etc.) lead to adaptively inferior individuals. During the course of a generation, the favored genotypes will tend to increase in frequency under natural selection from early to late stages in the life cycle. However, under random outcrossing, recombination at gametogenesis will tend to undo the effects of selection in maintaining the favored multigene complexes across generations. This process of decay in gametic phase disequilibrium (D) can be inhibited if the loci are physically linked on a chromosome because intergenic recombination then may be restricted (Box. 5.3). Inbreeding similarly retards the rate at which D converges to zero, by limiting *effective* recombination. For two loci under a model of mixed self-mating and random mating, this rate is given by

$$1 - 0.5 \{0.5 (1 + \lambda + s) + [(0.5 (1 + \lambda + s))^2 - 2s\lambda]^{0.5}\}, \quad (6.9)$$

where s is the selfing probability and λ is the amount of linkage (for $\lambda = 0$, the recombination fraction is $c = 0.5$) (Weir and Cockerham, 1973). For example, with $s = 0.98$ and $\lambda = 0.00$ (or with $s = 0.00$ and $\lambda = 0.98$), the rate of decay of D is 1.0% per generation. The point is that linkage and selfing can be seen to enter this equation similarly and, therefore, they retard disequilibrium decay in similar fashion.

One important difference between the effects of linkage versus inbreeding in restricting recombination is that the former acts locally among physically linked loci, whereas the latter acts globally, restricting recombination between all loci whether on the same or different chromosomes. Thus, in theory, inbreeding in conjunction with selection could serve to organize entire genomes into integrated multilocus systems (Allard, 1975).

---

subsequently has spread across much of North America. In its native range, *R. decollata* exists as a complex of inbred monogenic or mildly polygenic strains characterized by different suites of allozymic and morphological markers. Two strains ("light" and "dark") predominate in intensively surveyed areas in southern France and show fixed differences at 13 of 26 allozyme loci. The dark form typically occupies protected mesic environments such as under logs or rocks, whereas the light form is associated with open xeric habitat. Occasional outcrossing between these strains releases extensive recombinational variation that otherwise is expressed as between-strain genetic differences. Nevertheless, the two forms tend to retain their separate identities and habitat correlations in nature, suggesting that, as in *Avena barbata*, strong multilocus associations and pronounced population genetic structures probably stem from natural selection [as well as stochastic population factors (Selander, 1975)] operating in conjunction with the self-fertilizing breeding system that reduces effective recombination (Selander and Hudson, 1976). As a further note of interest, some inbred strains of *R. decollata* apparently have achieved extensive geographic distributions and high population numbers in the absence of appreciable genetic variation; all assayed populations in North America are allozymically identical and, thus, probably derive from a single strain introduced from Europe (Selander and Kaufman, 1973b).

Gametic and Zygotic Dispersal

POLLEN AND SEEDS

In outcrossing plants, mobile male gametes (pollen) are transferred to sedentary eggs in the female flower by wind or by animal vectors such as insects, birds, or mammals, some of which may be capable of moving long distances. The zygotes (seeds) produced then are dispersed from the maternal parent either by gravity, wind (sometimes assisted by a winged or plumose seed condition), or by animals (via seed attachment to the vector's body or passage through the digestive system). To what extent do varying dispersal mechanisms for gametes and seeds influence gene flow regimes and genetic structures of plant populations?

Most tropical trees outcross predominantly, with pollination and seed dispersal usually animal-mediated (Hamrick and Murawski, 1990). Using allozyme data, Hamrick and Loveless (1989) compared the population genetic structures in several outcrossing species of trees and shrubs in Panama. The authors rank-ordered 14 species with respect to a priori predictions about expected interpopulational genetic structure based primarily on considerations of pollen and seed dispersal mechanisms and found that the empirical estimates of gene flow [Nm (Box 6.4)] were correlated significantly with this ranking (Fig. 6.3). They concluded that geographic population structure appeared greater in species with weak-flying pollinators and anemic means of seed dispersal. The few wind-pollinated species examined also had relatively strong population structures, consistent with earlier suggestions that wind may not be a particularly effective agent for pollen flow in the low-density populations of conspecifics in tropical forests. Nonetheless, the correlations between reproductive biology and genetic structure were moderate at best, explaining far less than 50% of the overall variance in outcomes. Furthermore, over the spatial scales of several kilometers monitored, all species exhibited $Nm > 1.0$. This raises the more fundamental point that gene flow appears moderate to high in most populations of these tropical forest trees.

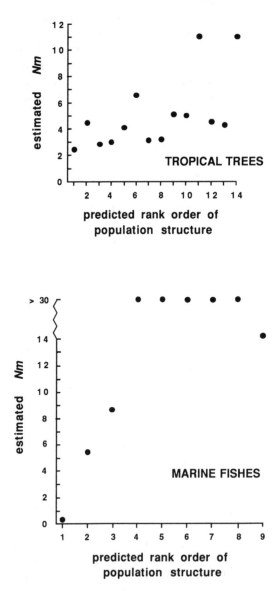

Figure 6.3. Relationship between gene flow and life history mediated dispersal potential. **Top:** Fourteen species of tropical trees and shrubs (data from Hamrick and Loveless, 1989); **Bottom:** Nine species of marine shore fishes (Waples, 1987). Species are plotted in rank order according to the predicted magnitude of population structure based on suspected dispersal capabilities of pollen and seeds (plants) or larvae (fishes). Gene flow (Nm) values were estimated from allozyme data using Wright's F_{ST} approach (Box 6.4). The Spearman's coefficients of correlation between Nm and predicted rank were $r = 0.68$ ($P < 0.01$, plants) and $r = 0.88$ ($P < 0.01$, fishes). Caution is indicated, however, because Nm values estimated from Slatkin's private-allele approach for these same species were not significantly correlated with dispersal rank.

Temperate trees, more of which are wind pollinated, also commonly yield moderate to high estimates of gene flow. For example, among populations of pitch pine (*Pinus rigida*) throughout the species' range in the eastern United States, mean F_{ST} across allozyme loci was only 0.024 (Guries and Ledig, 1982), a value associated with $Nm = 10.2$. Other wind-pollinated pines that exhibit low among-population differentiation across broad and continuous ranges include *P. banksiana* [$Nm = 6.7$ (Dancik and Yeh, 1983)], *P. contorta* [$Nm = 8.1$ (Wheeler and Guries, 1982)], and *P. ponderosa* [$Nm = 16.4$ (Hamrick et al., 1989)]. On the other hand, pine species whose populations are distributed as scattered isolates tend to show far greater spatial structure, as might be expected, and this is reflected in lower estimates of gene flow: *P. torreyana*, $Nm = 0$ (Ledig and Conkle, 1983); *P. halapensis*, $Nm = 0.6$ (Scheller et al., 1985); and *P. muricata*, $Nm = 1.0$ (Millar, 1983).

Hamrick et al. (1992) reviewed published allozyme data on conspecific population structures within 322 woody plant species. The overall mean estimate for the interpopulational component of genetic diversity was $G_{ST} = 0.085$ ($Nm = 2.7$). However, only 16% of the heterogeneity in genetic structure across species could be accounted for by differences in seven life history and ecological traits considered. The authors conclude that other influences, including the specific and idiosyncratic evolutionary histories of species, must have played important roles in determining how genetic diversity is partitioned. Most woody species are long-lived and self-fertilization is rare. Another review of the broader allozyme literature for 449 herbaceous and woody plants permitted breeding system and life form to be added to the list of life history influences considered (Hamrick and Godt, 1989). Across this wider taxonomic scale of comparison, the most important predictors of the magnitude of interpopulational structure were the selfing versus outcrossing breeding system and annual versus perennial life form. On average, 146 annual species showed $G_{ST} = 0.357$ ($Nm = 0.45$) and 78 selfing species showed $G_{ST} = 0.510$ ($Nm = 0.24$). Thus, plant species that live for short periods and self-fertilize tend to exhibit far greater spatial structure than those with contrasting features.

MARINE GAMETES AND LARVAE

Many marine invertebrates and fishes shed gametes and larvae into the water column, in analogous fashion to the atmospheric release of pollen and seed by many land plants. The planktonic duration varies widely for different species. For example, larvae of the sessile polychaete *Spirorbis borealis* remain free-living for a few hours at most and are competent to settle immediately on release from their parents, whereas larvae in another polychaete genus *Phyllocha-etopterus* can delay settlement and metamorphosis in excess of one year (Scheltema, 1986). More usually, the planktonic larval duration of benthic marine invertebrates lasts from a few days to weeks. Among marine fishes, larvae commonly remain in the plankton for weeks or months (e.g., Victor, 1986), but the remarkable leptocephalus larvae of some eels may remain pelagic for three years or more (Castle, 1984). On the other hand, other marine invertebrates and fishes produce nonplanktotrophic eggs or larvae. These less-dispersive propagules may be demersal or may be brooded by parents [e.g., in oral cavities (marine catfishes), abdominal pouches (pipefishes and seahorses), or other storage sites (many invertebrates)].

Are these widely varying potentials for gene flow via eggs and larvae correlated with population genetic structures in marine species? Some evidence suggests that they are. Thus, for invertebrates, several genetic studies have reported a correspondence between increased potential for larval dispersal and diminished genetic differentiation among geographic populations (Berger, 1973; Crisp, 1978; Gooch, 1975; Liu et al., 1991). For example, the larval-brooding snail *Littorina saxatilis* exhibited greater interpopulational variation in allozymic and morphologic characters than did a congeneric free-spawning species *L. littorea* (Janson, 1987). In sea urchins of the genus *Heliocidaris*, one species (*H. tuberculata*) with a several-week planktonic larval stage showed little differentiation in mtDNA genotypes between populations separated by 1000 km of open ocean, whereas populations of a congener (*H. erythrogramma*) with only a three- to four-day planktonic larval duration were strongly partitioned over comparable geographic scales (McMillan et al., 1992). Among the vertebrates, Ehrlich (1975) noted that Pacific damselfishes with pelagic larvae show allozyme uniformity over huge areas, whereas the one assayed species that lacks a pelagic larval phase (*Acanthochromis polyacanthus*) was highly structured genetically. Waples (1987) assessed allozyme differentiation in several species of marine shore fishes sampled along the same geographic transect in the eastern Pacific and reported a strong negative correlation with dispersal capability inferred from planktonic larval durations (Fig. 6.3). The species with the lowest potential for dispersal (a livebearer with no pelagic larval stage—*Embiotoca jacksoni*) exhibited the highest spatial genetic structure, whereas a species with the highest dispersal potential (a fish associated with drifting kelp and characterized by an extended larval duration—*Medialuna californiensis*) exhibited no detectable spatial genetic differentiation. Such results appear generally consistent also with the long-standing observation that marine species with dispersive larvae show a greater tendency to colonize oceanic islands and to exhibit broader geographic ranges than those with sedentary larvae (Jablonski, 1986; Thorson, 1961; however, see Thresher and Brothers, 1985 for counterexamples).

Population genetic structures in North Atlantic eels have attracted particular interest because of the extraordinary catadromous life histories of these species. Juvenile eels (*Anguilla rostrata* in the Americas and *A. anguilla* in Europe) inhabit coastal and inland waters, but during sexual maturation migrate to the

Sargasso Sea (western tropical mid-Atlantic Ocean) where spawning takes place. Conventional wisdom (reviewed by Williams and Koehn, 1984) is that the pelagic larvae produced from this suspected mass spawn disperse passively to continental margins via ocean currents, perhaps settling at locales randomly oriented with respect to the homesteads of their parents. If mating is indeed panmictic, and larval dispersal passive, all continental populations could represent random draws from a single gene pool and lack spatial structure accordingly. To a first approximation, genetic data for *A. rostrata* collected throughout eastern North America appear consistent with this scenario. Thus, Williams et al. (1973) and Koehn and Williams (1978) observed only very mild spatial structure at polymorphic allozyme loci (which they took as evidence for clinal selection in a panmictic species), and Avise et al. (1986) detected no geographic structure in mtDNA with the available sample sizes. However, American and European eels proved to be clearly distinct genetically, confirming the much-debated presence of at least two largely independent gene pools in the North Atlantic (Avise et al., 1986). Further genetic analysis also revealed a low-frequency presence of hybrids between *A. rostrata* and *A. anguilla* in Iceland, an island longitudinally intermediate to North America and Europe (Avise et al., 1990b). This unexpected appearance of hybrid eels in Iceland, thousands of kilometers from where the zygotes presumably were produced in the Sargasso Sea, raises the intriguing possibility of hybrid intermediacy in larval migratory behavior.

In general, a long-duration planktotrophic larval stage certainly affords the opportunity for extensive gene flow, and this potential appears to have been realized in many marine species as evidenced by a near absence of genetic (allozyme or mtDNA) differentiation over vast areas: for example, among populations of the sea urchin (*Strongylocentrotus purpuratus*) along the 2500-km west coast of North America (Palumbi and Wilson, 1990); among populations of the red rock lobster (*Jasus edwardsii*) across 4600 km of Australasian habitat (Ovenden et al., 1992); within each of six species of reef fishes from locales 1000 km apart in the Caribbean (Lacson, 1992); among red drum populations from the Atlantic and Gulf of Mexico seaboards in the southeastern United States (Bohlmeyer and Gold, 1991; Gold and Richardson, 1991); among damselfish (*Stegastes fasciolatus*) populations throughout the 2500-km length of the Hawaiian archipelago (Shaklee, 1984); among milkfish (*Chanos chanos*) populations from localities up to 10,000 km apart in the Pacific (Winans, 1980); between populations of tuna (*Katsuwonus pelamis* and *Thunnus alalunga*) in the Atlantic versus Pacific Oceans (Graves et al., 1984; Graves and Dizon, 1989); and among global populations of the orange roughy fish (*Hoplostethus atlanticus*) (Smith, 1986). Extensive movement of adults no doubt also contributes to the geographic uniformity in some of these fishes.

On the other hand, several other marine species with pelagic larvae *do* exhibit dramatic population differentiation over microgeographic or macrogeographic scales (Avise, 1987; Burton, 1983, 1986; Hedgecock, 1986). For example, rock-pool inhabiting populations of the copepod *Tigriopus californicus* show strong genetic (allozymic) differentiation both regionally and locally, notwithstanding a natural history that includes free-swimming adult and larval forms (Burton et al., 1979; Burton and Feldman, 1981). Populations of the American lobster *Homarus americanus*, whose pelagic larval stage lasts two to eight weeks, show genetic differences between the Atlantic Ocean and the Gulf of St. Lawrence in the northeastern United States (Tracey et al., 1975). In the horseshoe crab *Limulus polyphemus*, a dramatic genetic distinction in mtDNA exists between continuously-distributed adult populations along the Gulf of Mexico and Atlantic coasts in the southeastern United States (Saunders et al., 1986), despite the presence in this species of a trilobite larval stage that is "specialized for dispersal" (Rudloe, 1979).

Significant but ephemeral genetic patchiness also characterizes local populations of an intertidal limpet (*Siphonaria* sp.) and a sea urchin (*Echinometra mathaei*) in western Australia (Johnson and Black, 1982; Watts et al., 1990), despite macrogeographic similarities in allele frequencies and the presence of potentially dispersive planktonic larval stages. Such "chaotic patchiness" was attributed to idiosyncrasies among local sites in histories of larval recruitment. From a compilation of such examples, Burton (1983) concluded that although invertebrate species with planktotrophic larvae tend to show less spatial heterogeneity than those with nonmotile larvae, "previously described relationships between length of planktonic larval life and the geographic boundaries of panmictic populations are not strongly supported. In particular, substantial differentiation has been observed in several species that appear to have high dispersal capabilities."

There are several reasons why high dispersal potential of gametes or larvae may not always translate into spatial population genetic homogeneity and the attendant high estimates of gene flow (Hedgecock, 1986). First, actual levels of gene flow may be lower than presumed because of physical impediments to larval dispersal. Thus, the influences of particular oceanic currents in New England and in the southeastern United States may account in part for the genetic differences reported between regional populations of the American lobster and the horseshoe crab, respectively (see above). Second, larvae may not always be the passive dispersal propagules commonly assumed, but rather may adopt more active migrational behaviors and settlement choices in some species. Crisp (1976) and Woodin (1986) review the available evidence for discriminatory larval settlement in benthic intertidal and infaunal marine invertebrates, respectively. Some larvae fall far short of their dispersal potential. For example, in the shrimp *Alphaeus immaculatus* that has an extended free-swimming larval stage but whose adults live symbiotically with sea anemones, a detectable proportion of successful recruits settles on anemone colonies within a few meters of the parents (Knowlton and Keller, 1986).

Diversifying natural selection operating on particular loci via differential survival or mating success also could convey a false impression of low gene flow among populations. For example, in the blue mussel *Mytilus edulis,* allele frequencies at a leucine amino-peptidase (*Lap*) allozyme locus are significantly heterogeneous spatially, but strongly correlated with environmental salinity. Physiological and biochemical studies indicate that the alleles involved function differentially in relieving osmotic stress in varying salinity environments, via their influence on the free amino acid pools and volumes of cells (Hilbish et al., 1982). Thus, frequencies of the non-neutral *Lap* alleles probably say more about environmental conditions than about the gene flow regime of the species (Boyer, 1974; Koehn, 1978; Theisen, 1978). At other polymorphic allozyme loci, mussel populations exhibit large, moderate, or small interpopulational variances in allelic frequencies (Koehn et al., 1976), such that estimates of gene flow under assumptions of neutrality differ considerably across genes.

A particularly sobering example of how different genetic markers can sometimes yield contradictory pictures of gene flow involves populations of the American oyster (*Crassostrea virginica*) from the Gulf of Mexico and Atlantic coasts of the southeastern United States. Surveys of polymorphic allozymes revealed a near uniformity of allele frequencies throughout this range (Fig. 6.4), a result that understandably was attributed to high interpopulational gene flow resulting from "the rather long planktonic stage of larval development, since this species has the ability to disperse zygotes over great distances when facilitated by tidal cycles and oceanic currents" (Buroker, 1983). However, mtDNA genotypes revealed a dramatic genetic "break," involving cumulative and nearly fixed mutational differences that cleanly distinguished most Atlantic from Gulf oyster populations (Reeb and Avise, 1990). Subsequent surveys of nuclear RFLPs tended to support the dramatic Atlantic/Gulf mtDNA dichotomy (Karl and Avise, 1992; Fig. 6.4) and, thus, appear to eliminate possible differences in dispersal of male gametes versus female gametes as a potential explanation for the contrasting allozymic and mitochondrial population structures. One remaining possibility is that several of the allozyme loci may be under uniform balancing selection, and thus do not record the population subdivision that seems evidenced so clearly by mtDNA and by some of the nuclear RFLP distributions (Karl and Avise, 1992). This suggestion appears consistent also with the long-standing observation that multilocus allozyme heterozygosities in mollusks are associated strongly with presumed fitness components such as metabolic efficiency and growth rate (Garton et al., 1984; Zouros et al., 1980). Whether this explanation or its converse is correct (that allozymes faithfully register high gene flow in oysters, but mtDNA and some scnRFLPs differ in the Atlantic versus Gulf because of diversifying selection), the conclusion remains that natural selection must have acted on at least some of the genetic markers. This finding underlines the need for caution in inferring population structure and gene flow from any single gene or perhaps even class of marker loci.

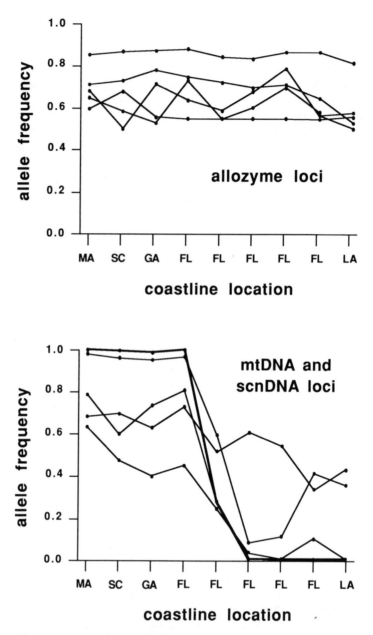

Figure 6.4. Allele frequencies in oyster populations. Shown are frequencies of the most common alleles at five polymorphic allozyme loci (**top**) and five DNA loci (**bottom**) along a coastline transect running from Massachusetts through South Carolina, Georgia, Florida, and Louisiana (after Karl and Avise, 1992). The allozyme loci are *Est1, Lap1, 6Pgd, Pgi,* and *Pgm* (data from Buroker, 1983); loci assayed at the DNA level are mtDNA (heavy line; from Reeb and Avise, 1990) and each of four anonymous single-copy nuclear genes.

Direct Estimates of Dispersal Distances

In some cases, molecular markers that are rare or unique have been employed to monitor gene dispersal from known point sources over one or a few generations. Such "genetic-branding" approaches (Ferris and Berg, 1986) are similar in concept to traditional labeling and tracking studies based on nonheritable markers [such as physical tags, fluorescent dyes, radiotracers, morphological characters, parasite loads, etc. (see the review in Levin, 1990)], with the added benefit that actual gene movement may, in some cases, be monitored over multiple generations.

For example, Burton and Swisher (1984) introduced copepods (*Tigriopus californicus*) carrying rare allozyme alleles into several natural tide-pool populations on rocky California headlands. There was an initial decline in the frequencies of the introduced alleles, but within six weeks the alleles had spread to nonrecipient pools on the same outcrop, and within eight months all subpopulations on an outcrop were nearly homogeneous in allelic frequencies, indicating rather extensive interpool gene flow over these small spatial scales. Similarly, Dillon (1988) introduced freshwater snails (*Goniobasis proxima*) carrying unique allozyme alleles into isolated streams in the southern Appalachians of Virginia. Few introductions were successful (2 of 12), but among those that were, introduced alleles spread at a rate of about 15–20 m per year upstream and only 5–10 m per year downstream. Thus, one of the peculiar natural history features confirmed for this species is its behavioral tendency to migrate by crawling against the current.

Gametic dispersal can be investigated in a similar fashion. Grosberg (1991) monitored the effective dispersal of sperm by placing allozymically marked colonies of the marine ascidian *Botryllus schlosseri* onto the pilings of a harbor dock and later assaying brooded embryos within nearby natural colonies for the presence of the introduced allele (signaling fertilization by dispersed sperm). From the rapid decline in fertilization success observed beyond about 50 cm, Grosberg concluded that effective sperm dispersal in this species is extremely limited. Similarly, Schaal (1980) monitored effective pollen flow in the Texas bluebonnet (*Lupinus texensis*) by assaying for heterozygotes among the F_1 progeny in an experimental population where an allozymically marked pollen donor was surrounded by plants of alternate genotype. Schaal found that gene dispersal via pollen was quite restricted (a few meters), with most gene movement being to neighboring plants. Nonetheless, this movement of pollen genes was significantly greater than would have been inferred from interplant flight distances by the pollinators (bees), apparently because of "pollen carryover" wherein some fraction of the pollen a bee deposits on a stigma comes from flowers visited before the most recent. Such studies of allelic dispersal that involve screening of progeny arrays for introduced genetic markers are allied closely to the parentage analyses discussed in Chapter 5.

Nonetheless, several limitations attend attempts to monitor dispersal directly by genetic markers. First is the problem of generating or finding unique alleles that with assurance are not found also in the natural population to which the markers are introduced. Second is the possibility that the markers themselves, or perhaps the selective breeding programs often used in their generation, may impair or otherwise modify the dispersal capabilities of the propagules in comparison to the natural state. A related possibility is that the dynamics of the introduced markers might be influenced primarily by natural selection acting on the markers or their genetic backgrounds rather than by the dispersal and gene flow regimes that usually are of primary interest in such studies. Finally, the greatest problem is the infeasibility of monitoring long-distance movement of potentially vagile markers, due to large spatial scales and the inevitable effects of dilution from the point source (Jones et al., 1981). Such undetected long-distance gene flow, even if rare, could have a significant homogenizing influence on the genetic structure of a species.

Vagility, Philopatry, and Dispersal Scale

Particularly in animals with nondispersive gametes and larvae, adult vagility in conjunction with habitat "grain" [the geographic scale at which organisms "perceive" environmental patchiness (Levins, 1968)] no doubt assume major roles influencing gene flow and population structure. To a land snail such as *Helix aspersa*, parking lots and streets may be formidable barriers to movement, and colonies within and among city blocks indeed do exhibit significant population structure as registered by allozymes (Selander and Kaufman, 1975b). At the other extreme, some turtles are oceanic mariners that swim many thousands of kilometers during their lifetimes; surveys of scnRFLPs in green turtles (*Chelonia mydas*) indicate population genetic structures throughout entire ocean basins comparable in magnitude to those among *Helix* populations on adjacent city blocks within a town [mean $F_{ST} \cong 0.13$ (Karl et al., 1992)].

Although the spatial scales of potential gene flow clearly are influenced by the mobilities of animals, apparent population genetic structure is not inextricably tied to an organism's vagility for several reasons, including (a) the presence of physical or ecological barriers that overmatch dispersal capability, (b) behavioral considerations such as habitat choice, or philopatry (site faithfulness) with regard to reproduction, (c) gender-biased patterns of dispersal and gene flow (applying particularly to genetic markers with sex-limited transmission), (d) influences of natural selection on particular genetic markers (or on loci linked to them), or (e) historical demographic events that have removed populations from equilibrium expectations between gene flow and genetic drift. These will be illustrated in turn.

PHYSICAL DISPERSAL BARRIERS

Bluegill sunfish (*Lepomis macrochirus*) are active and mobile swimmers abundant throughout their freshwater range in North America. An allozyme survey of 2560 specimens divided equally among 64 localities (8 sites per reservoir, 4 reservoirs in each of 2 adjacent river drainages) revealed that nearly 90% of the overall variance in allele frequency occurred between reservoirs of a drainage (Avise and Felley, 1979). Among locales within reservoirs (which ranged in size to more than 100,000 acres), allele frequencies seldom were significantly heterogeneous. Some of the reservoirs surveyed were geographically proximate, but intervening dams no doubt constitute formidable barriers to bluegill movement. Clearly, the subdivided structure of the physical environment has imposed a corresponding genetic structure on these otherwise highly mobile freshwater fishes.

Gyllensten (1985) reviewed allozyme literature on population structures for 19 species of fishes characterized according to habitat and life-style: strictly freshwater, anadromous (migrating from saltwater to freshwater to spawn), and marine. The average proportion of total intraspecific gene diversity allocated between locales increased considerably from the marine species and anadromous forms (1.6% and 3.7%, respectively) to the freshwater taxa (29.4%). Thus, the differences in the distribution of genetic variability coincided generally with qualitative differences in the occurrence of obvious geographic barriers to migration. Nevertheless, some marine and anadromous fish do exhibit levels of population structure across their ranges that are more or less comparable to those of some freshwater fish species (Avise et al., 1987b; Bowen and Avise, 1990).

In a flightless waterstrider (*Aquarius remigis*) that migrates by rowing on water surfaces, an allozyme survey by Preziosi and Fairbairn (1992) revealed that populations from different streams of a watershed are highly structured ($F_{ST} = 0.46$), whereas samples within a stream are undifferentiated ($F_{ST} = 0.009$). By contrast, another waterstrider with functional wings (*Limnoporus canaliculatus*) exhibited nearly homogeneous allele frequencies throughout several Atlantic seaboard states (Zera, 1981), suggesting that the differences in dispersal capacity of these two Hemipteran species exert an important influence on their population genetic structures. On the other hand, a comparison of population structures in five species of carabid ground beetles revealed no correlation with degree of flight-wing development (ranging from vestigial to fully winged), but a positive correlation was noted between F_{ST} values and the elevations of collecting sites (Liebherr, 1988). Perhaps the greater fragmentation of highland habitats was a more important factor than inherent dispersal capability in influencing magnitudes of genetic structure in these beetles, or perhaps different populational histories were involved.

Numerous other species likewise occupy discontinuous habitats and, depending on dispersal capabilities, may show significant population genetic structure related to environmental patchiness. To troglobitic (obligate cave dwelling) crickets in the genera *Hadenoecus* and *Euhadenoecus,* different cave systems represent isolated pockets of habitat, and these species tend to exhibit greater allozymic population structure than do their epigean (surface dwelling) counterparts (Caccone and Sbordoni, 1987). To *Peromyscus* mice on islands, ocean channels severely constrain dispersal, such that small island populations often show diminished within-population variability and exaggerated between-island genetic differences relative to their mainland counterparts (Ashley and Wills, 1987, 1989; Avise et al., 1974; Selander et al., 1971). To any habitat specialist, suitable environments may be scattered. In the beetle *Collops georgianus,* which is endemic to granitic outcrops in the southeastern United States, differentiation among distant outcrops appeared pronounced (F_{ST} = 0.19) relative to the spatial genetic heterogeneity exhibited over microgeographic scales (F_{ST} = 0.01) (King, 1987). These same granitic outcrops also serve as rather isolated island habitats for endemic plant populations, where allozyme data have indicated severe restrictions on gene flow (Wyatt et al., 1992).

PHILOPATRY TO NATAL SITE

Each reproductive season, many marine turtles migrate hundreds or thousands of kilometers from foraging grounds to particular nesting locales, where eggs are deposited on sandy beaches. For example, female green turtles (*Chelonia mydas*) that nest on Ascension (a small isolated island on the mid-Atlantic oceanic ridge) otherwise inhabit feeding grounds along the coast of Brazil, some 2000 km distant. From repeated captures of physically tagged adults, it long has been known that green turtles exhibit strong nest-site fidelity, i.e., Ascension females nest there and nowhere else, Costa Rican and Venezuelan nesters are faithful to their respective rookeries, and so on. What had remained unknown is whether the site to which a female is fidelic as an adult is also her natal rookery. If female "natal homing" prevails, rookeries within an ocean basin should exhibit clear genetic differences from one another with regard to maternally-transmitted genetic traits (such as mtDNA), even if appreciable interrookery exchange of nuclear genes occurs via the mating system and male-mediated gene flow (Karl et al., 1992). In recent surveys of mtDNA from green turtle rookeries around the world (Bowen et al., 1992; Meylan et al., 1990), it was discovered that (a) a fundamental phylogenetic split distinguishes all specimens in the Atlantic–Mediterranean from those in the Indo-Pacific Oceans and (b) additional genetic substructure characterizes populations within each ocean basin, as evidenced by fixed or nearly fixed genotypic differences between most rookeries (Fig. 6.5). Evidently, female green turtles have a strong propensity for natal homing. Re-

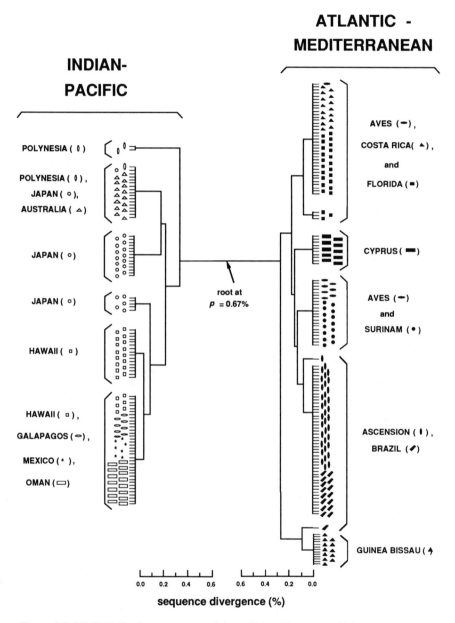

Figure 6.5. UPGMA dendrogram summarizing relationships among 226 sampled nests of the green turtle (after Bowen et al., 1992). To conserve space, the sequence divergence (*p*) axes on the bottom are presented as mirror images centered around the root leading to two distinct clonal assemblages (Atlantic–Mediterranean versus Indian–Pacific ocean basins) that also were evident in qualitative parsimony analyses of the same data.

sults exemplify how populations of a highly mobile species nonetheless can be structured dramatically, due in this instance both to geographic constraints (physical separations between oceans) as well as to a philopatric reproductive behavior (natal homing within oceans).

Whales too are impressive oceanic travelers, normally moving many thousands of kilometers seasonally. Baker et al. (1990) analyzed mtDNA from skin biopsies of 84 humpback whales (*Megaptera novaeangliae*), and found clear haplotype differences between groups previously reported to show distinct patterns of migration between summer feeding grounds in subpolar or temperate environs and winter breeding areas in the tropics. The authors interpreted this spatial segregation of genotypes, that occurred within as well as between ocean basins, "to be the consequence of maternally directed fidelity to migratory destinations," again illustrating the importance of site-fidelic behavior on population structure in a highly mobile animal.

Salmon are active and powerful swimmers, but also are notorious for suspected natal-homing propensity. In anadromous forms of these species, parr (juveniles) spawned in freshwater streams migrate to sea before eventually returning as adults to natal streams to complete the life cycle. Several allozyme and mtDNA studies of both Atlantic and Pacific species have revealed significant genetic differences among spawning populations at various spatial scales (Billington and Hebert, 1991; Ferguson, 1989; Gyllensten and Wilson, 1987a; Ryman, 1983; Ståhl, 1987). For example, Atlantic salmon (*Salmo salar*) from North America and Europe could be distinguished cleanly in genotypic composition (Bermingham et al., 1991; Davidson et al., 1989). However, on the smaller spatial scales of nearby drainages, genetic differences at most loci tend to involve allele frequency shifts that are less pronounced (e.g., G.M. Wilson et al., 1987). Presumably, the lack of greater local differentiation may be due to occasional "mistakes" in natal homing that provide gene flow between otherwise isolated spawning areas, as well as to the fact that most high-latitude streams currently utilized by salmon must have been colonized within the last few thousand years (following glacial retreats) and, hence, the populations no doubt are tightly connected in a historical sense.

Birds present a special enigma with regard to magnitude and pattern of population structure. On the one hand, most birds have high dispersal potential (because of flight, and often a migratory propensity), and many species have broad geographic distributions, suggesting that gene flow may be high and population differentiation minimal. On the other hand, many species also exhibit strong tendencies for nest-site philopatry and show obvious geographic variation in body size, song, or plumage (often leading to recognition of subspecies) suggesting that interpopulational gene flow may be low. Barrowclough (1983) reviewed the allozyme literature on geographic variation among populations within 57 vertebrate species. In comparison to most nonavian vertebrates, the surveyed birds (primarily temperate zone passerines) exhibited only minor geographic differentiation (mean $F_{ST} = 0.02$, compared to mean F_{ST} estimates of 0.11, 0.23, 0.30, and 0.38 for fishes, mammals, reptiles, and amphibians, respectively). Results were interpreted as "consistent with a presumed generally greater vagility in birds than in animals such as salamanders and rodents." Barrowclough (1983) also noted, however, that the results might reflect differences in the ages of populations.

Subsequent research employing the more discriminatory mtDNA molecule (Avise and Zink, 1988) has revealed that avian species exhibit a wide variety of population genetic structures (see the review in Avise and Ball, 1991). Some species such as the red-winged blackbird (*Agelaius phoeniceus*) and downy woodpecker (*Picoides pubescens*) show little differentiation in terms of mtDNA phylogeny across the entire North American continent (Ball et al., 1988; Ball and Avise, 1992), whereas others such as the seaside sparrow (*Ammodramus maritimus*), Canada goose (*Branta canadensis*), and fox sparrow (*Passerella iliaca*) show deep divisions in intraspecific mtDNA phylogeny that tend to align with geographic or morphologic partitions (Avise and Nelson, 1989; Van Wagner and Baker, 1990; Zink, 1991). Most of these studies involved restriction site analyses of total mtDNA. Recently, Wenink et al. (1993) employed sequences from the hypervariable mtDNA control region to demonstrate significant regional population structure in the dunlin (*Calidris alpina*), a long-distance migrant shorebird with Holarctic nesting distribution. Thus, high-resolution molecular assays likely will reveal significant geographic structure in many avian species, perhaps even over relatively small spatial scales [as has been demonstrated for example in the song sparrow *Melospiza melodia* (Hare and Shields, 1992; Zink, 1991)].

GENDER-BIASED DISPERSAL AND GENE FLOW

The degree of faithfulness to natal site or to social group often is gender-dependent. For example, as described earlier, lion prides normally consist of matrilineally related females and most interpride movement is by males (Packer et al., 1991; Schaller, 1972). Conversely, Florida scrub jays (*Aphelocoma coerulescens*) live in extended family groups built around patrilineal kinship—the young often remain at the nest as helpers, accession of territories is patrilineal, and most dispersal is by females (Woolfenden and Fitzpatrick, 1984). In general, most mammalian species with asymmetric philopatry exhibit male-biased dispersal, whereas most such avian species exhibit female-biased dispersal (Greenwood, 1980; Greenwood and Harvey, 1982). One likely consequence of such asymmetric dispersal is that a species may exhibit qualitatively different patterns of geographic population structure at genes with biparental transmission (most nuclear loci) versus those at which transmission occurs through only one sex (e.g., mtDNA, the Y chromosome of mammals and insects, or the W chromosome of birds). Although several Y- or W-linked genes have been identified in various mammalian and avian species, respectively (Casanova et al., 1985; Ellis et al., 1990; Page et al., 1985; Rabenold et al., 1991; Rasheed et al., 1991; Whisenant et al., 1991), few as yet have proven sufficiently variable to be of particular phylogenetic utility at the intraspecific level (Maynard Smith, 1990; but see Bishop et al., 1985). Thus, most molecular studies of gender-biased dispersal have relied on data from mtDNA in conjunction with those from allozymes or nuclear RFLPs.

A case in point involves macaque (*Macaca*) monkeys, where mirror-image patterns of geographic variation have been reported in nuclear-encoded allozymes versus mitochondrial DNA. Male macaques typically leave their natal group before sexual maturity, whereas females remain for life. Melnick and Hoelzer (1992) reviewed the literature on molecular variation in several macaque species (*M. fascicularis, M. mulatta, M. nemestrina,* and *M. sinica*) and reported that the observed patterns of geographic population structure appear consistent with these gender-specific behaviors. For example, in the nuclear genome of *M. mulatta,* only 9% of the total intraspecific diversity proved attributable to variation among geographic locales, whereas 91% of the overall diversity in mtDNA occurred between populations. Thus, the population genetic structures registered by the two genomes are "very different from one another . . . (and) . . . intimately linked to the asymmetrical dispersal patterns of males and females and the maternal inheritance of mtDNA" (Melnick and Hoelzer, 1992).

Another example of distinctive genetic signatures resulting from gender-based differences in behavior may involve the green turtle (*Chelonia mydas*). As already mentioned, most rookeries within an ocean basin are strongly isolated with regard to mtDNA lineages (mean inferred $Nm \cong 0.3$), indicating a strong propensity for natal homing by females (Bowen et al., 1992). However, these same rookeries are somewhat less differentiated at assayed scnRFLPs (mean $Nm = 1.7$; Karl et al., 1992), perhaps because of occasional male-mediated gene flow operating through the mating system. Green turtles are known to mate at sea, often on feeding grounds or other locales spatially removed from the nesting sites. Thus, interrookery matings may provide an avenue for the exchange of nuclear genes that largely is closed to mtDNA because of female natal philopatry.

The lesser snow goose (*Chen caerulescens*) provides an exception to the prevalent pattern of male-biased philopatry in birds. In this species, like many other migratory waterfowl, pair formation occurs on wintering grounds where considerable mixing of birds from different nesting areas can take place. Yet a mated pair normally returns to the female's natal or prior nesting area. Among all avian species for which direct banding returns are available (Cooke et al., 1975), according to Greenwood (1980) "The lesser snow goose is the best documented example of male biased natal and breeding dispersal. . . ." This natural history pattern suggests considerable intercolony gene flow mediated by males, an expectation consistent with results of both allozymic (Cooke et al., 1988) and scnRFLP studies (Quinn, 1988; Quinn and White, 1987). The behavior also suggests that colonies should be isolated with regard to matriarchal lineages, but surprisingly this has not proved to be the case. In a mtDNA survey of 160 geese from colonies across their breeding range (from Wrangel Island, Russia to Baffin Island in the eastern Canadian Arctic), no significant differences were observed in the spatial frequencies of two major mtDNA clades, a result strongly indicative of considerable population connectedness and gene flow involving females (Avise et al., 1992b; see also Quinn, 1992). One likelihood is that the entire current range of the snow goose was colonized recently from expansion out of Pleistocene refugia where the separation between the two mtDNA clades may have been initiated. A related possibility is that of ongoing gene flow, either via occasional lapses in philopatry by individual females (a phenomenon indeed documented from banding returns) or by episodic pulses of gene flow during periods of colony perturbation (also suspected from field observations). Whatever the process, snow goose colonies clearly have been in recent matrilineal contact, notwithstanding the propensity for natal philopatry by females.

Important object lessons are to be gained from these comparisons of banding and genetic data for snow geese. First, direct behavioral or marking studies on contemporary populations can, in some cases, provide a misleading picture of the geographic distributions of genetic traits because they fail to access the important evolutionary aspects of population connectedness revealed in the genes. Second, and conversely, geographic distributions of genetic markers can, in some cases, provide a misleading picture of contemporary dispersal and gene flow because they retain a record of evolutionary events and demographic parameters that may differ from those of the present. Thus, a full appreciation of the geographic population structure of a species requires an integration of evolutionary (genetic) and contemporary (behavioral) perspectives.

Africanized bees, or "killer bees" of the popular press, are aggressive forms of *Apis mellifera* that have spread rapidly in the New World following the introduction of African honeybees into Brazil in the late 1950s. The mode of spread and composition of colonies have been much debated, and at least two hypotheses were prevalent. First, perhaps queens are relatively sedentary, and most of the geographic expansion in aggressive behavior results from gene flow mediated by drones. Under this hypothesis, these males might travel considerable distances and mate with the more docile honeybees of European ancestry that formerly comprised the domesticated hives in the Americas. Alternatively, perhaps the gene flow results from colony swarming, a mechanism of maternal migration whereby a queen and some of the workers leave a hive and fly elsewhere to establish a new colony. Under this scenario, hybridization with the domesticated European forms is not necessarily required. Two independent research groups (Hall and Muralidharan, 1989; D.R. Smith et al., 1989) applied mtDNA assays to Africanized honeybees (mostly feral hives) in the neotropics and showed that these colonies usually carry African-type as opposed to European-type mtDNA. These results document the spread of African honeybees as continuous matrilines (the swarming hypothesis), thus eliminating the possibility that drones alone are responsible for the gene flow (see also Hall and Smith, 1991). Subsequent studies based on nuclear RFLPs and allozymes showed that African and European honeybees also hybridize at least occasionally in the neotropics and that this has led to introgression of nuclear genes as a part of the Africanization process, although to an argued degree (Hall, 1990; Lobo et al., 1989; Rinderer et al., 1991; Sheppard et al., 1991).

NON-NEUTRALITY OF SOME MOLECULAR MARKERS

Lewontin and Krakauer (1973) pointed out that one expected signature of natural selection on genetic markers is the appearance of significant heterogeneity across loci in the variances of allele frequencies among geographic populations. Because genetic drift, gene flow, and the breeding structure of a species should, in principle, affect all loci in a similar fashion, different population genetic structures across loci might indicate either that (a) allele frequencies at geographically variable loci are under diversifying selection (despite high gene flow as evidenced by geographically uniform genes) or (b) allele frequencies at geographically uniform loci are under stabilizing or equilibrium selection (despite low gene flow as evidenced by heterogeneous allele frequencies at geographically variable loci). Lewontin and Krakauer applied this reasoning to suggest that natural selection acted on at least some human-blood-group polymorphisms (Cavalli-Sforza, 1966), which on a global scale showed interpopulational allele frequency variances spanning a wide range (from $F_{ST} = 0.03$ to $F_{ST} = 0.38$)! The "Lewontin-Krakauer" test subsequently was criticized on the grounds that their proposed statistical methods seriously underestimated the variances in gene frequencies expected under the null (neutral) theory (Nei and Maruyama, 1975; Robertson, 1975; see also Lewontin and Krakauer, 1975). Nevertheless, it remains true that different loci within a species can sometimes paint very different pictures of population structure and gene flow when interpreted under models of selective neutrality.

An extreme example involves the deer mouse *Peromyscus maniculatus*. In allozyme surveys of populations from across North America, F_{ST} values at six polymorphic loci ranged from 0.04 (inferred $Nm = 6$) to 0.38 ($Nm = 0.4$) (Avise et al., 1979c). Particularly noteworthy was the observation that populations from central Mexico to northern Canada, and from the east to the west coasts of the United States, invariably exhibited the same two electromorphs at the aspartate aminotransferase (*Got-1* or *Aat-1*) locus, and in roughly similar frequencies ($F_{ST} = 0.05$). Subsequent molecular screening by varied electrophoretic techniques and other discriminatory protein assays failed to reveal any appreciable "hidden variation" within these two *Aat-1* electromorph classes (Aquadro and Avise, 1982b). Yet this relative geographic homogeneity at *Aat-1* contrasts sharply with the extreme geographic heterogeneity exhibited by this species in morphology, ecology, karyotype, and mtDNA sequence (Blair, 1950; Baker, 1968; Bowers et al., 1973; Lansman et al., 1983). In particular, the number of acrocentric chromosomes ranges from 4 to 20 across populations (Bowers et al., 1973), and regional populations often show deep divisions in a mtDNA genealogy involving cumulative and fixed mutational differences (Lansman et al., 1983). It is difficult to escape the conclusion that *Aat-1* provides a serious underestimate of the magnitude of geographic population structure in this species, perhaps because geographically uniform selection balances *Aat-1* allele frequencies despite the severe historical and contemporary restrictions on gene flow that appear to be registered by many other genetic traits.

The converse of this situation may apply to a well-studied allozyme polymorphism in *Drosophila melanogaster* for alcohol dehydrogenase (*Adh*), an enzyme whose main biochemical function is to catabolize ethanol that is abundant in fermented fruits in the fruit flies' natural environment. Several studies have shown that the Adh^F allele has significantly higher enzymatic activity than Adh^S, but is less heat resistant, and that these and other biochemical and physiological attributes can translate into fitness differences between *Adh* genotypes in particular experimental regimes (Sampsell and Sims, 1982; van Delden, 1982). In natural populations, frequencies of these two *Adh* alleles sometimes vary locally [e.g., inside vs. outside wine cellars! (Hickey and McLean, 1980)], and also show strong latitudinal clines with Adh^F more common with increasing latitude in both the northern and southern hemispheres (Oakeshott et al., 1982). Such evidence for environmental selection on *Adh* implies that *prima facie* conclusions about gene flow based on this polymorphism alone could be misleadingly low. From numerous other genetic traits, Singh and Rhomberg (1987) conclude that gene flow in *D. melanogaster* indeed is moderately high even on continental scales (*Nm* ≅ 1–3), but overall that natural selection in combination with migration [as well as population history (Hale and Singh, 1991)] must be considered in studies of geographic variation in molecular markers.

Based on similar arguments from comparative geographic patterns, natural selection has been implicated for various molecular polymorphisms in other species as well (e.g., Ayala et al., 1974). When coupled with further lines of evidence for balancing or other forms of selection acting on particular protein systems (Chapter 2), as well as on some DNA characters (e.g., Hughes and Nei, 1988; Kreitman, 1991; MacRae and Anderson, 1988; Nei and Hughes, 1991), it becomes clear that interpretations of geographic population structure under the assumption of strict neutrality are made with some peril. At the very least, conclusions about the forces shaping population structure in any species should be based on information from multiple independent loci.

HISTORICAL DEMOGRAPHIC EVENTS

Because of mathematical tractability, many theoretical models in population genetics yield only the equilibrium expectations between counteracting evolutionary forces [e.g., the diversifying influence of genetic drift in small populations versus the homogenizing influence of gene flow under an island or stepping-stone model (Box 6.4)]. Seldom is it feasible to consider formally the idiosyncratic histories of particular species or to treat nonequilibrium situations. Yet the demographic histories and phylogenies of real species *are* highly idiosyncratic and likely to produce nonequilibrium population structures. Hence, interpretation of empirical genetic structures of natural populations is especially challenging.

For example, in comparative analyses of three anadromous species of fish along the same coastline transect in the southeastern United States, Bowen and Avise (1990) were led to consider several historical demographic and zoogeographic factors that might have produced the observed differences in geographic population structure as registered by mtDNA. All species showed significant haplotype differences between the Atlantic and Gulf of Mexico, but the magnitude and pattern of the genetic variation differed greatly among the taxa. The black sea bass (*Centropristis striata*) showed little within-region polymorphism and a clear phylogenetic distinction between Atlantic and Gulf populations; menhaden (*Brevoortia tyrannus* and *B. patronus*) showed extensive within-region polymorphism and a paraphyletic (Fig. 4.10) genealogical relationship between the Atlantic and Gulf; and sturgeon (*Acipenser oxyrhynchus*) exhibited extremely low mtDNA variation both within regions and overall. From the mtDNA variabilities observed, estimates of evolutionary effective population sizes (Chapter 2) proved to vary by more than four orders of magnitude, from $N_{f(e)} = 50$ (Gulf of Mexico sturgeon) to $N_{f(e)} = 800,000$ (Atlantic menhaden), and to be rank-order correlated with current-day census sizes. These differences in $N_{f(e)}$, which presumably reflect widely different demographic histories of the three species, may help to explain some of the phylogenetic features, including the clean distinction between Atlantic and Gulf forms of the sea bass versus the paraphyletic pattern in menhaden. From prior population genetic analyses of several other maritime species (see upcoming section on phylogeography), Atlantic versus Gulf populations commonly exhibit genetic differences that appear to trace to Pleistocene events. All else being equal, smaller populations isolated by such events should evolve to a status of reciprocal monophyly in a gene tree more rapidly than large populations (Chapter 4). However, historical separations between the Atlantic and Gulf cannot explain all features of the population genetic structure in these three fishes. Thus, for the menhaden and sturgeon (but not the sea bass), recent gene flow between the Atlantic and Gulf also was implicated strongly by the shared presence of a few nearly identical mtDNA haplotypes (that would not likely have remained unaltered over long evolutionary periods).

Whether these particular inferences are correct or not, they serve to introduce some of the historical demographic considerations and nonequilibrium environmental conditions that *must* have impacted the genetic structures of real populations. In interpreting empirical data on population structures, a challenge involves deciding how far to pursue idiosyncratic demographic explanations, particularly because these can seldom be tested critically in controlled or replicated settings (however, see Fos et al., 1990; Scribner and Avise, 1993b; Wade and McCauley, 1984) and because alternative scenarios might account for the data. Nonetheless, cognizance of the limitations of equilibrium theory, and of the potential impact of historical demographic factors on population-genetic structures, must represent steps toward greater realism.

PHYLOGEOGRAPHY

Background

The introduction of mtDNA data to population genetics in the late 1970s prompted a revolutionary shift in attitude toward such historical, phylogenetic perspectives on intraspecific population structure. Because of the maternal, nonrecombining mode of mtDNA inheritance and rapid evolution in mtDNA sequence, the molecule often provides multiple alleles or haplotypes that can be ordered phylogenetically within a species, yielding intraspecific phylogenies (gene genealogies) interpretable as a matriarchal component of the organismal pedigree. MtDNA transmission in animal species constitutes the female analogue of male surname transmission in many human societies (Avise, 1989b)—both sons and daughters inherit their mother's mtDNA genotype, which only daughters normally transmit to the next generation. Thus, mtDNA lineages reflect mutationally interrelatable "female family names" of a species, and their historical dynamics can be interpreted according to the types of theoretical models long used by human demographers to analyze surname distributions (Lasker, 1985; Lotka, 1931; Chapter 4). Furthermore, mtDNA clones and clades within many species have proved to be geographically localized. Such observations prompted introduction of the word "phylogeography" (Avise et al., 1987a), which refers to the study of the principles and processes governing the geographic distributions of genealogical lineages, including those at the intraspecific level.

One of the first phylogeographic applications of mtDNA data still serves as a useful introduction to the types of population structures frequently revealed. The pocket gopher (*Geomys pinetis*) is a fossorial rodent that inhabits a three-state area in the southeastern United States. Analysis of 87 individuals from across this range by 6 restriction enzymes revealed 23 different mtDNA genotypes whose phylogenetic relationships and distributions are presented in Figure 6.6. Clearly, most mtDNA haplotypes in these gophers are localized geographically,

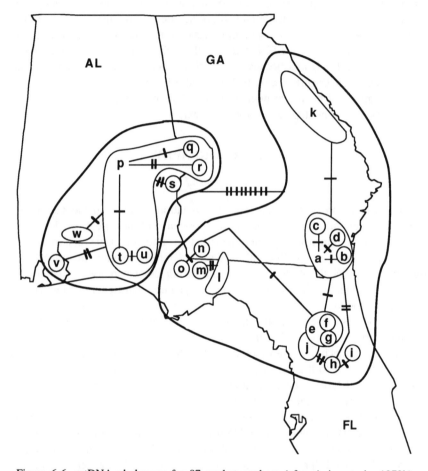

Figure 6.6. mtDNA phylogeny for 87 pocket gophers (after Avise et al., 1979b). Lowercase letters represent different mtDNA genotypes, which are connected by branches in a parsimony network that is superimposed over the geographic sources of the collections in Alabama, Georgia, and Florida. Slashes across network branches reflect the numbers of inferred mutational steps along a pathway. Heavier lines encompass two distinctive mtDNA clades that differ by at least nine mutational steps.

appearing only at one or a few adjacent collection sites. Furthermore, genetically related clones tended to be geographically contiguous or overlapping, and a major gap in the matriarchal phylogeny exhibited a strong geographic orientation cleanly distinguishing eastern from western populations.

Population subdivision characterized by localized genealogical structure and/or significant mtDNA phylogenetic gaps across a species' range subsequently have been reported in a wide variety of animal species: mammals ranging from voles and mice to some whales (Carr et al., 1986; Cronin et al., 1991a; Cronin, 1992; MacNeil and Strobeck, 1987; Plante et al., 1989; Prinsloo and Robinson, 1992; Riddle and Honeycutt, 1990; Wada et al., 1991); birds ranging from sparrows to geese (Avise and Nelson, 1989; Shields and Wilson, 1987b; Van Wagner and Baker, 1990; Zink, 1991); reptiles ranging from gekkos to tortoises (Densmore et al., 1989a; Lamb and Avise, 1992; Lamb et al., 1989; Moritz, 1991); amphibians (Wallis and Arntzen, 1989); freshwater and some marine fishes (Avise, 1987; Bermingham and Avise, 1986; Crosetti et al., 1993); insects (Hale and Singh, 1987, 1991; Harrison et al., 1987); snails (Murray et al., 1991); horseshoe crabs (Saunders et al., 1986); and many others (reviews in Avise et al., 1987a; Moritz et al., 1987; Wilson et al., 1985). On the other hand, several species have proved to exhibit little or no mtDNA phylogeographic structure across broad ranges; for example, some large mammals such as black bears (Cronin et al., 1991b) and coyotes (Lehman and Wayne, 1991; Lehman et al., 1991); some birds (Ball et al., 1988; Tegelström, 1987a); several marine invertebrates and fishes (Árnason et al., 1992; Avise, 1987; Ovenden et al., 1992; Palumbi and Wilson, 1990); and the nematode *Ostertagia ostertagi* [a parasite of cattle that may have been spread widely by transport of livestock (Blouin et al., 1992)]. In general, differences in organismal mobility and in environmental fragmentation appear to exert important influences on patterns of mtDNA phylogeographic structure. Thus, low-vagility pocket gophers and tortoises exhibit pronounced spatial-genetic structure, as do many freshwater fishes across isolated drainages, whereas mobile species that occupy more or less continuously suitable habitat—e.g., some birds, marine fishes, and large terrestrial mammals—often exhibit less spatial differentiation.

Presumably, the localization of related mtDNA clones in most species reflects limited contemporary gene flow (at least via females), and the deeper genetic breaks sometimes observed may evidence longer-term historical population separations. Several authors have discussed the philosophical distinctions between contemporary gene flow versus historical connectedness between populations in a genealogical sense (Avise, 1989a; Larson et al., 1984; Slatkin, 1987). What follows are a few illustrations of how mtDNA analyses have added a phylogenetic dimension to perspectives on intraspecific population structure.

Case Histories

THE ORIGIN OF GREEN TURTLES ON ASCENSION ISLAND

This major rookery for green turtles (*Chelonia mydas*) is a tiny (8 km in diameter) island situated on the mid-Atlantic ridge, halfway between Brazil and Liberia. From direct tagging studies, it is known that the females that nest on Ascension otherwise inhabit shallow-water feeding pastures along the South American coastline. Thus, for each nesting episode (every two to three years for an individual), females embark on a 5000-km migration to Ascension Island and back, in a several-month odyssey involving navigational and exertional feats that nearly defy human comprehension. How might Ascension turtles have established such an unlikely migratory circuit, particularly since suitable nesting beaches along the coast of South America are utilized by other green turtles? Carr and Coleman (1974) proposed an historical biogeographic scenario involving plate tectonics and natal homing. Under their hypothesis, the ancestors of Ascension Island green turtles nested on islands adjacent to South America in the late Cretaceous, soon after the opening of the equatorial Atlantic Ocean. Over the past 70 million years, these volcanic islands have been displaced from South America by sea-floor spreading (at a rate of about 2 cm per year). A population-specific instinct to migrate to the present-day Ascension Island thus was proposed to have evolved over tens of millions of years of genetic isolation (at least with regard to matriarchal lineages) from other green turtle rookeries in the Atlantic.

Bowen et al. (1989) critically tested the Carr–Coleman hypothesis by comparing mtDNA genotypes of Ascension Island nesters with those of other green turtles. They found fixed or nearly fixed mtDNA differences between many Atlantic rookeries, consistent with severe restrictions on contemporary interrookery gene flow by females and, thus, with the natal homing aspects of the Carr–Coleman hypothesis. However, the *magnitude* of sequence divergence of the Ascension rookery from others in the Atlantic [$p < 0.002$ (Fig. 6.5)] indicates a very recent separation, within at most the last 100,000 years (under a conventional vertebrate mtDNA clock) or 800,000 years [under a slower mtDNA clock calibration proposed for turtles (Avise et al., 1992a)]. Indeed, these are upper estimates of divergence times because in available assays the Ascension mtDNA haplotype proved indistinguishable from that in a geographically proximate Brazilian rookery (Bowen et al., 1992). In any event, the genetic results clearly are incompatible with the Carr–Coleman scenario—colonization of Ascension Island, or at least extensive matriarchal gene flow into the population, has been evolutionarily recent.

MIGRATORY HISTORIES OF MONARCH BUTTERFLIES

Each year, monarch butterflies (*Danaus plexippus*) from across North America undertake massive migrations that culminate in huge overwintering aggregations. Monarchs from east of the Rocky Mountains migrate to refugia in the Transvolcanic Range of Central Mexico, and those from west of the Rockies winter along the central California coast. In terms of allozymes, Eanes and Koehn (1978b) first noted that monarchs from across the eastern range showed very low geographic differentiation when compared to other species, and a recent survey based on mtDNA restriction sites was unable to document any genetic differences between populations east versus west of the rockies (Brower and Boyce, 1991). Although it is impossible to prove a null hypothesis (in this case that there are no genetic differences between eastern and western populations), the results do suggest that historical contact between the two areas has been extensive and recent. The monarch is the only temperate representative of an otherwise exclusively tropical subfamily of the Nymphalidae, so on these grounds too it appears plausible that the current migrational phenomenon may represent a post-Pleistocene colonization of North America from one or a few nearly homogeneous ancestral sources.

INTERBROOD SWITCHING IN PERIODICAL CICADAS

Species of *Magicicada* have an unusual life cycle in which larvae remain in the ground for long periods, only to emerge *en masse* and mate at periodic intervals (e.g., every 17 years in *M. septendecim,* and every 13 years in *M. tredecim*). Each year class is referred to as a brood. Life-cycle length commonly is assumed to be a genetically fixed species-specific trait, and differences in emergent periodicity supposedly impose severe temporal reproductive isolation between broods (13 and 17 year broods should co-emerge only once every 13 × 17 = 221 years!). Alternatively however, periodical cicadas might switch between life-cycle lengths in response to genetic or environmental factors (Lloyd and White, 1976; Lloyd et al., 1983). Some such switches must have occurred evolutionarily to account for the origins of broods with different periodicities.

Martin and Simon (1988, 1990) observed an anomalous distribution of mtDNA, allozyme, and morphological markers in *M. septendecim* and *M. tredecim* in the central and eastern United States; in all data sets, 13-year cicadas from a large region of their northern range proved more similar to 17-year cicadas than to 13-year cicadas collected elsewhere in the same year. This distribution of phylogenetic markers, in conjunction with geographical information and other aspects of the data, were interpreted to evidence an historical switch in maturation time, in which a large number of 17-year cicadas underwent a 4-year acceleration in development to become 13-year cicadas, thus isolating the new population temporally from the parent brood. Furthermore, in this case, the developmental acceleration led to an overlap in emergence time with preexisting 13-year broods, thus opening new opportunities for gene flow between former temporal isolates.

NATURAL SELECTION AND BIOGEOGRAPHIC HISTORY IN THE KILLIFISH

In *Fundulus heteroclitus*, two common alleles at a lactate dehydrogenase (LDH) nuclear gene exhibit a pronounced clinal shift in frequency along the east coast of the United States (Fig. 6.7). Detailed laboratory studies have revealed kinetic and biochemical differences between these LDH alleles that predict significant differences among individuals in metabolism, oxygen transport, swimming performance, developmental rate, and relative fitness (see the review in Powers et al., 1991a). The nature of these differences is such that latitudinal shifts in environmental temperature have been posited as directly responsible for the clinal allelic structure (Mitton and Koehn, 1975; Powers et al., 1986). Does adaptation to local conditions provide the entire story for the genetic architecture of these killifish populations?

González-Villaseñor and Powers (1990) surveyed populations along the same coastal transect with regard to restriction site variation in mtDNA and demonstrated a pronounced phylogenetic subdivision of *F. heteroclitus* into northern versus southern units. Quite likely, these populations were isolated during the Pleistocene, and now hybridize secondarily along the mid-Atlantic coast in such a way as to contribute to the clinal structure observed in LDH and in some other nuclear genes. This example demonstrates how phylogenetic and selective mechanisms need not be opposing influences, but may in some cases act in concert to achieve an observed population structure (Powers et al., 1991b). Although the differences in LDH allele frequency are mediated to a significant degree by environmental selection, the historical context in which this selection has taken place adds an important dimension to our understanding of the contemporary population genetic architecture.

HUMAN POPULATIONS—OUT OF AFRICA?

Not surprisingly, a great deal of attention has been directed toward estimating and interpreting mtDNA phylogeny within *Homo sapiens* (e.g., Ballinger et al., 1992; Cann et al., 1984; Denaro et al., 1981; DiRienzo and Wilson, 1991; Excoffier, 1990; Johnson et al., 1983; Merriwether et al., 1991; Stoneking et al., 1986; Torroni et al., 1992; Ward et al., 1991; Whittam et al., 1986). However, two publications stand out as having had major historical and conceptual impact.

First, Brown (1980) provided an early glimpse of global mtDNA diversity by assaying mtDNA RFLPs from 21 humans of diverse racial and geographic origin. Genetic differentiation was rather limited, with mean sequence divergence estimated at only $p = 0.0036$. Using standard mtDNA clock calibrations, Brown (1980) concluded that this level of mtDNA sequence heterogeneity "could have been generated from a single mating pair that existed $180-360 \times 10^3$ years ago, suggesting the possibility that present-day humans evolved from a small mito-

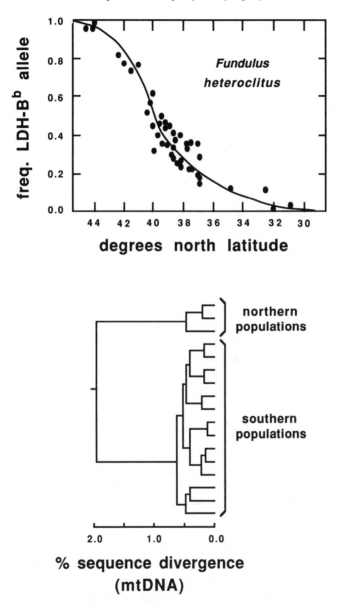

Figure 6.7. Molecular geographic patterns in the killifish (after Powers et al., 1991a). **Top:** Latitudinal cline in population frequencies of the nuclear "b" allele of a lactate dehydrogenase polymorphism. **Bottom:** Phenogram of mtDNA haplotypes in these same populations, showing the major phylogenetic distinction between genotypes characteristic of northern versus southern populations.

chondrially monomorphic population that existed at that time.'' The magnitudes of these estimated sequence divergences and separation times have changed relatively little with the addition in subsequent studies of many more samples as well as direct mtDNA sequence information (e.g., Hasegawa and Horai, 1991; Vigilant et al., 1991). Thus, the overall mtDNA picture for humans remains one of fairly shallow lineage separations, at least relative to conspecific populations of many other species similarly assayed. [In this regard, the mtDNA results also parallel findings from nuclear genomes that human populations and races are remarkably similar in molecular composition, notwithstanding obvious phenotypic differences in traits such as hair texture and skin color (Nei and Livshits, 1990; Nei and Roychoudhury, 1982). For example, in an early summary of the protein electrophoretic literature, Nei (1985) concluded that ''the net gene differences between the three races of man, Caucasoid, Negroid, and Mongoloid, are much smaller than the differences between individuals of the same races, but this small amount of gene differences corresponds to a divergence time of 50,000 to 100,000 years.'']

In any event, the provocative conclusion of the Brown (1980) study was that the coalescence of extant human mtDNA lineages might trace to a single female (dubbed ''Eve'' by the popular press) within the last few hundred thousand years, and furthermore that this indicated a severe bottleneck in absolute human numbers (the ''Garden of Eden'' scenario). This *latter aspect* of the scenario soon was challenged with results of models and computer simulations of population lineage sorting as a function of population demography (Chapter 4). From such a theory on gene genealogies, Avise et al. (1984a) concluded that '' 'Eve' could have belonged to a population of many thousands or tens of thousands of females, the remainder of whom left no descendants to the present day, due simply to the stochastic lineage extinction associated with reproduction.'' In other words, the observation that the genealogy of mtDNA (or any other locus) coalesces does not necessarily imply an extreme bottleneck in absolute population numbers (see also Hartl and Clark, 1989, p. 90). Latorre et al. (1986) were more blunt: Even if it is true that all extant mtDNA lineages do trace to a single female some 200,000 years ago, ''the Mother Eve hypothesis is . . . fallacious.'' Recent analyses of allelic genealogies at nuclear loci have bolstered these conclusions. Thus, as summarized by Takahata (1993), ''Genetic variation at most loci examined in human populations indicates that the (effective) population size has been $\cong 10^4$ for the past 1 Myr . . . (and) that the population size has never dropped to a few individuals, even in a single generation.''

A second influential study extended the mtDNA survey to 147 humans from around the world and produced a parsimony tree whose root traced to the African continent (Cann et al., 1987). The findings led to the ''out of Africa'' hypothesis, whereby human maternal lineages originated in Africa and spread within the last few hundred thousand years to the rest of the world, replacing those of other archaic populations. This conclusion also has proved highly controversial. One criticism has come from some paleontologists who on the basis of fossil or other evidence favor a multiregional origin for humans that far predates the apparent time scale of the mtDNA spread (e.g., Wolpoff, 1989; Wolpoff et al., 1984; however, see also Stringer and Andrews, 1988; Wilson et al., 1991). Another criticism has come from some geneticists who found that the postulated African root of the molecular phylogeny was not strongly supported (nor refuted) when additional tree-building analyses were applied to the mtDNA data (Hedges et al., 1992c; Maddison, 1991; Templeton, 1992). [Evidence remains unchallenged that extant populations in Africa house the highest level of mtDNA polymorphism and this has provided a second argument for an African mtDNA root (Cann et al. 1987).] Perhaps the truth lies somewhere between the extreme versions of the out-of-Africa and the multiregional models. Thus, a third possible scenario is that humans with modern anatomical features appeared first in Africa and then spread throughout the world, not completely replacing archaic populations but rather interbreeding with them to some extent (Li and Sadler, 1992). If so, the gene genealogies for some fraction of nuclear loci might well retain branches tracing to early roots on continents other than Africa.

In any event, most of the recent discussions of human origins based explicitly on gene trees have involved only the matrilineal component of our genetic history, as recorded by mtDNA. At first thought, it might be supposed that knowledge of the analogous patrilineal component (i.e., the transmission pathway of the Y chromosome) would complete the story by revealing an "Adam" ["the father of us all" (Gibbons, 1991)], but this is fallacious. The vast majority of our genetic heritage involves nuclear loci whose alleles trace through both sexes during human evolutionary history. In theory, such nuclear genealogies can, of course, differ from that of mtDNA as well as from one gene to the next. As yet, few such nuclear genealogies have been estimated empirically, but analyses of haplotypes at the loci encoding apolipoprotein and B-globin also were interpreted to support an African origin for *Homo sapiens sapiens,* with later dispersal to Europe and the Pacific (Rapacz et al., 1991; Wainscoat et al., 1986; however, see Giles and Ambrose, 1986). On the other hand, recent analyses of the apolipoprotein C-II gene led Xiong et al. (1991) to suggest an Asian origin for these sequences. Nonetheless, because of the limited amount of genealogical data currently available for nuclear loci and the great potential for such information gain through additional sequencing, at this time only one aspect of human origins seems absolutely certain—that interest and debate will continue (Goldman and Barton, 1992).

GENEALOGICAL CONCORDANCE AND THE SHARP-TAILED SPARROW

Because gene trees within an organismal phylogeny can differ considerably from locus to locus (Chapter 4), under what circumstances might the lineages within gene genealogies faithfully reveal significant phylogenetic subdivisions within conspecific pedigrees? One initial consideration is whether a particular gene tree receives statistical support by criteria such as bootstrapping. Given that a clade in a gene tree has been verified, the next consideration is whether concordant genealogical support exists at other, independent loci. Ideally, an examination of intraspecific phylogenies (of the sort provided by mtDNA) for each of several unlinked nuclear genes would be desirable. In the absence of such extensive genealogical data, surrogate genetic information must be employed.

For example, a recent molecular analysis of sharp-tailed sparrows (*Ammodramus caudacutus*) uncovered a highly significant mtDNA distinction that divides the species into two phylogeographic groups: a "northern" assemblage currently inhabiting the interior prairies of the United States and Canada, the Hudson Bay lowlands, the St. Lawrence River Valley, and the Canadian maritime provinces to Maine; and a "southern" group that occurs along the Atlantic coast from southern Maine to Virginia (Rising and Avise, 1993). Populations belonging to these two mtDNA clades proved to be recognizable concordantly by multivariate analyses of morphological attributes, as well as by behavioral distinctions including song and flight displays (Greenlaw, 1993; Montagna, 1942). Such concordance among independent lines of evidence suggests strongly that, in this case, the split in the mtDNA gene tree also reflects a phylogenetic distinction in the organismal phylogeny (likely tracing to Pleistocene population separations). This example is of further interest because the subdivision of the sharp-tailed sparrow into two distinctive phylogenetic units does not coincide with the existing subspecific designations for the complex, nor with the present-day geographic subdivisions upon which this taxonomy in part is based.

PHYLOGEOGRAPHY OF A REGIONAL FAUNA

Another potential class of genealogical concordance involves shared phylogeographic patterns across unrelated taxa, presumably due to similar historical influences on intraspecific genetic architectures. A remarkable case in point involves the maritime and freshwater faunas of the southeastern United States, where a total of 19 species has been compared for phylogeographic patterning in mitochondrial (and in some cases nuclear) markers (see the review in Avise, 1992). Among 10 estuarine species or species complexes surveyed throughout most of their respective distributions [including creatures as diverse as the horseshoe crab (*Limulus polyphemus*), American oyster (*Crassostrea virginica*), black sea bass (*Centropristis striata*), diamondback terrapin (*Malaclemys terrapin*), and seaside sparrow (*Ammodramus maritimus*)], populations of at least five and probably eight species evidence a fundamental mtDNA phylogenetic break between the Atlantic and Gulf of Mexico shorelines (Figs. 6.8 and 6.9). The striking concordance in geographic patterns (though not always in the magni-

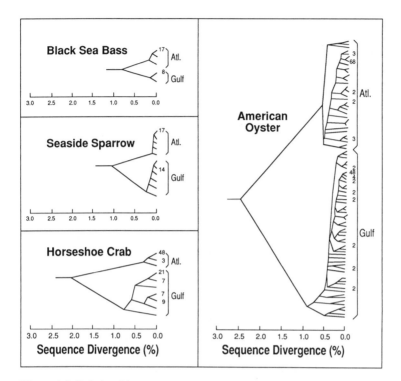

Figure 6.8. Relationships among mtDNA haplotypes in four maritime species of the southeastern United States (from Avise, 1992). Numbers of individuals belonging to various mtDNA clones are indicated to the right, whereas terminal branches without numbers were represented by single assayed individuals. Note that all UPGMA phenograms are plotted on the same scale of estimated mtDNA sequence divergence.

tudes of separation) strongly suggests an overriding influence of shared historical biogeographic forces. Most likely, ancestral forms of the various taxa were separated during the Pleistocene into disjunct Atlantic and Gulf isolates (perhaps via changes in sea level and the associated alterations in estuarine habitat), with subsequent dispersal and contemporary ecological influences leading to the present-day distributions of these clades. The different depths in the mtDNA gene trees might reflect heterogeneity in molecular evolutionary rates across taxa and/or isolations that date to different Pleistocene glacial or interglacial episodes. The Gulf Stream, a major oceanic current that flows out of the Gulf of Mexico and hugs the southern Florida coastline, is most likely an important contemporary influence on gene flow, perhaps facilitating in some species the inferred "leakage" of Gulf mtDNA haplotypes into southeastern Florida (Fig. 6.9). The

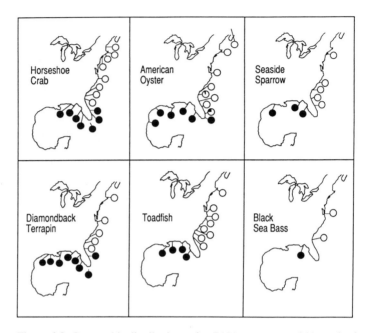

Figure 6.9. Geographic distributions of mtDNA genotypes within each of six maritime taxa (from Avise, 1992). Shown are pie diagrams summarizing frequencies of the two fundamental clades in populations of each species.

zone of demarcation between conspecific populations near Cape Canaveral (in eastern Florida) is also a long-recognized boundary between marine zoogeographic provinces, as identified by a more traditional type of zoogeographic evidence—concentrations of species' distributional limits (Briggs, 1958, 1974). Similar historical processes may be evidenced by these intraspecific and interspecific phenomena; for example, if either of the distinctive conspecific units recognized in the mtDNA analyses had become extinct, current northern or southern range limits of the extant sister taxon likely would occur along eastern Florida.

A similar pronounced pattern of intraspecific mtDNA genealogical concordance across taxa characterizes freshwater fishes in the southeastern United States (Avise, 1992; Bermingham and Avise, 1986; Scribner and Avise, 1993a). Within each of six species or species complexes examined genetically, ranging from the bowfin (*Amia calva*) to mosquitofish (*Gambusia affinis/holbrooki*) and spotted sunfish (*Lepomis punctatus*), a sharp phylogenetic subdivision distinguishes populations in eastern (primarily Atlantic coast) from western (primarily Gulf coast) drainages (Figs. 6.10 and 6.11). The geographic dividing line differs

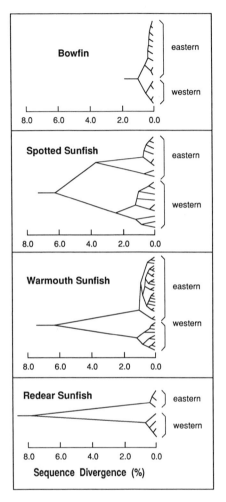

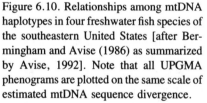

Figure 6.10. Relationships among mtDNA haplotypes in four freshwater fish species of the southeastern United States [after Bermingham and Avise (1986) as summarized by Avise, 1992]. Note that all UPGMA phenograms are plotted on the same scale of estimated mtDNA sequence divergence.

somewhat from one species to the next (as do the magnitudes of genetic distance), but considering the number and variety of species examined, shared aspects of the phylogeographic patterns remain striking. As in the maritime realm, these concordant features presumably reflect shared elements in biogeographic histories, likely involving Pleistocene refugia in eastern and western drainages with subsequent dispersal routes influenced by subsequent drainage interconnections. The general zone of demarcation between the eastern and western freshwater populations also corresponds well to a contemporary boundary between piscine zoogeographic provinces as identified by concentrations of species' distributional limits (Swift et al., 1985).

It remains to be seen how often multispecies assemblages within other regional

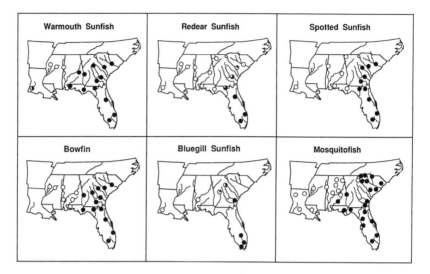

Figure 6.11. Geographic distributions of mtDNA genotypes within each of six fresh-water fish species (from Avise, 1992). Shown are pie diagrams summarizing frequencies of the two fundamental clades in populations of each species.

biotas will prove to exhibit such pronounced patterns of phylogeographic concordance. In the American southwest, populations of the desert tortoise (*Xerobates agassizi*) east versus west of the Colorado River exhibit pronounced mtDNA differentiation that was interpreted provisionally as stemming from the isolating effects of historical marine incursions in the area (Lamb et al., 1989). However, subsequent mtDNA analyses of two other reptile species with similar ranges (the desert iguana *Dipsosaurus dorsalis* and the chuckwalla *Sauromalus obesus*) failed to reveal any phylogeographic concordance with the desert tortoise pattern (Lamb et al., 1992). Each species displays a unique signature of genetic differentiation, but there is no evidence across species for the overriding impact of any singular biogeographic factor.

General Conclusions About Intraspecific Phylogeography

Experience with the intraspecific mtDNA architectures of numerous species has led to the formulation of several phylogeographic hypotheses and their corollaries (Box 6.6). Whether or not these particular hypotheses are confirmed with additional data from mtDNA and nDNA genealogies, it can be argued that a more explicit concern with the historical aspects of population structure should assume a place in evolutionary study at least commensurate with that of "ecogeography," a long-established discipline that focuses on selection-mediated trends in geographic distributions of organismal attributes. For example, Bergmann's eco-

---

Box 6.6. Phylogeographic Hypotheses Derived from Studies of mtDNA Gene Genealogies Within Species (after Avise et al., 1987a)

I. Most species are composed of geographic populations whose members occupy different branches of an intraspecific, phylogenetic tree (pedigree).

II. Species with limited or "shallow" phylogeographic population structure have life histories conducive to dispersal and have occupied ranges free of long-standing impediments to gene flow.

III. Monophyletic groups distinguished by large phylogenetic gaps usually arise from long-term extrinsic (biogeographic) barriers to gene flow.

Corollaries
   (a) As time since isolation increases, the degree of phylogeographic concordance across separate gene genealogies increases.
   (b) The geographic placements of phylogenetic gaps may tend to be concordant across species.
   (c) Phylogenetic gaps within species tend to be geographically concordant with boundaries between traditionally-recognized biogeographic provinces.

---

geographic rule notes the tendency in homeotherms for larger body sizes at higher latitudes, presumably a surface/volume adaptation for heat conservation in colder climates; Allen's rule notes a latitudinal trend in lengths of limbs, perhaps because shorter extremities serve similarly to conserve heat in cold climates; Gloger's ecogeographic rule notes a tendency for populations in humid areas to be more heavily pigmented, perhaps a manifestation of selection for background-matching related to predation and competition. Emphasis on these and related ecogeographic trends perhaps reflects the prior dominance of an "adaptationist paradigm" (Gould and Lewontin, 1979) for viewing intraspecific evolution, but this perspective should now be balanced with the additional insight to be gained from a phylogeographic perspective. The proposed place for phylogeography within the broader discipline of biogeography is summarized in Figure 6.12.

A gene genealogy within any species represents but one realization of the process of lineage sorting through an organismal pedigree and, hence, must be interpreted with caution as an indicator of overall genomic history. In other words, in the absence of concordant support from the phylogenies of independent loci, significant partitions in a gene tree cannot necessarily be assumed to evidence fundamental historical subdivisions at the population level (Chapter 4; Avise and Ball, 1990). However, additional support for long-standing population subdivisions may take any of several forms, including (a) a correspondence between significant phylogenetic subdivisions in a gene tree and the boundaries between historical biogeographic provinces as evidenced by nonmolecular data, (b) a concordance between subdivisions in a gene genealogy and subdivisions

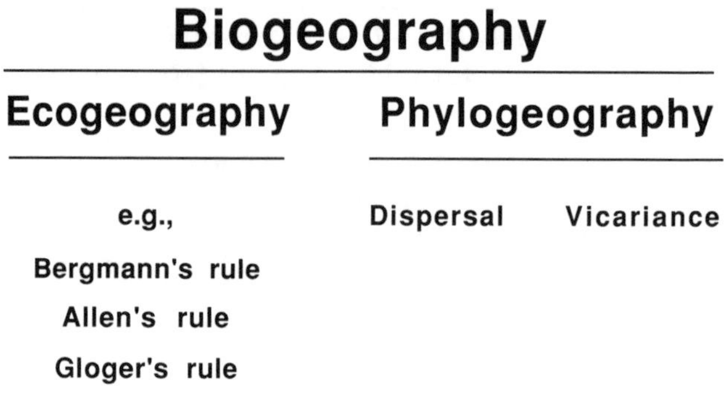

Biogeography

Ecogeography	Phylogeography

	Dispersal Vicariance
e.g.,	
Bergmann's rule	
Allen's rule	
Gloger's rule	

Figure 6.12. The proposed status of "phylogeography" within the broader discipline of biogeographic study. An explicit concern with phylogenetic aspects of intraspecific population structure should help to broaden and enrich the selectionist perspectives of "eco-geography" and encompass the realization that both past demographic conditions (including dispersal and gene flow), as well as historical environmental disjunctions [vicariance (see Chapter 8)], no doubt have played influencing roles in generating present-day lineage distributions.

registered by other attributes such as morphology or behavior (as in the sharp-tailed sparrow), or (c) geographic concordance in significant genealogical partitions among the populations of several independent species (as for the freshwater and maritime faunas of the southeastern United States).

Thus, in interpreting gene phylogenies, a distinction should be drawn between generalized conclusions about the magnitude of population structure and gene flow, and the evidence for particular historical population separations. In species whose populations exhibit contemporary isolation by distance in the absence of long-standing biogeographic barriers to dispersal, phylogenies of independent genes each may reveal significant population subdivision, yet exhibit little concordance in the particular population units identified. Such results imply generalized contemporary restrictions on gene flow, but the particular populations distinguished by any gene may be of little phylogenetic consequence. In contrast, when populations or metapopulations are recognized concordantly by multiple lines of molecular or other genetic evidence, significant historical separations in the organismal pedigree are implicated.

MICROTEMPORAL PHYLOGENY

Most DNA sequences evolve far too slowly to permit the direct temporal monitoring of significant phylogenetic changes in populations over yearly or decade-long time scales. One valiant attempt to describe such changes involved the comparison of mtDNA sequences in extant and museum-preserved samples of the Panamint kangaroo rat (*Dipodomys panamintinus*) taken from the same locales in California (Thomas et al., 1990). The museum samples were dried skins prepared in 1911, 1917, and 1937, from which a mtDNA segment was amplified by the PCR. The results showed a temporal stability in genotype composition over this century, with relationships among three geographic populations being the same for the modern as for the museum collections. However, even if genetic changes had been observed, presumably most would have involved shifts in frequencies of preexisting ancestral haplotypes (as can occur rapidly under genetic drift or shifting family structure, for example) rather than the evolutionary accumulation of de novo mutations over such a short time scale.

Molecular systems which apparently *do* evolve rapidly enough to permit the monitoring of short-term mutational changes involve RNA viruses including HIV, the human immunodeficiency retrovirus that produces AIDS (acquired immune deficiency syndrome). The mean rate of synonymous nucleotide substitution for the HIV genome is approximately 10×10^{-3} per site per year (W.-H. Li et al., 1988), or about 1,000,000 times greater than typical rates in the nuclear genomes of most higher organisms (Chapter 2). Presumably, a rapid pace of change in particular regions of these RNAs (along with the possibility of interstrain recombination) underlies the observed rapid rates of change in viral pathogenicity and antigenicity (Coffin et al., 1986; Gallo, 1987). Another consequence of the rapid mutational process is that HIVs from different individuals [and sometimes within the same individual through time (Holmes et al., 1992)] usually are found to be genetically distinct but related, and such heterogeneity has found epidemiologic and forensic applications. For example, in sequence comparisons of a HIV from a dentist, 7 of his infected patients, and 35 other HIV carriers from the local geographic area, it was shown that the viruses from the dentist and 5 of his patients were related closely (Ou et al., 1992). These molecular markers provided the first genetic confirmation for transmission of HIV from an infected health care worker to clients.

The human AIDS virus appears to have originated in Africa, perhaps within the past few decades following a postulated transfer from wild monkeys (Diamond, 1992) and, subsequently, has infected millions of people worldwide. Studies of HIV nucleotide sequences have played an important role in documenting the histories and patterns of migration of the virus (Desai et al., 1986; Gallo, 1987; Yokayama and Gojobori, 1987). For example, W.-H. Li et al. (1988) analyzed the sequences from 15 HIV isolates from the United States, Haiti, and Zaire to generate the estimated phylogeny reproduced in Figure 6.13. Results are consistent with the hypothesized African ancestry of HIV, its subsequent spread to Haiti, and later to the United States. Most remarkable about these phylogenetic reconstructions are the micro-time-scales involved. From the

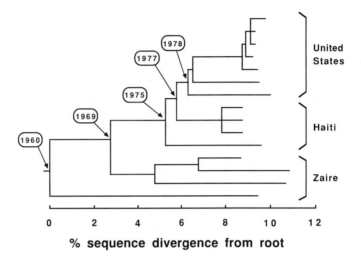

Figure 6.13. Phylogenetic relationships among 15 HIV isolates based on nucleotide sequence data (after W.-H. Li et al., 1988). Note the micro-time-scales as reflected by recent dates for various nodes in the tree.

synonymous substitution rates in HIV, the inferred separation date of the New World from the African isolates was 1969, and the separation date of the United States from the Haitian isolates was 1977!

SUMMARY

1.   All conspecific individuals are related genealogically through an extended pedigree that constitutes the intraspecific phylogeny of a species.

2.   Molecular approaches for assessment of kinship within a species normally require highly-polymorphic qualitative markers with known transmission patterns. However, because of the complexity of potential transmission pathways between relatives and the indeterminate nature of Mendelian inheritance, *exact* kinship relations between particular pairs of individuals seldom can be established, and analyses normally are confined to assessment of *mean* genetic relatedness within groups.

3.   In eusocial species such as many haplo-diploid hymenopterans, mean intracolony relatedness values sometimes have proved to be high, but numerous exceptions present an enigma for sociobiological theories on the evolution of reproductive altruism.

4.   Within noneusocial groups, a variety of mean genetic relatedness values

has been observed using molecular markers, and these often appear interpretable in terms of the suspected behaviors and natural histories of the species assayed. Genetic markers also have helped to address questions regarding mechanisms and genetic consequences of kin recognition.

5. Most populations are genetically structured geographically. These genetic architectures have been revealed for numerous species using molecular markers and appear to be influenced by a variety of factors including the nature of the mating system and the gene flow regime. These, in turn, are influenced by the species-specific dispersal capabilities of gametes and zygotes and by the vagilities as well as philopatries of postzygotic life-history stages.

6. The genetic structure of any species also has been influenced by historical biogeographic and demographic factors. In large measure, historical perspectives on population genetic structure were stimulated by the relatively unambiguous genealogical reconstructions made possible by development of assays for nonrecombining mtDNA haplotypes. A new discipline termed ''phylogeography'' has enriched biogeographic study and provides a balance to traditional ecogeographic perspectives.

7. Some viruses evolve so rapidly that genetic changes can be observed directly over time scales of years or decades. Molecular phylogenetic appraisals of the HIV virus have revealed details of its origin and spread within this century.

7

Speciation and Hybridization

*Without gene flow, it is inevitable that there will
be speciation.*

M.H. Wolpoff, 1989

With regard to sexually-reproducing organisms, several "species concepts" have been advanced (Box 7.1). Most of these entail the perception of conspecific populations as a field for gene recombination—in other words, as an extended reproductive community within which genetic exchange potentially takes place. For example, under the popular "biological species concept" (BSC) championed by Dobzhansky (1937), species are characterized as "groups of actually or potentially interbreeding natural populations which are reproductively isolated from other such groups" (Mayr, 1963). Many authors have expressed sentiments on the BSC similar to those of Ayala (1976b): "among cladogenetic processes, the most decisive one is speciation—the process by which one species splits into two or more . . . Species are, therefore, independent evolutionary units. Adaptive changes occurring in an individual or population may be extended to all members of the species by natural selection; they cannot, however, be passed on to different species." Thus, under the BSC and related concepts, species are perceived as biological and evolutionary entities that are far more meaningful and less arbitrary than other taxonomic categories such as subspecies, genera, or orders (Dobzhansky, 1970). Nonetheless, several complications can attend the application of BSC principles.

One of these difficulties involves the discretionary judgments required about the specific status of related extant forms in allopatry (as well as of extant forms

Box 7.1. Examples of Various Species Concepts and Definitions

1. Biological species concept (BSC) (Dobzhansky, 1937): "Species are systems of populations: the gene exchange between these systems is limited or prevented by a reproductive isolating mechanism or perhaps by a combination of several such mechanisms."

   Comment: Unquestionably the most influential species concept, and one that remains popular today.

2. Evolutionary species concept (ESC) (Simpson, 1951): "a lineage (ancestral-descendant sequence of populations) evolving separately from others and with its own unitary evolutionary role and tendencies."

   Comment: Applicable both to living and extinct groups, and to sexual and asexual organisms. However, this concept is vague operationally in what is meant by "unitary evolutionary role and tendencies."

3. Recognition species concept (RSC) (Paterson, 1985): the most inclusive population of biparental organisms which share a common fertilization system.

   Comment: Similar to the BSC in viewing conspecific populations as a field for gene recombination. However, this concept shifts attention from isolating mechanisms as barriers to gene exchange, to a focus instead on the positive function of these mechanisms in facilitating reproduction among members of a species. Although reproductive barriers can indeed arise as a by-product of speciation, they are not viewed as an active part of the speciation process.

4. Cohesion species concept (CSC) (Templeton, 1989): "the most inclusive population of individuals having the potential for cohesion through intrinsic cohesion mechanisms."

   Comment: Attempts to incorporate strengths of the BSC, ESC, and RSC, and avoid their weaknesses. The major classes of cohesion mechanisms are genetic exchangeability (factors that define the limits of spread of new genetic variants through gene flow) and demographic exchangeability (factors that define the fundamental niche and the limits of spread of new genetic variants through genetic drift and natural selection).

5. Phylogenetic species concept (PSC) (Cracraft, 1983): a monophyletic group composed of "the smallest diagnosable cluster of individual organisms within which there is a parental pattern of ancestry and descent."

   Comment: Explicitly avoids all reference to reproductive isolation and focuses instead on phylogenetic histories of populations. A serious problem involves how monophyly is to be recognized and how to distinguish histories of traits (e.g., gene trees) from histories of organisms (pedigrees).

6. Concordance principles (CP) (Avise and Ball, 1990): a suggested means of recognizing species by the evidence of concordant phylogenetic partitions at multiple independent genetic attributes.

   Comment: Attempts to incorporate strengths of the BSC and PSC and avoid their weaknesses. This approach accepts the basic premise of the BSC, with the understanding that the reproductive barriers are to be interpreted as intrinsic as opposed to extrinsic (purely geographic) factors. When phylogenetic concordance is exhibited across genetic characters solely because of extrinsic barriers to reproduction, subspecies status is suggested.

to their evolutionary ancestors). Inevitably, reproductive isolating barriers [RIBs (Box 7.2)] develop between geographically separated populations as a nonadaptive by-product of genomic divergence, but the time scales involved and the magnitudes of differentiation are matters for study in particular instances. The "acid test" for biological species status—whether the populations retain separate identities in sympatry—often has not been carried out in nature. A second practical difficulty involves the issue of how much genetic exchange disqualifies populations from status as separate biological species. Thus, the study of speciation conceptually links the topic of gene flow (Chapter 6) with that of hybridization and introgression. Under the BSC, there are no black-and-white solutions to either of these two difficulties because evolutionary divergence and speciation normally are gradual processes, and because levels of genetic exchange can vary along a continuum (Dobzhansky, 1976).

Another challenge in applying the BSC involves the need to distinguish the evolutionary origins of RIBs from their genetic consequences. Normally, reproductive barriers under the BSC are considered *intrinsic* biological factors rather than purely *extrinsic* limits to reproduction resulting from geographic separation alone. However, this distinction blurs when syntopic populations (those occupying the same macrohabitat) are isolated via preferences for different microhabitats, particularly when these habitat proclivities are coupled with differences in mate choice (Diehl and Bush, 1989). In such situations, one substantive as well as semantic issue is whether speciation occurred sympatrically versus allopatrically followed by secondary range overlap. Another issue is whether certain types of RIBs arise in direct response to selection pressures favoring homotypic matings (for example) or whether they reflect nonselected by-products of genomic differentiation that occurred for other reasons (Box 7.2).

Associated with the speciation process, under any definition, is the conversion of genetic variability within a species to between-species genetic differences. However, because RIBs retain primacy in demarcating species under the BSC, no arbitrary magnitude of molecular genetic divergence can provide an infallible metric to establish specific status, particularly among allopatric forms. Furthermore, as noted by Patton and Smith (1989), almost "all mechanisms of speciation that are currently advocated by evolutionary biologists . . . will result in paraphyletic taxa as long as reproductive isolation forms the basis for species definition." How then may molecular markers inform speciation studies? First, molecular patterns might provide distinctive genetic signatures relatable to demographic events during speciation or to the geographic settings in which speciation took place (Neigel and Avise, 1986; Templeton, 1980a; Fig. 7.1). Second, estimates of genetic differentiation between populations at various stages of RIB acquisition might be useful in assessing temporal aspects of the speciation process (Coyne, 1992). Finally, molecular markers can prove invaluable for assessing the magnitude and pattern of genetic exchange among related forms, and thereby can contribute to an understanding of the intensity and nature of RIBs.

Box 7.2. Classification of Reproductive Isolating Barriers (RIBs)

1. Prezygotic Barriers

   (a) Ecological or habitat isolation: Populations occupy different habitats in the same general region, and most matings take place within these microhabitat types.

   (b) Temporal isolation: Matings take place at different times, e.g. seasonally or diurnally.

   (c) Ethological isolation: Individuals from different populations meet but do not mate.

   (d) Mechanical isolation: Interpopulational matings occur but no transfer of male gametes takes place.

   (e) Gametic mortality or incompatibility: Transfer of male gametes occurs but eggs are not fertilized.

2. Postzygotic Barriers

   (a) F_1 inviability: F_1 hybrids have reduced viability.

   (b) F_1 sterility: F_1 hybrids have reduced fertility.

   (c) Hybrid breakdown: F_2, backcross, or later-generation hybrids have reduced viability or fertility.

One rationale for distinguishing between prezygotic and postzygotic RIBs is that in principle only the former are selectable directly. Under the "reinforcement" scenario of Dobzhansky (1940; Blair, 1955), natural selection can act to superimpose prezygotic RIBs over any preexisting postzygotic RIBs (that may have arisen, for example, in former allopatry). As stated by Dobzhansky [1951, as quoted from Butlin (1989)], "Assume that incipient species, A and B, are in contact in a certain territory. Mutations arise in either or both species which make their carriers less likely to mate with the other species. The nonmutant individuals of A which cross to B will produce a progeny which is adaptively inferior to the pure species. Since the mutants breed only or mostly within the species, their progeny will be adaptively superior to that of the non-mutants. Consequently, natural selection will favour the spread and establishment of the mutant condition."

Notwithstanding its conceptual appeal, Dobzhanksy's suggestion has proven difficult to verify observationally or experimentally (see the review in Butlin, 1989). Koopman (1950) and Thoday and Gibson (1962) provide widely-quoted examples of selective reinforcement of prezygotic RIBs, but other such experimental studies have produced equivocal outcomes (e.g., Spiess and Wilke, 1984).

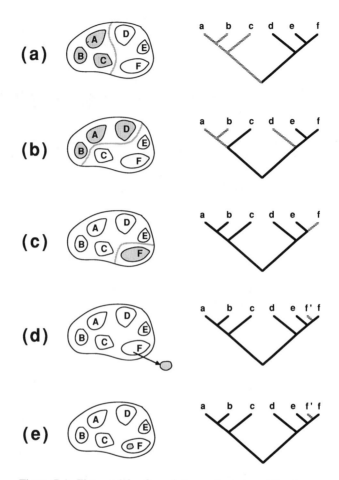

Figure 7.1. Five models of speciation and corresponding gene trees (after Harrison, 1991). Shown are distributions of alleles in the two daughter species. For simplicity, each population is represented as monomorphic, and the gene genealogy in each case is (((a,b)(c))((d)(e,f))). In reality, most populations are likely to be polymorphic and, hence upon separation, are expected to evolve through intermediate states of polyphyly and paraphyly in the gene tree (Chapter 4). The five models are (a) speciation by geographic subdivision with the physical partition congruent with an existing phylogenetic discontinuity; (b) speciation by subdivision with the partition not congruent with an existing phylogenetic discontinuity; (c) speciation in a peripheral population; (d) speciation via colonization of a new habitat by propagule(s) from a single population; and (e) local sympatric speciation. In the allelic phylogenies, solid versus hatched lines represent lineages occurring in the two respective daughter species.

THE SPECIATION PROCESS

What follows are examples of some questions in speciation theory that have been readdressed through use of molecular genetic markers.

How Much Genetic Change Accompanies Speciation?

TRADITIONAL PERSPECTIVES

One long-standing view of species differences, stated clearly by Morgan early in this century (1919), is that species differ from each other "not by a single Mendelian difference, but by a number of small differences." A counterproposal frequently expressed was that new species or even genera might arise by single mutations of a special kind—"macromutations" or "systemic mutations" (deVries, 1910; Goldschmidt, 1940)—that suddenly transform one kind of organism into another. Although such suggestions for saltational speciation probably are untenable in their original formulation, more recent theories have stressed plausible routes by which species might arise rapidly, perhaps in some cases with minimal molecular genetic divergence overall. Examples of such avenues to saltational speciation in plants and animals are summarized in Box 7.3. But apart from these "special cases" (which nonetheless may be quite widespread), can species arise quickly and with little genetic alteration?

One class of argumentation for sudden speciation comes from paleontology. Based on a reinterpretation of "gaps" in the fossil record, Eldredge and Gould (1972) proposed that diagrams of the "tree of life," in which divergence is plotted on one axis and time on the other, are better represented as "rectangular" branching patterns (Stanley, 1975) reflecting evolution through "punctuated equilibria." According to this view, a new species arises rapidly and, once formed, represents a well-buffered homeostatic system, resistant to within-lineage change (anagenesis) until speciation again is triggered, perhaps by an alteration in ontogenetic (developmental) pattern (Gould, 1977). A second class of argumentation for sudden speciation has come from molecular genetics: The molecular events responsible might involve changes in gene regulation perhaps mediated by relatively few control elements that could have a highly disproportionate influence on organismal evolution, including speciation (Britten and Davidson, 1969, 1971; Krieber and Rose, 1986; McDonald, 1989, 1990; Rose and Doolittle, 1983; Wilson, 1976). A third class of argumentation involves demographic and population-genetic considerations. For example, Mayr (1954) suggested that "founder effects" in small, geographically isolated populations might produce "genetic revolutions" leading to new species; Carson (1968) advanced a "founder-flush" model in which rapid population expansion and relaxed selection following a severe founder event might facilitate the appear-

Box 7.3. Sudden Speciation

There are several known or suspected pathways to rapid speciation that entail little or no change in genetic composition at the allelic level (beyond rearrangements of genetic variation from the ancestral forms). These include the following:

(a) Polyploidization. The origin of stable polyploids usually is associated with hybridization between populations or species differing in chromosome constitution. If the hybrid is sterile only because its parental chromosomes are too dissimilar to pair properly during meiosis, this difficulty is removed by a doubling of chromosomes producing a polyploid hybrid. Furthermore, such a polyploid species spontaneously exhibits reproductive isolation from its progenitors because any cross with the parental species produces progeny with unbalanced chromosome sets (e.g., a cross between a tetraploid and a diploid progenitor produces mostly sterile triploids).

Examples: The treefrog *Hyla versicolor* is a tetraploid that on the basis of allozyme and immunological comparisons arose recently from hybridization between distinct eastern and western populations of its cryptic diploid relative *H. chrysoscelis* (Maxson et al., 1977; Ralin, 1976). However, polyploidy is relatively uncommon in animals and confined primarily to forms that reproduce asexually (see the section on hybridization in this chapter). On the other hand, recent estimates indicate that 70–80% of angiosperm plant species may be of polyploid origin (Lewis, 1980), and molecular genetic data often provide definitive evidence identifying the ancestral parental species. For example, allozyme analyses established that the tetraploid goatsbeard species *Tragopogon mirus* and *T. miscellus* arose recently from crosses between the diploids *T. dubius* and *T. porrifolius,* and *T. dubius* and *T. pratensis,* respectively (Roose and Gottlieb, 1976). These allopolyploids (polyploids arising from combinations of genetically distinct chromosome sets) expressed additively all examined allozyme alleles inherited from their respective progenitors. In several cases, similar molecular genetic analyses involving allozymes or cpDNA have demonstrated that particular polyploid forms are of polyphyletic (multihybridization in this case) origin, e.g., *Plagiomnium* bryophytes (Wyatt et al., 1988), *Asplenium* ferns (Werth et al., 1985), *Glycine* soybeans (Doyle et al., 1990), *Heuchera* alumroots (Soltis et al., 1989) and *Senecio* composites (Ashton and Abbott, 1992). New autopolyploid taxa (polyploids that arise by the multiplication of one basic set of chromosomes) also have been described through molecular assays (Rieseberg and Doyle, 1989). An especially ingenious application of genetic markers involved documentation of a complicated cytological pathway leading to a new tetraploid species of fern, *Asplenium plenum.* Using protein electrophoresis, Gastony (1986) showed that *A. plenum* must have arisen through a cross between triploid *A. curtissii* (which had produced an unreduced spore) and a diploid *A. abscissum* (which had produced a normal haploid spore). The nearly-sterile triploid *A. curtissii* itself was shown to have arisen through a cross between a tetraploid species *A. verecundum* and diploid *A. abscissum.*

(b) Chromosomal rearrangements. Closely related taxa differing in a variety of structural chromosomal features—including translocations, inversions, or chromosome numbers—may exhibit reproductive isolation for at least two reasons. First, the structural differences themselves may cause difficulties in chromosomal pairing and proper disjunction during meiosis in hybrids, resulting in partial or complete sterility. Second, the gene rearrangements may otherwise diminish fitness in hybrid classes through disruptions of gene expression patterns resulting from position effects.

Examples: White (1978a) compiled evidence for animals that chromosomal rearrangements commonly are involved in the speciation process (see also Sites and Moritz, 1987), as did Grant (1981) for plants. When chromosomal rearrangements have conferred reproductive isolation recently, allelic differentiation between the descendant species may yet be minimal. Three examples in which the magnitude of allozymic differences between chromosomally differentiated forms are reported to be about the same as those among populations within a species involve subterranean *Thomomys* and *Spalax* rodents (Nevo and Shaw, 1972; Nevo et al., 1974; Patton and Smith, 1981) and *Sceloporus* lizards (Sites and Greenbaum, 1983). In some of these cases, however, the chromosomal differences do not provide complete barriers to reproduction.

(c) Changes in the mating system. Many plant species exhibit self-incompatibility whereby pollen fail to fertilize ova from the same individual. The mechanisms may involve alleles at a self-incompatibility locus known to be highly polymorphic within some species (Ioerger et al., 1990) or to a physical barrier such as a difference in the lengths of styles and stamens (heterostyly) that inhibits self-pollination. A switch in mating system [for example, from self-incompatibility to self-compatibility (autogamy) as mediated by a change from heterostyly to homostyly] could, in principle, precipitate a rapid "speciation" event with little overall change in genic composition. Other alterations of the breeding system such as the timing of reproduction similarly might generate reproductive isolation rapidly.

Examples: In many plant groups, closely related taxa exhibit contrasting reproductive modes suggesting that "the evolution of floral syndromes, and their influence on mating patterns, is intimately associated with the development of reproductive isolation and speciation" (Barrett, 1989). For example, self-compatible *Stephanomeria malheurensis* apparently arose from a self-incompatible progenitor *S. exigua* ssp. *coronaria* and also differs from it by chromosomal rearrangements which are the principle cause of the hybrid sterility (Stebbins, 1989). As judged by high allozymic similarities (Gottlieb, 1973b), the process took place recently such that the derivative species "was extracted from the repertoire of genetic polymorphisms already present in the progenitor" (Gottlieb, 1981). Such evolution of self-fertilization probably favors the establishment of chromosomal rearrangements contributing to reproductive isolation of the selfing derivative (Barrett, 1989).

ance and survival of novel recombinants leading to a new species; and, Templeton (1980b) introduced a "transilience model" in which speciation involves a rapid shift to a new adaptive peak under conditions where a founder event causes a rapid but temporary accumulation of inbreeding without severe depletion of genetic variability [see Carson and Templeton (1984) for comparisons of these models, and Provine (1989) for the history of concepts]. The generality of such rapid-speciation scenarios has proved difficult to document because speciation no doubt is a highly eclectic process and because many of the proposed genetic and demographic events likely occur also at the populational level *without* producing new species.

On the other hand, many authors have viewed the speciation process (though not necessarily its products) as a rather unexceptional continuation of the microevolutionary processes generating geographic population structure, with the added factor of the evolutionary acquisition of intrinsic reproductive isolation (see the reviews in Barton and Charlesworth, 1984; Charlesworth et al., 1982). This view was termed "phyletic gradualism" by Eldredge and Gould (1972). However, Wright (1931) and some others who interpret speciation mostly as a continuation of microevolution (Provine, 1986) nonetheless emphasize how episodic shifts in evolution can result from genetic drift (in conjunction with selection) in facilitating rapid leaps across adaptive peaks. Thus, the crucial distinction is not whether evolutionary change is gradual versus episodic, but whether speciation as a process is somehow decoupled from processes of intraspecific population differentiation [as Gould (1980) suggested]. As an approach to addressing these issues, many assessments have been made of the magnitude and pattern of genetic differentiation associated with species formation.

Traditional nonmolecular approaches to quantifying genetic divergence between closely related species involve study of phenotypic traits among the later-generation progeny of hybrid crosses. One method is to measure the variance among F_2 hybrids for particular behavioral or morphological characters. Frequently, it was observed that such variances greatly exceed those in the parental populations and in the F_1's, and that few F_2's fall into the parental classes (DeWinter, 1992; Lamb and Avise, 1987; Rick and Smith, 1953). Such results appear attributable to recombinationally derived variation and indicate that for these assayed characters the parental species must differ in multiple genes, each with small effect [although only the minimum number of such polygenes actually can be estimated by this approach (Lande, 1981)]. Another traditional method of assessing the genetic differences between species involves chromosomal mapping of RIB genes through searches for consistent patterns of cosegregation in experimental backcross progeny (see the review in Charlesworth et al., 1987). For example, in sibling species of *Drosophila*, partial hybrid sterility and inviability have proven attributable to differences at several (usually anonymous) loci on each chromosome (Dobzhansky, 1970, 1974; Orr, 1987, 1992), with X-linked genes typically having the greatest effects (Coyne and Orr, 1989a).

However, there are at least two serious limitations to these classical Mendelian approaches. First, they can be applied only to hybridizable taxa. Second, patterns of allelic assortment can be inferred only for loci distinguishing the parental species—genes identical in the parents escape detection. But to determine the *proportion* of genes distinguishing species, both divergent and nondivergent loci must be monitored. Thus, following the introduction of allozyme methods in the mid-1960s, many researchers reexamined the issue of genetic differentiation during speciation, under the rationale that direct molecular assays permitted, for the first time, an examination of a large sample of gene products unbiased with regard to the divergence issue (see the reviews in Avise, 1976; Ayala, 1975; Gottlieb, 1977).

The most extensive allozyme survey of genetic differentiation accompanying RIB acquisition involved the *Drosophila willistoni* complex (Ayala et al., 1975b), which includes populations at several stages of the speciation process as gauged by reproductive relationships and geographic distributions. This complex is distributed widely in northern South America, Central America, and the Caribbean, and provides a paradigm example of gradual speciation involving (a) geographic populations that are fully compatible reproductively, (b) different "subspecies" that are allopatric and exhibit incipient reproductive isolation in the form of postzygotic RIBs (hybrid male sterility in laboratory crosses), (c) "semispecies" that overlap in geographic distribution and show both postzygotic RIBs and prezygotic RIBs (homotypic mating preferences), the latter presumably having evolved under the influence of natural selection after sympatry was achieved between formerly separated subspecies (Box 7.2), (d) sibling species that show complete reproductive isolation but remain nearly identical morphologically, and (e) nonsibling species that diverged earlier. Frequency distributions of genetic similarities across 36 allozyme loci are summarized in Figure 7.2. Between subspecies (or semispecies), nearly 15% of these genes showed substantial or fixed allelic frequency differences involving detected replacement substitutions. Results were interpreted to indicate that "a substantial degree of genetic differentiation occurs during the first stage of speciation" (Ayala et al., 1975b). The subsequent allozyme literature on 119 pairs of closely related *Drosophila* taxa was reviewed by Coyne and Orr (1989b). These taxa provided a cross section of populations at various stages of the speciation process, as defined by geographic distributions and experimentally determined levels of prezygotic and postzygotic reproductive isolation. Results demonstrated that large genetic distances often are associated with various levels of partial reproductive isolation (Table 7.1).

Other noteworthy early studies demonstrating moderate to large allozymic distances between populations at various stages of speciation involved *Lepomis* sunfishes (Fig. 7.2), *Peromyscus* mice (Zimmerman et al., 1978), and *Helianthus* sunflowers (Wain, 1983). In a sense, these and the *Drosophila* studies merely affirm what already was emphasized in Chapter 6—that considerable genetic differentiation among geographic populations can accumulate prior to the completion of intrinsic reproductive isolation.

Among the vertebrates, perhaps the current record for magnitude of genetic differentiation prior to the completion of speciation involves the salamander *Ensatina eschscholtzii*. This complex of morphologically differentiated subspe-

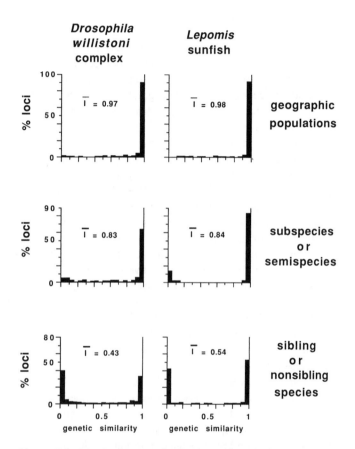

Figure 7.2. Distributions of allozyme loci with respect to genetic
similarity (Nei's 1972 measure) in comparisons among populations
at various stages of evolutionary divergence. Shown are results from
Drosophila willistoni complex fruit flies (after Ayala et al., 1975b)
and *Lepomis* sunfish (after Avise and Smith, 1977). For the fruit
flies, some categories such as "subspecies" and "semispecies" that
were treated separately in the original paper are grouped here, with
equal weighting.

Table 7.1. Means and standard errors of genetic distance (Nei's *D*, allozymes) characterizing *Drosophila* taxa at indicated levels of prezygotic and postzygotic reproductive isolation[a]

Reproductive Isolation Index	No. of Comparisons	Mean Genetic Distance (SE)	
		Prezygotic	Postzygotic
0.00	13	0.122 (0.046)	0.138 (0.058)
0.25	8	0.370 (0.078)	0.251 (0.083)
0.50	21	0.257 (0.080)	0.249 (0.032)
0.75	29	0.578 (0.098)	0.722 (0.198)
1.00	13	0.523 (0.089)	0.991 (0.127)

[a]Prezygotic isolation index: $1 - \dfrac{\text{(frequency heterotypic matings)}}{\text{(frequency homotypic matings)}}$.

Postzygotic isolation index: A measure of hybrid inviability and hybrid sterility, scaled from zero to one.

Source: After Coyne and Orr (1989b)

cies encircles the Central Valley of California in ring-like fashion, with adjacent populations along the chain normally capable of genetic exchange. However, at the southern end of the distribution, populations stemming from coastal areas and the interior overlap sympatrically with limited or no hybridization and thus appear to have achieved the status of biological species in this area (Wake et al., 1986). Remarkably, various populations in this ring complex show profound allozymic distances, often greater than $D = 0.5$ [far greater than values for many congeneric vertebrate species (Fig. 1.2)] (Wake and Yanev, 1986). Huge genetic distances ($p > 0.12$) are apparent in the mtDNA genomes as well (Moritz et al., 1992). Wake et al. (1989) interpret the results to evidence "several stages of speciation in what appears to be a continuous process of gradual allopatric, adaptive divergence." If genetic distances provide even a crude measure of evolutionary time, speciation in this salamander complex must be extremely slow.

On the other hand, by these same allozymic criteria, other recognized species of animals and plants are characterized by much smaller genetic distances, often within the range of values more normally associated with conspecific populations, i.e., *D* less than about 0.05 (Fig. 1.2). Examples can be found within herbaceous plants (Ganders, 1989; Witter and Carr, 1988), insects (Harrison, 1979; Simon, 1979), snakes (Gartside et al., 1977), mammals (Apfelbaum and Reig, 1989; Hafner et al., 1987), birds (Thorpe, 1982), and many others. Presumably, the presence of few or no differentiating electromorphs between related biological species indicates that insufficient time has elapsed for the accumulation of greater de novo mutational differences in these assays. Indeed, this time-dependent aspect of allozyme divergence has proved useful in reassessing speciation dates in several instances. For example, two minnow species in California that had been placed in different genera (*Hesperoleucus* and *Lavinia*) proved to exhibit an allozyme distance of only $D \cong 0.05$, suggesting a far more recent separation than had been implied from their generic assignments (Avise et al., 1975). In the plant genera *Clarkia, Lycopersicon, Erythronium,* and *Gaura,* particular progenitor-derivative species pairs with low allozymic distances now are interpreted to be of relatively recent origin (Gottlieb, 1974; Gottlieb and Pilz, 1976; Pleasants and Wendel, 1989; Rick et al., 1976). Conversely, the self-pollinating plant *Clarkia franciscana* was thought to have evolved recently by rapid reorganization of chromosomes from the morphologically similar *C. rubicunda,* but has proved to be totally divergent from that species at 75% of allozyme loci, indicating that the phylogenetic separation of the two species occurred much longer ago than formerly supposed (Gottlieb, 1973a).

Apart from allozyme methods, few molecular techniques have been brought to bear *explicitly* on the issue of magnitude of genetic differentiation during speciation (Harrison, 1991). Multilocus DNA fingerprinting approaches are too sensitive, and DNA–DNA hybridization methods are generally too insensitive (however, see Caccone et al., 1987), for addressing these meso-evolutionary issues. Single-locus sequencing or RFLP approaches applied to mtDNA or scnDNA may offer the proper window of resolution, but normally provide only individual gene genealogies that in theory (Chapter 4) and practice (see next section) may differ from one locus to another and from the composite species phylogeny. Although it will be interesting to reexamine "genetic differentiation during speciation" using more refined molecular approaches, any enhanced resolution of species differences might well be matched by increased resolution of populational differences. Thus, in any future studies using refined genetic markers, a significant challenge again will be to obtain a suitable frame of reference by extensive sampling of intraspecific variation and by assessment of the magnitude and pattern of interlocus variation.

Do Speciations Entail Severe Population Bottlenecks?

All sudden modes of speciation described in Box 7.3 no doubt are initiated by very small numbers of individuals who first acquire the relevant chromosomal or reproductive alterations. Apart from such situations, do founder events underlie speciations in many other animal and plant groups? If so, significant shifts in frequencies of ancestral polymorphisms might be entailed, but at the outset probably little de novo sequence divergence will have accumulated. A severe and prolonged population bottleneck accompanying speciation should also greatly diminish genetic variability in the neospecies.

A remarkable radiation of drosophilid flies has occurred in the Hawaiian archipelago, with about 800 species endemic to the islands (compared to about 2000 species in the remainder of the world) (Carson and Kaneshiro, 1976; Wheeler, 1986). Founder-induced speciation models have figured prominently in discussions of the prolific speciation among Hawaiian *Drosophila* (Giddings et al., 1989), where species formation is postulated to follow the colonization of new islands perhaps by one or a small number of gravid females. However, molecular genetic data are equivocal on these scenarios. Some sister species such as *D. silvestris* and *D. heteroneura* do indeed exhibit high allozymic similarities suggestive of recent speciation (Sene and Carson, 1977). On the other hand, many recently derived Hawaiian species appear no less variable allozymically than do typical continental *Drosophila,* a result used by Barton and Charlesworth (1984) to dispute founder events, but defended by Carson and Templeton (1984) as consistent with the founder-flush and transilience models of speciation. The Hawaiian endemics *D. silvestris* and *D. heteroneura* also show relatively high genotypic and nucleotide diversities in mtDNA (DeSalle et al., 1986a, 1986b), further suggesting to Barton (1989) that founder-induced speciations were not involved.

Ovenden and White (1990) surveyed both allozyme and mtDNA variation in the Australian fish *Galaxias truttaceus,* a landlocked form of which constitutes an incipient species separated from coastal ancestors within the last 3000–7000 years (based on geologic evidence). Among the total of 58 mtDNA haplotypes observed in coastal populations, only two characterized the landlocked forms. Heterozygosities at allozyme loci nonetheless were nearly identical in the freshwater and coastal populations. The genetic results were interpreted to indicate that a severe but transitory population bottleneck accompanied the speciational transition to lacustrine habitat (because in principle, such bottlenecks might affect genotypic diversity of mtDNA more greatly than that of nDNA).

In terms of gene genealogy, a founder-induced speciation should initially produce a paraphyletic relationship between the ancestral and descendant species (Fig. 7.1; Chapter 4). Indeed, several examples of paraphyly for mtDNA or scnDNA gene trees recently have been reported for related species (Powell, 1991). For example, in terms of mtDNA phylogeny, the deer mouse (*Peromyscus maniculatus*), which occupies most of North America, exhibits a paraphyletic relationship to the old-field mouse (*P. polionotus*) (Fig. 7.3), a species confined to the southeastern United States. Similarly, the mallard duck (*Anas platyrhynchos*), with broad Holarctic distribution, appears paraphyletic in mtDNA genealogy to the American black duck (*Anas rubripes*), which inhabits eastern North America only (Fig. 7.3). However, the mere appearance of genealogical paraphyly is insufficient for concluding that founder-induced speciations necessarily were involved for these or other species, for several reasons. First, paraphyly is expected even when a derivative, geographically restricted species

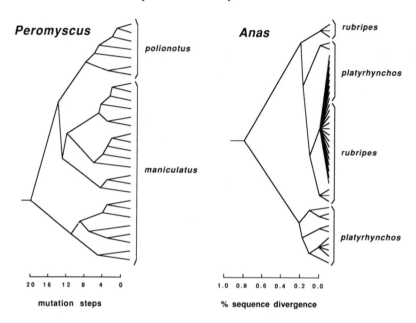

Figure 7.3. Empirical examples of paraphyletic relationships in the mtDNA gene genealogies of closely related species. **Left:** *Peromyscus* mice (after Avise et al., 1983); **right:** *Anas* ducks (after Avise et al., 1990a).

arises via gradual allopatric divergence in the absence of population bottlenecks (Fig. 7.1c). Second, under most geographic modes of speciation entailing moderate or large populations, paraphyly in gene trees is an anticipated stage temporally intermediate to polyphyly and reciprocal monophyly (Chapter 4). Finally, the appearance of paraphyly also could result from secondary introgressive hybridization that has transferred some allelic lineages from one species to another (see section on hybridization and introgression later in this chapter).

The influences of founder events on the magnitude of genic variability within species, and on the pattern of genetic differentiation among species, are in theory also functions both of the size and duration of the population bottleneck and the subsequent rate of population growth when variation recovers (Nei et al., 1975). All else being equal, mtDNA haplotypes might register founder effects more clearly than autosomal nuclear loci because of their expected fourfold lower effective population size (Wilson et al., 1985). However, in any given instance much stochasticity must exist with regard to which particular lineages happen to survive a bottleneck and thereby contribute to genetic diversity during population recovery. Thus, population genetic inferences about the demographic events accompanying past speciations can be complicated and are confounded further by the possibility that nonspeciational bottlenecks can either predate or postdate erection of the RIBS themselves.

Are Speciation Rates and Divergence Rates Correlated?

One intriguing possibility is that speciation events themselves greatly accelerate evolutionary differentiation within clades. If so, the magnitude of divergence between extant species should be proportional to the numbers of speciation events in their evolutionary histories rather than to the sidereal time elapsed since the species last shared a common ancestor. With regard to morphological divergence, this, indeed, is a logical consequence of the original model of punctuated equilibrium (Eldredge and Gould, 1972), which proposes a general stasis for organismal lineages except during speciation events. To test this possibility at the genetic level, Avise and Ayala (1975) introduced a conceptual approach that involves comparing mean genetic distances within clades (phylads) that are of similar evolutionary age but have experienced different rates of speciation. If genetic divergence is proportional to time, mean distances should be similar among extant members of rapidly speciating (speciose) and slowly speciating (depauperate) phylads. On the other hand, if genetic divergence is a function of the number of speciation events, mean genetic distance among extant forms should be much greater in the speciose clade (Fig. 7.4).

One set of empirical tests utilizing this approach involved the speciose North American minnows (approximately 200 species, more than 100 within *Notropis* alone) versus the relatively depauperate sunfish (approximately 30 species, 11 within *Lepomis*). From fossil evidence, these groups appear to be of similar evolutionary age on the continent (Miocene), and an important assumption underlying the test is that the greater number of minnow species is a consequence of a more rapid pace of speciation (rather than a lower rate of species extinction, for example). Pairwise comparisons of more than 80 species based on multilocus protein electrophoretic methods revealed that mean genetic distances within the minnows and sunfish were similar (Avise, 1977b; Avise and Ayala, 1976): for example, the mean genetic distance among 47 assayed species of *Notropis* was $D = 0.62$, and among 10 species of *Lepomis* was $D = 0.63$ (thus, $D_R/D_P \cong 1.0$). Furthermore, the minnow and sunfish species had comparable levels of mean allozymic heterozygosity ($H = 0.052$ and $H = 0.049$, respectively), suggesting that the apparent differences in rates of speciation were not attributable to differing levels of intraspecific genetic variability available for conversion to between-species differences (Avise, 1977c). Overall, the results for minnows and sunfish were interpreted as consistent with phyletic gradualism and inconsistent with punctuated equilibrium, at least with regard to allozyme evolution.

Gould and Eldredge (1977) justifiably questioned the relevance of these findings to the broader punctuated equilibrium/phyletic gradualism debate. They

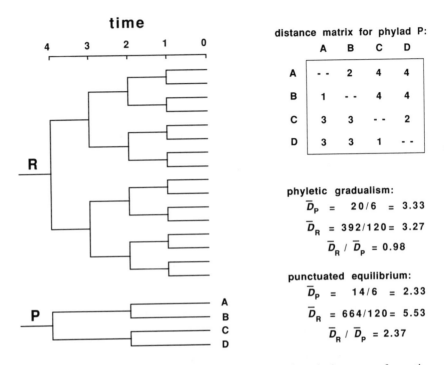

Figure 7.4. Explanation of models underlying the test for whether rates of genetic divergence and speciation are correlated (after Avise and Ayala, 1975). R and P are species-rich and species-poor phylads, respectively, of comparable evolutionary age. The distance matrix in the upper right applies to phylad P (the larger matrix for R is not presented) and shows expected distances between species pairs when differentiation is time dependent (phyletic gradualism, above diagonal) versus speciation dependent (punctuated equilibrium, below diagonal). In the lower right are shown expected ratios of *mean* distances (D values) for extant species within R and P under these competing models. Under phyletic gradualism, $D_R/D_P \cong 1.0$, whereas under punctuated equilibrium, $D_R/D_P \gg 1.0$.

pointed out that the controversy refers primarily to patterns of morphological divergence, whereas allozymes might merely be neutral molecular characters and, hence, irrelevant to the discussion. Douglas and Avise (1982) therefore reexamined these fishes in terms of quantitative morphological differences and showed by multivariate character analyses that the magnitude of phenotypic divergence was approximately the same in the rapidly speciating minnows and the slowly speciating sunfish, a result again inconsistent with some of the predictions of punctuated equilibrium.

Mayden (1986) also criticized the original genetic tests on minnows and sunfish, on grounds that the groups examined may not be monophyletic and that their ages were poorly known from fossils. However, if the minnows do not constitute a clade, they should be older than assumed and, therefore, display even larger distances under the time-divergence model (and group ages are irrelevant to predictions of punctuated equilibrium). Nonetheless, several reservations do remain about the validity and generality of this case study. For example, the minnow and sunfish taxa originally were recognized on the basis of qualitative morphological appraisals, so the quantitative reassessment by Douglas and Avise (1982) likely includes an element of "circular" reasoning in this test of "rectangular" evolution!

In any event, considering the importance of whether most change in evolution arises via anagenesis or cladogenesis, it is surprising that few other tests of this sort involving neontological data (those from extant species) have been attempted (Lemen and Freeman, 1981, 1989; Ricklefs, 1980). Recently, Mindell et al. (1990) summarized the literature on allozyme genetic distances within 111 vertebrate genera and reported a positive correlation between genetic divergence and species richness that they attributed to the accelerating influence of speciation on molecular differentiation. However, among taxa this diverse, one cannot be certain that many other uncontrolled variables do not also correlate with and possibly influence molecular rates.

Can Speciation Occur Sympatrically?

A long-standing controversy in evolutionary biology concerns how often new species arise sympatrically, most likely under the influences of diversifying natural selection on resource utilization (Levine, 1953; Maynard Smith, 1966; Tauber and Tauber, 1989). Some biologists question the occurrence of sympatric speciation (e.g., Felsenstein, 1981; Futuyma and Mayer, 1980; Mayr, 1963), whereas others perceive it as a prevalent mode at least in some groups such as insects (e.g., Bush, 1975; Kondrashov and Mina, 1986). This debate will not be rehashed here; instead, a few examples will be presented in which molecular methods have shed some light on evolutionary relationships in settings where sympatric speciation had been suspected.

FLOCKS OF FISHES

Within each of several isolated lakes or drainages scattered around the world, closely related fishes comprise "species flocks" (Echelle and Kornfield, 1984) that might have arisen through sympatric speciations. The numbers of named "species" in such flocks range from just a few as in salmonid complexes in high-latitude lakes of the Northern Hemisphere to more than 500 for cichlid fishes in the rift-valley lakes of east Africa (Fryer and Iles, 1972; Greenwood, 1981).

For such flocks, a primary question is whether the differences among sympatric morphotypes do indeed reflect the presence of distinct species, or alternatively whether some or all of the differences might reflect phenotypic plasticity within species [perhaps due to developmental switches triggered by environmental conditions such as diet (Meyer, 1987)]. Polymorphic nuclear markers such as those provided by allozymes are well suited for addressing this issue. For example, among *Cichlasoma* fishes in an isolated basin near Coahuila Mexico, three trophic morphs described as separate species are present: a snail-eating form with molariform (crushing) teeth and a short gut; a detritus- or algae-eating form with papilliform teeth and a long gut; and a fish-eating form with fusiform body. At 27 monomorphic and polymorphic allozyme loci, these morphs proved indistinguishable genetically (Kornfield and Koehn, 1975; Kornfield et al., 1982; Sage and Selander, 1975), a result which led Sage and Selander (1975) to conclude that "trophic radiation in the Cuatro Cienegas cichlids has been achieved through ecological polymorphism rather than speciation." Subsequent studies confirmed this conclusion—different morphotypes among progeny within a brood can be generated by altering the rearing conditions (Meyer, 1987).

Another reported example of pronounced trophic polymorphism within a species involves stream-dwelling fishes in the genus *Ilyodon*. In certain rivers in Mexico, two sharply dichotomous trophic forms formerly considered distinct species were shown to be indistinguishable at several polymorphic allozyme loci, with pooled genotypic frequencies in agreement with Hardy–Weinberg expectations for single random-mating populations (Grudzien and Turner, 1984; Turner and Grosse, 1980). "Conspecificity of the trophic types appears to be the most straightforward interpretation of the data" (Turner and Grosse, 1980).

In many trout and salmon, coexisting forms often exhibit contrasting life histories: nonanadromous individuals spend their entire lives in freshwater, whereas anadromous individuals hatched in the same streams or lakes migrate to sea before returning to freshwater to spawn. In addition, some landlocked populations (those that lack present-day access to the ocean) include both stream-resident and lake-migratory individuals. A long-standing question is whether these coexisting life history types represent genetically separate populations (i.e., species). Protein-electrophoretic studies of several such fish complexes— rainbow trout *Salmo gairdneri* (now *Oncorhynchus mykiss*) (Allendorf and Utter, 1979), cutthroat trout *O. clarki* (Campton and Utter, 1987), brown trout *S. sutta* (Hindar et al., 1991), Atlantic salmon *S. salar* (Ståhl, 1987), and sockeye salmon *O. nerka* (Foote et al., 1989)—have revealed that freshwater-resident forms often are genetically close to or indistinguishable from anadromous populations in the same regions, whereas by these same assays, populations from geographically separate spawning localities frequently show highly significant differentiation (Ferguson, 1989; Ryman, 1983; see Chapter 6). On the other hand, several other studies on Salmoniformes have reported mild but significant molecular genetic differences among nearby or sympatric forms that differ in lifestyle (Baby et al., 1991; Birt et al., 1991; Krueger and May, 1987; Skaala and Nævdal, 1989; Vuorinen and Berg, 1989). In most of these latter cases, it has remained unclear whether the genetic differences were attributable to extrinsic environmental structure per se, or to innate behavioral preferences for different microhabitats or spawning times.

Whether or not these nonanadromous fishes at a given locale are fully isolated reproductively from their anadromous counterparts at the present time, the small genetic distances normally involved, and the polyphyletic appearance of particular life history patterns suggest that such isolations might well be ephemeral evolutionarily. Switches between life-styles must have occurred commonly in the past, and any contemporary phylogenetic separations are probably of recent origin (as indicated also by the fact that most of the locales under consideration were covered by glacial ice as recently as 10,000 years ago). Furthermore, rearing and tagging studies of brown trout have shown that freshwater-resident individuals can develop from anadromous parents, and vice versa (Skrochowska, 1969), indicating a considerable element of phenotypic plasticity in life-style. In this case, a possibility is that the freshwater-resident behavior is associated with a slow growth rate of parr (Hindar et al., 1991), which itself probably is influenced by both genetic and environmental factors.

Molecular analyses of several other putative fish species flocks *have* revealed significant differences in allozymes or mtDNA among strictly sympatric forms and, thus, confirmed the suspected status of separate biological species. Examples include representatives of the atherinids of central Mexico (Echelle and Echelle, 1984), cyprinodontids of eastern Mexico (Humphries, 1984), coregonids of Canada and the northern United States (Kirkpatrick and Selander, 1979), certain of the cichlids in the African rift-valley lakes (Sage et al., 1984; Sturmbauer and Meyer, 1992), and the now-extinct cyprinids of Lake Lanao in the Philippines (Kornfield and Carpenter, 1984). However, despite significant allelic differences at particular loci, in several of these cases the overall genetic distances among species remained small (Nei's $D < 0.05$ in the allozyme studies), suggesting that the evolutionary separations were recent.

A follow-up question for such validated species flocks is whether the speciations took place sympatrically, within the lake or drainage basin. In the Allegash basin of eastern Canada and northern Maine, coexisting dwarf and normal-sized lake whitefish (*Coregonus clupeaformis*) represent independent gene pools, albeit at a small overall level of genetic distance (Kirkpatrick and Selander, 1979). Might this represent a case of sympatric speciation (perhaps involving changes at a few loci regulating rate of maturation and time of spawning)? If so, populations of dwarf and normal-sized fish within the Allegash should be one another's closest relatives (i.e., sister taxa), whereas if speciation was allopatric and sympatry achieved secondarily, closer genealogical ties might exist with populations outside this drainage. Bernatchez and Dodson (1990) discovered from RFLP analyses of mtDNA that the Allegash populations probably represent a secondary overlap of two monophyletic groups that evolved allopatrically in separate refugia during the last (Wisconsin) Pleistocene glaciation. The evidence was that nearby western populations outside the Allegash belong to one mtDNA clade, eastern populations to another, and only in the Allegash basin do the two clades overlap and appear alternately fixed in the dwarf and normal-sized forms. (A competing hypothesis—that the Allegash basin was the original, continuously occupied homeland for whitefish and still retains ancestral polymorphisms that have been fixed elsewhere—appears unlikely because this drainage was glaciated entirely during the Wisconsin.) An extension of the mtDNA survey to populations from Alaska to Labrador revealed additional mtDNA clones and clades whose distributions appear interpretable in terms of the suspected Pleistocene glacial histories of these drainages (Bernatchez and Dodson, 1991).

A different phylogenetic outcome was revealed in mtDNA analyses of the African rift-valley cichlids in lakes Victoria and Malawi. Comparisons of some 800 nucleotide positions in the cytochrome *b* gene, two transfer RNA genes, and a normally highly variable portion of the control region revealed almost no differentiation among species within Lake Victoria, yet a large genetic distinction (> 50 base substitutions) from assayed fishes in Lake Malawi (Meyer et al., 1990). Results strongly support a recent monophyletic origin for these cichlid fishes within lakes and conflict diametrically with a taxonomic reappraisal based on morphology in which Greenwood (1980) had concluded that "the overall picture is one of a super-flock comprised of several lineages whose members cut across the boundaries imposed by the present-day lake shores." The genetic results leave open the possibility of sympatric speciation in these African cichlids and raise a number of additional questions such as how so many species can arise so rapidly and coexist ecologically within a lake (Avise, 1990).

Another example of the variety of genetic outcomes observed among fish species flocks involves the sculpin (*Cottus*) radiation in Lake Baikal in Russia. Grachev et al. (1992) recently reported rather extensive mtDNA sequence differences among three species, indicating that by whatever pathway speciation in this flock took place, "the cottoid fish of Lake Baikal are much more ancient than cichlid fish of the great lakes of Africa."

Host- or Habitat-Switching in Insects

Changes in host utilization by phytophagous and zoophagous parasites might, in principle, give rise quickly to new species reproductively isolated from their sympatric progenitors (Bush, 1975). If such speciations took place recently, the genetic footprints again should involve a paraphyletic relationship for each genetically similar progenitor/derivative parasite pair and probably a reduction of variation in the neospecies. Furthermore, if such speciations were common in the past, the phylogenies of parasites and their hosts could be conspicuously decoupled, the host-shifts, in effect, producing reticulations of parasite lineages across host lineages (e.g., Baverstock et al., 1985). However, caution is indicated because a lack of congruence between the phylogenies of parasite and host also could result from asynchronized speciations and lineage sorting processes, such that surviving lineages in the parasite trace to phylogenetic splits either predating or postdating nodes in the host phylogeny (Fig. 7.5). Such patterns of parasite

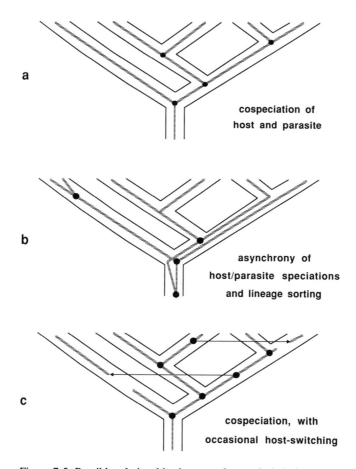

Figure 7.5. Possible relationships between the species' phylogeny of a parasite (inner branches) and that of its host taxon (thick branches). Black dots denote parasite speciations. (a) Cospeciation, wherein host and parasite phylogenies are perfectly concordant; (b) asynchronous speciation, wherein nonconcordant lineage sorting characterizes host and parasite lineages; (c) host-switching by the parasite.

lineage sorting within a host phylogeny bear some analogies to similar processes that can distinguish a gene tree from a species tree (Chapter 4) and also are reminiscent of Fitch's (1970) distinction between orthology and paralogy (Chapter 1).

One of the classic examples of host-switching involves frugivorous *Rhagoletis* flies, where several host-specific forms such as the "apple race" and the "hawthorne race" of *R. polmonella* are postulated to be in the process of sympatric speciation currently (Bush, 1969). Available genetic data are not inconsistent with this possibility. In allozymic composition, collections from the apple and hawthorne forms in local sympatry show small but significant differentiation (Feder et al., 1988; McPheron et al., 1988), and such differences between paired samples extend across the eastern United States and Canada (Feder et al., 1990a, 1990b). Furthermore, Berlocher and Bush (1982) concluded from allozymic data that the phylogeny for many *Rhagoletis* flies and relatives differs from that of the hosts, a result at least consistent with host-switching aspects of the sympatric speciation scenario. Of course, such consistencies do not necessarily eliminate all competing scenarios.

Can Related Species Be Diagnosed Reliably?

Regardless of origin, molecular differences between species can be of great utility in diagnosing closely related forms, even where morphological or other traditional markers have failed or are ambiguous. For example, despite a near identity in morphological appearance, several sibling species of *Drosophila* proved readily separable with a battery of allozyme markers (Ayala and Powell, 1972b; Ayala et al., 1970; Hubby and Throckmorton, 1968; Fig. 7.2), as did sibling species of *Trachyphloeus* weevils (Jermiin et al., 1991) and *Chthamalus* barnacles (Hedgecock, 1979). Mitochondrial sequences in morphologically similar *Penaeus* shrimp show huge genetic differences, greater than those distinguishing some orders of mammals (Palumbi and Benzie, 1991). A polychaete worm (*Capitella capitata*) widely used as an indicator of marine pollution and environmental disturbance was thought to be a single cosmopolitan species, but allozyme analyses indicate the presence of at least six sibling species (Grassle and Grassle, 1976). In corals assigned to "*Montastraea annularis,*" enormous phenotypic variation of questionable genetic basis proved to involve at least three unnamed species, as judged by significant differences in allozymic composition among sympatric morphotypes (Knowlton et al., 1992).

Among the vertebrates also, numerous problematic species have been diagnosed using molecular characters. Morphologically cryptic fish species have been revealed through isozymic or other molecular studies (see the reviews in Powers, 1991; Shaklee, 1983; Shaklee et al., 1982). For example, several closely related species of tuna (*Thunnus*) can be distinguished by allozyme and mtDNA markers (Bartlett and Davidson, 1991; Dotson and Graves, 1984; Sharp and Pirages, 1978). In frogs of the genus *Gastrotheca,* immunological and protein electrophoretic evidence revealed that at least six species in two different groups formerly were masquerading under the name *G. riobambae* (Duellman and Hillis, 1987; Scanlan et al., 1980). Remarkable examples of morphological stasis in the face of extensive speciation involve lungless salamanders in the family Plethodontidae, where multiple fixed allozymic differences are the rule rather than the exception among populations formerly thought to be conspecific (Larson, 1984, 1989). [For example, the slimy salamander in the eastern United States had been considered a single species (*Plethodon glutinosus*), but dramatic differences in allozymes and albumin (Highton, 1984; Maha et al., 1989), including those among sympatric populations, revealed at least 16 groups that appear to have achieved the species or semispecies level of divergence (Highton et al., 1989).] Among avian taxa, sibling species in the flycatcher genera *Empidonax* and *Contopus* are difficult to distinguish morphologically, but often exhibit diagnostic allozyme alleles (Zink and Johnson, 1984). In general, birds are among the best known of organisms at the species level, but a new species of skrike (*Lanarius liberatus*) recently was discovered that differs from congeners by large mtDNA distances and represents the first avian species described solely on the basis of molecular data (Hughes, 1992; E.F.G. Smith et al., 1991).

In many species, adults are distinguishable morphologically, whereas their juveniles or larvae are not. By comparing molecular characters in unknown larvae against known adults, species assignments often can be made. In examples of this approach, researchers have employed protein electrophoresis or DNA restriction mapping to identify the larvae of a variety of marine organisms, including oysters (Hu et al., 1992), white perch and striped bass (Morgan, 1975; Sidell et al., 1978), other basses (Graves et al., 1989), flounders (Smith and Benson, 1980), tunas (Graves et al., 1988), and snappers (Smith and Crossland, 1977). In the near future, PCR-based methods should prove of particular utility in identifying small larval forms or eggs. Silberman and Walsh (1992) used the PCR to amplify nuclear 28S rRNA genes from phyllosome larvae of three species of spiny lobsters (*Panulirus argus, P. guttatus,* and *P. laevicauda*). The amplified products then were subjected to restriction digests that revealed species-diagnostic banding patterns. In an interesting problem of larval identification involving sea cucumbers (Echinodermata), Olson et al. (1991) used PCR-based sequences from the 16S ribosomal RNA gene in mtDNA to assign a collection of bright red pentacula larvae to *Cucumaria frondosa*. This assignment came as a surprise—from the coloration and other morphological considerations, the larvae had been assumed to belong to a distantly related sea cucumber, *Psolus fabricii.*

Economic or medical exigencies often require the correct identification of sibling taxa, and molecular markers can play an important role. For example, the two principle mosquito vectors for malaria in Africa, *Anopheles gambiae* and *A.*

arabiensis, differ in behavior and preferred habitat but are indistinguishable morphologically. Finnerty and Collins (1988) isolated a diagnostic RFLP probe from rDNA that can be used to assay single dried adults, thus extending earlier species-discriminating capabilities for the *gambiae* complex based on allozymes (Miles, 1978). In the eastern United States, another *Anopheles* complex (formerly considered a single species *A. quadrimaculatus*) has proved from molecular and cytological evidence to consist of at least four morphologically cryptic species. Numerous molecular markers [including those from allozymes (Narang et al., 1989a, 1989b, 1989c), rRNA genes (Mitchell et al., 1993), and mtDNA (Kim and Narang, 1990)] show concordant genetic partitions that agree also with reproductive boundaries as revealed through experimental crosses. The genetic differences indeed are large, such that many molecular characters are diagnostic, and dichotomous keys to species identification in *Anopheles* can be constructed (Table 7.2). In another mosquito genus (*Aedes*), similar work based on allozymes was used to discriminate morphologically cryptic and sometimes sympatric forms (Munstermann, 1988), including behavioral types within the *Aedes aegypti* complex that is the primary vector to humans of the viruses for dengue fever and yellow fever (Powell et al., 1980; Tabachnick and Powell, 1978; Tabachnick et al., 1979).

Further development of PCR-based methods for DNA amplification likely will revolutionize abilities to discriminate among microbes and other small organisms. For example, Wimpee et al. (1991) employed two highly conserved regions of the *luxA* gene as PCR primers to prepare species-specific probes from four major groups of marine luminous bacteria. The probes subsequently were used to identify various bacterial species from field isolates. Schmidt et al. (1991) examined the species composition represented in bulk genomic DNA isolated from picoplankton collected in the central Pacific Ocean. The approach involved cloning the mixed population of DNA into phage, screening these clones for presence of the 16S rRNA genes, and subsequent nucleotide sequencing of these isolates. The data were used to establish the identities of the picoplankton by comparison with an established information base of rRNA gene sequences. This method allowed identification of 15 bacterial sequences that could be related to cyanobacteria and proteobacteria. Similar molecular approaches have been employed to characterize the phylogenetic positions of previously unknown microbial taxa from the Sargasso Sea (Giovannoni et al., 1990). With the accumulation of larger data bases, such approaches might also allow much finer taxonomic assignments (Weller and Ward, 1989; Weller et al., 1991). For example, Fell et al. (1992) used rRNA gene sequences from 117 species representing 23 genera of basidiomycetous yeasts to identify and characterize marine collections of these micro-eukaryotes.

Zooxanthellae are unicellular algae that occur as endosymbionts in many species of marine invertebrates. Due to a paucity of informative morphological

Table 7.2. Diagnostic allozyme loci (**A**) and dichotomous biochemical key (**B**) to four sibling species in the *Anopheles quadrimaculatus* complex of mosquito species.[a]

A. Diagnostic Loci for Species Pairs					
A:B	A:C	A:D	B:C	B:D	C:D
Idh-1	*Acon-1*	*Acon-1*	*Acon-1*	*Acon-1*	*Got-1*
Idh-2	*Idh-2*	*Idh-2*	*Idh-1*	*Idh-1*	*Had-1*
Est-2	*Had-1*	*Got-1*	*Had-1*	*Got-1*	*Had-3*
Est-5	*Had-3*	*Got-2*	*Had-3*	*Got-2*	*Pep-4*
Est-7	*Pep-2*	*Pep-2*	*Got-2*	*Pep-2*	*Pgi-1*
Had-1	*Got-2*	*Pep-4*	*Pep-2*	*Pep-4*	*Me-1*
6Pgd-1	*Pgi-1*	*Me-1*	*Pgi-1*	*Me-1*	*Est-2*
	Est-2	*Mpi-1*	*Est-4*	*Est-2*	*Mpi-1*
	Est-6		*Est-5*	*Est-7*	
	Mpi-1		*Est-6*	*Mpi-1*	
	6Pgd-1		*Est-7*		
	Xdh-3		*Mpi-1*		
	Ao-1		*Xdh-3*		

B. Biochemical Key

1. *Mpi-1* slow (62 allele, rarely with 52 as heterozygote) . . . . species D
   Mpi-1 faster (78 or greater) . . . . go to 2
2. *Idh-1* slow (86) and *Idh-2* fast (162) . . . . species B
   Idh-1 faster (≥ 100, sometimes with 86 as heterozygote);
      Idh-2 fast or slower (100, 132, 162) . . . . go to 3
3. *Had-3* slow (45); *Pgi-1* slow (95) . . . . species C
   Had-3 faster (100, sometimes with 45 as heterozygote);
      Pgi-1 faster (100, rarely with 95 as heterozygote) . . . . species A

[a]The diagnostic loci provide correct identification with probability greater than 99%. In the key shown (one of many that could be generated), the numbers indicate electromorph mobilities relative to a standard strain.

Source: After Narang et al. (1989b).

characters, most zooxanthellae have been placed provisionally in the genus *Symbiodinium,* but the numbers of species are unknown as are their evolutionary relationships and possible host-specificities. Rowan and Powers (1991) assayed RFLPs and sequences in the small-subunit nuclear ribosomal RNA gene from zooxanthellae isolated from 22 host taxa and found several genetically distinct forms whose estimated phylogenetic relationships bore little resemblance to the taxonomies of their hosts. For example, some genetically similar zooxanthellae were isolated from hosts of ordinal or greater taxonomic distance, such as anemones, corals, and gorgonians; some dissimilar zooxanthellae were isolated from congeneric hosts. Overall, results suggest that many zooxanthellae species exist and that the symbioses can arise by a shuffling of symbionts among even unrelated host taxa. The conclusion that *Symbiodinium* contains many species was supported further by the finding that genetic diversity in ribosomal RNA gene sequences within this genus is comparable to that among different orders of nonsymbiotic dinoflagellates (Rowan and Powers, 1992)!

Not surprisingly, such molecular-based identifications of microbial taxa also have found applications in more applied areas, including medicine and the food industry. For example, Regnery et al. (1991) employed PCR-amplified DNA sequences from the citrate synthase and the 190-kDa antigen genes as substrates for RFLP assays that proved to differentiate among various rickettsial species causing spotted fever; Salama et al. (1991) employed subspecies-specific rRNA gene probes to identify *Lactococcus lactis cremoris,* a bacterium whose few available strains are relied on by the dairy industry for the manufacture of cheddar cheese free of fermented and fruity flavors.

Should a Phylogenetic Species Concept Replace the BSC?

For more than 50 years, the biological species concept (BSC) has been the major theoretical framework orienting research on the origins of biological diversity. Recently, a serious challenge has come from some systematists who argue that the BSC lacks a sufficient phylogenetic perspective and, hence, provides an inappropriate guide to the origins and products of evolutionary diversification (de Queiroz and Donoghue, 1988; Donoghue, 1985; Eldredge and Cracraft, 1980; Mishler and Donoghue, 1982; Nelson and Platnick, 1981). Many critics of the BSC argue that "reproductive isolation should not be part of species concepts" (McKitrick and Zink, 1988). This has led to a call for replacement of the BSC with a "phylogenetic-species concept" (PSC—Box 7.1) under which a species has been defined as a monophyletic group composed of "the smallest diagnosable cluster of individual organisms within which there is a parental pattern of ancestry and descent" (Cracraft, 1983). Because molecular data provide unprecedented power for phylogeny estimation, it might be supposed that molecular evolutionists would be strong advocates for the PSC, but this has not necessarily been the case.

One fundamental difficulty with existing PSC proposals concerns the nature of evidence required to justifiably diagnose a monophyletic group warranting species recognition. Molecular technologies have made it abundantly clear that multitudinous derived traits often can be employed to subdivide named species into diagnosable subunits (Chapter 6). Indeed, most individuals within sexually-reproducing species can be distinguished from one another with high-resolution assays such as DNA fingerprinting (Chapter 5). If each individual is genetically unique, then the grouping of individuals into phylogenetic "species" requires that distinctions below some arbitrary threshold be ignored. The evolutionary significance of any such threshold surely must be questionable. For these and other reasons, Avise and Ball (1990) suggest that if a broader framework of the PSC is to contribute to a significant advance in systematic practice (and they believe that it can), a shift from issues of diagnostics to those of magnitudes and patterns of phylogenetic differentiation, and of the historical and reproductive reasons for such patterns, will be required. Toward this end, they suggest how principles of "genealogical concordance" might be employed to combine desirable elements of the PSC and the BSC.

Within any organismal pedigree, allelic phylogenies can differ greatly from locus to locus (Ball et al., 1990), due to the inevitable Mendelian vagaries of meiotic segregation and syngamy and to the reproductive successes of individuals through which the alleles happen to have been transmitted. An array of individuals phylogenetically grouped by one locus may differ from an array of individuals grouped by another locus, unless some overriding evolutionary forces may have concordantly shaped the phylogenetic structures of the independent genes. One such force expected to generate genealogical concordance across loci is intrinsic reproductive isolation (the focal point of the BSC). Through time, due to processes of lineage turnover, populations isolated from one another by intrinsic RIBS (i.e., biological species) inevitably tend to evolve toward a status of reciprocal monophyly with respect to particular gene genealogies (Chapter 4); furthermore, through time, the genealogical tracings of independent loci almost inevitably sort in such a way as to partition these isolated populations concordantly (Avise and Ball, 1990). Thus, genealogical concordance per se becomes a deciding criterion by which to recognize significant evolutionary partitions at the level of organismal phylogeny.

However, genealogical concordance across loci also can arise from extrinsic barriers to reproduction, as among populations geographically isolated for sufficient lengths of time [relative to effective population size (Chapter 4)]. As emphasized in Chapter 6, dramatic phylogenetic partitions commonly are observed among populations considered conspecific under the BSC. It might be argued that such populations also warrant formal taxonomic recognition, on the grounds that they represent significant biotic partitions of relevance in such areas as biogeographic reconstruction (Chapter 6) and conservation biology (Chapter 9).

From consideration of these and other factors, Avise and Ball (1990) suggested the following guidelines for biological taxonomy, based on applications of genealogical concordance principles. The biological and taxonomic category "species" should continue to refer to groups of actually or potentially interbreeding populations isolated by intrinsic RIBS from other such groups. In other words, a retention of the philosophical framework of the BSC is warranted, in no small part because reproductive barriers generate significant (concordant) genealogic partitions at multiple loci in an organismal phylogeny. At the intraspecific level, "subspecies" warranting formal recognition then may be defined as groups of actually or potentially interbreeding populations phylogenetically distinguishable from, but reproductively compatible with, other such groups. Importantly, the evidence for phylogenetic distinction must come, in principle, from concordant genetic partitions across multiple, independent, genetic-based traits (be they molecular or phenotypic). This phylogenetic approach to taxonomic recognition represents a novel and direct outgrowth of molecular-based perspectives on the evolutionary process.

HYBRIDIZATION AND INTROGRESSION

The term "hybridization" is as challenging to define as is "speciation," and for the same reasons. In the early literature of systematics, a hybrid was defined as an offspring resulting from a cross between species, whereas "intergrade" was reserved for any product of a cross between distinctive conspecific populations or subspecies. But as we have seen, this distinction can be rather arbitrary, and in the more recent literature hybridization often is employed in a broad sense to include crosses between genetically "differentiated" forms regardless of their current taxonomic status. Introgression refers to the movement of genes between species (or between well-marked genetic populations) mediated by backcrossing.

Background

FREQUENCIES AND GEOGRAPHIC SETTINGS OF HYBRIDIZATION

Hybridization and introgression are common phenomena in many plant and animal groups. More than 20 years ago, Knoblock (1972) compiled a list of 23,675 reported instances of interspecific or intergeneric plant hybridization, despite the availability of detailed studies on only a small fraction of the botanical world. Introgression is more challenging to assess, but Rieseberg and Wendel (1993) provided a compilation of 155 "noteworthy" cases of plant introgression, many of which include molecular documentation. Similarly, hybridization and introgression have been noted involving numerous animal taxa (Harrison, 1993). For example, Schwartz (1972, 1981) compiled a list of 3759 references dealing with natural and artificial hybridization in fishes, many cases of which have been verified and characterized further using molecular markers (Campton, 1987; Verspoor and Hammar, 1991). Among the vertebrates, fishes appear most prone to hybridization, but the phenomenon is widespread.

The frequency of hybridization and the extent of introgression both can vary along a continuum, and molecular markers are invaluable for assessing where a given situation falls. At one extreme, hybridization may be rare and confined primarily to production of F_1 hybrids, a situation likely to arise when such progeny have greatly reduced viability or fertility. For example, the two largest mammalian species, the blue whale (*Balaenoptera musculus*) and fin whale (*B. physalus*), normally are distinct but also are suspected of occasional hybridization. Recent molecular analyses of three anomalous specimens by nuclear and mitochondrial DNA revealed that these individuals were indeed interspecific hybrids resulting from crosses that had taken place in both directions with respect to sex; one hybrid female also carried a backcross fetus with a blue whale father (Árnason et al., 1991; Spilliaert et al., 1991). In milkweed bugs (*Oncopeltus*), genetically divergent species that are capable of producing hybrids under laboratory conditions failed to show significant allozymic exchange in nature due to strong prezygotic and postzygotic RIBs (Leslie and Dingle, 1983). At the other extreme, hybridization can be extensive and the hybridizing taxa may merge completely into one panmictic gene pool. This situation is exemplified by a local hybrid swarm between two genetically well-marked subspecies of bluegill sunfish (*Lepomis macrochirus macrochirus* and *L.m. purpurescens*). These subspecies inhabit different regions of the southeastern United States, but overlap in parts of Georgia and the Carolinas where hybridization takes place (Avise and Smith, 1974). In one well-characterized hybrid population, mtDNA and allozyme markers normally diagnostic for the parental subspecies proved to be shuffled thoroughly: Gametic phase disequilibria (Box 5.3) involving nuclear and cytonuclear allelic associations all were negligible (Asmussen et al., 1987; Avise et al., 1984b).

In many taxonomic groups, organisms separated for long periods of evolutionary time nonetheless may retain the anatomical and physiological capacity for hybrid production. Using microcomplement fixation to assay the albumins of 50 pairs of frog species known to be capable of generating viable hybrids, Wilson et al. (1974a) demonstrated a mean immunological distance of 36 units, which corresponds to an average separation time of about 21 million years under a conventional albumin clock (Prager and Wilson, 1975; Wilson et al., 1977). From similar molecular assays, hybridizable species-pairs of birds were estimated to have last shared common ancestors on average 20 mya, whereas hybridizable pairs of mammals separated on average only 2 mya (Fig. 7.6). From these dramatic differences, it was concluded that birds and frogs have lost the potential for interspecific hybridization slowly in comparison to mammals, perhaps because of a slower pace of chromosomal evolution and/or a hypothesized lower rate of change in regulatory genes (Prager and Wilson, 1975; Wilson et al., 1974a, 1974b). In any event, some organisms clearly retain a potential for hybridization over extremely long evolutionary periods. How often such potential is realized in nature is, of course, another issue, and one that also can be addressed powerfully through molecular approaches.

Frequently, hybridization follows human-mediated transplantations. In the early 1980s, the pupfish *Cyprinodon variegatus* was introduced to the Pecos River in Texas where it then hybridized with an endemic species *C. pecosensis*.

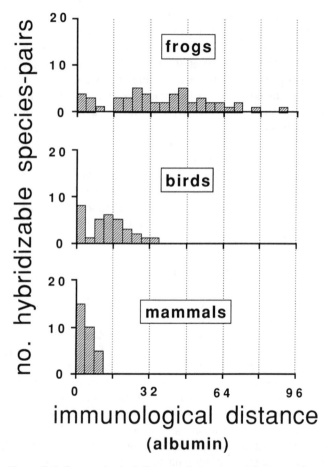

Figure 7.6. Immunological distances between vertebrate species capable of producing viable hybrids [data from Prager and Wilson (1975) and Wilson et al. (1974a)]. Data are shown for more than 100 species-pairs, many of which appear quite old. Albumin clocks have been calibrated at about 1.7 immunological distance (ID) units per million years in frogs and mammals (Prager and Wilson, 1975) and about 0.6 ID units per million years in birds (Prager et al., 1974).

Protein-electrophoretic data revealed that within five years, panmictic admixtures of the two pupfishes occupied approximately 430 river-kilometers, or roughly one-half the historic range of the endemic species (Echelle and Connor, 1989; Echelle et al., 1987). In the land snail genus *Cerion*, Bahamian individuals of *C. casablancae* introduced in 1915 into the range of *C. incanum* on Bahia Honda Key, Florida, have been hybridizing with the latter during this century. Analyses of allozymes (and morphology) revealed that these snails now are panmictic, that no pure *C. casablancae* remain, and that there has been a 30% reduction in frequency of the introduced genome (Woodruff and Gould, 1987). Other examples of extensive introgression following artificial transplantations involve trout species in the western United States, where repeated introductions of millions of hatchery-reared rainbow trout into endemic cutthroat trout habitats, and of cutthroat populations from one locale to another, have resulted in extensive introgressive hybridization well documented by molecular markers (Allendorf and Leary, 1988; Busack and Gall, 1981; Forbes and Allendorf, 1991; Gyllensten et al., 1985b; Leary et al., 1984).

Natural hybridization may occur sporadically between broadly sympatric species or be confined to particular geographic contact areas. Hybrid zones, which often appear linear (Hewitt, 1989) or mosaic (Harrison and Rand, 1989; Rand and Harrison, 1989), are regions in which genetically distinct populations meet and produce progeny of mixed ancestry (Barton and Hewitt, 1989; Harrison, 1990). Most hybrid zones represent a secondary overlap between formerly allopatric or parapatric taxa. [However, the usual evidence for secondary contact, concordance in character clines across the zone, could in theory also be a primary outcome of intense diversifying selection within a single continuously-distributed population (Endler, 1977).] If secondary hybrid zones are not ephemeral, their persistence might be explained by the hypothesis of "bounded hybrid superiority" wherein hybrids have superior fitness in areas of presumed ecological transition, or by the "dynamic-equilibrium" hypothesis wherein hybrid zone maintenance is achieved through a balance between dispersal of parental types into the zone and hybrid inferiority (Moore and Buchanan, 1985).

In any event, hybrid zones are marvelous settings to apply molecular markers for several reasons (Hewitt, 1988). First, the populations or species involved are genetically differentiated (by definition), such that multiple markers normally can be uncovered for characterizing the hybrid gene pool. Second, because true hybrid zones involve amalgamations of independently evolved genomes, exaggerated effects of intergenomic interactions might be anticipated, thus magnifying the impact of such processes as recombination and selection, and perhaps making these evolutionary forces easier to study. Third, various sexual asymmetries frequently are involved in hybrid zones, and powerful means now exist for dissecting these factors by utilizing joint data from cytoplasmic and nuclear markers.

To introduce the power of molecular genetics in hybridization analysis, consider an example involving the cytonuclear dissection of an introgressed population of treefrogs in ponds near Auburn, Alabama. *Hyla cinerea* and *H. gratiosa* are distinctive species distributed widely and sympatrically throughout the southeastern United States. From morphological evidence, they are known to hybridize at least sporadically, and at the Auburn site extensive introgressive hybridization has continued for more than 25 years since its initial discovery. One reason for particular interest in this population stems from behavioral observations suggesting the potential for a sexual bias in the direction of interspecific matings. During the breeding season, *H. gratiosa* males call from the water surface, whereas *H. cinerea* males call from perches along the shoreline (Fig. 7.7). In the evenings, gravid females of both species approach the ponds from surrounding woods and become amplexed. Thus, the hypothesis was raised that most interspecific matings might involve *H. cinerea* males with *H. gratiosa* females, rather than the converse: *H. gratiosa* females presumably must "hop the gauntlet" of *H. cinerea* males before reaching conspecific pairing partners!

Lamb and Avise (1986) employed five species-diagnostic allozyme loci plus mtDNA to characterize 305 individuals from this hybrid population. The allozyme loci were chosen because they exhibit fixed allelic differences between the species, thus allowing provisional assignment of each assayed individual to one of the following six categories: pure *H. cinerea,* pure *H. gratiosa,* F_1 hybrid, progeny from a backcross to *H. cinerea,* backcross to *H. gratiosa,* or later-generation hybrid. For example, a F_1 hybrid should be heterozygous at all marker loci, and a *H. cinerea* backcross progeny likely would appear heterozygous at some loci and homozygous for *H. cinerea* alleles at others. [With multiple diagnostic markers, the probabilities of misclassifying an individual into an alternative category are low and can be calculated readily from Mendelian considerations; for example, a true first-generation *H. cinerea* backcross progeny would be mistaken for a pure *H. cinerea* with probability $k = (0.5)^n$, where n is the number of fixed marker loci, so that in this case $k = 0.031$.] The mtDNA genotypes then allowed assignment of the female (and hence male) parent for each allozymically characterized specimen.

The molecular data revealed a striking genetic architecture for the hybrid population that generally proved consistent with suspected mating behaviors of the parental species (Table 7.3; Fig. 7.7). Thus, all 20 F_1 hybrids carried "*gratiosa*"-type mtDNA, indicating that they had *H. gratiosa* mothers. Furthermore, 52 of 53 individuals identified as backcross progeny to *H. gratiosa* possessed "*gratiosa*"-type mtDNA (as predicted, because their mothers were either F_1's or pure *H. gratiosa*). Furthermore, among the progeny of backcrosses to *H. cinerea,* individuals carrying either "*gratiosa*"-type or "*cinerea*"-type mtDNA

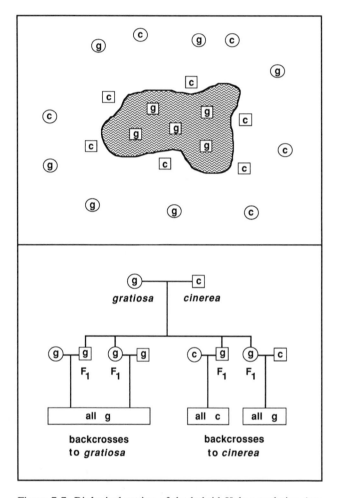

Figure 7.7. Biological setting of the hybrid *Hyla* population (see text). **Top:** Diagrammatic aerial view of an Auburn pond, with calling male treefrogs denoted by c (*H. cinerea*) and g (*H. gratiosa*). **Bottom:** Expected pedigree involved in production of F_1 hybrids and various backcross classes, under the behavior-motivated assumption that hybridizations primarily entail matings of male *H. cinerea* with female *H. gratiosa*. Each letter (c or g) in the pedigree refers to the species origin of the mtDNA genotype; circles and squares indicate females and males, respectively.

Table 7.3. Genetic architecture of a hybrid population involving the tree frogs *Hyla cinerea* and *H. gratiosa* (after Lamb and Avise, 1986).[a]

Allozyme Category	"gratiosa" mtDNA		"cinerea" mtDNA	
	Observed	Expected	Observed	Expected
Pure *H. gratiosa*	103	—	0	—
Pure *H. cinerea*	0	—	60	—
*F*₁ hybrid	20	20	0	0
H. cinerea backcross	22	29	36	29
H. gratiosa backcross	52	53	1	0
Later-generation hybrids	9	Some[b]	2	Some[b]

[a]For each category of frogs as identified allozymically, numbers of observed individuals with indicated mtDNA genotypes are presented. Also shown are these expected numbers based on the behaviorally motivated hypothesis that interspecific crosses are in the direction *H. cinerea* male X *H. gratiosa* female (see text), and that to a given backcross category, *F*₁ hybrids of both sexes (who thus have *H. gratiosa* mtDNA) have contributed equally.

[b]Both "cinerea" and "gratiosa" mtDNA genotypes are expected among later-generation hybrids, but relative frequencies are dependent on additional factors and are, thus, hard to predict.

were both well represented (also as predicted, because the mtDNA genome transmitted in a given mating would depend on whether the F_1 hybrid parent was a male or female—Fig. 7.7). Nevertheless, the asymmetric mating pattern alone probably cannot explain all aspects of the data, because individuals with pure *H. cinerea* and *H. gratiosa* genotypes remained present in high frequency (Table 7.3). Additional factors at work may involve selection against hybrids and/or continued migration of the parent species into the area. In formal models that allowed variation in parental immigration rates and included tendencies for positive assortative mating between conspecifics, Asmussen et al. (1989) found an excellent fit to the empirical cytonuclear data when, at equilibrium, about 32% of the inhabitants of the hybrid zone were pure-species immigrants each generation. Of course, other unexplored scenarios (e.g., involving selection against hybrids) also might explain the data.

How much of this pronounced genetic structure in the *Hyla* population would have been uncovered from a traditional morphological assessment alone? Lamb and Avise (1987) applied multivariate analyses to numerous phenotypic characters in the same *Hyla* individuals and compared results against those obtained from the molecular genetic assessments. Although the pure *H. gratiosa* and *H. cinerea* specimens (as classified by molecular genotype) could be distinguished cleanly by discriminate analyses of morphological traits, various hybrid classes proved less recognizable. For example, in terms of morphology 18% of true F_1's were indistinguishable from pure parental species. Other percentages of misclassification by morphology were as follows: 27% of backcrosses in either direction were not distinguished from F_1's; 50% of *gratiosa* backcross progeny were misidentified as pure *H. gratiosa;* and 56% of *cinerea* backcross progeny were misidentified as pure *H. cinerea.* By contrast, the expected levels of misclassification based on the surveyed molecular genotypes were always less than 4% (based on straightforward Mendelian considerations). Furthermore, the pronounced asymmetry in the mating behavior that apparently exerted exceptional influence on the genetic architecture of the hybrid population would have remained completely undetected by morphological assessment alone.

Numerous molecular genetic analyses of hybridization and speciation have appeared in the past two decades, and thorough recent reviews can be found in Arnold (1992), Barton and Hewitt (1985), and Harrison (1993), among others. A few select examples will be presented here to illustrate the diversity of issues addressed.

Hybrid Zone Asymmetries

Various types of genetic asymmetries frequently attend hybridization and introgression processes. Some of the asymmetries can stem from differential compatibilities of various alleles when placed in heterologous genetic backgrounds via introgression and may be revealed by comparisons among multiple independent molecular markers. Other asymmetries can stem from different behaviors or fitnesses between the sexes and may be revealed by comparisons of cytoplasmic and nuclear genomes.

DIFFERENTIAL INTROGRESSION ACROSS A HYBRID ZONE

A classic case of differential introgression across a well-defined hybrid zone involves the house mice *Mus musculus* and *M. domesticus.* These forms, previously considered subspecies, meet and hybridize along a narrow line that extends throughout central Europe (Fig. 7.8). From a molecular genetics perspective, this situation first was studied by Selander et al. (1969), who documented pronounced divergence between the taxa (fixed or nearly fixed allelic differences at 6 of 40 allozyme loci). Subsequent genetic analysis of nearly 2700 mice from the contact zone across Denmark revealed the following (Hunt and Selander, 1973): (a) free interbreeding within the hybrid zone (as indicated by agreement of genotype frequencies with random-mating expectations); (b) an asymmetry of introgression adjacent to the zone, with extensive introgression of some of the *domesticus* alleles into *musculus* but little gene movement in the other direction; and (c) a marked increase in the width of the zone in western Denmark, compared to the east where 90% of the transition in genic character occurred over a distance of only 20 km. Furthermore, different slopes and patterns of allelic clines across loci were observed and interpreted as evidence for

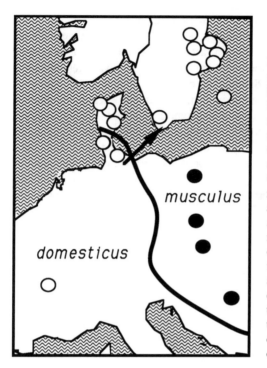

Figure 7.8. Distributions of two commensal species of house mice in central Europe. The heavy line indicates the position of a hybrid zone at the contact between *Mus musculus* (to the north and east, including Scandinavia) and *M. domesticus* (to the south and west). Open and closed circles indicate mtDNA genotypes normally characteristic of *M. domesticus* and *M. musculus,* respectively, and the arrow indicates the postulated route of colonization of Scandinavia by female *M. domesticus* (Gyllensten and Wilson, 1987b). Note that the distributions of the mtDNA genotypes are strikingly discordant with the ranges of the two species as defined by morphology and nuclear genes.

different selective values for alleles as determined in part by the internal genetic environment. In other words, "selection against introgression of the genes studied (or the chromosomal segments that they mark) is presumed to involve reduced fitness in backcross generations caused by disruption of co-adapted parental gene complexes" (Hunt and Selander, 1973). Another molecular study of the 20-km-wide hybrid zone in southern Germany revealed that about 98% of the mice had backcross genotypes. Furthermore, these hybrids were unusually susceptible to parasitic pinworms, other nematodes, and tapeworms, leading to the conclusion that their reduced fitnesses "act as a genetic sink, interfering with the flow of genes between the two species" (Sage et al., 1986). Subsequent study of Y-chromosome allelic frequencies (based on Southern-blot assays using diagnostic Y-specific probes) through portions of the hybrid zone in Bulgaria and Denmark revealed sharp clines indicative of severe restrictions on Y-introgression as well (Vanlerberghe et al., 1986).

Throughout most of Europe, *M. musculus* and *M. domesticus* also differ clearly in mtDNA composition (mean $p \cong 0.05$; Ferris et al., 1983a,b). However, an unexpected pattern emerged in parts of Scandinavia, where mice that by the evidence of nDNA and morphology appear to be *M. musculus* nevertheless carried *M. domesticus*-type mtDNA exclusively (Fig. 7.8). Gyllensten and Wilson (1987b) proposed that the "foreign" mtDNA originated from a small number of female *M. domesticus* that colonized Sweden from a southern source, perhaps in association with the spread of farming from northern Germany to Sweden some 4000 years ago. Continued backcrossing to *M. musculus* males might thereby have introduced mtDNA from *M. domesticus* into populations that retain a predominant *M. musculus* nuclear genetic background. Further study of the hybrid zone in central Denmark revealed that mtDNA genotypes in northern Danish populations, though also characteristic of *M. domesticus*, nonetheless differ detectably from present-day *M. domesticus* mtDNA in southern Denmark, a finding that adds support to the Gyllensten-Wilson scenario by eliminating the possibility that contemporary introgression alone can account for the mtDNA transfer (Vanlerberghe et al., 1988).

In some hybrid zones, patterns of variation for cytoplasmic and nuclear markers are highly concordant (Baker et al., 1989; Nelson et al., 1987; Szymura et al., 1985), but in several other instances pronounced discordances across loci (or between molecular and morphological data) appear to reflect differing historical patterns of introgression (Table 7.4). One phenomenon frequently reported is that of cytoplasmic "capture," wherein the mtDNA or cpDNA genotypes normally characteristic of one species sometimes occur against a predominant nuclear background of another species, presumably because of current or past introgressive hybridization. One speculation has been that mtDNA tends to introgress more readily than nuclear DNA because genes contributing to reproductive isolation might be housed primarily in the nucleus (Barton and Jones, 1983). However, even if true, many additional factors governing introgression must be involved. The following sections provide examples of how asymmetries stemming from different behaviors and/or fitnesses of males and females can lead to patterns of differential exchange for nuclear and cytoplasmic loci.

HALDANE'S RULE

An empirical generality first noticed by Haldane (1922) is that "When in the F_1 offspring of two different animal races one sex is absent, rare, or sterile, that sex is the heterozygous [heterogametic] sex." Thus, in species with heterogametic females (such as birds and butterflies), female hybrids more often show decreased fitness, whereas in species with heterogametic males (such as fruit flies and mammals), male hybrids tend to show more severe reductions in viability or fertility (Coyne and Orr, 1989a).

Tegelström and Gelter (1990) compared nuclear and cytoplasmic markers in a contact zone between two hybridizing flycatcher species in Europe, *Ficedula albicollis* and *F. hypoleuca*. Allozyme differentiation was negligible, due perhaps to extensive ongoing introgression of nuclear genes via male hybrids [which from DNA fingerprint analyses of parentage are known to be fertile (Gelter et al.,

Table 7.4. Examples of mtDNA and cpDNA "capture" reportedly due to introgressive hybridization between related species of animals and plants, respectively.[a]

Genus	Common Name	Reference
Animal mtDNA		
Caledia	Grasshoppers	Marchant, 1988
Clethrionomys	Voles	Tegelström et al., 1988
Drosophila	Fruit flies	Solignac and Monnerot, 1986
Gryllus	Crickets	Harrison et al., 1987
Hyla	Treefrogs	Lamb and Avise, 1986
Mus	House mice	Ferris et al., 1983a
Notropis	Minnows	Dowling et al., 1989; Dowling and Hoeh, 1991
Odocoileus	Deer	Carr et al., 1986
Rana	Frogs	Spolsky and Uzzell, 1984, 1986
Plant cpDNA		
Argyroxiphium	Silverswords	Baldwin et al., 1990
Brassica	Cabbages and allies	Palmer et al., 1983
Dubautia	Silverswords	Baldwin et al., 1990
Gossypium	Cotton	Wendel et al., 1991; Wendel and Albert, 1992
Helianthus	Sunflowers	Rieseberg et al., 1990b
Heuchera	Heucheras	Soltis et al., 1991
Persea	Avacados	Furnier et al., 1990
Pisum	Peas	Palmer et al., 1985
Populus	Poplars	Smith and Sytsma, 1990
Quercus	Oaks	Whittemore and Schaal, 1991
Salix	Willows	Brunsfeld et al., 1992
Tellima	(herb. perennial)	Soltis et al., 1991
Zea	Teosintes, maize	Doebley, 1989

[a]See text for discussion, and consult Harrison (1989) and Rieseberg and Soltis (1991) for additional examples.

1992)]. However, mtDNA divergence between these avian species remained high, a result interpreted by the authors to evidence a lack of interspecific cytoplasmic genetic exchange (the heterogametic hybrid females appear sterile, in accord with Haldane's rule). However, samples from outside the zone of sympatry were not compared, so the geographic extent of nuclear introgression remains unknown (and indeed, alternatives to the biased introgression scenario might be entertained). In voles (*Clethrionomys*), interspecific crosses result in fertile female but sterile male offspring (Tegelström et al., 1988). In northern Scandinavia, an observed discordance between species boundary and mtDNA phylogeny led to the conclusion that mtDNA stemming from *C. rutilus* has passed the species boundary and now is found in *C. glareolus,* perhaps as a result of a limited hybridization episode perhaps dating to the postglacial colonization of the region some 10,000 years ago (Tegelström, 1987b). Thus, this scenario is very similar to that for the unidirectional introgression of mtDNA in the Scandinavian *Mus* populations discussed above.

In *Drosophila* hybrids, the heterogametic males frequently show partial or complete sterility (sometimes in one direction of a cross only), whereas the homogametic females normally remain fertile. Male sterility has proved to be polygenic, with responsible loci mapped (using chromosomal or molecular markers) to the X and Y chromosomes and to various autosomes (Dobzhansky, 1974; Vigneault and Zouros, 1986). This asymmetry of hybrid fertility has been invoked to explain the apparent ease with which mtDNA appears to cross some species' boundaries (because the fertile females leave open an avenue for interspecific cytoplasmic exchange). For example, in *D. mauritiana,* a high proportion of individuals carries a mtDNA genotype also found in nearby populations of *D. simulans,* and these observations were interpreted to indicate a recent introgression of *D. simulans* mtDNA into the Mauritiana population (Solignac and Monnerot, 1986). This hypothesis subsequently gained support from population cage studies in which the predicted takeover by *D. simulans* mtDNA was documented experimentally over a few generations of introgressive hybridization (Aubert and Solignac, 1990). Powell (1983) reported a similar situation where suspected hybridization was postulated to account for a greater mtDNA similarity between populations of *D. persimilis* and *D. pseudoobscura* in sympatry than in allopatry [however, see Powell (1991) for possible reinterpretations]. On the other hand, DeSalle and Giddings (1986) report that the mtDNA phylogeny for several closely related species of Hawaiian *Drosophila* matches the suspected species phylogeny quite well, despite a postulated historical introgression that has complicated phylogenetic reconstructions based on nuclear genes.

DIFFERENTIAL MATING BEHAVIORS

The *Hyla* example discussed earlier provides a powerful illustration of how a behavioral asymmetry has influenced the genetic architecture of a hybrid zone. Another example involves a contact zone in France between the hybridizing newts *Triturus cristatus* and *T. marmoratus* (Arntzen and Wallis, 1991). Allozymes again were employed to characterize the hybrid status of individuals, and mtDNA genotypes were used to identify the female parents. All F_1 hybrids possessed the *T. cristatus*-type mtDNA, perhaps due to a strong asymmetry in mate choice. Furthermore, an observed absence of introgression of mtDNA in areas where *T. cristatus* replaced *T. marmoratus* is consistent with this interpretation.

Sunfish in the genus *Lepomis* are renowned for propensity to hybridize, both in artificial pond situations and in nature. As stated by Breder (1936), ''There is probably no group of fishes, North American at least, in which there would seem to be a concatenation of reproductive and other events so well arranged as to lead to extensive hybridizing; *i.e.*, the species are numerous; there is less geographic separation than usual; spawning occurs at about the same temperature threshold; spawning sites are limited and similar for most species; nests are exchanged among species.'' From the observation of diminished hybrid fertility in both sexes, Hubbs and Hubbs (1933; see also Hubbs, 1955) concluded that natural hybridization probably was limited to the F_1 generation, but later work with experimental populations revealed that ''a number of different kinds of hybrid sunfishes . . . are not sterile, are fully capable of producing abundant F_2 and F_3 generations, and can be successfully backcrossed to parent species and even outcrossed to nonparental species'' (Childers, 1967). Nonetheless, the 11 recognized species within the genus normally exhibit large genetic distances at protein-coding loci (Fig. 7.2) and, for the most part, retain distinctive morphological identities throughout their respective ranges. Thus, questions remain about the degree of *Lepomis* hybridization in nature and whether introgression has played a significant role in the evolution of the group. Avise and Saunders (1984) characterized a total of 277 sunfish from two locations in north Georgia for species-diagnostic allozymes and mtDNA. The genetic data revealed the following: (a) a low frequency (5%) of interspecific hybrids, all of which appeared to be F_1's; (b) the involvement of five sympatric *Lepomis* species in the production of these hybrids; and (c) no evidence for introgression in these study locales. With respect to the current discussion, two additional points were of interest: There was a strong tendency for the hybridizations to take place between parental species differing greatly in abundance, and a tendency for the rare species in the hybrid cross to provide the female parent. Although conclusions are tentative because the number of hybrids discovered was small, the data suggest a density-dependent mating pattern in which the absence of conspecific pairing partners and spawning stimuli for females of rarer species might be important factors in increasing the likelihood of interspecific hybridization.

Another molecular study of hybridization in sunfish involved bluegills (*Lepomis macrochirus*) and pumpkinseeds (*L. gibbosus*) in a lake in southern Canada. Among 44 hybrids examined by allozymes and mtDNA, all appeared to be F_1's, and to have had pumpkinseed mothers (Konkle and Philipp, 1992). From prior studies, male bluegill are known to exhibit two alternative life history strategies with respect to reproduction: a ''parental pathway,'' in which sexual maturation is delayed until about eight years of age, at which time the male builds and defends a nest, courts and spawns with females, and guards eggs and young after fertilization; and a ''precocial pathway,'' in which a male matures at about two years of age, fails to build a nest or court females, and instead ''steals'' fertilizations from conspecific parental males through sneaky behavior or female mimicry (Gross, 1979; Gross and Charnov, 1980). The highly asymmetrical genetic results in the hybrid population studied are consistent with the postulate that precocial male bluegills may also be cuckolding heterospecific pumpkinseed males, and at a higher frequency than is true for crosses in the converse direction.

DIFFERENTIAL GAMETIC EXCHANGE

Particularly in plants, a pronounced decoupling of male and female components of gene flow across species is possible due to the two distinct avenues for genetic movement—pollen and seeds (Arnold, 1992; Paige et al., 1991). When genetic transfer is mediated solely by pollen, maternally inherited genetic markers (e.g., cpDNA in most angiosperms) will not introgress. On the other hand, immigration of seeds into a foreign population might lead to the introgression of both nuclear and cytoplasmic genes when the resulting plants are fertilized by pollen from resident individuals. Clearly, alternative modes of gene transfer can differentially influence cytonuclear associations (Asmussen and Schnabel, 1991).

Arnold et al. (1990a, 1990b, 1991, 1992) have employed molecular markers from allozymes, nDNA, and cpDNA to document asymmetrical gene flow between the Louisiana irises *Iris fulva* and *I. hexagona,* part of a species complex representing a classic example of plant hybridization and introgression (Anderson, 1949). At particular locales, nuclear alleles that normally are species-diagnostic have been shuffled into a variety of recombinant genotypes, confirming suspicions from morphological analyses that introgressive hybridization has occurred. Furthermore, many individuals of hybrid ancestry (as gauged by nuclear genes) nonetheless retain the cpDNA of *I. hexagona* (Fig. 7.9), suggesting that they are the products of pollen transfer from *I. fulva* onto *I. hexagona* flowers. Thus, introgression of nDNA markers has occurred in the absence of cpDNA introgression. Results appear consistent with the natural history of these species, which includes pollination by highly mobile bumblebees and an absence of special adaptations for seed dispersal.

Might long-distance pollen movement lead to the exceptionally wide areas of introgression postulated for some plant species complexes? In the southeastern United States, three parapatric species of buckeye trees (*Aesculus sylvatica, A. flava,* and *A. pavia*) appear to be engaged in introgression over a broad region at least 200 km wide, as inferred from patterns of morphology, geographic distribution, and meiotic irregularities associated with decreased germinability of pollen from putative interspecific hybrids. Allozyme data gathered for these species also are consistent with the introgression scenario, and further raise the possibility that long distance movement of genes beyond the morphology-recognized hybrid area may have taken place (dePamphilis and Wyatt, 1990). For example, one of the hybrid zones appears highly asymmetrical, with alleles characteristic of coastal plain *A. pavia* also found in Piedmont populations where *A. sylvatica* normally occurs. The authors hypothesized that an important polli-

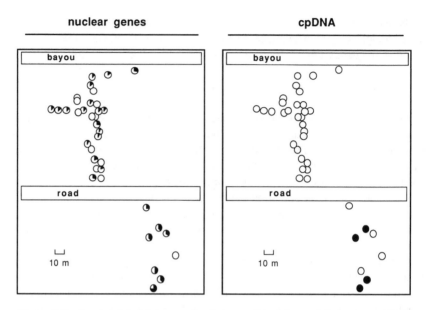

Figure 7.9. Asymmetrical introgression between *Iris fulva* and *I. hexagona* likely resulting from pollen flow (after Arnold, 1992). Each circle represents a single plant. **Left:** Relative proportion of *I. fulva* (shaded) and *I. hexagona* (unshaded) nuclear markers. **Right:** Similar representation for maternally-transmitted chloroplast DNA markers. Note, in particular, the population between the road and bayou, where multilocus nuclear genotypes suggest the presence of advanced generation hybrids or backcrosses, despite the apparent absence of seed dispersal that would be registered by cpDNA from *I. fulva* in this area.

nation agent for buckeyes, the ruby-throated hummingbird, may have effected long-distance pollen flow during its northward spring migrations, thus accounting for both the width and asymmetry of the hybrid zone (dePamphilis and Wyatt, 1989).

Once pollen have arrived on a heterospecific style, additional challenges await that might lead to asymmetric barriers to successful hybridization. For example, in interspecific crosses among *Eucalyptus* species, *E. nitens* pollen tubes grow slowly and never reach full length in the larger *E. globulus* styles, whereas *E. globulus* pollen tubes grow rapidly in *E. nitens* styles and enter the ovary (Gore et al., 1990). Hybridization between several species of *Eucalyptus* occurs rather commonly in nature (Griffin et al., 1988), and unilateral cross-incompatibility might play an important role in structuring cytonuclear associations. A variety of other selective mechanisms operating at prezygotic or postzygotic stages also could lead to asymmetrical introgression in *Eucalyptus* species and elsewhere (Gore et al., 1990; Potts and Reid, 1985).

One such postzygotic mechanism is "cytoplasmic male sterility" (CMS), a widespread phenomenon in plants (Edwardson, 1970) involving pollen abortion in hybrid progeny due to an interaction of cytoplasmically-transmitted (usually mtDNA) mutations with a foreign nuclear background (Hanson, 1991). Male sterility in first-generation hybrids or backcross progeny could have the effect of attenuating the introgression of nuclear genes, while nonetheless leaving open an avenue for cytoplasmic transfer via fertile females. CMS is known to occur, for example, in hybrids between some of the *Helianthus* species that have figured prominently in discussions of intertaxon exchange of cpDNA, as described next.

RETICULATE EVOLUTION EVIDENCED BY CPDNA PHYLOGENIES

Owing to a strong propensity for hybridization in many plant taxa, botanists have been particularly concerned with the possibility of widespread reticulate evolution (Grant, 1981; Stebbins, 1950), wherein phylogenies of some groups might be characterized as anastomotic or "netlike," rather than strictly dichotomous and branching. Recent molecular analyses based on comparisons of nDNA and cpDNA have uncovered considerable evidence for this phenomenon.

In *Helianthus* sunflowers, gross incongruities between phylogenies estimated from morphological and chromosomal variation, experimental crossing success, and various molecular markers have led to the conclusion that both recent and relatively ancient episodes of intertaxon gene exchange have produced a reticulate pattern of relationships among these species (Rieseberg, 1991; Rieseberg et al., 1988, 1991). For example, the cpDNA-based phylogeny for numerous *Helianthus* taxa shown in Figure 7.10 contrasts significantly at the indicated positions with suspected relationships based on morphological characters and on the nuclear genes for ribosomal RNA. Thus, each of five species (*H. anomalus, H. annuus, H. debilis, H. neglectus,* and *H. petiolaris*) possesses two or more highly distinct cpDNA genotypes otherwise characteristic of different phylogenetic groups within the genus. Furthermore, some species appear to have captured the cytoplasm of other species on multiple occasions. For example, *H. petiolaris* has acquired the cytoplasm of *H. annuus* at least three times, as evidenced by additional information on geographic locations of the introgressed populations and by the particular *H. annuus* cpDNA genotypes exhibited (Rieseberg and Soltis, 1991).

As with animal mtDNA, the cpDNA molecule appears especially helpful (as an adjunct to nuclear analyses) in revealing such cases of reticulation because its clonal transmission allows particular ancestral sources to be identified without the complication of recombination (including that at the intragenic scale) that can lead to a mosaic ancestry for the nuclear genome. In addition, there may be several biological factors including CMS that facilitate cytoplasmic (relative to nuclear) exchange across species (see earlier in this Chapter). Rieseberg and

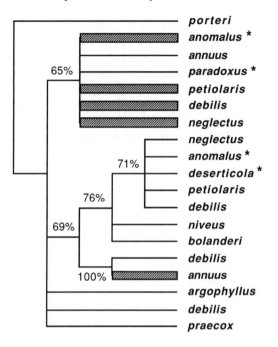

Figure 7.10. Evidence for cytoplasmic introgression in *Helianthus* sunflowers (after Rieseberg and Soltis, 1991). Shown is a most-parsimonious tree (Wagner network) based on cpDNA data and indicating areas of gross discrepancy (hatched branches) with a morphological classification. These discrepancies were interpreted to be the result of interspecies transfer of cpDNA mediated by hybridization. Numbers indicate levels of bootstrap support for putative clades. Asterisks specify taxa that are stabilized hybrid derivatives.

Soltis (1991) reviewed the rapidly growing literature on instances of interspecific cpDNA "capture" attributable to introgressive hybridization (Table 7.4) and concluded that reticulate evolution is indeed a widespread phenomenon in plants. Such reticulations can seriously compromise phylogeny reconstructions based on small numbers of characters or those generated under the assumption that phylogenies are strictly branching and hierarchical. Furthermore, no taxonomic zone may be completely safe from the phylogenetic consequences of reticulate evolution, because gene transfer ". . . between major evolutionary lines during early stages of their divergence has the potential to impact cpDNA phylogenetic reconstruction at all taxonomic levels. . . ." (Rieseberg and Soltis, 1991).

In summarizing this general section, it is now abundantly clear that hybridization phenomena are best analyzed through multiple lines of evidence involving numerous molecular (and other) markers. Barriers to reproduction between closely related taxa seldom are absolute and, in addition, often appear differentially "semipermeable" to cytoplasmic and various nuclear alleles. Thus, a rich and varied fabric of gene genealogies (seldom evident from traditional morphological assessment alone) characterizes many hybrid contacts, revealing varying degrees of reticulation among the phylogenetic branches connecting related species.

Speciation by Hybridization

Can differentiated genomes brought together through hybridization sometimes produce a new species? One such mechanism already has been described (Box 7.3)—allopolyploidization. From surveys of molecular markers, other hybridization-mediated routes to speciation have been revealed as well (Abbott, 1992).

DIPLOID OR HOMOPLOID SPECIATION

In the plant literature, a traditional proposal has been that species isolated by a chromosomal sterility barrier might give rise via hybridization to new fertile diploid species that are at least partially isolated reproductively from both parents (Grant, 1963; Stebbins, 1950). Such "recombinational speciation" (Grant, 1981) has been tested under artificial conditions by experimental synthesis of new hybrid species (see the review in Rieseberg et al., 1990a), but questions remained as to the prevalence of this speciational mode in nature. Numerous candidates for hybrid species status were identified from early studies of morphology, ecology, and geographic distributions of plant taxa in nature, but final confirmation of hybridity awaited the application of molecular markers.

Stephanomeria diegensis is a diploid annual plant native to southern California, purportedly derived following stabilization of hybrid segregants from a natural cross between two divergent diploid relatives (*S. exigua* and *S. virgata*) with the same chromosome number. Gallez and Gottlieb (1982) demonstrated that *S. diegensis* indeed displays an additive profile of the allozyme alleles of its presumed relatives, a finding consistent with the plants' intermediate morphology and karyotype and, thus, supportive of the postulated hybrid origin. Similarly, analyses of allozymes and other nuclear markers in the putative hybrid *Iris nelsoni* indicate that this species possesses a combination of nuclear genes characteristic of three species that thus appear involved in its formation—*I. fulva, I. hexagona,* and *I. brevicaudis* (Arnold et al., 1990b, 1991). However, not all molecular genetic reappraisals have confirmed the suspected hybrid origins of problematic plant taxa. The diploid annual *Lasthenia burkei* proved not to possess a combination of allozyme alleles present in *L. conjugens* and *L. fremontii,* thus disputing earlier hypotheses that *L. burkei* is a stabilized hybrid derivative of the latter two species (Crawford and Ornduff, 1989).

Rieseberg et al. (1990a) noted that, by hard criteria, apparent additivity of alleles in the nuclear genome is insufficient to fully confirm the hybrid origin of a diploid species because a nonexcluded alternative is that the taxon in question might be ancestral to its putative parents. One solution to this problem is to use additional markers (such as those from cpDNA) to establish the polarity of relationships. By adopting this philosophy as applied to allozymes and cpDNA genotypes in diploid sunflowers, Rieseberg et al. (1990a) concluded that two problematic taxa, appropriately named *Helianthus paradoxus* and *H. neglectus,* had dissimilar pathways of origin: The former is indeed a hybrid species having arisen from *H. annuus* and *H. petiolaris,* whereas the latter appears to be a recent nonhybrid derivative of *H. petiolaris.*

Somewhat less attention has been devoted to the possibility of homoploid speciation via hybridization in animals. On the basis of morphological and allozyme data, Highton et al. (1989) suggested that the salamander *Plethodon teyahalee* had a hybrid origin as a result of interbreeding between white-spotted *P. glutinosus* and red-legged *P. jordani.* Among fishes, several examples of stabilized hybrid forms have been suspected from morphological or distributional considerations. For example, the "zuni sucker" in the Little Colorado drainage of the western United States was suggested to be a hybridization-derived intermediate between *Catostomus discobolus* of the Colorado drainage and *C. plebeius* of the Rio Grande; and the "white shiner" in the Roanoke and adjacent drainages of the eastern United States was proposed to have arisen from hybridization between nearby *Luxilus cornutus* and *L. cerasinus.* However, molecular reevaluations of these forms either did not support the hybrid origin scenario [in the case of the zuni sucker (Crabtree and Buth, 1987)] or were equivocal [in the case of the white shiner (Meagher and Dowling, 1991)] due to the difficulty of discriminating among the possibilities of ancestral polymorphism, convergence, and past hybridization to account for the observed allozymic distributions.

Molecular reappraisals *have* provided support for the postulated hybrid origin of another recognized fish species. On the basis of an intermediate morphology, *Gila seminuda* in the Virgin River of the western United States had been proposed as a hybrid derivative between the roundtail chub (*Gila robusta*) and bonytail chub (*Gila elegans*). In terms of allozymes, *G. seminuda* proved to be polymorphic for alleles at two allozyme loci otherwise diagnostic for the putative parental taxa; as judged by mtDNA, the matriarchal lineage retained by *G. seminuda* derives from *G. elegans* (DeMarais et al., 1992). As noted by the authors, such stabilized hybrid derivatives might be relatively common in some groups of fishes, but remain unrecognized due to a lack of detailed molecular studies. As also noted by the authors, the formal taxonomic status of such introgressed forms likely will remain a point of contention, particularly when the introgressed population currently is isolated from its parental species by extrinsic (geographic) barriers to reproduction. Should the hybrid form be considered a distinct population, subspecies, or species? This question is not merely academic, but also is relevant to the implementation of conservation programs (Chapter 9).

ORIGINS OF UNISEXUAL BIOTYPES

Unisexual biotypes that reproduce by parthenogenesis, gynogenesis, or hybridogenesis (Fig. 5.3) are not biological species in the usual sense applied to sexually reproducing taxa, but nonetheless they are isolated genetically from their sexual relatives (as well as from other unisexual lineages) and typically are afforded formal taxonomic recognition. Among the vertebrates, essentially all of the 70 known unisexual taxa arose through hybridizations between related sexual species. In a number of studies, molecular markers have been highly informative in revealing the mode of origin and parentage of these all-female biotypes.

In most cases, the particular bisexual progenitors of various unisexual vertebrates had been suspected from earlier comparisons of morphology, karyotype, geographic range, or other information, but molecular surveys of allozymes and mtDNA have confirmed the suspected hybrid origins in several instances, and for the first time revealed the directions of the original crosses. For example, the gynogenetic live-bearing fish *Poecilia formosa* in northeastern Mexico exhibits nearly fixed heterozygosity at numerous protein and allozyme loci that distinguish or are polymorphic in *P. latipinna* and *P. mexicana* (Abramoff et al., 1968; Balsano et al., 1972; Turner, 1982), thus confirming evidence from morphology and geography that *P. formosa* arose via hybridization between these sexual species. The gynogen *P. formosa* also carries the mtDNA of *P. mexicana*, which is highly divergent ($p \cong 0.07$) from that of *P. latipinna*, indicating that the direction of the formational hybrid cross(es) was *P. mexicana* female $\times$ *P. latipinna* male (Avise et al., 1991). To date, similar molecular inspections have allowed unambiguous determination of the sexual progenitors for more than 25 unisexual biotypes (Table 7.5).

A few unisexuals carry genomic contributions from more than two sexual ancestors. The gynogenetic fish *Poeciliopsis monacha-lucida-viriosa*, for example, includes *P. viriosa* nuclear genes apparently introgressed as a result of occasional matings of *P. monacha-lucida* females with *P. viriosa* males rather than with males of their usual sexual host *P. lucida* (Vrijenhoek and Schultz, 1974). As judged by allozymes and other evidence, several triploid parthenogenetic lizards in the genus *Cnemidophorus* also carry genes from three sexual progenitors (Dessauer and Cole, 1989; Good and Wright, 1984), likely as a result of multiple hybridization events involving these species.

In many cases, molecular analyses have further pinpointed the geographic and

Table 7.5. Examples of species parentage determined for unisexual vertebrates (for an extended list, see Avise et al., 1992c).[a]

Unisexual Biotype	Ploidy Level	Reprod. Mode[b]	Bisexual Parental Species		Ref.[c]
			Male	Female	
Cnemidophorus lizards					
uniparens	3n	P	*burti*	*inornatus* (2)	a
tesselatus	2n	P	*septemvittatus*	*marmoratus*	b
velox	3n	P	*inornatus* (2)	*burti* or *costatus*	c
laredoensis	2n	P	*sexlineatus*	*gularis*	d
Heteronotia lizards					
binoei (widespread form)	3n	P	*binoei*	sp. "CA6"	e
Menidia fish					
clarkhubbsi complex	2n	G	*beryllina*	*peninsulae*	f
Phoxinus fish					
eos-neogaeus	2n,3n	G	*eos*	*neogaeus*	g
Poecilia fish					
formosa	2n	G	*latipinna*	*mexicana*	h
Poeciliopsis fish					
monacha-lucida	2n	H	*lucida*	*monacha*	i
monacha-occidentalis	2n	H	*occidentalis*	*monacha*	j

[a]The bisexual parental species were identified from comparisons of allozymes, morphology, karyotype, geographic ranges, or other information, and the female parent was identified by mtDNA comparisons in the references indicated.

[b]P = parthenogenetic; G = gynogenetic; H = hybridogenetic (see Fig. 5.3).

[c]References: (a) Densmore et al., 1989a; (b) Brown and Wright, 1979; Densmore et al., 1989b; (c) Moritz et al., 1989; (d) Wright et al., 1983; (e) Moritz, 1991; (f) A.A. Echelle et al., 1989; (g) Goddard et al., 1989; (h) Avise et al., 1991; (i) Quattro et al., 1991; (j) Quattro et al., 1992a.

genetic source of particular unisexuals. For example, from mtDNA comparisons, the matriarchal components of nine unisexual biotypes in the *sexlineatus* group of *Cnemidophorus* lizards all appear to stem from females within one of the four nominate geographic subspecies of *C. inornatus*—*C.i. arizonae* (Densmore et al., 1989a); the maternal ancestry of five triploid unisexual strains in the *Poeciliopsis monacha-lucida* fish complex trace phylogenetically to an extant bisexual *P. monacha* from the Río Fuerte in northwestern Mexico (Quattro et al., 1992b).

Typically, the bisexual relatives of unisexuals have proved highly distinct in mtDNA genotype, whereas mtDNAs of the unisexuals are closely related to or indistinguishable from those of only one of the sexual progenitors. Thus, an emerging generalization is that most extant unisexual biotypes originated through asymmetrical hybridization events, occurring in one direction only (e.g., *A* female × *B* male versus *B* female × *A* male). Whether this reflects some asymmetrical mechanistic constraints on the origin of unisexuals, or merely the survival of a limited subset of lineages from crosses in both directions, generally remains unclear. However, in the case of the *Poeciliopsis* hybridogens, laboratory crosses of *P. monacha* females × *P. lucida* males sometimes result in the spontaneous production of viable hybridogenetic lineages, whereas the reciprocal matings do not (Schultz, 1973). This direction of cross is consistent with the molecular-inferred origins of natural hybridogenetic strains, all of which possess *P. monacha*-type mtDNA (Quattro et al., 1991). Furthermore, these extant natural hybridogens have arisen multiple times through separate hybridization events, as gauged by their links to several different branches in the mtDNA phylogeny of their maternal ancestor *P. monacha* (Fig. 7.11).

One exception to such straightforward hybrid origins involves the hybridogenetic frog *Rana esculenta* of Europe, in which individuals exhibit mtDNA geno-

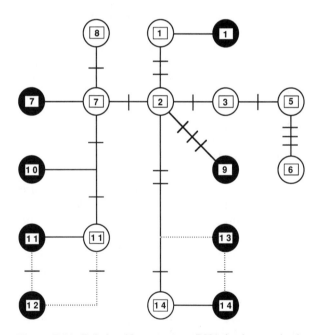

Figure 7.11. Relationships among mtDNA haplotypes in the sexual fish *Poeciliopsis monacha* (open circles) and its unisexual derivative *P. monacha-lucida* (shaded circles) (after Quattro et al., 1991). Slashes represent inferred mutations along the parsimony network; dotted lines indicate alternative network pathways.

types normally characteristic of either *R. lessonae* or *R. ridibunda* (Spolsky and Uzzell, 1986). The hybridogen *R. esculenta* is unique among the assayed "asexual" biotypes in consisting of high frequencies of both males and females. From behavioral considerations, the initial hybridizations producing *R. esculenta* were postulated to involve male *R. lessonae* × female *R. ridibunda*. Once the hybridogen was formed, occasional matings of male *R. esculenta* with female *R. lessonae* secondarily may have introduced *R. lessonae*-type mtDNA into *R. esculenta*. Furthermore, females belonging to such *R. esculenta* lineages appear to have served as a natural bridge for the interspecific transfer of *R. lessonae* mtDNA into certain *R. ridibunda* populations via matings with *R. ridibunda* males (Spolsky and Uzzell, 1984). Such crosses apparently produced "*R. ridibunda*" frogs with normal nuclear genomes (because the *R. lessonae* chromosomes are excluded during meiosis), but with *R. lessonae*-type mtDNA.

Another complex scenario surrounds the hypothesized maternal ancestry of the triploid salamander *Ambystoma 2-laterale-jeffersonianum*, which by allozyme evidence contains nuclear genomes of the bisexual species *A. laterale* and *A. jeffersonianum*, but reportedly carries mtDNA from *A. texanum* (Kraus and Miyamoto, 1990). These authors favor an explanation in which an original *A. laterale-texanum* hybrid female produced an ovum with primarily *A. laterale* nuclear chromosomes, but the female-determining sex chromosome (W) and the mtDNA of *A. texanum*. When fertilized by a male *A. laterale*, female progeny with two *A. laterale* nuclear genomes and the mtDNA of *A. texanum* would result. Subsequent hybridization with male *A. jeffersonianum* could then produce the observed *A. 2-laterale-jeffersonianum* biotypes carrying *A. texanum* mtDNA. Although this scenario remains highly speculative, its mere feasibility suggests that distinct reticulate histories could characterize different genomic elements in some hybridogenetic taxa.

Another question about unisexual vertebrates addressed by molecular markers concerns mechanistic modes of polyploid formation. More than 60% of known unisexual biotypes are polyploid (Vrijenhoek et al., 1989) and two competing hypotheses have been advanced to account for the origins of these forms (Fig. 7.12). Under the "primary hybrid origin" hypothesis, a disruption of meiotic processes in an F_1 interspecific hybrid leads to the production of unreduced diploid gametes, whose subsequent fertilization by sperm leads to a triploid condition (Schultz, 1969). Alternatively, under the "spontaneous origin" hypothesis, parthenogenetic triploids might have arisen when unreduced oocytes from a diploid nonhybrid were fertilized by sperm from a second bisexual species (Cuellar, 1974, 1977). As diagrammed in Figure 7.12, joint comparisons of mitochondrial and nuclear markers permit an empirical test of these competing possibilities. If a unisexual biotype arose spontaneously from sexual ancestors and hybridization was involved only secondarily, the paired homospecific nuclear genomes should derive from the maternal parent, and thus should be coupled with mtDNA derived

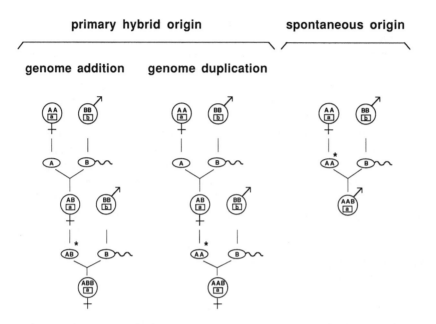

Figure 7.12. Competing scenarios for the origin of triploid unisexual taxa (from Avise et al., 1992c; see text). Each uppercase letter represents one nuclear gene set (A or B) from the respective parental species, and the lowercase letters in boxes similarly refer to the maternally transmitted mtDNA genomes. Smaller ovals indicate sperm and eggs, respectively, the latter being unreduced where indicated by stars. In the genome duplication scenario, this suppression of reduction occurs during an equational division such that the AB hybrid produces AA (or BB) ova.

from the same species. Conversely, under a model of primary hybrid origin, the paired homospecific nuclear genomes could be coupled with the mtDNA type from either of the sexual ancestors, depending on additional details by which the nuclear genome was duplicated or added (see below).

Cytonuclear genetic analyses for several unisexual taxa have provided support for the "primary hybrid origin" hypothesis. For example, the triploid parthenogen *Cnemidophorus flagellicaudus* possesses the mtDNA of *C. inornatus* but two homospecific nuclear genomes from *C. burti*, and a similar type of cytonuclear pattern was observed for 8 of 10 parthenogenetic *Cnemidophorus* biotypes examined (Densmore et al., 1989a; Moritz et al., 1989). Similarly, the triploid gynogenetic fish *Poeciliopsis monacha-2 lucida* possesses the mtDNA of *P. monacha* but two nuclear genomes from *P. lucida* (Quattro et al., 1992b). Thus, these results appear to refute the "spontaneous origin" scenario (unless the diploid nonhybrid that produced unreduced gametes was a male, in which case all bets are off!).

Assuming correctness of the primary hybrid origin scenario, two further cytogenetic pathways to triploidy can be distinguished (Fig. 7.12). Under the "genomic addition" scenario (Schultz, 1969), interspecific F_1 hybrids produce unreduced ova (AB) that on backcrossing to one of the sexual ancestors leads to allotriploid biotypes AAB or ABB. Under the "genomic duplication" scenario (Cimino, 1972), suppression of an equational division in an F_1 hybrid could produce unreduced AA or BB ova, which following a backcross to species A or B would produce AAB or ABB offspring (autopolyploid AAA or BBB progeny could also result from this process, but no self-sustaining populations of autopolyploid unisexual vertebrates are known). An important distinction between these pathways involves the predicted level of heterozygosity in the homospecific nuclear genomes. Heterozygosity should be extremely low under the genome duplication pathway (the only variation being derived from postformational mutations), whereas normal heterozygosity is predicted under the genomic addition pathway. At least one test of these scenarios is available—in triploid *Poeciliopsis* gynogens, all assayed strains proved to be heterozygous for homospecific nuclear markers at one or more allozyme loci, a result which effectively excludes the genome duplication hypothesis for these fishes (Quattro et al., 1992b).

Recently, molecular markers contributed to the discovery of the first instance in nature of "androgenesis" in an animal species. Androgenesis is a sort of male analogue of gynogenesis (Fig. 5.3), involving the development of an individual solely under the influence of its paternally derived chromosome set (i.e., in the absence of instructions from the mother's genetic material). The process had been demonstrated experimentally in the laboratory (see Giorgi, 1992) and was uncovered recently in nature as well. Using protein electrophoresis in conjunction with chromosomal markers, Mantovani and Scali (1992; see also Mantovani et al., 1991) first identified two hybridogenetic strains of stick insects in Italy (*Bacillus rossius-grandii benazzii* and *B. rossius-grandii grandii*) which had arisen from hybridization between *B. rossius* females and two subspecies of *B. grandii* males. The males of these hybridogenetic strains are infertile, whereas females can reproduce both by hybridogenesis or gynogenesis. The most surprising discovery, however, was that when a female *B. rossius-grandii benassii* is fertilized by a *B. rossius* male, up to 20% of the offspring have the nuclear genetic makeup solely of the father. Thus, these individuals are androgenetic. They also have proved to be diploid (through duplication of the male chromosomes or by fusion of two sperm) and fertile.

Overall, the detailed understanding of the evolutionary genetics of "unisexual" taxa provided by molecular markers was unimaginable even a short time ago. As recently as 1978, in referring to a hybrid-derived parthenogenetic grasshopper, a leading student of the speciation process lamented that "we are never likely to know which species was the female parent" (White, 1978b). Since then, parentage determinations for taxa with parthenogenetic or related modes of reproduction have become routine, and indeed are viewed as but a starting point for more refined genetic analyses of evolutionary origins and pathways.

SUMMARY

1. Various speciation patterns are expected to leave characteristic phylogenetic signatures on the genomes of recently separated species. By revealing such signatures, molecular markers have prompted reexamination of long-standing issues in speciation theory, including: How much genetic change accompanies speciation? Do most speciations entail severe population bottlenecks? Are rates of speciation correlated with rates of genetic evolution? Have speciations occurred in sympatry? These and related questions have been answered for several studied groups.

2. Molecular markers have pragmatic utility in distinguishing closely related taxa, including sibling species that may have gone unrecognized by non-molecular appraisals. Molecular diagnoses sometimes involve sibling species of medical or economic importance.

3. Molecular approaches have enriched the traditional biological species concept (BSC) by adding a phylogenetic perspective to discussions of population relationships. The distinction between gene trees and species trees has led to the development and elaboration of principles of genealogical concordance for the recognition of subspecies and species.

4. Molecular markers provide powerful means for identifying hybrid organisms and for characterizing patterns of introgression. Degrees of hybridization and introgression have proved to vary along a continuum, from instances of sporadic production of F_1's to extensive introgression leading to genetic merging of formerly separate taxa.

5. Through joint examination of multiple nuclear loci and cytoplasmic genes, several sources of genetic asymmetry in hybrid zones have been recognized. These differential patterns of introgression can be due to interlocus variation in selection intensity against alleles on heterologous genetic backgrounds, Haldane's rule whereby gender-specific fitness differences characterize hybrid organisms, differential mating behaviors of the hybridizing taxa, or other sources of differential gametic exchange.

6. Phylogenetic studies of cpDNA suggest that reticulate evolution has been common in many plant groups.

7. Hybrid origins for several recognized taxa have been revealed through molecular markers. Particularly informative have been genetic characterizations of both the parentage and the cytological pathways leading to production of hybridization-derived unisexual biotypes.

8

Species Phylogenies
and Macroevolution

*All the organic beings which have ever lived on this
Earth may be descended from some one primordial form.*

C. Darwin, 1859

*. . . study of the gene at the most fundamental level will
soon tell us more about the phylogenetic relationships of
organisms than we have managed to learn in all the 173
years since Lamarck.*

R.K. Selander, 1982

We proceed now to the provenance traditionally equated with molecular phylogenetics—estimation of evolutionary relationships among species and higher
taxa. After reproductive barriers have been erected and the speciation process
completed, molecular characters continue to evolve in a more or less time-
dependent fashion (Chapter 4), such that the overall genetic distance between
species under study provides a compelling guide to the general magnitude of
evolutionary time since their common ancestry. Furthermore, many qualitative
molecular markers considered individually or in combination can provide powerful characters for clade delineation. In assessing relationships among species,
a variety of molecular techniques has been brought to bear, including protein
electrophoresis, immunological assays, DNA restriction analysis, DNA–DNA
hybridization, and others. In recent years, nucleotide sequencing has become
widespread and currently is revolutionizing the field. Collectively, these methods
have been applied to phylogenetic estimation in many hundreds of taxonomic

groups, at evolutionary depths ranging from closely related congeners to the deepest branches in the tree of life. No attempt will be made here to summarize this vast literature exhaustively. Rather, the purpose of this chapter is to illustrate by chosen examples some of the wide variety of problems and approaches in supraspecific phylogeny tackled through molecular markers.

RATIONALES FOR PHYLOGENY ESTIMATION

Phylogenetic Mapping of Organismal Traits

Phylogenetic hypotheses (explicit or implicit) underlie virtually all conclusions in comparative evolution (Harvey and Purvis, 1991). For example, the inference that powered flight in mammals is a derived rather than ancestral condition stems ultimately from the restricted phylogenetic position of bats (Chiroptera) within a group (Mammalia) whose ancestral forms unquestionably were terrestrial. More challenging in this case is whether wings and flight evolved once or more than once (convergently) in bat evolution, a contentious issue whose resolution depends on whether Chiroptera is a monophyletic or polyphyletic assemblage (see below and Fig. 8.1).

An important task in evolutionary biology is to understand adaptations. Unfortunately, direct experimental analyses often are not possible, so evolutionists rely heavily on the comparative method to infer underlying evolutionary processes. A significant complication in this general approach is the potential lack

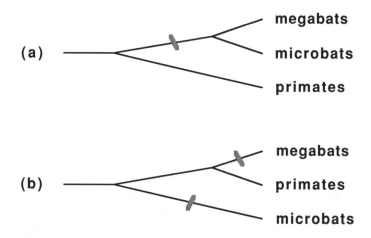

Figure 8.1. Competing hypotheses concerning the phylogenetic relationships of microchiropteran bats, megachiropteran bats, and primates (simplified from R.J. Baker et al., 1991). Slashes across tree branches indicate hypothesized origins of powered flight. Molecular data provide significant support for scenario (a).

of statistical independence among data points due to phylogenetic legacy (Felsenstein, 1985b; Richman and Price, 1992). Thus, in comparative phylogenetic analyses, if it is erroneously assumed that each taxon constitutes an independent unit, the degrees of freedom in a statistical test could be grossly overestimated. For example, many passerine birds tend to have more or less monogamous mating systems (however, see Chapter 5), whereas large herbivorous mammals usually are highly polygynous; how many of the examples of these contrasting mating systems are to be interpreted as independent adaptive responses to different ecological circumstances faced by birds and mammals (as opposed to the mere retentions of these respective ancestral conditions across the evolutionary radiations of avian and mammalian species)? By seeking independent evolutionary origins of organismal characteristics through phylogenetic analyses, conclusions from comparative studies can be placed on a firmer empirical and statistical grounding.

In other words, suppose that species' phylogenies were known with certainty. Taxonomic distributions of morphological, behavioral, or life history characters superimposed on these phylogenetic trees should then illuminate the evolutionary origin(s) and directions of change in each organismal feature. "Phylogenetic character mapping" involves the matching of particular traits with their associated species on a cladogram, with the purpose of revealing evolutionary patterns in those traits. Clearly, the phylogeny or cladogram itself must be estimated using data that are different from and independent of the attributes to be mapped. With the advent of molecular approaches, such independent appraisals of phylogeny have become commonplace.

ANATOMICAL FEATURES

Because bats are such unusual and distinctive creatures, it might seem unlikely that they could have evolved multiple times from independent ancestors. Nonetheless, on the basis of newly-observed neuroanatomical features, a "diphyletic hypothesis" was proposed in which megabats (the large "flying foxes," suborder Megachiroptera) are closer phylogenetically to primates than to microbats (typical bats, Microchiroptera) (Pettigrew, 1986, 1991). If so, mammalian wings and powered flight must have evolved from nonflying ancestral conditions on at least two separate occasions: once in an early ancestor of the Microchiroptera, and again in a separate line leading to the Megachiroptera after their ancestors separated from the Primates (Fig. 8.1). On the contrary, under the traditional monophyletic view, microbats and megabats are one another's closest relatives. Thus, a clear dilemma exists: either the neuroanatomical characters shared by primates and megabats are homoplasious (show similarity due to convergence or reversal) or the shared features of wing structure and powered flight are homoplasious in megabats and microbats.

Recent molecular analyses have led to a rejection of the "flying primate" hypothesis for megabats. Nucleotide sequences from the 12S ribosomal RNA gene and the cytochrome oxidase gene in mitochondria (Adkins and Honeycutt,

1991; Bennett et al., 1988; Mindell et al., 1991), and from the ε-globin gene in the nucleus (Bailey et al., 1992), all support a monophyletic scenario for Chiroptera. For example, 39 derived nucleotide sequence changes observed at the ε-globin locus were shared uniquely by microbats and megabats, whereas only two or three such changes were shared by megabats with primates; and at the 12S locus, about 10 more synapomorphies united megabats and microbats than was true for less parsimonious, alternative tree topologies. Thus, from molecular appraisals of phylogeny, the anatomical features associated with powered mammalian flight appear to be of a single evolutionary origin.

The king crabs of Alaska (genera *Lithodes* and *Paralithodes*) are well-known decapod crustaceans that look like typical "crabs" (apart from extraordinarily large size), with a strongly calcified exoskeleton and a greatly reduced abdomen that is folded up under the body. By contrast, hermit crabs (some 800 species in more than 80 genera) have a long, decalcified abdomen that the animals coil into adopted gastropod shells which provide the hermits with a protective home. Thus, at least superficially, the morphology of hermit crabs is quite different from that of king crabs, being intermediate between true crabs and the other major groups of decapod crustaceans—lobsters and shrimps. Nonetheless, morphologists long have suspected a close genealogical tie between hermit crabs and king crabs, for the following reasons (Gould, 1992): The abdomen of king crabs, though reduced, is asymmetrical as in hermit crabs; one or two pairs of legs are reduced greatly in the king crabs and hermit crabs, respectively, whereas all 10 legs are developed fully in typical crabs; larval forms of the two groups are remarkably alike; and carcinization (the evolution of crablike features) appears to have been a recurring theme in hermit crab evolution under ecological circumstances where the shells of gastropod snails are of limited availability (as for example in the deep sea).

Recent molecular data from a ribosomal RNA gene in mtDNA appear to clinch the case for a close genetic link between hermit and king crabs and, in addition, allow placement of a provisional time scale on the evolutionary separation (Fig. 8.2). From phylogenetic analyses of DNA sequences of 12 species of hermit and king crabs, plus an outgroup the brine shrimp (*Artemia salina*), the king crabs appear to have branched off from a restricted genealogical subset of hermit crabs and, indeed, are nested within the hermit crab genus *Pagurus*. Furthermore, this split from hermit ancestors was estimated to have occurred about 13 to 25 million years ago, thus placing an upper bound on the time transpired during the evolutionary loss of shell-living habit and the complete carcinization of king crabs (Cunningham et al., 1992). Evolutionary changes in the timing of organismal development (heterochrony) likely account for these dramatic morphological shifts.

Living cetaceans traditionally have been divided into two distinctive suborders, the Odontoceti (echolocating toothed whales and dolphins) and the Mysticeti (filter-feeding baleen whales). Surprisingly, recent sequence analyses of mitochondrial ribosomal RNA genes have suggested that the carnivorous sperm whales are related more closely to baleen whales than to other toothed whales (Fig. 8.3),

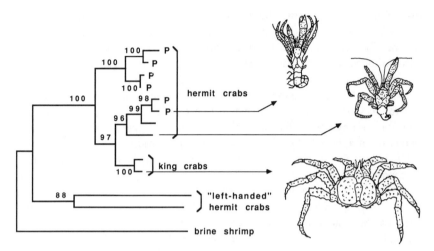

Figure 8.2. Molecular phylogenetic tree for brine shrimp (*Artemia*), 2 species of king crabs (genera *Lithodes* and *Paralithodes*), and 10 species of hermit crabs (after Cunningham et al., 1992). The letter P indicates species in the genus *Pagurus*. Both distance-based and parsimony methods applied to mtDNA ribosomal RNA gene sequences produced essentially the same phylogeny. Numbers indicate percentages of bootstrap support for the various nodes in the parsimony analysis. From molecular clock considerations and extrapolation downward from nodes dated by independent fossil and geographic evidence, the phylogenetic separation of king crabs from hermit crabs was estimated at about 13–25 mya.

such that the evolutionary relationship between Mysticeti and Odontoceti is one of paraphyly rather than reciprocal monophyly. This suggested phylogeny for the two whale groups, if correct, leads to a parsimonious inference that baleen whales probably lost the echolocation capability secondarily [alternatively, echolocation would have to have been gained independently by sperm whales and other toothed cetaceans (Milinkovitch et al., 1993)]. In other regards, the phylogeny estimated from mtDNA sequences is consistent with conventional wisdom about relationships within Cetacea, thus providing increased confidence that the mtDNA sequences are informative phylogenetically. Important areas of agreement between molecular and traditional systematic data include an inferred monophyletic status for each of the following: beaked whales (Ziphiidae), baleen whales (Balaenopteridae), sperm whales (Physeteroidea), dolphins (Delphinidae), and porpoises (Phocoenidae) (Fig. 8.3).

Molecular data also have led to a reassessment of the phylogenetic position of cetaceans within mammals. The transition from a terrestrial to a fully aquatic life-style required a gross remodeling of several biological systems and has led to specialized cetacean anatomies, physiologies, and behaviors that have com-

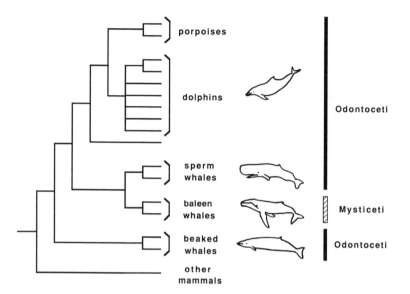

Figure 8.3. Molecular phylogenetic tree for 16 cetacean species based on rRNA gene sequences in mtDNA (after Milinkovitch et al., 1993). Levels of bootstrap support were near 100% for most (but not all) of the putative clades identified by parsimony analyses and neighbor-joining, including the hypothesized relationship of sperm whales to baleen whales which was supported at about the 90% level of confidence.

plicated attempts to establish genealogical ties to other mammalian groups. Early paleontological data suggested that condylarths (archaic ungulates, now extinct) might provide the ancestral link between extant cetaceans and the modern ungulates (hoof-bearing mammals), two prominent orders of which are Perissodactyla (odd-toed forms, including horses, tapirs, and rhinoceroses) and Artiodactyla (even-toed forms, including pigs, camels, and deer). A variety of molecular data are either consistent with or positively supportive of the view that cetaceans indeed are related more closely to ungulates than to other mammals—these include information from protein sequences (Goodman et al., 1985; Miyamoto and Goodman, 1986), mtDNA sequences (Irwin et al., 1991; Milinkovitch et al., 1993; Southern et al., 1988), and DNA-DNA hybridization (Milinkovitch, 1992). Among the extant ungulates, recent molecular as well as paleontological evidence suggest a close association of cetaceans with Artiodactyla (Milinkovitch et al., 1993; Novacek, 1992).

Earlier in this century, the startling discovery of a living coelacanth (a member of an otherwise extinct subset of crossopterygian fishes) created hope that this "living fossil" would provide a long-sought missing link morphologically close to the ancestral form of tetrapods (land vertebrates). However, even with extant material available, systematists have disagreed on the exact placement of the evolutionary root for tetrapods along the candidate phylogenetic branches that interconnect the land vertebrates, ray-finned fishes (Actinopterygii), and lobe-finned fishes (which in addition to the coelacanth include extant lungfishes). Three plausible resolutions have been suggested (Fig. 8.4): (a) lungfishes as the sister group to coelacanths plus tetrapods; (b) tetrapods as the sister group to lungfishes plus coelacanths; and (c) coelacanths as the sister group to lungfishes plus tetrapods. A recent analysis based on mtDNA sequences (slowly-evolving positions at the *cytb* and 12S rRNA loci) provided initial support for phylogenetic arrangement (c) (Meyer and Dolven, 1992; Meyer and Wilson, 1990). Onto this molecular phylogenetic backdrop, numerous morphological features then were mapped (Table 8.1). Through this approach, 14 head and body traits were postulated to have undergone a single change on the lungfish-tetrapod stem. These include acquisition of internal nostrils and controlled access to the trachea through a glottis (both of which facilitate the tetrapod mode of feeding while breathing), a division of the heart auricle leading to separation of oxygenated

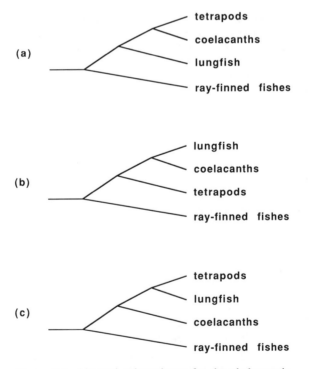

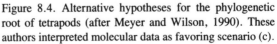

Figure 8.4. Alternative hypotheses for the phylogenetic root of tetrapods (after Meyer and Wilson, 1990). These authors interpreted molecular data as favoring scenario (c).

Table 8.1. Phylogenetic mapping of 22 morphological traits along the molecular-inferred tree (Fig. 8.4c) for tetrapods, lobe-finned, and ray-finned fishes.

	Presence (+) Versus Absence (−) in			
Trait	Lungfish	Tetrapods	Coelacanth	Ray-Finned Fish
Single Hypothesized Origin				
1. Internal nostrils	+	+	−	−
2. Palate fused with neurocranium	+	+	−	−
3. Glottis	+	+	−	−
4. Pharyngobranchial gill arches	−	−	+	+
5. Autopalatine bone	−	−	+	+
6. Depressor mandibulae muscle	+	+	−	−
7. Free hyomandibular bone	+	+	−	−
8. Ethmoid sensory canal	−	−	+	+
9. Saccus vasculosus of pituitary	−	−	+	+
10. Pars nervosa of pituitary	+	+	−	−
11. Truncus anteriosus of heart	+	+	−	−
12. Divided auricle of heart	+	+	−	−
13. Limbs with >4 mesomeres	+	+	−	−
14. Pelvic girdles joined	+	+	−	−
Parallelisms or Reversals				
1. Labial pit and muscular lip-fold	+	−	+	−
2. Electroreceptors mostly on snout	+	−	+	−
3. Septum dividing brain hemispheres	+	−	+	−
4. Thickened dorsal thalamus	+	−	+	−
5. Maxilla bone	−	+	−	+
6. Short dentary bone	+	−	+	−
7. Glenoid convex	+	−	+	−
8. Endolymphatic commissure	−	+	+	−

Source: After Meyer and Wilson (1990).

from unoxygenated blood, the fusion of pelvic girdles, and the incorporation of more mesomeres into the limbs. By contrast, the distributions of eight other morphological traits along the molecular-inferred phylogeny called for more complex explanations involving parallelisms or reversals along the tree.

This study is among the most ambitious and explicit of available attempts to phylogenetically map large numbers of organismal characters. However, the conclusions also have been called into question, primarily on the grounds that the molecular phylogeny may not be well supported. For example, Gorr et al. (1991) interpreted hemoglobin sequences as indicative that the coelacanth rather than the lungfish was the closest living relative of tetrapods, and reanalyses of these same data by Stock and Swofford (1991) and Sharp et al. (1991) suggested that the phylogenetic placement of the coelacanth simply was inconclusive. Further phylogenetic analyses of *cytb* and other mitochondrial genes (Normark et al., 1991), and of 18S rRNA sequences (Stock et al., 1991), likewise found inconclusive support for the postulated lungfish-tetrapod clade. Whether the phylogenetic arrangement proposed by Meyer and Wilson ultimately proves correct remains to be seen, but, in any event, the important take-home message is that phylogenetic character mapping depends critically on a correct tree topology. If the lungfish and tetrapods do not constitute a clade, phylogenetic scenarios for many of the characters listed in Table 8.1 would change dramatically.

Since first described in 1869, the giant panda of China (*Ailuropoda melanoleuca*) has been a phylogenetic enigma. It certainly looks like a bear (family Ursidae), but also has many traits that are not at all bearlike: flattened teeth and various other adaptations associated with a bamboo diet; an opposable ''thumb''; lack of hibernation; a bleating voice like a sheep; and a karyotype consisting of 21 pairs of chromosomes that appears superficially unlike that of bears with 37 chromosome pairs. In terms of chromosomal numbers, the giant panda more closely resembles the red panda (*Ailurus fulgens,* 22 chromosome pairs), a very different-looking species also from China that traditionally had been considered a member of the raccoon family Procyonidae. Over 40 morphological treatises on the subject of giant panda ancestry have reached no consensus on whether *A. melanoleuca* is allied more closely to bears, raccoons, or to neither (O'Brien, 1987). However, recent molecular appraisals appear to have solved the mystery, showing that the giant panda stemmed ($\cong$20 mya) from an early offshoot of the lineage leading to modern Ursidae [whereas the ancestor of the red panda separated ($\cong$30 mya) from a different line leading perhaps to the modern Procyonidae or other carnivores (Fig. 8.5; O'Brien et al., 1985a)].

This study illustrates two general points. First, molecular phylogenies are most convincing when supported concordantly by multiple lines of evidence. The molecular assays applied to pandas and relatives included multilocus protein electrophoresis (Goldman et al., 1989), protein-immunological methods (Sarich, 1973), and DNA–DNA hybridization, and these several data sets converged on

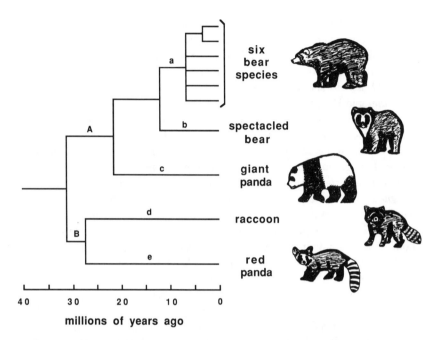

Figure 8.5. Consensus molecular phylogeny showing the genealogical position of the giant panda relative to the taxonomic families of bears (Ursidae, node A) and raccoons (Procyonidae, node B) (after O'Brien, 1987). Lowercase letters represent suggested subfamily designations.

the consensus phylogeny depicted in Figure 8.5. A second point is that consensus molecular phylogenies can be informative as a backdrop for interpreting the evolutionary histories of problematic molecular or cellular features, just as they are for morphological traits. In this case, the consensus molecular phylogeny prompted a reexamination of the bizarre karyotype of the giant panda, using refined methods that reveal details of chromosomal banding patterns. Results showed conclusively that previously noted differences between the gross karyotypes of giant pandas and bears (42 versus 74 chromosomes, respectively) were superficial, attributable to simple centromeric fusions in the line leading to the giant panda (O'Brien et al., 1985a).

BEHAVIORAL, PHYSIOLOGICAL, AND LIFE-HISTORY FEATURES

Molecular-based phylogenies also can provide useful backdrop for the recovery of the evolutionary histories of other classes of organismal characteristics. An available case in point involves the phenomenon of interspecific brood parasitism in birds (the laying of eggs by individuals of one species into nests of another species). Most New World blackbirds (Icterinae) exhibit conventional parental life-styles, but several cowbirds in this subfamily display varying degrees of specialization for brood parasitism: *Molothrus badius* takes over the nests of other species but rears its own young; *M. rufoaxillaris* specializes on a single host species; *M. aeneus* and *Scaphidura oryzivora* parasitize confamilial genera only; and *M. ater* and *M. bonariensis* are generalists, utilizing a wide assortment of host taxa as foster parents. Recent parsimony analyses applied to sequence data from an 852-bp region of the cytochrome b gene in mtDNA revealed that among 26 species of Icterinae examined, all of the brood parasites (listed above) formed a monophyletic group (Lanyon, 1992). Furthermore, the two generalist taxa (*M. ater* and *M. bonariensis*) proved to constitute a molecular clade whose sister taxon (*M. aeneus*) is a brood parasite with intermediate host specificity. These results led to the conclusion that brood parasitism probably evolved a single time within Icterinae, and that the generalized form of brood parasitism is the derived condition. This finding is important because it appears to conflict with a prevailing hypothesis that as hosts develop defense mechanisms against brood parasitism, parasites might evolve toward specialization on fewer host taxa.

Another example in which the direction of evolutionary change in behavior was inferred from molecular phylogenetic analysis involves *Lasioglossum* sweat bees of the subgenus *Evylaeus,* a hymenopteran group that contains both solitary and eusocial species exhibiting a wide range of nest architectures. Packer (1991) employed multilocus allozyme approaches to estimate a cladogram for eight Old World species, and then mapped a variety of behavioral characteristics onto the molecular phylogeny (Fig. 8.6). Among the major conclusions drawn from this exercise in phylogenetic mapping were the following: most *Evylaeus* species share sociality by descent from a eusocial common ancestor; multiple-foundress associations are a derived condition within *Evylaeus;* and an architectural trait featuring an extended opening of brood cells during development probably originated twice independently among the species considered.

Endothermy, the ability to maintain elevated body temperature by metabolic means, is rare among fishes, the only examples having been documented within large oceanic teleosts of the suborder Scombroidei (including tunas, mackerels, and billfishes). Did endothermy evolve once or multiple times within this assemblage? A recent mapping of this physiological adaptation on a phylogeny estimated from the cytochrome b sequences in mtDNA suggests that endothermy arose at least three times independently within the Scombroidei (Fig. 8.7). These convergent appearances of endothermy have involved diverse physiological and morphological pathways (Block et al., 1993).

Tunicates (ascidians or sea squirts) have been thought to be primitive chordates, due largely to the presence a well-organized notochord in their tadpole-type larvae. Recent phylogenetic analyses based on sequences from rRNA genes (Field et al., 1988) and on molecular features of muscle actins (Kusakabe et al.,

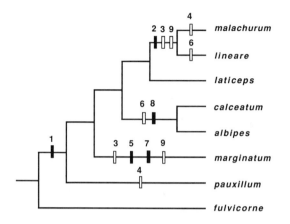

1) eusociality

2) nonoverlapping caste sizes

3) less than 1% workers mate

4) more than one worker brood per year

5) perennial societies

6) multiple foundress nests

7) nests without cavity

8) lateroid present

9) brood cells open during juvenile development

Figure 8.6. Allozyme-based cladogram for sweat bees, onto which have been mapped the various behavioral evolutionary changes listed (after Packer, 1991). Solid bars crossing branches indicate traits inferred to be of monophyletic origin within the group; open bars indicate traits with suggested multiple origins.

1992) tend to support this view by placing the ascidians somewhat closer to the vertebrates than to the invertebrates. Among the Ascidiacea (of which about 2300 species are known), two distinctive reproductive/developmental modes are exhibited: some (solitary ascidians) live as individuals, whereas others (colonial ascidians) form colonies and can propagate asexually by budding, strobilation, or regeneration. Under an orthodox classification, based primarily on morphological features of the branchial sac and gonad, ascidians have been divided into two taxonomic orders (Enterogona and Pleurogona), irrespective of whether the lifestyle is solitary or colonial. If this view is correct, and reflective of phylogeny, then one or both life history modes probably arose multiple times in evolution. Alternatively, the distinctive life-styles themselves might register a fundamental

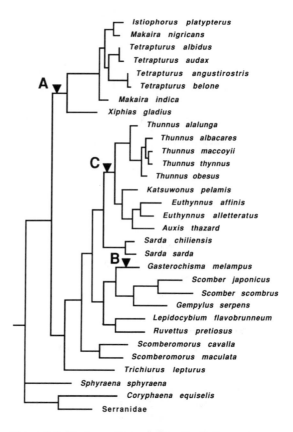

Figure 8.7. Phylogenetic mapping of endothermy among marine fishes in the suborder Scombroidei (after Block et al., 1993). The phylogeny was estimated from nucleotide sequences of a portion of the cytochrome b gene in mtDNA. Letters indicate three separate origins of endothermy, each with a different physiological basis: A—modification of the superior rectus muscle into a thermogenic organ; B—modification of the lateral rectus muscle into a thermogenic organ; and C—use of vascular countercurrent heat exchangers in the muscle, viscera, and brain.

phylogenetic split, such that the branchial sac and gonadal characters would be homoplasious. To test these possibilities, Wada et al. (1992) sequenced portions of the 18S rRNA gene in 8 diverse species of ascidians representing both solitary and colonial forms. Phylogenetic analyses of these sequences grouped the ascidians into two distinct lineages that agreed with traditional ordinal assignments. Thus, the solitary and colonial life-styles likely evolved independently after a true phylogenetic split between the Enterogona and Pleurogona.

Of course, plants also have morphologies and behaviors that can be subjected to phylogenetic mapping using molecular markers. Androdioecy (in which male and hermaphroditic flowers occur on separate plants) traditionally has been viewed as an intermediate step in the evolution of dioecy from hermaphroditism. However, in the tree genus Datiscaceae which contains three dioecious and one androdioecious species, androdioecy (rather than dioecy) has been postulated to represent the derived condition. To test this hypothesis, Rieseberg et al. (1992) utilized restriction sites from PCR-amplified cpDNA fragments to construct molecular phylogenies for 14 species (Datiscaceae and 10 outgroups) against which to interpret the taxonomic distributions of reproductive traits. Results supported the contention that the androdioecious species (*Datisca glomerata*) occupies a derived position relative to dioecious members of the family and, hence, that androdioecy evolved from dioecy in this plant family.

Carnivory in flowering plants involves a large suite of morphological and physiological features associated with the attraction, retention, trapping, killing, and digestion of animals, and absorption of their products. On this basis, it might be supposed that the evolutionary acquisition of a carnivorous habit would be extremely difficult and, hence, rare. Nonetheless, mapping of this life-style against a molecular-based estimate of phylogeny for more than 70 taxonomic plant families [from nucleotide sequences of the *rbcL* gene (see section on chloroplast DNA later in this chapter)] has revealed that carnivory undoubtedly is polyphyletic. Furthermore, "flypaper traps" appeared to have had at least five separate origins among dicotyledons, and "pitcher traps" had at least three independent evolutionary origins (Albert et al., 1992). However, some of the more detailed components of carnivory did appear to be clustered phylogenetically, such that the overall syndrome displayed a mixture of homologous and analogous elements. As stated by the authors, "form is not a reliable indictor of phylogenetic relationships among carnivorous plants at highly inclusive levels (such as trapping mechanism), whereas it appears to be at less inclusive ones (such as glandular anatomy)." Figure 8.8 provides a schematic view of how the polygenic determinants of complex phenotypic features might wax and wane along phylogenetic branches, occasionally crossing thresholds where the more inclusive syndromes are exhibited.

Even the simplest of eukaryotic organisms also have provided subject material for phylogenetic character mapping (Knoll, 1992). Molecular data, particularly

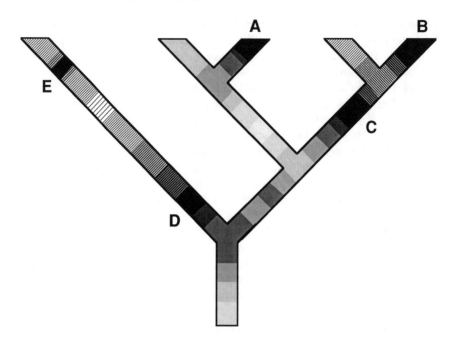

Figure 8.8. Concept of gradients and thresholds in the phylogeny of quantitative traits. Consider a polygenic trait whose level of expression is a function of appropriate alleles at numerous unlinked loci (as is no doubt true for many phenotypic attributes). The diagram shows how genetic changes at these loci can produce gradual evolutionary shifts in the level of expression (intensity of shading) of the final trait. Furthermore, different suites of underlying alleles may be responsible for a given level of trait expression (as indicated by various patterns of barring and stippling). Suppose now that the phenotypic trait requires some threshold number of alleles for expression. For example, assume that black in the diagram is above the required threshold (indicating ''presence'' of the trait) and non-black is below the threshold (indicating trait ''absence''). ''Trait presence'' would then have arisen polyphyletically (at positions A–E), due to shifts in levels of polygenic support. Thus, a quantitative trait could have a complex mixture of both homologous and homoplasious elements.

from ribosomal RNA genes and some conservative protein-coding loci, have suggested that a basal split in eukaryotic evolution was between a group dubbed the Archaezoa (see Cavalier-Smith, 1991) and all other eukaryotes (including most protists). All examined members of the Archaezoa lack mitochondria, peroxisomes (membrane-bound organelles containing enzymes that metabolize hydrogen peroxide), Golgi apparatus (a cell organelle that sequesters substances synthesized by the membraneous endoplasmic reticulum), and spliceosomal introns, all features that also are lacking in noneukaryotes. From such evidence, it now seems likely that ancestral eukaryote(s) may have been far simpler in cellular and molecular organization than are most extant eukaryotic forms.

Before leaving this general section, it should again be emphasized that the phylogenies derived from molecular markers, like those from other systematic characters, represent *estimates* (rather than definitive documentations) of phylogeny. Thus, the conclusions stemming from phylogenetic character mapping must be viewed as provisional evolutionary hypotheses, subject to further evaluation. Nonetheless, the continued exploration of the comparative evolutionary distributions of different classes of attributes (molecular and organismal) holds great promise.

Biogeographic Assessment

A second rationale for phylogeny estimation is in biogeographic reconstruction. Just as phylogeographic relationships among conspecific populations have been revealed through molecular analyses (Chapter 6), so too have phylogeographic relationships among species and higher taxa.

VICARIANCE VERSUS DISPERSAL

The potential geographic range of any taxon, of course, is limited by the suitability of environmental conditions. Within that zone of ecological tolerance, the realized geographic distribution is influenced additionally by historical factors. The "success" of numerous human-mediated introductions of plants and animals around the world is testimony to the fact that not all habitats suitable for a species are occupied naturally (or, perhaps, that humans have so disturbed native habitats that introductions can succeed). Thus, whether a species occurs in a particular region is a function also of historical demographic factors and dispersal patterns, which themselves are influenced by the proximity and spatial relationships of the environments potentially habitable. The study of such historical factors is the focus of molecular phylogeography.

In cases where related taxa show disjunct geographic distributions, two competing hypotheses often are advanced to account for the spatial arrangements (Box 8.1). Under "dispersalist" scenarios, such a taxonomic group came to occupy its current range through active or passive dispersal across a preexisting geographic or ecologic barrier. Alternatively, under "vicariance" scenarios, the more or less continuous ranges of ancestral forms may have been split by particular vicariant geographic events, e.g., the rise of a mountain range that sundered lowland taxa, a continental breakup that partitioned populations of terrestrial organisms, or subdivision of a body of water that split aquatic or marine populations. Both theories subscribe to the proposition that speciation is predominantly allopatric. However, an important prediction of vicariance biogeography (not shared by dispersalist scenarios) is that the cladogram for a group of related species should match the historical "area-cladogram" of environments occupied (Box 8.1).

Most tests of particular vicariance scenarios have involved phylogenetic assessments using traditional nonmolecular characters (e.g., Cracraft, 1986), but in

---

Box 8.1. Vicariance Versus Dispersalist Biogeography

Disjunct distributions of related taxa pose special challenges for biogeographic interpretation. Traditionally, disjunct ranges were interpreted to evidence organismal dispersal across preexisting geographic or ecological barriers, usually from a biogeographic "center of origin" where a given taxonomic group presumably originated (Darlington, 1957, 1965). Dispersalist explanations sometimes became quite strained, however, as for example in accounting for the global distributions of relatively sedentary creatures such as the large flightless birds (ostriches, rheas, emus, and their presumed relatives). In recent decades, vicariance biogeographers have challenged the dispersalist paradigm by proposing that disjunct distributions also can result from environmentally-mediated separations of widespread ancestral forms, without the need to invoke long-distance or improbable dispersal events ["sweepstakes" dispersal (Simpson, 1940)].

Vicariance biogeography as a formal discipline (Humphries and Parenti, 1986; Nelson and Platnick, 1981; Nelson and Rosen, 1981; Rosen, 1978) grew out of several developments in the 1960s and 1970s, prominent among which were the following (Wiley, 1988): (a) a growing appreciation from the study of plate tectonics and other geological processes that the earth's features were not fixed, but rather showed tremendous historical dynamism; and (b) the growth of cladistics (Chapter 2), which armed researchers with new conceptual outlooks and analytical tools for recovering the historical, phylogenetic component of evolution.

Under strict vicariance hypotheses (in contrast to dispersalist expectations), cladistic relationships among related disjunct taxa should mirror faithfully the historical relationships among the geographic regions occupied. Indeed, one major goal of vicariance biogeography is to reveal relationships among the units of real estate themselves [e.g., "Is Cuba more closely related to Hispaniola than to Jamaica?" (Wiley, 1988)], through comparative searches for congruent patterns ("generalized tracks") in organismal phylogeography.

geographic phylogeny:

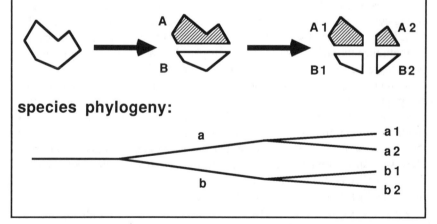

species phylogeny:

recent years molecular methods have played an increasingly important role. For example, Bermingham et al. (1992) used data from mtDNA restriction sites to test a vicariance model for the evolution of North American birds in the black-throated green warbler complex. It had been proposed that episodic glacial advances during the Pleistocene repeatedly fragmented the ranges of forest-dwelling birds into eastern and western populations, in such a way that subsequent speciations produced a series of western endemics each linked phylogenetically to the widespread eastern form (*Dendroica virens* in this case), but at different evolutionary depths (Mengel, 1964). However, molecular data appeared inconsistent with certain cladistic aspects of this scenario, suggesting instead that some of the endemic western warblers in the black-throated green complex budded off from one another (perhaps via intermontane isolations) rather than directly from *D. virens*. Nonetheless, other aspects of Mengel's scenario *were* supported by the molecular data, including the basal branching of *D. nigrescens* from the *D. virens* lineage and a probable Pleistocene origin for several avian species-pairs.

A similar Pleistocene scenario constituted the conventional wisdom for anuran evolution in southwestern Australia, where several western endemics were hypothesized to have arisen following multiple invasions from eastern source stocks. However, microcomplement fixation studies of albumin evolution in more than 20 species of frogs representing 6 genera revealed several discrepancies with phylogenetic predictions of this model and led to a rejection of the multiple-invasion scenario in favor of a model that includes speciation events within southwestern Australia (Maxson and Roberts, 1984; Roberts and Maxson, 1985). Furthermore, by molecular evidence, many of these speciation events appeared to predate the Pleistocene significantly.

Allozyme data have been used to test competing vicariance versus dispersalist models for the trans-Pacific distributions of many tropical marine shore fishes (Rosenblatt and Waples, 1986). More than 50 such taxa occur on both sides of this ocean basin, despite 5000 km separating the closest islands of the central versus eastern Pacific. One possibility is that recent long-distance dispersal (probably west to east, via free-swimming adults or planktonic larvae utilizing equatorial currents) connects these trans-Pacific fishes. Alternatively, as had been proposed for certain invertebrate taxa with similar distributions, these fish populations in the eastern Pacific might conceivably represent vicariant relicts of former worldwide ancestors in the Tethyan Sea (an ocean body that separated Laurasia from Gondwana following the breakup of the supercontinent Pangaea during the Mesozoic) (McCoy and Heck, 1976). The molecular results proved consistent with the hypothesis of recent or ongoing dispersal from the west Pacific. Thus, in each of 12 comparisons attempted, allozyme distances between trans-Pacific forms were small and markedly lower than genetic distances observed between amphi-American sister taxa that were separated into Atlantic and Pacific units by the Isthmus of Panama, which arose only 3 mya.

Thus, one important advantage of molecular approaches is that temporal issues (based on molecular clock considerations) can be examined in addition to cladistic assessments per se. In another example, Hedges et al. (1992b) utilized immunological distances in albumins to assess the evolutionary times of separation between a wide variety of amphibians and reptiles in the Caribbean region. Geologic evidence indicates that the Greater Antilles islands were formed in close proximity to North and South America about 110–130 mya, but under the influence of plate tectonics began separating from the mainlands by the late Cretaceous (80 mya). It had been proposed previously that much of the present West Indian terrestrial biota reflects these ancient proto-Antillean vicariant separations. Alternatively, postvicariant overwater dispersal might account for the presence of related taxa on the various islands. The molecular data for 38 pairs of terrestrial vertebrates proved inconsistent with the hypothesis of ancient vicariance, in two regards (Fig. 8.9): (a) in comparisons involving particular island pairs or an island versus the mainland, genetic distances between independent pairs of taxa showed a large variance, suggesting widely differing times of colonization as might be expected under a dispersalist scenario; (b) nearly all estimates of divergence time (assuming an albumin evolutionary clock as calibrated from independent data) postdated considerably the vicariant geographic events hypothesized to have isolated the respective populations. Many additional phylogeographic details also have been revealed through these and other protein and DNA comparisons of the Caribbean herpetofauna (Burnell and Hedges, 1990; Hedges, 1989; Hedges et al., 1991). For example, the six species of *Anolis* lizards currently on Jamaica apparently represent a monophyletic group that radiated mostly during the Pliocene from a colonizer that arrived in the mid-Miocene, about 14 mya (Hedges and Burnell, 1990).

A remarkable instance of long-distance historical dispersal documented by molecular markers involves two species of annual plant, the diploid *Senecio flavus* of the Saharo-Arabian and Namibian deserts in Africa and the tetraploid *S. mohavensis* of the Mojave and Sonoran deserts of North America. Using multilocus allozyme assays, Liston et al. (1989) showed that these two species indeed are remarkably similar genetically (Nei's $I \cong 0.95$). The genetic results not only affirm the traditional congeneric status of these taxa (based on gross morphology), but also imply strongly that recent intercontinental dispersal must account for their highly disjunct distributions. How this "sweepstakes dispersal" (Box 8.1) occurred remains a mystery, but one possibility is that the sticky seeds were transported by migrating or lost birds.

Not all molecular appraisals have cast doubt on vicariance scenarios. One long-standing evolutionary enigma concerns how the large flightless birds (the "ratites," which include the ostriches of Africa, rheas of South America, and the emus, cassowaries, and kiwis of Australia, New Guinea, and New Zealand, respectively) came to possess such a wide distribution in the Southern Hemi-

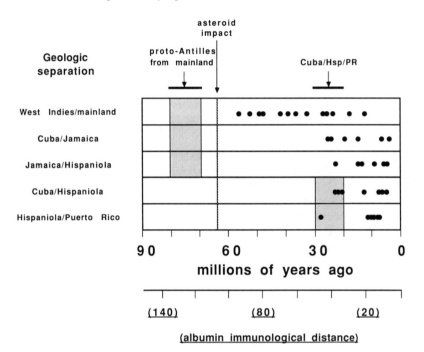

Figure 8.9. Empirical molecular tests of dispersalist versus vicariance hypotheses for the origins of terrestrial vertebrates in the Caribbean region (after Hedges et al., 1992b). Shown are immunological distances in albumin proteins, and the associated time scales under an evolutionary clock, between various island and mainland species. Shaded areas indicate dates of faunal separation predicted under vicariance scenarios based on geologic events.

sphere. One possibility is that these birds are not linked phylogenetically, their similarities in morphology being attributable instead to convergent evolution from unrelated ancestors on separate continents. Alternatively, under a vicariance model, lineage separations eventually leading to the extant ratites may have stemmed from the Mesozoic breakup of Gondwana and subsequent drift of the southern continents via plate tectonics. This vicariance model carries at least two testable predictions: ratites should be monophyletic; and the separations among extant species should be ancient (e.g., beginning > 70 mya). Remarkably, available molecular data support these predictions. Using immunological distances in transferrins, Prager et al. (1976) concluded that the ratites indeed are monophyletic and that the molecular-based estimates of separation times are congruent with the suspected dates of particular continental separations. Data from DNA-DNA hybridizations (e.g., Fig. 3.5) (Sibley and Ahlquist, 1981, 1990), and from amino acid sequences of an α-crystallin eye lens protein (Stapel et al., 1984), similarly indicate that the living ratites are both monophyletic and old. For example, ostriches and rheas appeared to be sister taxa that separated about 80 mya, a date generally consistent with the opening of the South Atlantic Ocean between Africa and South America (Sibley and Ahlquist, 1981).

Except for purposes of organizing thought, it is probably unwise to dichotomize dispersalist versus vicariance scenarios too strongly, because both factors probably have played a role in many instances. For example, although the global phylogeography of the ratite birds has been interpreted as consistent with vicariance explanations, recent molecular findings also suggest a role for dispersal in the colonization of New Zealand. Both kiwis and the now-extinct moas inhabited New Zealand during the Pleistocene, and conventional wisdom is that these groups shared a common ancestor on the island. However, a phylogeny estimated from 12S rRNA gene sequences suggests that the kiwis are related more closely to Australian and African ratites than to the moas and, thus, that New Zealand probably was colonized at least twice by ratite ancestors (Cooper et al., 1992).

COMMON ANCESTRY VERSUS CONVERGENCE

In phylogenetic mapping, a general goal is to distinguish whether shared organismal features arose through common ancestry or through convergent evolution from unrelated ancestors. These issues often come into particularly sharp focus in a geographic context. For example, various marsupial mammals in Australia bear at least superficial resemblance in behavioral or morphological features to particular placental mammals elsewhere in the world. These ecological similarities (e.g., between the Tasmanian wolf and placental carnivores, and between some bandicoots and placental rabbits) are "known" to reflect convergent evolution because the marsupials retain other detailed and distinctive evolutionary signatures interpreted as conclusive evidence of common ancestry (such as a marsupium or pouch). Furthermore, the suspected monophyly of the marsupials (almost all of which occur in the Southern Hemisphere) makes geographic sense. In particular, Australia was isolated from other landmasses during the early and middle Tertiary period (from about 30–60 mya), during which time a remarkable marsupial radiation took place that filled many ecological niches on that continent that elsewhere in the world were occupied by "ecological equivalents" among the placentals. Many temporal and cladistic details of this phylogenetic radiation of Australian marsupials have been worked out using molecular methods including protein immunology (Baverstock et al., 1987; Kirsch, 1977) and DNA-DNA hybridization (Kirsch et al., 1990; Springer and Kirsch, 1989, 1991; Springer et al., 1990).

One of the most remarkable and provocative scenarios to emerge in all of molecular phylogenetics concerns a similar in situ adaptive radiation recently proposed for Australian songbirds. Most of the birds of Australia were discovered and named after European ornithologists already had classified other elements of the world's avifauna. Many of the Passeriformes ("perching birds") in Australia appeared to fit neatly into taxonomic categories previously established, e.g., Australian warbler-like birds into Sylviinae (true warblers), Australian flycatchers into Muscicapidae (Afro-Eurasian flycatchers), treecreepers into Certhiidae (Eurasian-American creepers), and sitellas into Sittidae (Holarctic nuthatches). Nonetheless, evidence from DNA–DNA hybridization has suggested that these traditional taxonomies are incorrect when interpreted as phylogenetic arrangements. Instead, many of the Australian songbirds may stem from a common ancestor on the continent (Sibley, 1991; Sibley and Ahlquist, 1986). From molecular comparisons of many hundreds of avian species, Sibley and Ahlquist (1990) conclude that the songbirds of the world should be divided (phylogenetically) into two groups—the suggested suborder Passerida whose members appear to have evolved in Africa, Eurasia, and North America, and the Corvida which originated in Australia. The latter includes numerous species thought previously to be unrelated to one another (Fig. 8.10). If this fascinating

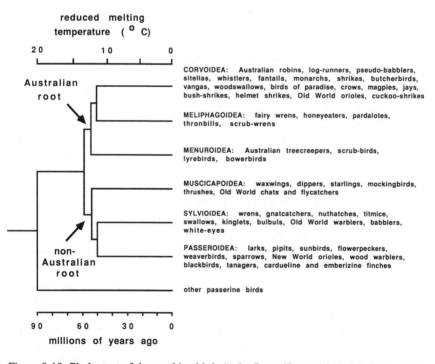

Figure 8.10. Phylogeny of the perching birds (order Passeriformes) based on DNA-DNA hybridization data (after Sibley and Ahlquist, 1986). Two major groups of the suborder Passeres were postulated: one tracing to an Australian root (some of whose members such as crows and jays secondarily radiated elsewhere in the world) and the other tracing to a non-Australian origin.

proposal for a monophyletic origin of Corvida indeed is correct, then the phylogeographic history of a major segment of the Australian avifauna parallels that of the native Australian mammals.

RECENT ISLANDS, ANCIENT INHABITANTS

Another spectacular evolutionary radiation studied through molecular markers involves representatives of the Drosophilidae flies of the Hawaiian Islands, home to an estimated 800+ species (a remarkable count, because the entire archipelago accounts for only 0.01% of the earth's land area). These flies traditionally are divided into two recognizable groups, the "drosophiloids" and the "scaptomyzoids," that have been postulated to derive from only one or perhaps two founder populations of unknown continental source (Throckmorton, 1975). The oldest of the major, present-day Hawaiian Islands dates by geological evidence to only 5 mya. Was the incredible proliferation of drosophiline species on the Hawaiian Islands truly accomplished within such a short evolutionary time span?

From immunological comparisons of larval hemolymph proteins, Beverley and Wilson (1985) provided the first extensive molecular evidence that some of the lineage separations in Hawaiian drosophilines vastly predate the volcanic emergences of the present-day islands. According to their molecular clock calibrations, the Hawaiian Drosophilidae stem from a colonist that landed on the archipelago about 42 mya. To explain the paradox, they note that the current Hawaiian Islands are merely the latest in a series of former islands that date back over 70 million years, the remnants of which remain today as eroded seamounts or low atolls to the northwest of the current chain. Thomas and Hunt (1991) and DeSalle (1992a, 1992b) have reexamined this issue using DNA sequences from the alcohol dehydrogenase nuclear locus and mtDNA, respectively, and their findings strongly support the contention that *Drosophila* inhibited the Hawaiian chain well before 5 mya (although estimates of the exact divergence times differ somewhat). Thus, according to molecular evidence, many speciation events probably occurred on islands no longer in existence (as flies island-hopped to newly arisen terrain), such that the drosophiline radiation was accomplished over a considerably longer temporal framework than suspected previously. Extensive molecular and other data on particular *Drosophila* subgroups (such as the "picture-winged" flies) demonstrate that at least some speciations also have taken place recently on the present-day Hawaiian Islands (e.g., Carson, 1976, 1992).

Analogous questions apply to the biota of another volcanic archipelago that has figured prominently in evolutionary studies—the Galapagos Islands. All of the present-day Galapagos are less than three million years old, and some researchers have considered this to represent the maximum age over which the exuberant evolution on the island chain must have taken place (Hickman and Lipps, 1985). Indeed, the small genetic distances among several species of

Darwin's finches (Geospizinae) are consistent with the hypothesis of a very recent radiation within this avian clade, perhaps within the last one million years (Polans, 1983; Yang and Patton, 1981). On the other hand, much larger genetic distances have been found between the marine iguana (*Amblyrhynchus cristatus*) and the two land iguanas of the Galapagos (*Conolophus pallidus* and *C. subcristatus*), suggesting separation dates of perhaps 15–20 mya (Wyles and Sarich, 1983). One possibility is that these genera stem from different (but unknown) ancestral stocks on the South American mainland [such scenarios also have been advanced to account for large genetic distances observed among some other native Galapagos lizards and rodents (Lopez et al., 1992; Patton and Hafner, 1983; Wright, 1983)]. However, another possibility is that speciations took place on former islands before they sank beneath the ocean surface. For this reason, considerable excitement attended the recent geological discovery of drowned islands downstream from the Galapagos volcanic hotspot (Christie et al., 1992). These geological observations and radiometric data indicate that islands have been present over the Galapagos area for at least nine million years, and perhaps much longer. Whether this temporal extension of available habitat actually accounts for ancient speciation events in the Galapagos iguanas or its other biotas remains to be determined.

Academic Pursuit of Genealogical Roots

In truth, most studies in molecular phylogeny probably are initiated out of sheer intellectual curiosity about the ancestry of a particular group. Most evolutionists have a favorite taxon (be it fishes or fungi), the phylogenetic understanding of which can become an obsession. Although it would be presumptuous here to choose particular examples in molecular systematics as being of special inherent interest, one case history stands out as worthy of discussion—the phylogenetic position of humans relative to the great apes and other primates. Probably no other topic in molecular evolution has attracted so much interest, or controversy.

Traditionally, *Homo sapiens* has been placed alone (in terms of extant species) in the Hominidae, a taxonomic family belonging to the superfamily Hominoidea which also includes the Asiatic apes [gibbons (*Hylobates*), siamangs (*Symphalangus*), and orangutans (*Pongo*)] and the African apes [gorillas (*Gorilla*) and chimpanzees (*Pan*)]. Closest relatives outside the Hominoidea are the Old World monkeys (Cercopithecoidea). Within Hominoidea, conventional (but not universal) wisdom has been that humans' closest living relatives are the great apes of Africa (Pongidae). Beyond these points, agreement usually ended.

Prior to the availability of molecular data, a common paleontological scenario was that the line leading to humans split from a line leading to gorillas and chimpanzees about 15–30 mya (see Patterson, 1987). In 1967, a stunning report by Sarich and Wilson based on immunological studies of albumin challenged this belief in two major respects (Fig. 8.11). First, the molecular data were interpreted to indicate that the phylogenetic split leading to *Homo* occurred much more recently than formerly supposed—only about 5 mya. Second, the molecular data suggested that the African apes might not form a distinct clade—rather, chimpanzees, gorillas, and humans constituted a close-knit but unresolved phylogenetic "trichotomy." Nearly 30 years and dozens of molecular studies later, the two major conclusions of Sarich and Wilson both have been vindicated. It now is accepted widely that the human lineage separated from apes about 4–8 mya, and that humans, chimpanzees, and gorillas are related more or less equidistantly to one another. These conclusions represent an impressive consensus of information from immunological assays, protein electrophoresis, amino acid sequencing, DNA-DNA hybridization, DNA restriction-site analyses, and nucleotide sequencing from mtDNA and several nuclear genes and noncoding regions (e.g., Bruce and Ayala, 1979; Caccone and Powell, 1989; Goodman et al., 1990; Hasegawa, 1990; Miyamoto and Goodman, 1990; Nei and Tajima, 1985; Sibley et al., 1990; Williams and Goodman, 1989; and references therein).

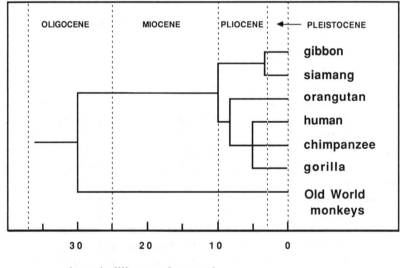

Figure 8.11. One of the first molecular-based estimates of the phylogenetic position of *Homo sapiens* within the primates (after Sarich and Wilson, 1967; see also Goodman, 1962). This phylogeny, based on immunological distances in albumins, revolutionized thought about human origins (see text), and its major features have been confirmed with much additional molecular evidence.

Earlier competing scenarios based on morphological and paleontological evidence also have been reinterpreted to accommodate (at least partially!) this overwhelming molecular evidence (Andrews, 1987; Pilbeam, 1984).

In recent years, most attention has focused on resolving the human–chimp–gorilla trichotomy, and this has included discussions of the suitability of various classes of molecular data as well as methods of statistical analysis (e.g., Kishino and Hasegawa, 1989; Nei et al., 1985; Saitou and Nei, 1986; Templeton, 1983). Although the issue is not yet settled to everyone's satisfaction, a developing consensus favors a human-chimpanzee clade as a sister group to the gorilla (Hasegawa, 1990; Horai et al., 1992; Li and Graur, 1991; Sibley and Ahlquist, 1984; Williams and Goodman, 1989). In any event, the major point is that molecular data have revealed just how tightly the phylogenetic ties bind humans to our primate relatives. The traditional taxonomic placement of *Homo sapiens* within the monotypic Hominidae probably says much more about anthropocentric bias than it does about objective phylogenetic reality.

SPECIAL APPROACHES TO PHYLOGENY ESTIMATION

Another way to organize thought about the available plethora of molecular phylogenetic studies is to focus on the principle methodologies employed. This section will highlight additional major approaches that have been or promise to be of special importance in phylogeny estimation.

DNA–DNA Hybridization and Avian Systematics

At the time of this writing, the clear record for number of published taxa assayed in a molecular phylogenetic study belongs to Charles Sibley and Jon Ahlquist, who over a 12-year period applied DNA-DNA hybridization methods to more than 1700 avian species representing all but 3 of the 171 taxonomic families of birds conventionally recognized. The result was an extended estimate of avian phylogeny [a printed version of which spans 42 pages in their summary tome (Sibley and Ahlquist, 1990)] that in ornithological circles has become known simply as "the Tapestry." Some examples of conclusions derived from their work already have been cited (on ratite birds and on the Australian avifauna, this chapter; New World and Old World vultures, Chapter 1). In general, most of their molecular-based phylogenetic conclusions agree quite well with traditional ornithological thought, thus providing considerable confidence in the approach. However, many problematic and contentious results also emerged (Table 8.2), and the Tapestry and its underlying data have by no means been met with universal approval (e.g., Cracraft, 1992; Cracraft and Mindell, 1989; Sarich et al., 1989).

Table 8.2. Examples of surprising and provocative conclusions from DNA–DNA hybridization studies of birds.

Example	Explanation
1. Barbets and toucans	Traditional view: African and South American barbets a monophyletic group within Piciformes Molecular view: New World barbets and toucans are related more closely to one another than either is to Old World barbets
2. Hoatzin	Traditional view: a specialized form that has been a complete taxonomic puzzle Molecular view: a highly modified cuckoo related to roadrunners and anis (Cuculiformes)
3. Owls	Traditional view: allied either to diurnal birds of prey (Falconiformes) or to "nightjars" (Caprimulgiformes) Molecular view: related to nightjars, and not to diurnal birds of prey
4. Totipalmate swimmers	Traditional view: fully-webbed toes and other shared features indicative of monophyly for Pelecaniformes (pelicans, boobies, gannets, cormorants, anhingas, frigatebirds, and tropicbirds) Molecular view: Pelecaniformes likely a polyphyletic assemblage
5. Grebes	Traditional view: closely related to loons (Gaviiformes) Molecular view: not related closely to loons, and indeed have no close living relatives
6. Sandgrouse	Traditional view: related perhaps to pigeons (Columbiformes), plovers (Charadriiformes), or chickens (Galliformes) Molecular view: closest allies among the Charadriiformes
7. Vultures	Traditional view: New World and Old World vultures (in Falconiformes) closely related Molecular view: New World vultures related more closely to storks (Ciconiiformes) than to Old World vultures
8. Shoebill	Traditional view: related to storks or herons (Ciconiiformes) Molecular view: related more closely to pelicans (Pelecaniformes)
9. Starlings	Traditional view: related closely to crows (Corvidae) Molecular view: a sister group to mockingbirds (Mimidae), unrelated to crows

Source: Sibley and Ahlquist (1990).

Whether the more provocative of the Sibley–Ahlquist conclusions are proved correct remains to be determined by comparative analyses using additional molecular (and other) appraisals. In any event, two important points regarding this immense effort warrant attention here. First, this study is among the most broadly based of molecular attempts to capitalize explicitly on the "common yardstick" rationale in phylogenetic reconstruction (Chapter 1). Sibley and Ahlquist promulgated a molecular metric (ΔT, that presumably is related to evolutionary time) that could be applied as a standardized measure of the magnitude of evolutionary separation among any avian (or other) taxa, be they hummingbirds or ostriches and rheas. Second, they boldly advocated a direct translation of these metrics into a formal taxonomy that is intended to encompass the concept of "categorical equivalency" (Table 8.3). For example, they suggest that species be classified at the family level when they exhibit a ΔT in the range 9–11°C, and at the subordinal level when a ΔT of 18–20°C is observed. Whether or not the particular details of this translation are adopted widely, the Sibley-Ahlquist study remains a monumental and pioneering empirical effort toward the revolutionary goal of a universal systematics—an aspiration that *is* one of the ultimate promises of molecular phylogenetics.

Table 8.3. Suggested levels of genetic divergence (as measured by DNA–DNA hybridization) to be associated with indicated levels of taxonomic recognition in birds.

Taxonomic Category	Ending	$\Delta T_{50}H$ Range	Example
Class	—	31–33	Aves
Subclass	-ornithes	29–31	Neornithes
Infraclass	-aves	27–29	Neoaves
Parvclass	-ae	24.5–27	Galloanserae
Superorder	-morphae	22–24.5	Anserimorphae
Order	-iformes	20–22	Anseriformes
Suborder	-i	18–20	—
Infraorder	-ides	15.5–18	Anserides
Parvorder	-ida	13–15.5	—
Superfamily	-oidea	11–13	—
Family	-idae	9–11	Anatidae
Subfamily	-inae	7–9	Anatinae
Tribe	-ini	4.5–7	Anatini
Subtribe	-ina	2.2–4.5	—
Congeneric spp.	—	0–2.2	*Anas* (puddle ducks)

Source: After Sibley et al. (1988).

Mitochondrial DNA and the Higher Systematics of Animals

In addition to its utility as an intraspecific, micro-evolutionary marker (Chapter 6), mtDNA also is employed widely as an informative guide to phylogenetic relationships among higher animal taxa. For these latter purposes, unique structural alterations in the molecule, or sequence regions that evolve more slowly than average, are monitored.

ECCENTRIC mtDNA MARKERS

With respect to the intended property of providing a numerical "average" of genetic distances over a large but unspecified number of (single-copy) genes, the DNA–DNA hybridization studies discussed earlier represent one end of a philosophical continuum in molecular systematics. Near the other end of this continuum are analyses that focus on idiosyncratic molecular traits which, because of supposed singularities of origin, should be of special significance in a cladistic sense. One promising suite of such characters involves gene orders and other unusual compositional properties of mtDNA.

The same ensemble of about 37 genes comprises the mtDNA of most animal species, but gene orders sometimes differ. For example, identical mtDNA gene arrangements are displayed by assayed placental mammals, amphibians, and fishes, but these differ from the gene alignment in quail by a transposition involving five loci (Desjardins and Morais, 1991). Also, three assayed marsupials have rearranged tRNA clusters in mtDNA relative to the common vertebrate theme (Pääbo et al., 1991). More comprehensive surveys of additional avian and marsupial taxa should reveal precisely where these apparently unique and derived transpositions arose and, hence, the clades that they delineate.

Distinctive mtDNA gene arrangements recently have been employed as phylogenetic markers in the Echinodermata, an invertebrate phylum traditionally divided into five classes: Asteroidea (sea stars), Echinoidea (sea urchins), Ophiuroidea (brittle stars), Holothuroidea (sea cucumbers), and Crinoidea (crinoids). From fossil evidence, these classes are believed to have separated from one another early in the Paleozoic (some 450–550 mya), but the exact phylogenetic branching sequence remained controversial (Smith, 1992). Recent studies of mtDNA gene order provided an additional clue: ophiuroids and asteroids have a similar mtDNA gene arrangement that contrasts sharply with the gene order common to echinoids and holothuroids (crinoids were not assayed) (Jacobs et al., 1988, 1989; M.J. Smith et al., 1989, 1990, 1993). Outgroup comparisons (against vertebrates) suggest that the asteroid-ophiuroid condition is synapomorphic and, therefore, defines a clade.

Rearrangements in mtDNA gene order also have been employed successfully as phylogenetic markers among *Suillus* and related mushroom genera (Bruns and Palmer, 1989; Bruns et al., 1989). More comprehensive surveys of mtDNA gene order across metazoan phyla currently are underway (e.g., Okimoto et al., 1992). Rare mtDNA structural rearrangements appear to hold special phylogenetic promise (Sankoff et al., 1992), as, for example, in characterizing higher invertebrate taxa where evolutionary relationships heretofore have been highly conjectural.

Ascertainment of the phylogenetic relationships among the taxonomic classes of Cnidaria (corals, anemones, jellyfishes, and allies) has been a classic problem in invertebrate zoology, due to a paucity of morphological characters that are independent of the dramatic differences in the lifecycles of various forms. The recent discovery that some cnidarians possess native mtDNA in linear (rather than circular) form has prompted use of this structural feature as a phylogenetic marker. All surveyed members of the classes Cubozoa, Scyphozoa, and Hydrozoa proved to possess linear mtDNA, whereas all Anthozoa, Ctenophora (the supposed outgroup), and most other surveyed metazoan phyla have circular mtDNA (Bridge et al., 1992). Thus, the derived (linear) condition appears to define a clade, and if so, a mapping of life-cycle characters on this rooted phylogeny implies that the benthic polyp stage (rather than the pelagic medussa stage) probably came first in cnidarian evolution (Bridge et al., 1992).

Several other kinds of genetic novelties (e.g., differences in secondary structures of tRNAs, sizes of rRNA genes, and peculiar features of the control region) also distinguish the mtDNA genomes of various metazoan groups (Wolstenholme, 1992). One particularly intriguing class of potential phylogenetic markers involves different patterns of codon assignment in protein translation. Following the initial discovery of modified codon assignments in mammalian mtDNA (Barrell et al., 1979), it was postulated that "drift" in codon usage away from the "universal" code of most nuclear genomes had occurred during mtDNA evolution. Wolstenholme (1992) has summarized available information on mtDNA codon assignments across 19 metazoan animals representing six phyla and superimposed the differences on a suspected phylogenetic tree (Fig. 8.12). Several of the differences appeared to be informative phylogenetically. For example, in the mtDNAs of all assayed invertebrate phyla (with the exception of Cnidaria), AGA and AGG specify serine, whereas in vertebrate mitochondria they cause chain termination. In the Cnidaria, as in the universal nuclear code, these same codons specify arginine. Thus, from the shared-derived codon conditions, vertebrates appear to constitute a clade, as do the assayed invertebrates exclusive of Cnidaria.

On the other hand, some evidence for evolutionary convergence in codon assignments also was uncovered (Fig. 8.12). The best example involved the nucleotide triplet AAA, which in the Echinodermata and Platyhelminthes (phyla otherwise thought to be unrelated) appears to specify asparagine, rather than lysine as in other metazoan mtDNAs and in the universal code. If these unusual codon assignments for asparagine are confirmed (the evidence is not yet conclusive), the derived condition probably arose twice in evolution, independently. Similarly, an apparent reversion to the presumed ancestral condition in which

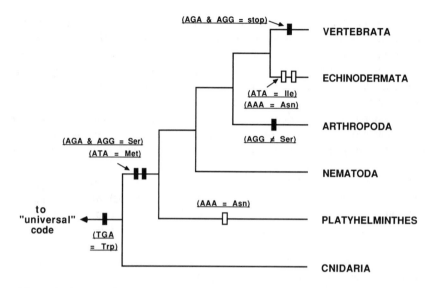

Figure 8.12. Evolutionary scenario for alterations of the genetic code in metazoan mtDNA (after Wolstenholme, 1992). Observed changes from the "universal genetic code" are plotted on a presumed phylogeny for metazoan animals based mostly on other evidence. Solid bars across branches indicate probable synapomorphies indicative of various clades; open bars indicate probable homoplasious changes (convergences or reversals).

ATA specifies isoleucine (rather than methionine) may have occurred in an ancestor of the echinoderms.

These latter observations raise a cautionary note for all attempts to define clades by any single genetic marker. No matter how secure a synapomorphy might appear, the possibilities of evolutionary convergence or reversal seldom can be eliminated entirely. Thus, confirmation from independent sources of data should always be attempted before definitive phylogenetic conclusions are drawn.

MtDNA Sequences

A second approach to higher animal systematics through mtDNA involves direct comparisons of more slowly-evolving nucleotide sequences or portions thereof (such as transversional changes, or nonsynonymous substitutions in protein-coding regions). The identification of "universal" primers that can be used to PCR-amplify particular segments of mtDNA from numerous species (Box 3.3) has facilitated greatly such sequencing studies. Prominent among the mtDNA gene regions now sequenced routinely are those encoding the ribosomal RNA subunits, cytochrome oxidases, and cytochrome b. Several phylogenetic examples already have been presented (see the earlier sections on bats, crabs, whales, coelacanths, and blackbirds), and more will follow later.

Chloroplast DNA and the Higher Systematics of Plants

The molecular appraisal of cpDNA recently has blossomed into a major industry within the field of plant phylogenetics. The chloroplast genome is well suited for higher systematic studies because (a) it is a widely distributed and relatively abundant component of plant total DNA, (b) much background information is available (including complete sequences from at least three distantly related species), which facilitates experimental and comparative work, (c) distinctive structural features of cladistic utility characterize the cpDNAs of some taxa, and (d) the molecule generally exhibits a conservative rate of nucleotide substitution (Clegg and Zurawski, 1992). Two distinct phylogenetic approaches again have been employed (Olmstead et al., 1990). The first involves monitoring the taxonomic distributions of idiosyncratic, often fortuitously discovered structural characteristics of cpDNA, the rationale being that unusual or singularly-arisen features are especially powerful for clade delineation. The second approach involves either RFLP analyses or direct nucleotide sequencing of particular cpDNA genes or regions. These latter studies provide a much larger number of potentially informative characters, but the data sets are generally afflicted with a higher level of homoplasy.

Eccentric cpDNA Markers

Relatively unusual or idiosyncratic features of cpDNA that have been employed as phylogenetic markers include inversions, losses of genes and introns, and losses of a large inverted repeat region (Downie and Palmer, 1992). These will be discussed in turn.

With a few notable exceptions [involving *Pisum* (Palmer et al., 1988b), *Trifolium* (Milligan et al., 1989), and conifers (Strauss et al., 1988)], gene order normally is a conservative feature of cpDNA in vascular plants. This is illustrated by the fact that the gene arrangement in tobacco (*Nicotiana tabacum*) is similar to the presumed ancestral vascular plant gene order also found in most other examined angiosperms, ferns, and the *Ginkgo* (Palmer et al., 1988a). Among those cpDNA genomes that do differ in gene order, one or a few responsible inversions often can be deduced. For example, cpDNAs from bryophytes (*Marchantia*), mosses (*Physcomitrella*), and lycopsids (*Lycopodium, Selaginella, Isoetes*) differ from that of most vascular plants by a 30-kb inversion (Calie and Hughes, 1987; Ohyama et al., 1986; Raubeson and Jansen, 1992a), one of the few large structural alterations in cpDNA accepted over the 400–500 million years of evolution involved. This inversion appears to be a synapomorphy for vascular plants minus the lycopsids and identifies a basal split among vascular plants.

One of the first and most comprehensive of phylogenetic studies of a cpDNA rearrangement involved a 22-kb inversion found to be shared by 57 genera representing all tribes of the Asteraceae (sunflowers), a large plant family with more than 20,000 species and 1100 genera (Jansen and Palmer, 1987). The absence of this inversion from the subtribe Barnadesiinae of the Mutisieae tribe, and from all families allied to Asteraceae, suggested that Barnadesiinae represents the most basal lineage in the Asteraceae and that, contrary to earlier opinion, Mutisieae is not monophyletic. These conclusions subsequently were supported by congruent results obtained from phylogenetic analyses of cpDNA restriction sites (Jansen and Palmer, 1988) and sequences (Kim et al., 1992; see review in Jansen et al., 1992).

Unlike the case for mtDNA in most higher animals, cpDNA genes carry numerous introns [more than 20 have been identified at least tentatively in *N. tabacum* (Shinozaki et al., 1986)]. Systematic surveys have revealed that losses of entire introns (in contrast to the more frequently observed mutations in intron length) are relatively rare events that can be quite informative phylogenetically. For example, the absence of an intron in the *rpl2* gene marks all examined members of the Caryophyllales (Downie et al., 1991). However, caution in the use of such markers alone is indicated also, because a loss of this intron apparently has occurred independently in at least five other unrelated dicot lineages (Downie et al., 1991). Occasional evolutionary losses of particular genes from cpDNA (perhaps to the nucleus) also have been employed as phylogenetic signals (Downie and Palmer, 1992).

Most land plant cpDNAs possess a 20–30-kb inverted repeat (IR) region, the rare deletion of one copy of which has been employed to infer monophyly for six tribes and one putatively allied genus *Wisteria* within the subfamily Papilionoideae (Lavin et al., 1990). However, in a phylogenetic sense the character state "IR loss" again may be informative locally but potentially misleading as a guide to global phylogeny because IR losses have occurred independently in more than one plant group (Doyle et al., 1992). In particular, conifers also possess only one IR element (Lidholm et al., 1988; Raubeson and Jansen, 1992b).

Generally speaking, convergent evolutionary gains of rare features are even less likely than convergent losses (although both types of events can be troubled by homoplasy), and hence the shared possessions of de novo genomic additions should be of special significance in clade identification. The relationship of green algae to land plants long has intrigued botanists. One green algal group from which land plants were suspected to have arisen is the Charophyceae, a suggestion that gained recent additional support from molecular observations on cpDNA. All previously examined algae as well as eubacteria lack introns in their tRNA^{Ala} and tRNA[Ile] genes, whereas all assayed land plants possess them. A recent discovery of homologous introns in representatives of the Charophyceae (*Coleochaete, Nitella,* and *Spirogyra*) was interpreted to indicate the evolutionary acquisition of a genetic novelty, presumably some 400–500 mya, marking a clade that indeed does link the land plants to the charophyceans (Manhart and Palmer, 1990).

cpDNA RESTRICTION SITES AND SEQUENCES

Because the cpDNA genome is large (relative to animal mtDNA, for example), typical whole-genome studies based on Southern blotting procedures often include a large number of six-cutter restriction sites (typically 200–1000). An example of one such study, which deals with the magnificent woody giant-rosette plants in the genus *Lobelia,* is summarized in Figure 8.13. This group has a nearly pantropical distribution with spectacular evolutionary radiations in such places as the mountains of eastern Africa and the Hawaiian Islands, but phylogeographic

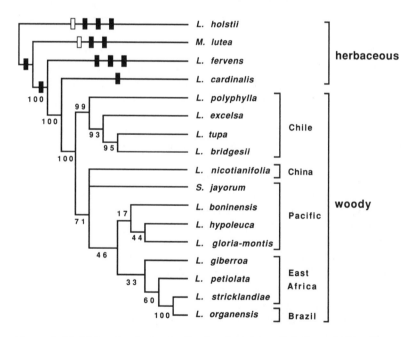

Figure 8.13. Molecular phylogeny for the plant genus *Lobelia* and allies (*S.* = *Sclerotheca*; *M.* = *Monopsis*) (after Knox et al., 1993). The tree is derived from 132 informative cpDNA restriction-site mutations. Numbers indicate levels of bootstrap support for putative clades. Superimposed on the phylogeny are structural rearrangements in cpDNA that also were monitored: deletions (open bars) and inversions (closed bars).

relationships in the assemblage have remained controversial. Phylogenetic analyses of 17 species using restriction-site mutations revealed that the woody lobelias (a) are derived monophyletically from diploid herbaceous ancestors, (b) include a hexaploid group in Chile that is a sister clade to a pan-tropical tetraploid clade, (c) include the genus *Sclerotheca,* which now appears to be derived from a woody *Lobelia* ancestor, and (d) contain a clade consisting of the giant lobelias of eastern Africa plus a Brazilian species (Knox et al., 1993). Apart from these phylogenetic insights, this study is important because it included also a phylogenetic analysis of cpDNA structural rearrangements (deletions and inversions) that occur within the group. These structural features could be overlaid on the restriction-site tree topology without conflict (Fig. 8.13), indicating that the two sets of data yield congruent (though not equally detailed) phylogenetic estimates.

In recent years, sequence analyses of cpDNA have become popular. A favorite sequencing target is *rbcL,* a gene encoding the large subunit of ribulose-1,5-biphosphate carboxylase. Reasons for this preference include the following (D.E. Soltis et al., 1990): (a) much comparative information has accumulated [at least 500 *rbcL* sequences are available at the time of this writing (Chase et al., 1993)]; (b) *rbcL* is a large gene (>1400 bp) that provides numerous characters for phylogenetic studies; and (c) the rate of *rbcL* evolution has proved especially appropriate for addressing questions of plant phylogeny at the subfamilial level or higher. One example of a tree generated from *rbcL* sequences (plus restriction-site data) is presented in Figure 8.14. Interestingly, in this case, restriction-site data from the entire cpDNA genome proved more useful than the *rbcL* sequences, presumably due to the sampling of more variation in the restriction-site comparisons and the higher incidence of homoplasy in the sequence data (Kim et al., 1992).

One advantage of sequence (or site) analyses over particular eccentric markers (see above) in phylogenetic assessment is that general estimates of divergence times (as well as branching patterns) can be attempted, using appeals to molecular clock considerations. For example, from analyses of published sequences of numerous chloroplast and nuclear genes in a variety of plant species, Wolfe et al. (1989) estimated that the cycad-angiosperm split occurred about 340 mya and that a monocot-dicot divergence occurred about 200 mya (with an uncertainty of about 40 my). Within the dicots, an evolutionary radiation of higher lineages was estimated from *rbcL* sequences and the fossil record to have occurred during a relatively short period about 85–90 mya (Olmstead et al., 1992). Within the monocots, analysis of *rbcL* sequences from 104 species representing more than 50 taxonomic families produced a tree indicating a relatively rapid radiation of major lineages within this group also, although the authors did not hazard a guess as to the absolute age of this event (Duvall et al., 1993). In general, caution should, of course, be exercised in interpreting conclusions about specific divergence times from *rbcL* (or other) sequences because suspected rate differences of at least fivefold have been reported for this gene across plant lineages (Gaut et al., 1992).

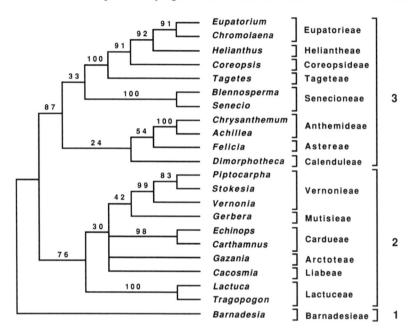

Figure 8.14. Molecular phylogeny for the angiosperm family Asteraceae (after Kim et al., 1992). The tree is a consensus parsimony network based on more than 750 cpDNA characters (restriction sites and *rbcL* mutations). Numbers indicate levels of bootstrap support for putative clades.

It has not gone unnoticed (Rieseberg and Brunsfeld, 1992) that much of the current interest in the molecular systematics of higher plant taxa is based on the same molecule (cpDNA) that has been used to document reticulate evolution due to interspecific hybridization and introgression between closely related plant species (Chapter 7). Could these or other phenomena that can lead to discrepancies between cpDNA gene trees and organismal trees also cause gross errors in higher phylogeny estimation? Clegg and Zurawski (1992) suggest that this is unlikely: "It is reasonable to assume that the approximation to organismal history will improve as time increases, because the biases introduced by interspecific hybridization or intraspecific polymorphism will diminish with an increase in time scale." Nonetheless, the organismal phylogenies estimated by cpDNA (or any other single-gene genealogy) will require corroboration from additional lines of evidence.

Slowly Evolving Gene Sequences and Deep Phylogenetic Branches

Phylogenetic analyses of nucleotide sequences have revolutionized the study of organismal relationships, particularly among higher taxa representing the deepest branchings in the tree of life. Particularly noteworthy have been the large number of studies conducted on ribosomal RNA genes (examples in Table 8.4). The main function of rRNA is protein synthesis, so it is not surprising that the genomes of all organisms contain sequences that code for these essential molecules. Analyses of rRNA genes have been facilitated by the fact that portions of the coding region evolve very slowly [some nucleotide stretches are conserved across all species examined, from microbes to higher plants and animals (Gerbi, 1985; Jorgensen and Cluster, 1988)], whereas others evolve somewhat more rapidly and provide phylogenetic markers at intermediate evolutionary depths. The occurrence of ribosomal RNA genes and other evolutionarily conserved loci across widely different organisms has opened the possibility for development of what Wheelis et al. (1992) refer to as a "global classification" for all of life.

Before turning to studies at such global biotic dimensions, two examples (one each from plants and animals) will be presented to illustrate the utility of rRNA gene sequence analyses at "meso"-evolutionary levels. Hamby and Zimmer (1992) phylogenetically analyzed sequences collected from the nuclear 18S and 26S ribosomal RNA genes of 60 diverse plant taxa. Among the many conclusions tentatively reached concerning the evolution of vascular seed plants were the following (Fig. 8.15): (a) the "flowering" plants (angiosperms) are monophyletic; (b) the "naked-seed" plants (gymnosperms) are not monophyletic, but rather exhibit a paraphyletic relationship to the angiosperms; (c) within angiosperms, the monocots do not form a strictly monophyletic group, due primarily to the inclusion of water lilies (*Nelumbo* and *Ceratophyllum*) along a monocot branch; and (d) dicots, as traditionally viewed, exhibit a paraphyletic relationship to the major monocot lineage(s). The latter conclusion, supported further by analyses of *rbcL* sequences from the cpDNA genome (Chase et al., 1993), is important because it suggests that the morphological traits characterizing most monocots as a markedly distinctive group are probably shared-derived (synapomorphic) conditions, rather than ancestral features of plants.

Another finding of the Hamby and Zimmer (1992) study was that in the shortest parsimony trees, the order Gnetales [genera *Welwitschia*, *Gnetum* and *Ephedra* (Fig. 8.15)] appeared to be the earliest diverging group of seed plants, and that the other gymnosperms (conifers, cycads, and *Ginkgo*) constituted a sister group to the angiosperms. However, this result is not in accord either with other molecular information [from *rbcL* sequences (Chase et al., 1993)], or with cladistic analyses of morphological data (Crane, 1985), both of which place the Gnetales as the sister group to the angiosperms. This result again emphasizes the importance of basing any final phylogenetic conclusions on multiple genes (as well as other sources of evidence).

Hedges et al. (1990) published a similar phylogenetic analysis of the ribosomal RNA gene sequences for the nuclear 18S and 28S molecules in more than 25 species of tetrapod vertebrates, with a particular emphasis on the amphibians

Table 8.4. A small sample of stimulating initial suggestions based on phylogenetic analyses of various rRNA gene sequences.[a]

Taxonomic Group	RNA Subunit	Reference	Authors' Conclusions
Bacteria	16S	Fox et al., 1980	Two deep clades in prokaryotes (archaebacteria and eubacteria)
Protists	Nuclear 5S	Kumazaki et al., 1983	Green algae share common ancestor with vascular plants
Plants	Nuclear 5S	Hori et al., 1985	*Cycas* is a gymnosperm; land plants related to charophyte algae
All living forms	16S	Woese, 1987	Three clades of life (archaebacteria, eubacteria, eukaryotes)
All living forms	5S and nuclear 5S	Hori and Osawa, 1987	Archaebacteria and eukaryotes split off after eubacteria
Protozoa and fungi	Nuclear 16S	Edman et al., 1988	*Pneumocystis carinii* is a fungus
Animals	Nuclear 18S	Field et al., 1988	Cnidarians are separate from other animal lineages.
Eukaryotes	Nuclear 18S	Nairn and Ferl, 1988	Angiosperms are monophyletic
Bacteria	16S	Lake, 1988, 1989	New data analyses suggest archaebacteria are paraphyletic
Bacteria	16S, 23S	Gouy and Li, 1989	Further statistical analyses support Woese (1987) above
Eukaryotes	Nuclear 18S	Sogin et al., 1989	Fungi, plants, and animals diverged relatively recently
Vertebrates	Nuclear 28S	Hillis and Dixon, 1989	Coelacanths allied to tetrapods; birds, mammals perhaps linked
Fishes	Nuclear 18S	Stock and Whitt, 1992	Cyclostome fishes (lampreys and hagfishes) monophyletic
Archaebacteria	16S	Fuhrman et al., 1992	Novel archaebacterial lineage discovered in marine plankton

Source: After Hamby and Zimmer (1992).

[a]These stated conclusions should be viewed as working hypotheses in need of further testing using data from other loci, rRNA gene data from additional taxa, and further statistical evaluations.

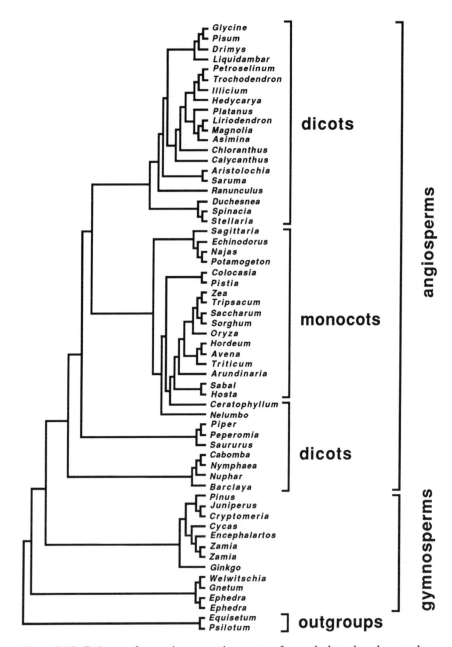

Figure 8.15. Estimate of a maximum parsimony tree for seed plants based on nuclear rRNA gene sequences (after Hamby and Zimmer, 1992). The original paper should be consulted for additional details and for alternative tree topologies, many of which were nearly as parsimonious as the one shown here.

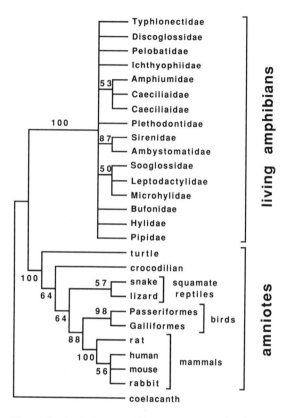

Figure 8.16. Consensus parsimony tree for amphibian families and other tetrapod vertebrates, with the coelacanth fish as outgroup, based on nuclear 18S rRNA gene sequences (after Hedges et al., 1990). Numbers indicate levels of bootstrap support for putative clades.

(Fig. 8.16). Among the major conclusions reached include the following: (a) the living amphibians represent a statistically significant monophyletic group; (b) so too do the Amniota (vertebrates with an extra embryonic membrane—the reptiles, birds, and mammals); and (c) most surprisingly, a sister-group relationship might exist between the birds and mammals. This latter suggestion by Hedges and colleagues contradicts conventional wisdom that the closest living relatives of birds are crocodiles, but nonetheless can be accommodated with molecular data from several other loci and thus will warrant further serious consideration (e.g., Hillis and Dixon, 1989; see discussion in Hedges et al., 1990).

The greatest impact of ribosomal RNA analyses in evolutionary biology has been in stimulating phylogenetic thought about the primary lineages of life. From

Greek antiquity to recent times, a common notion was that living organisms could be divided into two kingdoms, animals and plants. In this century, another primary division recognized widely was between the prokaryotes (microorganisms lacking a membrane-bound nucleus) and eukaryotes (organisms consisting of cells with true nuclei). In 1959, Whittaker proposed the existence of five kingdoms (animals, plants, fungi, unicellular eukaryotes, and prokaryotes), and this view gained widespread acceptance in the 1960s.

A breakthrough occurred in 1977, when Woese, Wolfe, and their colleagues conducted analyses of 16S rRNA gene sequences and concluded that all living systems should be divided in a different fashion—along what appeared to be distinct phylogenetic lines of descent (Fox et al., 1977; Woese and Fox, 1977). An updated version of this scenario (Woese et al., 1990), based on analyses of additional sequences from rRNA genes (and other loci), proposed that all forms of life should be classified into one of three "domains" above the rank of kingdom: Eucarya, including all eukaryotes (i.e., the basic eukaryotic cell stripped of any contribution from its organelles); Bacteria (previously called "eubacteria" or typical bacteria); and Archaea (formerly Archaebacteria), which includes methanogens and thermophilic forms (Fig. 8.17). The molecular differences separating extant representatives of these lineages appear to be "of a more profound nature than the differences that separate typical kingdoms, such as animals and plants" (Woese et al., 1990). Although there have been some alternative interpretations of the deep-branching structure of life as revealed by molecular data [see Day (1991), Lake (1991), and references therein], there is no doubt that major phylogenetic lineages, previously unrecognized, exist among prokaryotic life forms.

One question of special interest concerns where the root should be placed on the Eucarya-Bacteria-Archaea phylogeny. This issue has been addressed from a cladistic standpoint using phylogenetic analyses of the sequences of primordially duplicated genes: those encoding elongation factors and ATP-ase subunits (Gogarten et al., 1989; Iwabe et al., 1989). From observed distributions of shared-derived features post-dating the gene duplications, both studies concluded that the Archaea and Eucarya constitute an evolutionary clade to the exclusion of the Bacteria.

Some taxonomists disagree that the prokaryotes should be divided into two distinct groupings (Bacteria and Archaea), on the grounds that differences in levels of biological organization between prokaryotes and eukaryotes remain so profound that continued taxonomic recognition of these latter groups is desirable (Mayr, 1990). Woese et al. (1991) retort that Mayr's scheme would represent a return to an "artificial" and "flawed" classification based not on phylogeny. This debate raises a more general point. Should classifications reflect solely the phylogenetic branching order of taxa (as cladists suggest) or should perceived grades of organismal differentiation somehow be incorporated as well (as many

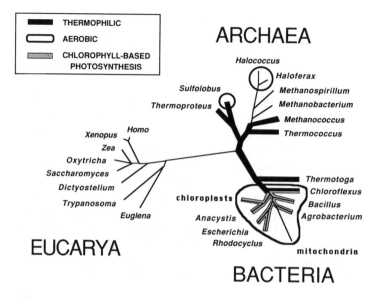

Figure 8.17. An updated version of one of the first reconstructions of the deep structure in the tree of life [inferred from sequences of the small subunit rRNA genes (G.J. Olsen and C.R. Woese, unpublished)]. Note the positions of chloroplast and mitochondrial genes, embedded deep within the Bacteria rather than the Eucarya where currently they are housed. Codings indicate the distributions of three physiological traits in the network (thermophilic adaptations, aerobic metabolism, and chlorophyll-based photosynthesis), as indicated by their occurrence within at least some members of the respective groups.

others would argue)? This is a subjective issue (though nonetheless important operationally), such that an answer depends on one's orientation about the nature of information that a formal classification should convey.

In any event, the original studies by Woese and colleagues provided a positive stimulus for much of the subsequent molecular work on the deep phylogenetic structure of life. One conclusion from analyses of many additional taxa is the astounding phylogenetic diversity among eukaryotic protists (Knoll, 1992; Sogin, 1991). Thus, extremely long evolutionary branches (inferred, for example, from 16S-like ribosomal gene sequences) lead separately to the extant entamoebas, euglenids, trichomonads, microsporidians, diplomonads, slime molds, and several other protistan groups that therefore must have had long and independent evolutionary histories (probably over a billion years).

Regardless of how the new molecular-based phylogenetic appraisals of life eventually are translated into a formal classification scheme, the branching diagrams themselves remain of considerable interest for the phylogenetic mapping of organismal traits (Figs. 8.17 and 8.18). For example, against the phylogenetic backdrop proposed by the Woese group, aerobic metabolism occurs in widely separated lineages of Bacteria and Archaea, suggesting a polyphyletic origin for this metabolic trait (provided that anaerobic metabolism was the ancestral condition, as most researchers now believe). On the other hand, the capacity for chlorophyll-based photosynthesis appears only in the Bacteria (and their chloroplast relatives), suggesting a single evolutionary origin for this capability (Fig. 8.17). Within the Bacteria, sulphur oxidation appears in at least two divergent lineages, suggesting that interpretation of this potential phylogenetic marker as a global synapomorph probably would be invalid (Fig. 8.18).

One of the most remarkable discoveries from the phylogenetic mapping of characters on a molecular backdrop concerns the evolutionary origins of cpDNA and mtDNA. Two theories previously were advanced to account for the existence of separate and distinctive nuclear and cytoplasmic genomes within eukaryotic cells. One theory stipulated that organellar genomes had an autogenous origin within eukaryotes, as fragments from the nuclear genome became incorporated into membrane-encased mitochondria or chloroplasts which subsequently as-

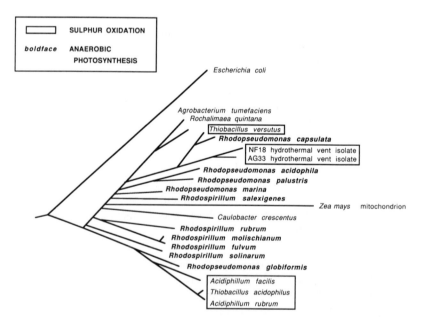

Figure 8.18. Phylogenetic tree for the α-subdivision of the purple Bacteria inferred from sequences of the small subunit rRNA genes (from C.R. Woese and G.J. Olsen, unpublished). Codings indicate the distributions of two physiological adaptations (sulphur oxidation and anaerobic photosynthesis) that had contributed to a more traditional taxonomy for these organisms.

sumed a quasi-independent existence (see Cavalier-Smith, 1975; Uzzell and Spolsky, 1981). The second theory stipulated that organellar genomes had exogenous origins, stemming from bacterial-like ancestors that invaded (or were engulfed by) proto-eukaryotic host cells bearing precursors of the nuclear genome (Margulis, 1981). The latter "endosymbiont theory" has received compelling support from multiple lines of molecular evidence (Gray, 1992).

Much of the early molecular evidence in favor of the endosymbiont theory came from c-type cytochromes (Schwartz and Dayhoff, 1978) and rRNA gene sequences. In terms of molecular phylogenetic affinities, the rRNA gene sequences of both mitochondria and chloroplasts appear allied to bacterial sequences rather than to the nuclear sequences of the Eucarya (Fig. 8.17). It is now accepted that most Eucarya really are genetic chimeras containing a mixture of truly distinctive evolutionary lineages.

Recent molecular attention has focused on closer details of the phylogenetic origins for cpDNA and mtDNA. From analyses of rRNA gene sequences, chloroplast sequences appear allied most closely to those of the photosynthetic cyanobacteria (Giovannovi et al., 1988), and mitochondrial sequences may derive from the α-subdivision of the purple bacteria (Proteobacteria) (Cedergren et al., 1988; Yang et al., 1985). Further proposals for the phylogenetic roots of mtDNA and cpDNA have been made by many authors (e.g., Bremer and Bremer, 1989; Gray et al., 1989; Howe et al., 1992; Kishino et al., 1990; Lockhart et al., 1992; W. Martin et al., 1992; Van den Eynde et al., 1988; Van de Peer et al., 1990; Villanueva et al., 1985). A recent reexamination of chloroplast origins using the inferred amino acid sequences from several protein-coding cpDNA loci (*psbA*, *rbcL*, *rbcS*, *tufA* and *atpB*) not only supported the hypothesis of a bacterial origin for chloroplasts but also suggested that the extant plastids in various algal and plant lineages probably trace to a single endosymbiotic event that was followed by at least one instance of a secondary transfer of an *rbc*LS operon from a purple bacterium into a plastid lineage (Morden et al., 1992).

The latter study by Morden et al. (1992) is especially noteworthy because, as has been argued repeatedly, final phylogenetic conclusions should be based on multiple lines of genetic evidence. A general difficulty in evaluating the deep-branching structure among life forms has been that until recently, relatively few other genetic characters had been monitored at these deep evolutionary levels. Thus, searches for other ancient molecular markers have been initiated. Among the loci identified thus far that have appeared most promising for the study of deep-branching topologies are those encoding ATP synthase subunits (Recipon et al., 1992), elongation factors EF-2/EF-G (Cammarano et al., 1992), and RNA polymerases (Pühler et al., 1989). Phylogenetic analyses of these sequences have further supported the alliance of cpDNA and mtDNA with Bacteria rather than Eucarya (Recipon et al., 1992), as well as confirmed the general distinctiveness of the Archaea as originally proposed by Woese and colleagues.

PROSPECTUS FOR A GLOBAL PHYLOGENY

In the foreseeable future, it should be possible to assemble molecular (and other) data into a grand phylogenetic encyclopedia—a universal Tapestry linking all life forms. This endeavor should include estimates of relationship at all phylogenetic levels, from intraspecific kinships to the most ancient of evolutionary separations. The hierarchical structure inherent in organismal phylogeny carries at least two ramifications for such an enterprise. First, life's hierarchical arrangement will facilitate information storage and retrieval. Sections of the global Tapestry could be stored and referenced as a nested series of phylogenies at increasingly greater evolutionary depths (as in Fig. 8.19). Second, although a complete Tapestry would necessitate the generation of millions of microphylogeny estimates (one for each species, clearly an impossible empirical task), the pyramidal structure of organismal relationships means that fewer and fewer such summaries will be required at increasingly higher taxonomic levels. Thus, it should be feasible to achieve a consensus phylogenetic picture of the major trunks and intermediate branches of life, plus occasional snapshots of microevolutionary relationships within selected species or genera of special interest.

A molecular phylogeny for fungi spanning all hierarchical levels (Fig. 8.19) provides a small illustration of the kind of information that a global Tapestry might include. This example also serves to illustrate at least two complications that will arise. One complication is that different molecules and assay procedures will have to be employed at different levels of the hierarchy, due to the varying windows of resolution provided. In this case, mtDNA restriction fragments were employed for the microevolutionary comparisons, and rRNA gene sequences for the meso- and macro-evolutionary trees. No single molecular approach normally will apply to all levels simultaneously. A second complication is that the separation times among branches, as well as the topologies of each tree, will be desirable. Purely cladistic appraisals aimed at resolving branching topology are important, particularly for issues such as phylogenetic character mapping, but they risk a diversion of attention from the equally important issue of divergence times. As is often the case, the mesophylogenetic and macrophylogenetic summaries for fungi shown in Figure 8.19 were based on procedures designed primarily to recover correct branching topology, but knowledge of absolute divergence times would enrich these phylogenetic pictures greatly.

As this book testifies, many scattered pieces of the grand Tapestry puzzle already have been gathered through molecular methods, and preliminary attempts have been made to assemble these elements into coherent summaries for particular taxonomic groups (e.g., Dutta, 1986; Fernholm et al., 1989; Sibley and Ahlquist, 1990; P.S. Soltis et al., 1990). In a way, molecular phylogenetic explorers are at a stage of biotic description analogous to that of the European naturalists (including Charles Darwin) in the 1800s, during the early global explorations of a newly discovered world—at that time too, many new bits of

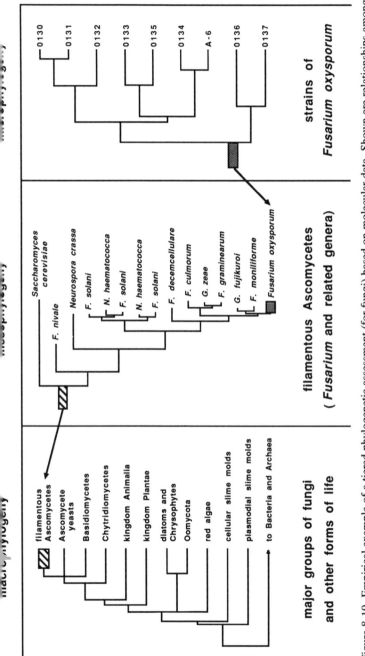

Figure 8.19. Empirical example of a tiered phylogenetic assessment (for fungi) based on molecular data. Shown are relationships among strains of the wilt fungus *Fusarium oxysporum* and their position within the broader phylogenetic hierarchy of life (compiled from diagrams and information in Bruns et al., 1991; Gaudet et al., 1989; and Jacobson and Gordon, 1990). For additional details about molecular relationships among the major fungal groups (Basidiomycetes, other Ascomycetes, and Chytridiomycetes), see Bowman et al. (1992).

biological information of relevance to systematics were coming in rapidly and an important challenge was to catalog, assemble, and interpret the findings.

SPECIAL TOPICS IN MOLECULAR PHYLOGENETICS

Horizontal Gene Transfer

Genetic transmission is overwhelmingly "vertical"; i.e., from parents to off-spring. Indeed, if this were not the case, phylogeny would have little meaning; and furthermore, evolutionary trees built from independent characters would seldom exhibit the coherent and consistent structures that have proved to characterize most well-studied taxonomic groups. Nonetheless, several reports of lateral transfer of genetic elements across seemingly inviolable taxonomic boundaries have appeared in recent years (Table 8.5), and most molecular evolutionists now accept that such "horizontal" movement occasionally does take place. Note that horizontal exchange between taxa should not be confused with hybridization-mediated introgression (Chapter 7), which is merely a special case of vertical genetic transmission. [However, some semantic difficulties can arise, as, for example, in the use of the word "conjugation" to describe the physical joining between distantly related organisms, such as that between bacteria and yeast which is known to result in the intertaxon transfer of DNA via conjugative plasmids (Heinemann and Sprague, 1989).]

The evidence for horizontal transmission usually comes from a gross inconsistency between the phylogenies inferred from different sets of molecular markers, or between a molecular marker and a traditionally accepted phylogeny (M.W. Smith et al., 1992). One of the first reported instances of a horizontal genetic transfer involved a supposed movement of the gene encoding copper-zinc superoxide dismutase (SOD), normally found predominantly in higher eukaryotes, from the ponyfish *Leiognathus splendens* to its symbiotic bacterium *Photobacterium leiognathi* (Bannister and Parker, 1985; Martin and Fridovich, 1981). The unusual phylogenetic occurrence of this form of SOD in a bacterium, and the observation that its amino acid composition resembled more closely the ponyfish SOD than it did known SODs from other sources, led to the horizontal-transfer scenario. However, further research over a broader phylogenetic scope has called this conclusion into serious question (Leunissen and de Jong, 1986; Steffens et al., 1983), due to two additional findings: the discovery of Cu-Zn SOD in other prokaryotes, and the lack of significant statistical support for the supposed phylogenetic grouping of the SOD of *P. leiognathi* with eukaryotic sequences [indeed, the *P. leiognathi* SOD now appears allied more closely to other prokaryotic SODs (Smith and Doolittle, 1992a)]. A similar controversy has centered on a glutamine synthetase II gene (*glnII*), which occurs in eukaryotes and in a small subset of bacteria (genera *Rhizobium, Agrobacterium,* and allies)

Table 8.5. Reported instances of horizontal gene transfer between different species and unrelated taxa. The first two cases in the list have been disputed seriously (see text).[a]

Gene Supposedly Transferred	Taxonomic Groups Involved and Direction of Movement (Where Postulated)	Reference
Prokaryote/Eukaryote Gene Transfers		
Copper/zinc superoxide dismutase	From fish to bacterium	Martin and Fridovich, 1981
Glutamine synthetase II	From plant to bacterium	Carlson and Chelm, 1986
Glyceraldehyde-3 phosphate dehydrogenase	From unspecified eukaryote to bacterium	Doolittle et al., 1990
Fibronectin type III	From animal to bacterium	Bork and Doolittle, 1992
Glucose-6-phosphate isomerase	From plant to bacterium	Smith and Doolittle, 1992b
Plasmid-mediated transfers	From bacterium to plant	Zambryski et al., 1989
Plasmid-mediated transfers	From bacterium to yeast	Heinemann and Sprague, 1989
Eukaryotic Transfers of Transposable Elements		
"mariner"	From *Drosophila* to *Zaprionus* flies	Maruyama and Hartl, 1991
"*p*"	From *D. willistoni* complex to *D. melanogaster*	Daniels et al., 1990
"hobo"	Between *Drosophila* species	Simmons, 1992
"jockey"	Between *Drosophila* species	Mizrokhi and Mazo, 1990
"hobo" and others	Between animals and plants	Calvi et al., 1991
"copia"-like	Between numerous plant, animal, and yeast hosts	Flavell, 1992

[a]For additional examples of lateral transfer, see Heinemann (1991), Kidwell (1992), Mazodier and Davies (1991), M.W. Smith et al. (1992), and Sprague (1991).

that are symbiotic with higher plants. Sequence similarities between the *glnII* loci in these bacteria and plants originally were interpreted to evidence a horizontal transfer between kingdoms (Carlson and Chelm, 1986), but subsequent analysis of a broader array of glutamine synthetases failed to support the contention that *glnII* was of plant origin or that gene transfers across large taxonomic gaps were required to explain the data (Shatters and Kahn, 1989).

Regardless of their final resolution, these case histories raise several important points. First, thorough phylogenetic analyses are required before putative cases of horizontal transfer can be confirmed (or refuted). An important step in such analyses should involve statistical tests of the significance of the discordant tree structures [for example, by bootstrapping procedures (Lawrence and Hartl, 1992)]. Second, it should be emphasized that a variety of evolutionary factors other than horizontal transfer, in principle, can lead to apparent phylogenetic discordancies among characters (Fig. 8.20). These include the shared retention of ancestral states by the taxa in question, extreme molecular rate heterogeneities across lineages, convergent evolution to a shared molecular condition, introgressive hybridization, and a mistaken assumption of orthology when the loci in question might truly be paralogous (Fig. 1.3). Third, the possibility of horizontal transfer of particular genes adds yet another compelling rationale for the desirability of including multiple lines of evidence in phylogenetic reconstructions.

Notwithstanding these considerable obstacles to the firm documentation of horizontal genetic transfer, several reported instances (Table 8.5) do appear convincing (or at least have not as yet been challenged seriously). Perhaps the most compelling case for the occurrence of horizontal transfer in any eukaryotic system involves the "*P* element" in fruit flies (Daniels et al., 1990). These transposable genetic elements (see below) have a patchy phylogenetic occurrence confined mostly to the genus *Drosophila* and related dipteran genera including *Lucilia* blowflies (Perkins and Howells, 1992). A remarkable genetic discovery was that *P*-element sequences in *D. melanogaster* are nearly identical to those in *D. willistoni*, despite a suspected evolutionary separation of these host species of more than 50 million years (Daniels et al., 1990). Furthermore, close relatives of *D. melanogaster* appear to lack *P* elements entirely, whereas the elements are widespread in species of the *D. willistoni* complex; and, there is strong circumstantial evidence for the recent spread of *P* elements within *D. melanogaster* (Kidwell, 1992). The overall conclusion is that *D. melanogaster* probably acquired *P* elements recently (within the last half-century!) via a horizontal genetic transfer from the *D. willistoni* complex. Recent evidence suggests that a semiparasitic mite (*Proctolaelaps regalis*) may have been the mediating vector (Houck et al., 1991).

Phylogeny of Retroviruses and Transposable Elements

Many of the suspected instances of horizontal genetic transfer involve transposable elements (TEs) (Table 8.5), a class of DNA sequences that may be

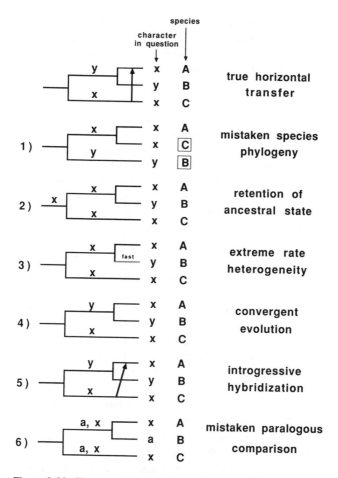

Figure 8.20. Six competing hypotheses to account for an apparent character-state discordance that otherwise might be attributable to a horizontal gene-transfer event.

predisposed to such movement due to a known ability to shift from one chromosomal position to another within cells. Among the mobile eukaryotic elements, some TEs transpose by reverse transcription of an RNA intermediate ("class I" retrotransposable elements), and others by a DNA to DNA transposition mechanism ("class II" elements) (Finnegan, 1989). Most TEs have characteristic structures that include genes coding the enzymes involved in the transposition process, usually flanked by terminal repeat sequences of varying length. In the retrotransposable elements (RTEs), one of these genes encodes reverse transcriptase (RT), which catalyzes the transcription of RNA to DNA.

Particularly intriguing are the biological and structural similarities between retrotransposable elements and retroviruses (RVs). Retroviruses are small, single-stranded RNA viruses found mostly in mammals that resemble RTEs in several features, including the production of a reverse transcriptase and the presence of long terminal repeats (LTRs) flanking the coding region. However, like other viruses and unlike RTEs, retroviruses can encase themselves in a protective envelope that facilitates independent infectious transport across the cells of the same or different organisms. These observations raise an interesting evolutionary question (Doolittle et al., 1989; Finnegan, 1983). Might RTEs represent "degenerate" retroviruses that secondarily lost much of this facility for autonomous intercellular transport? Or alternatively, did retroviruses evolve from ancestral RTEs by secondary acquisition of these capabilities?

In addition to its presence in both RTEs and RVs, a reverse transcriptase gene also is found in several other genetic elements, including the hepadnaviruses of animals and the caulimoviruses of plants. The RT gene also exhibits structural similarities (suggestive of shared ancestry) to the RNA-directed RNA polymerases of some other viruses. Xiong and Eickbush (1990) have taken advantage of these facts to estimate a molecular phylogeny for RT-containing genetic elements, based on nucleotide sequences from the RT and RNA polymerase genes. Results from this analysis for 82 sequenced retroelements are summarized in Figure 8.21. Major conclusions included the following: (a) there appear to be two fundamental branches in retroelement phylogeny, one leading to the non-LTR retrotransposons, and the other leading to the LTR retrotransposons, retroviruses, caulimoviruses, and hepadnaviruses; (b) most members within each of these five named assemblages group together in terms of RT phylogeny; and (c) due to the restricted position of retroviruses in the broader scheme of RT-containing elements, it seems likely that retroviruses evolved from retrotransposable elements, rather than the converse.

This example is presented not to suggest that the questions concerning RTE-RV evolution are solved to everyone's satisfaction, but rather to illustrate yet another arena (in this case involving some of the smallest and simplest forms of life) in which molecular phylogenetic appraisals are playing a novel and important role in evolutionary studies.

Molecular Paleontology

Most molecular appraisals have been directed toward extant organisms, with phylogenetic inferences representing extrapolations to the mutational changes and cladistic events of the past. A longstanding dream of molecular evolutionists has been to assess extinct biota more directly, through recovery of biological macromolecules from fossil material.

In 1980, Prager and co-workers reported a phylogenetic signal retained in the

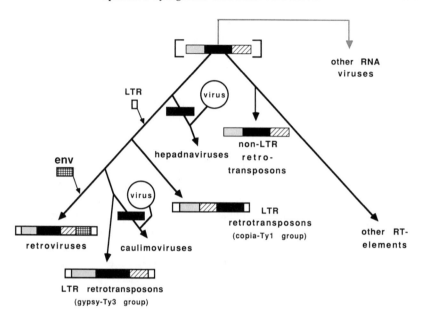

Figure 8.21. Phylogeny for retroelements based on RT sequences (after Xiong and Eickbush, 1990). Various structural features of the RT-containing elements are superimposed on the estimated phylogeny. Shaded boxes correspond to regions of the coding sequence: solid shading, RT region; stippled, *gag* gene region; diagonal shading, integrase region; cross-hatched shading, envelope gene. Unshaded boxes represent LTRs. The ancestral structure (in parentheses) is hypothesized.

serum albumin proteins of a 40,000-year-old mammoth (*Mammuthus primigenius*) whose carcass had been preserved in the frozen soil of eastern Siberia. In immunological tests, rabbits injected with ground mammoth muscle produced antibodies that reacted strongly with albumins from extant Indian and African elephants, weakly with sea cows (Sirenia, a taxonomic order thought to contain living relatives of elephants), and still more weakly or not at all with other mammalian albumins. Using similar assays, Lowenstein et al. (1981) showed that albumins from the extinct Tasmanian wolf (*Thylacinus cynocephalus*) produced phylogenetically informative levels of immunological reaction against the albumins of other extant Australian marsupials. The preserved tissue in this study was dried muscle that had adhered to museum specimens collected in the late 19th and early 20th centuries. Apart from a few such examples involving fortuitously well-preserved or recent tissue remains, most other attempts to extract significant genetic information from fossil proteins met with little success (Hare, 1980; Wyckoff, 1972).

Early studies of "ancient DNA" (Pääbo, 1989) fared somewhat better. In the first successful retrieval of phylogenetically informative DNA sequences from museum material, Higuchi et al. (1984, 1987) recovered short mtDNA segments from a salt-preserved study skin (140 years old) of the quagga (*Equus quagga*), an extinct member of the horse family. Fragments of DNA isolated from dried muscle and connective tissue were cloned into a λ vector, and sequences totaling 229 base pairs were obtained and compared against those of extant relatives. These and subsequent molecular studies (George and Ryder, 1986; Pääbo and Wilson, 1988) helped to solve a phylogenetic enigma regarding the quagga, which in terms of general morphology exhibited an odd mixture of horselike and zebralike features. Phylogenetic analyses of the molecular data indicated that the extinct quagga was related closely to the Burchell zebra, *Equus burchelli* (Fig. 8.22).

In 1985, a new record for the evolutionary age of recovered DNA was established with the isolation and biological cloning of nuclear DNA pieces from an Egyptian mummy 2400 years old (Pääbo, 1985). The record did not last long. One year later, a report appeared on the extraction of DNA from human brain tissue, 8000 years old, that had been buried in a swamp in central Florida (Doran et al., 1986). Notwithstanding these and a few other success stories, traditional isolation procedures seldom yielded ancient DNA sequences in a sufficient state of preservation to be of practical utility for phylogenetic comparisons.

Much excitement, therefore, has attended recent developments in the application of PCR-based methods to the recovery of ancient DNA (Pääbo et al., 1989). The amplification of DNA suitable for sequencing from exceedingly small amounts of fossil template, or from well-preserved museum materials, promises to revolutionize molecular paleontology. Indeed, during the years 1989—1991 alone, following the introduction of PCR, more publications appeared on fossil DNA than had accumulated over all prior decades (Brown, 1992).

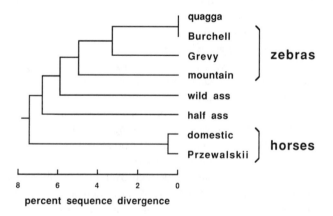

Figure 8.22. Phylogenetic tree based on mtDNA sequences from the extinct quagga and extant members of the genus *Equus* (after Pääbo et al., 1989).

Studies based on PCR approaches reportedly have extended the known temporal range of DNA preservation by more than 1000-fold. At the time of this writing, the record for age of fossil DNA from a plant belongs to Golenberg et al. (1990), who extracted, PCR-amplified, and sequenced an 820-bp fragment of the chloroplast *rbcL* gene from fossil *Magnolia* leaves dating to the Miocene (some 18 million years bp). Phylogenetic comparisons against homologous DNA sequences of modern magnolias and other plants indicated that the characterized molecules really did reflect fossil material, rather than more recent laboratory or field contamination.

Among animal fossils, even older records were established with the recovery of DNA from insects preserved in amber (tree sap): a 30 million-year-old termite [*Mastotermes electrodominicus* (DeSalle et al., 1992)]; a 25–40 million-year-old stingless bee [*Proplebeia dominicana* (Cano et al., 1992)]; and a 130 million-year-old weevil [undescribed species of Coleoptera (Cano et al., 1993)]. In the termite study, comparisons of ancient DNA fragments from two ribosomal RNA genes (16S from the mitochondrion, 18S from the nucleus) against homologous sequences from living termites, cockroaches, and mantids helped to confirm a morphological cladistic assessment of these same taxa and substantiated the suspected monophyly of this extinct species with an extant termite (*M. darwiniensis*) currently abundant in Australia. Apart from the impressive time scales over which dead DNA fragments were shown to persist, these studies also are of special interest because of the detailed phylogenetic comparisons made possible between molecules and morphology. Unlike most fossils, amber-preserved specimens are displayed beautifully in three dimensions, enabling detailed appraisals of "soft" morphological features such as mouthparts and genitalia.

Several other impressive examples of phylogenetic comparisons involving PCR amplifications of fossil DNA have appeared. For example, P.S. Soltis et al. (1992a) isolated *rbcL* sequences ($\cong$ 1300 bp long) from a Miocene species of bald cypress (*Taxodium*) and demonstrated that the fossil taxon is related more closely to *T. distichum* than to the other extant species in Taxodiaceae and Pinaceae. In an extension of the fossil-protein study cited above, Thomas et al. (1989) examined mtDNA sequences from the extinct Tasmanian wolf and demonstrated that *Thylacinus cynocephalus* was related more closely to other Australian marsupials than it was to a group of South American carnivorous marsupials. Janczewski et al. (1992) characterized mitochondrial and nuclear sequences from 14,000-year-old bones of the saber-toothed cat (*Smilodon fatalis*) from tar pits in Los Angeles, and thereby uncovered the phylogenetic position of this extinct species within the Felidae radiation.

SUMMARY

1. Phylogenetic hypotheses underlie virtually all conclusions in comparative organismal evolution. To make these hypotheses more explicit and testable, "phylogenetic character mapping" has become popular, whereby particular or-

ganismal features are matched with their associated species on a cladogram, with the purpose of revealing the evolutionary histories of those traits. The independent appraisals of phylogeny often are based on molecular markers.

2.　Phylogenetic mapping against a molecular backdrop has been accomplished for numerous anatomical, physiological, and behavioral characteristics in plants, animals, and microbes. Some organismal features have proved to have arisen monophyletically, others polyphyletically. Concepts of gradients and thresholds in the phylogeny of quantitative traits have been stimulated by the exercise of phylogenetic character mapping.

3.　Molecular markers are used widely in biogeographic assessment, for example to test dispersalist versus vicariance explanations for the appearance of related taxa in disjunct geographic regions, to distinguish between common ancestry and convergence for organismal similarities between regional biotas, and to reconstruct the biogeographic histories of island biotas.

4.　In macrophylogeny assessment, especially noteworthy have been broad-scale molecular studies based on DNA–DNA hybridization, nucleotide sequencing of slowly-evolving ribosomal RNA and other genes, and restriction-site and sequence analyses of animal mtDNA and plant cpDNA. From the latter two cytoplasmic systems, phylogenetic markers may involve sequence information itself, or eccentric molecular features (such as gene order, presence versus absence of particular introns, or patterns of codon assignment) that appear to have special significance for clade delineation.

5.　Horizontal gene transfer, whereby genetic information is exchanged among unrelated taxa, has been well-documented in several instances. However, discordance between a gene tree and a species phylogeny alone is not sufficient to document a horizontal transfer event because several alternative scenarios also can produce the discordance.

6.　Molecular studies have been extended to some of the simplest forms of life, including the quasi-independent retroviruses and transposable elements which have proved to be allied phylogenetically.

7.　With advent of the PCR, molecular phylogenetic appraisals have been extended to "ancient" DNA, in some cases extracted successfully from fossils up to tens of millions of years old.

8.　In the near future, prospects are great for developing a molecular-based global or universal phylogeny for all of life.

9

Conservation Genetics

Modern biology has produced a genuinely new way of looking at the world . . . to the degree that we come to understand other organisms, we will place a greater value on them, and on ourselves.

E.O. Wilson, 1984

In the final analysis, concerns about the conservation of biodiversity represent concerns about the conservation of genetic diversity. As we have seen, this genetic diversity is arranged hierarchically, from the family units, extended kinships, and geographic population structures within species, to a graded scale of genetic differences among reproductively isolated taxa that have been separated phylogenetically for various lengths of evolutionary time. As we have also seen, the visible external phenotypes of organisms are not an infallible guide to how this genetic diversity is partitioned. Ironically, even as we gain the molecular tools to assess genetic heterogeneity in new and exciting ways, the marvelous biodiversity that has carpeted our planet is being lost at a pace that is nearly unprecedented in the history of life, due to direct and indirect effects of explosive human population growth (Ehrlich and Ehrlich, 1991).

One goal of conservation biology is to preserve genetic diversity. Another goal should be to preserve evolutionary processes. Hybridization, introgression, and speciation are examples of natural and dynamic evolutionary processes that exert great influence on how genetic diversity is organized, yet these forces are sometimes viewed with considerable naivety in conservation legislation (such as that designed to protect "endangered species"). Not only must societies find ways to preserve existing genetic diversity, but they also must seek sustainable environments for life in which the evolutionary processes fostering biotic diversity are maintained.

How might studies in molecular phylogenetics contribute to the assessment of natural genetic diversity and evolutionary processes in ways that are serviceable to the field of conservation biology? Answers to this question are the subject of this concluding chapter.

ISSUES OF HETEROZYGOSITY

Most discussions of genetics in conservation biology have focused on how best to preserve variability within rare or threatened populations, a common assumption being that higher heterozygosity (*H*, one measure of the within-population component of genetic variation) enhances the probability of a population's survival over ecological or evolutionary time (Chapter 2). Traditional approaches to heterozygosity assessment and management have been indirect. Management for heterozygosity in captive populations (such as in zoos) normally occurs in de facto fashion through controlled breeding programs designed to avoid intense inbreeding, either by maintenance of closed populations above some "minimum viable population size," or, where feasible, by the exchange of breeding individuals among sites. For natural populations, the analogous concerns have been to ensure adequate habitat such that local effective population sizes remain above levels where inbreeding (and its associated fitness depression) might become pronounced and/or to maintain habitat corridors for purposes such as facilitating gene flow among local populations (Hobbs, 1992; Simberloff and Cox, 1987; however, see Simberloff et al., 1992 for a critical appraisal of such programs).

With the advent of molecular techniques, more direct estimates of heterozygosity have become possible through assays of multiple marker loci. From these new data bases, two major conservation-related issues regarding heterozygosity have arisen: Is molecular variability reduced significantly in rare or threatened populations? If so, is this reduction a cause for concern about that population's future?

Molecular Heterozygosity in Rare and Threatened Species

In the mid-1800s, indiscriminate commercial harvests of the northern elephant seal (*Mirounga angustirostris*) reduced this formerly abundant species to dangerously low levels. Fewer than 30 individuals may have survived through the 1890s (on a single remote island west of Baja California), but following legislative protection by the Mexican and U.S. governments, the species has rebounded and now numbers in the tens of thousands of individuals distributed among several rookeries. Bonnell and Selander (1974) surveyed 24 allozyme loci in 159 seals from five rookeries and observed absolutely no genetic variation, a striking finding given the high heterozygosity estimates reported for most other species assayed by similar protein-electrophoretic methods (Chapter 2). Recent molecular analyses by Hoelzel et al. (1993) have confirmed and extended these findings: At a total of 55 allozyme loci, no variability has as yet been detected in the northern elephant seal, and at the normally variable control region of mtDNA, only two haplotypes (sequence difference about 1%) were found. These results cannot be attributed to abnormally low genetic variation intrinsic to all seals because the southern elephant seal *M. leonina* also included in the assays displayed normal heterozygosity levels.

During the past decade, the isolated and endangered population of gray wolves (*Canis lupus*) on Isle Royale in Lake Superior declined from about 50 to fewer than 12 remaining individuals. Molecular studies have revealed that approximately 50% of allozyme heterozygosity was lost in this island population, relative to mainland samples (Wayne et al., 1991b). Furthermore, only a single mtDNA genotype was present. In terms of multilocus nuclear DNA fingerprints, the island wolves appeared to be about as similar genetically as are known full siblings in a captive wolf colony, suggesting that the Isle Royale population is inbred severely (Wayne et al., 1991b).

Hillis et al. (1991) employed allozyme markers to estimate genetic variability in the Florida tree snail (*Liguus fasciatus*), many of whose populations are threatened or already extinct. Among the 34 loci monitored in 60 individuals, only one was polymorphic, and mean heterozygosity overall was only $H = 0.002$. Perhaps a population bottleneck accompanied or followed the colonization of Florida by this species from Cuba; or perhaps the snail's habit of partial self-fertilization plays a role in heterozygosity loss (at least within local populations). Surprisingly, the lack of appreciable variation at the molecular (allozymic) level in this snail contrasts diametrically with the exuberant morphological variability exhibited, particularly with regard to the beautiful variants in shell pattern that are known to be genetically based.

Genetic variation in remnant populations of the Sonoran topminnow (*Poeciliopsis occidentalis*) in Arizona, where the species is endangered, has been compared to that in populations from Sonora, Mexico, where the fish is widespread and abundant. At 25 allozyme loci, the geographically peripheral populations in Arizona exhibited significantly less variation than did the Mexican populations near the center of the species' distribution (Vrijenhoek et al., 1985). The molecular analysis also revealed three major genetic groups within the species' range. The authors suggest that these groups should be maintained as discrete entities in nature, because the majority of overall genetic diversity in this species is attributable to the intergroup differences. They also recommend that any restocking efforts in Arizona employ local populations, whose mixing could increase within-population heterozygosity without compromising the genetic differences that characterize the broader geographic assemblages.

Another endangered species assayed extensively for molecular genetic variation is the cheetah (*Acinonyx jubatus*). The South African subspecies of this large cat first was surveyed for genetic variation at 47 allozyme loci, all of which proved monomorphic, and at 155 abundant soluble proteins revealed by two-dimensional gel electrophoresis, where heterozygosity also proved low [H = 0.013 (O'Brien et al., 1983)]. Subsequent assays of additional allozyme markers and of RFLP variation at the MHC (major histocompatibility complex) supported the contention that this population is genetically depauperate (O'Brien et al., 1985b; Yuhki and O'Brien, 1990), a result further confirmed by the failure of these cats to acutely reject skin grafts from "unrelated" conspecifics. Again, the low genetic variation observed in cheetahs cannot be attributed to some inherent property characteristic of all cats because most other species of Felidae exhibit normal to high levels of genic heterozygosity in each of the above assays. On the basis of the genetic evidence, O'Brien et al. (1987) proposed that at least two population bottlenecks occurred in cheetahs, one perhaps 10,000 years ago prior to geographic isolation of two recognized subspecies (which are highly similar genetically) and a second within the last century that may have led to the exceptional genetic impoverishment of the South African form. However, recent assays of rapidly-evolving molecular systems (mtDNA and VNTR nuclear loci) have uncovered moderate genetic variation in cheetahs, of a magnitude that Menotti-Raymond and O'Brien (1993) calculate could be due to postbottleneck mutational recovery over a time scale of about 6,000 to 20,000 years.

A similar scenario about bottleneck effects has emerged regarding the Asiatic lion (*Panthera leo persica*), which now occurs as a remnant population in the Gir Forest Sanctuary in western India. Allozyme surveys (46–50 loci) detected absolutely no variation in a sample of 28 individuals from this subspecies, whereas the Serengeti population of the African subspecies had much higher genetic variability (Wildt et al., 1987). The relict group of lions in the Gir forest descends from a population that was contracted to less than 20 animals in the first quarter of this century. The obvious interpretation is that the population reduction profoundly impacted genomic variation.

Table 9.1 lists additional examples of molecular studies on rare or endangered populations in which exceptionally low levels of molecular heterozygosity have been reported. In most of these instances, the authors provisionally attributed results to effects of genetic drift attending historical bottlenecks in population size. On the other hand, it also must be emphasized that many rare or endangered species have proved *not* to be unusually depauperate genetically. Examples of such species with more or less normal molecular heterozygosity levels include the following: a federally protected spring-dwelling fish (*Gambusia nobilis*) en-

protein-electrophoretic methods.

Species	Observation	Reference
Plants		
Bensoniella oregona (Saxifragaceae)	Complete absence of allozyme variation (24 loci) within or among populations of this endemic herbaceous perennial in southwest Oregon and northwest California	P.S. Soltis et al., 1992b
Pedicularis furbishiae (Scrophulariaceae)	Complete absence of allozyme variation (22 loci) within or among populations of this endangered hemiparasitic lousewort in northern Maine	Waller et al., 1987
Howellia aquaticus (Campanulaceae)	Complete absence of allozyme variation (18 loci) within or among populations of this rare and endangered aquatic plant in the Pacific Northwest	Lesica et al., 1988
Trifolium reflexum (Fabaceae)	Complete absence of allozyme variation (14 loci) in the only known population of this rare native clover in Ohio [however, allozyme assays (20 loci) of an endangered congener *T. stoloniferum* did reveal low to moderate levels of genetic variation]	Hickey et al., 1991
Animals		
Bison bison	Only one allozyme locus (among 24 tested) was polymorphic in a bison herd in South Dakota known to be descended from a small founder group	McClenaghan et al., 1990
Perameles gunnii	Complete absence of allozyme variation (27 loci) within an endangered, isolated population of the eastern barred bandicoot in Australia (however, a widespread and dense population of the same species in Tasmania also lacked genetic variation at these same loci)	Sherwin et al., 1991
Mustela nigripes	Only one allozyme locus (among 46 tested) was polymorphic in the one known remaining population of the highly endangered black-footed ferret	O'Brien et al., 1989
Strix occidentalis	Complete absence of allozyme variation (23 loci) in six populations of the endangered spotted owl from Oregon and California	Barrowclough and Gutiérrez, 1990

demic to the Chihuahuan desert (A.F. Echelle et al., 1989); some populations of a critically endangered flightless parrot (*Strigops habroptilus*) native to New Zealand (Triggs et al., 1989); most populations of the endangered red-cockaded woodpecker (*Picoides borealis*) in the southeastern United States (Stangel et al., 1992); Przewalski's horse (*Equus przewalskii*), which is extinct in the wild but survived by 600 animals in zoos (Bowling and Ryder, 1987); and the endangered manatee (*Trichechus manatus*) in Florida (McClenaghan and O'Shea, 1988).

In theory, the demographic details of population bottlenecks (such as their size, duration, and periodicity) should exert important influence on the severity of the expected reductions in neutral genetic variability. For example, the loss in mean heterozygosity can be rather small if population size increases rapidly following a single bottleneck of short duration (Nei et al., 1975). An empirical example of a severe population reduction that for suspected demographic reasons has not resulted in low heterozygosity involves the endangered one-horned rhinoceros (*Rhinoceros unicornis*). Prior to the 15th century, perhaps half a million of these rhinos ranged across a broad area from northwestern Burma to northern Pakistan. Land-clearing and human settlement then began to fragment and destroy rhino habitat, and by 1962 fewer than 80 animals remained, all in what is now the Royal Chitwan Park in Nepal. A surprising finding was that this herd exhibits among the highest allozyme heterozygosities reported for any vertebrate, near 10% (Dinerstein and McCracken, 1990). One possibility is that the loss of rhino habitat across the Indian subcontinent compressed surviving populations into the Chitwan area, thereby concentrating into a single locale genetic variation, some of which formerly had been distributed among regions. Whether or not this scenario is correct, these findings are significant because they again demonstrate that not all endangered taxa are genetic paupers.

Does Reduced Molecular Heterozygosity Matter?

The examples cited earlier indicate that genic heterozygosity, indeed, is reduced in populations of many (but not all) rare or threatened species. Do these findings carry any special significance for conservation efforts? Although it is tempting to assume that a paucity of genetic variation jeopardizes a species' future, the goal of firmly documenting a causal link between molecular heterozygosity and population viability remains elusive (Chapter 2). In general, there are several reasons for exercising caution in interpreting the low molecular heterozygosities reported for rare species: (a) most of the reductions in genetic variation presumably have been the outcomes rather than the causes of population bottlenecks; (b) at least a few widespread and successful species also appear to have low heterozygosities, as estimated by the same molecular methods (e.g., Fig. 2.2); (c) in some endangered species such as the northern elephant seal, low genetic variation appears not to have seriously inhibited population recovery from dangerously low levels (at least to this point in time); and (d) the fitness cost of inbreeding (Box 9.1) is known to differ widely among species, with some taxa highly susceptible but others relatively immune to fitness depression accompanying inbreeding (Laikre and Ryman, 1991; Price and Waser, 1979; Ralls et al., 1988).

An additional concern about interpreting the evolutionary significance of molecular variation is that published estimates based on any single class of markers (such as allozymes) may inadequately characterize genome-wide heterozygosity (Hedrick et al., 1986), including the variability that may underlie morphological or physiological traits of potential adaptive significance (e.g., recall the discussion of the Florida tree snails). Carson (1990) has gone further to suggest that "genetic variance available to natural selection may actually increase following a single severe bottleneck" and, thus, "character change in adaptation and speciation may, in some instances, be promoted by founder events." This conclusion stemmed from observations and experiments with bottlenecked populations of fruit flies and house flies.

For many of these and related reasons, Lande (1988) has argued that demographic and behavioral considerations should be of greater immediate importance than genetic (heterozygosity) concerns in the formulation of conservation plans for endangered populations. Lande emphasized that individuals in many species show decreased reproduction at low population densities for nongenetic reasons, such as lack of social interactions necessary for breeding, difficulties of finding a mate, or other density-dependent ecological factors collectively known as the "Allee effect" (Andrewartha and Birch, 1954). Furthermore, when populations are few in number and small in size, the possibility of species extinction through "stochastic" demographic fluctuations (irrespective of heterozygosity) may be of paramount immediate concern (Gilpin and Soulé, 1986).

On the other hand, several authors have argued forcefully that heterozygosity as measured by molecular markers *is* highly relevant to a population's health and continued survival probability and must be monitored accordingly in enlightened management programs (e.g., O'Brien and Evermann, 1988; Quattro and Vrijenhoek, 1989). In a few case studies, plausible arguments been been advanced for a direct association between observed molecular variability and the viability of an endangered taxon. For example, in the Sonoran topminnow, four measures of fitness (survival, growth, early fecundity, and developmental stability) were monitored experimentally among laboratory-reared progeny of fish representing natural populations that differed widely in heterozygosity levels as measured by allozymes (Quattro and Vrijenhoek, 1989). The authors found that all four fitness traits were correlated positively with mean heterozygosities in the populations from which they stemmed. For the Isle Royale population of gray wolves, Wayne et al. (1991b) speculated that an observed behavioral difficulty in the pair-bonding of adults might be due to a recognition-triggered instinct for incest-

Box 9.1. Inbreeding Depression

Inbreeding depression is the decrease in growth, survival, or fertility often observed following matings among relatives. The phenomenon is of special concern in conservation biology because individuals in small populations are likely to be inbred. Genetically, inbred populations have increased homozygosity (reduced heterozygosity) due to increased probabilities that individuals carry alleles that are "identical by descent" (stem from the same ancestral copy) in earlier generations of the pedigree. This probability for individual I is the inbreeding coefficient, which for known pedigrees can be calculated as:

$$F_I = \Sigma \, (1/2)^i \, (1 + F_A),$$

where the summation is over all possible paths through all common ancestors, i is the number of individuals in each path, and A is the common ancestor in each path [for computational details, see Ballou (1983) and Boyce (1983)].

Two competing hypotheses for the genetic basis of inbreeding depression have been debated for decades (see Charlesworth and Charlesworth, 1987). Under the "dominance" scenario, lowered fitness under inbreeding results from homozygosity at particular loci for deleterious recessive alleles that in outbred populations normally are masked in expression by their more common dominant counterparts. According to the "overdominance" or "heterozygous advantage" explanation, genome-wide heterozygosity levels per se are the critical influences on fitness. These two hypotheses make different predictions about the relative tolerance of populations to inbreeding (Lacy, 1992). If the expression of deleterious recessive alleles is the cause of inbreeding depression, then selection will have removed most such alleles from populations that have long histories of inbreeding, and those populations should be resistant to further inbreeding impacts. In other words, populations that survive severe inbreeding may be "purged" of deleterious recessive alleles. Under this scenario, mean heterozygosity (as estimated for example by molecular markers) should have little predictive value of a population's genetic "health." However, if inbreeding depression occurs because of a selective advantage to genome-wide heterozygosity, then previously inbred (and homozygous) populations would have reduced fitness and would fare no better under future inbreeding than would highly heterozygous populations.

In any event, different populations exhibit widely varying fitness costs associated with inbreeding. For example, in a survey of captive populations of a variety of mammalian species, relative reductions in survival in crosses between first-degree relatives (such as full sibs) varied across more than two orders of magnitude (Ralls et al., 1988, as summarized by Hedrick and Miller, 1992):

Species	Cost of Inbreeding
Sumatran tiger (*Panthera tigris sumatrae*)	0.003
Bush dog (*Speothos venaticus*)	0.06
Short bare-tailed opossum (*Monodelphis domestica*)	0.10
Gaur (*Bos gaurus*)	0.12
Pygmy hippopotamus (*Choeropsis liberiensis*)	0.33
Greater galago (*Galago c. crassicaudatus*)	0.34
Dorcas gazelle (*Gazella dorcas*)	0.37
Elephant shrew (*Elephantulus refuscens*)	0.41
Golden lion tamarin (*Leontopithecus r. rosalia*)	0.42
Brown lemur (*Lemur fulvus*)	0.90

avoidance, because the molecular findings suggest that as a result of inbreeding the extant wolves are related about as closely as siblings. For the isolated Gir forest population of Indian lions, O'Brien and Evermann (1988) concluded that the high levels of abnormal spermatozoa plus diminished testosterone concentrations observed in males (relative to lions of the African Serengeti) were attributable to intense inbreeding, because similar damaging effects on sperm development have been observed upon inbreeding of mice and livestock.

Perhaps the most intriguing case for a causal link between inbreeding, low molecular heterozygosity, and diminished population fitness involves the cheetah. As mentioned above, molecular evidence for severely reduced heterozygosity in the cheetah is multifaceted and includes RFLP assays of the major histocompatibility complex (a DNA region encoding cell-surface antigens involved in the immune response). Extreme monomorphism at the MHC is indicated further by the remarkable acceptance of skin grafts among "unrelated" individuals. Recently, an infectious disease (feline infectious peritonitis or FIP) caused by a coronavirus swept through several captive cheetah colonies and caused 50–60% mortality over a three-year period. The same virus in domestic cats (which have normal levels of MHC variation as indicated by graft rejections and molecular assays) has an average morbidity of only 1%. O'Brien and Evermann (1988) speculate that a FIP virus might have acclimated initially to one cheetah, and then spread rapidly to other individuals who were genetically uniform in their immunological defenses. In general, enhanced susceptibility to infectious diseases or parasitic agents probably constitutes one of the most serious of challenges faced by a genetically depauperate population (O'Brien and Evermann, 1988).

Interestingly, emphasis on a special adaptive significance for immunorecognition genes has led to the suggestion that captive breeding programs for endangered species should be designed with the "specific goal of maintaining diversity at MHC . . . loci," due to the possibility "that an individual heterozygous at all or most MHC loci will be protected against a wider variety of pathogens than a homozygous individual" (Hughes, 1991). Furthermore, according to Hughes (1991), "at most loci loss of diversity should not be a cause for concern, because the vast majority of genetic polymorphisms are selectively neutral." This provocative suggestion immediately was criticized on several grounds (Gilpin and Wills, 1991; Miller and Hedrick, 1991; Vrijenhoek and Leberg, 1991), not the least of which is the gross assumption that variability at genes other than the MHC is adaptively irrelevant. Several other loci are known to contribute to disease resistance itself, and polygenes underlying numerous other quantitative traits of potential adaptive relevance should hardly be ignored so cavalierly (Vrijenhoek and Leberg, 1991). Furthermore, selective breeding designed explicitly to maintain MHC diversity could have the counterproductive consequence of accelerating inbreeding, with concomitant accelerated loss of diversity elsewhere in the genome.

At this point, we are left with the conclusion that de facto management for genic heterozygosity through avoidance of unnecessary inbreeding and maintenance of large N_e probably is an important element in conservation programs for rare or threatened species. With sufficient effort, molecular markers can provide a useful index to genome-wide levels of genetic variation, particularly when multiple assays are employed and concordant results are discovered among the different methods. Thus, molecular approaches can help to identify those natural or captive populations that might be at special risk of extinction due to severe genetic impoverishment from past population bottlenecks and/or inbreeding. Nonetheless, implementation of heterozygosity guidelines for managed populations should not come at the expense of equally important behavioral or demographic considerations. Furthermore, when population histories are known (as is often the case in zoos), direct pedigree analyses rather than use of particular molecular markers should in theory provide a more robust guide to the true magnitude of heterozygosity loss through inbreeding, as well as to the breeding priorities of particular individuals for purposes of maximizing genetic variation within a population (Haig et al., 1990; Hedrick and Miller, 1992). For example, two breeding strategies that have been suggested for captive populations whose pedigrees are known involve the design of crosses to (a) equalize the expected genetic contributions from the original founder individuals (Lacy, 1989) or (b) emphasize reproduction by individuals who are judged to be of special genetic importance by virtue of unusual position in the pedigree (Geyer et al., 1989).

In the early literature of conservation genetics, Franklin (1980) and Soulé (1980) suggested that a minimum effective population size of 50 would be required to stem inbreeding depression, and Franklin (1980) added that an effective population size of 500 would prevent the long-term erosion of variability by genetic drift. These specific management guidelines (Franklin and Soulé, 1981), which became known as the "50/500 rule," have had great influence in the field of practical conservation science (Simberloff, 1988). Nonetheless, as should be clear from the above discussion, no single set of guidelines is likely to be valid universally. Furthermore, as pointed out by Varvio et al. (1986), such detailed instructions represent gross simplifications based on several untested assumptions, such that "any single principle should not be imposed as a general guideline for the management of small populations."

ISSUES OF PHYLOGENY

Debate over the relative importance of demographic versus genetic concerns in endangered species management (see above) has had at least two undesirable consequences: (a) it has tended to dichotomize genetics and demography, thereby placing them in an antagonistic relationship (when in reality both classes of concern are important, and indeed may sometimes be intimately interrelated) and (b) it has left the unfair impression that the only task for molecular genetics in conservation biology is in heterozygosity assessment (see Avise, 1989c). However, molecular markers can play many additional roles, involving phylogeny estimation at any of the hierarchical levels that have formed the organizational format for this book. The following sections provide examples that involve rare or endangered species.

Parentage and Kinship

Parentage assignments sometimes assume conservation or management relevance. For example, Longmire et al. (1992) used DNA fingerprinting to assess paternity within a small captive flock of the endangered whooping crane (*Grus americana*) that had been maintained at the Patuxent Wildlife Center in Maryland since 1965. A substantial investment is required to produce young cranes in captivity because of the species' long generation length and low reproductive output. In an attempt to maximize production of fertile eggs, adult females often had been inseminated with semen from several males. However, this procedure also made paternity uncertain, with the undesirable consequence that breeding plans based on maximizing N_e (i.e., avoidance of inbreeding) were compromised. The molecular genetic study partially rectified this situation by providing a posteriori knowledge of biological parentage within the flock. In turn, such pedigree information may prove useful in the design of subsequent matings.

Studies of parentage and kinship through molecular genetic markers often can provide useful background information about the biologies and natural histories of threatened species. The rare blue duck (*Hymenolaimus malacorhynchos*) of New Zealand inhabits isolated mountain streams, and recent DNA-fingerprinting assays have provided new insight into the family-unit structure of this species. From patterns of DNA band-sharing, significantly higher genetic relatedness was documented within than between blue duck populations from different rivers, likely due to a social system that includes limited dispersal from the natal site and matings among relatives (Triggs et al., 1992).

In some cases, molecular genetic findings on extended kinship have clear and immediate relevance for conservation strategies. For example, analyses of mtDNA have shown that rookeries of the endangered green turtle are characterized by distinctive maternal lineages, indicating a strong propensity for natal homing by the females of this highly migratory species (Chapter 6). Irrespective of the level of interrookery gene flow mediated by males, this structure of rookeries along matrilines indicates that green turtle colonies must be considered independent of one another demographically [because any extirpated colony would unlikely be reestablished naturally by females hatched elsewhere, at least over the ecological timescales relevant to human interests (Fig. 9.1)]. This genetics-based deduction is consistent also with the observation that rookeries exterminated by man over the past four centuries (including those on Grand Cayman, Bermuda, and Alto Velo) have yet to be recolonized. Because the continuing decline of many marine turtle colonies through overharvesting will not be ameliorated significantly by recruitment from foreign rookeries, each desired remaining colony will have to be protected individually (Bowen et al., 1992).

Population Structure and Intraspecific Phylogeny

The extended kinship structures of blue ducks and green turtles are a reminder that issues of kinship grade into those of geographic population structure and intraspecific phylogeny. At these levels too, many applications of molecular markers have been employed in a conservation context. For example, particu-

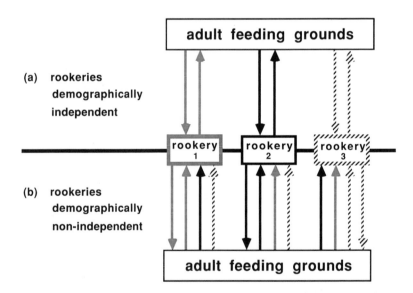

Figure 9.1. Alternative scenarios for the genetics and demography of green turtle rookeries (after Meylan et al., 1990). Coded arrows indicate possible migration pathways of females between natal and feeding grounds. (a) In the diagram above the heavy horizontal line, females are assumed to home to natal sites, in which case rookeries would be independent both genetically (with respect to mtDNA) and demographically. (b) In the diagram below the horizontal line, females commonly exchange among rookeries, yielding genetic and demographic non-independence. Molecular data from mtDNA are consistent with the former scenario.

larly in agronomically important plant species much effort has been devoted to the collection and long-term storage of "seed banks" from which future genetic withdrawals may prove invaluable in development of needed strains (Frankel and Hawkes, 1975). In organizing the deposits to such germplasm collections, genetic diversity may be maximized through knowledge of how natural variation is partitioned within and among plant populations, a task for which molecular genetic markers are well suited (Schoen and Brown, 1991). In general, by revealing how genetic variation is partitioned within any plant or animal species (Chapter 6), molecular methods can help to characterize the intraspecific genetic resources that conservation biology seeks to preserve.

GENETIC PARTITIONS WITHIN RARE AND THREATENED SPECIES

Numbers of the highly endangered black rhinoceros (*Diceros bicornis*) and white rhinoceros (*Ceratotherium simum*) in Africa have been decimated by poachers, who supply the lucrative markets that convert rhino horns to ornamental dagger handles and supposed medicines. Assuming that poaching may somehow be brought under control, additional genetic and demographic problems in the few remaining populations also could threaten these species' continued survival. Should all remaining conspecific rhinos be considered members of a single population for purposes of breeding and management? Or should the recognized extant forms (including the geographic subspecies *D.b. michaeli* and *D.b. minor* in the black rhino, and *C.m. cottoni* and *C.m. simum* in the white rhino) be managed as separate populational entities, perhaps because of strong genetic differentiation? The strategy of mixing and breeding conspecific rhinos from separate geographic sources might enhance each species' chance of survival by increasing effective population sizes, and thereby forestalling stochastic demographic extinctions and inbreeding depression. On the other hand, attempted amalgamations of well-differentiated genetic stocks could lead to outbreeding depression, or cause an overall erosion of genetic variability most of which might be distributed among rather than within geographic regions. Toward the goal of empirically assessing genetic differentiation among rhino populations, preliminary studies of proteins and DNA have been conducted. Merenlender et al. (1989) reported a small genetic distance ($D = 0.005$) between the two white rhino subspecies based on allozymes, and Ashley et al. (1990) reported only low levels of mtDNA sequence divergence ($p < 0.004$) for three black rhino populations based on restriction-site comparisons. However, due to the preliminary nature of the data, the lack of obvious genetic differentiation, and the fact that more extensive genetic surveys are underway, the available results should not be taken to provide an unambiguous guide to management strategy.

The piping plover (*Charadrius melodus*) of North America has been the subject of recent molecular analyses based on allozymes. Breeding birds have all but disappeared from the Great Lakes region in the past 50 years, but threatened or endangered populations persist in the interior Great Plains and along the Atlantic seaboard. Low variances in inter-populational allele frequencies at polymorphic loci (mean $F_{ST} \cong 0.02$), and low overall genetic distances ($D < 0.01$) indicate that geographically disjunct populations nonetheless have been in extensive and recent genetic contact (Haig and Oring, 1988). This finding may not be too surprising, given direct dispersal data suggesting low natal philopatry for members of this species, as well as a considerable mixing of birds from different breeding regions on the wintering grounds. Allozymic studies of the rare wood stork (*Mycteria americana*) in Florida reached similar conclusions. This species nests in spatially discrete colonies, but negligible differentiation among rookeries (mean $F_{ST} = 0.02$; $D < 0.001$) indicated that the colonies were connected by high gene flow and/or have separated only recently (Stangel et al., 1990).

Molecular studies on several other rare or endangered species *have* revealed significant population genetic structures apparently resulting from behavioral and/or historical geographic influences. The striking matriarchal population structure of endangered green turtle rookeries within ocean basins, attributable at least in part to natal homing, already has been mentioned, and a similar though less pronounced behavior-based genetic architecture also characterizes loggerhead turtles (*Caretta caretta*) in the North Atlantic Ocean and Mediterranean Sea (Bowen et al., 1993b). In the desert tortoise (*Xerobates agassizi*) of the American southwest, a dramatic difference in mtDNA haplotype composition between populations east versus west of the Colorado River presumably reflects a historical biogeographic separation of populations, as well as the limited vagility of members of this species (Lamb et al., 1989).

In a complex of endangered desert pupfishes (genus *Cyprinodon*) in and near Death Valley, California, allozymic studies have revealed little polymorphism within but considerable genetic divergence among most populations, which currently are confined to isolated springs and streams that are the remnants of inland lakes and connected watercourses in former pluvial times (Turner, 1974). A different pattern of geographic structure within one of the desert pupfish species may be an exception that proves the rule. In *C. macularius* populations of the Salton Sea area of southern California, polymorphism within remnant colonies accounted for 70% of the total genetic variance, with differences among colonies contributing only 30% [Echelle's et al. (1987) analysis of Turner's (1983) allozyme data]. The hydrologic history of this region suggests an explanation: The populations probably have been in repeated contact due to historical cycles of flooding of the lake basin, perhaps most recently earlier in this century when water broke out of the Colorado River irrigation system (Turner, 1983).

In general, fishes in the desert basins of North America are declining at an alarming rate, with more than 20 taxa having gone extinct in the last few decades and many more at risk of the same fate. Meffe and Vrijenhoek (1988) have considered management recommendations that might stem from two different types of population structure exhibited by the remaining species, and these scenarios should apply to other biological settings as well. In the "Death Valley model" (patterned after the desert pupfish), populations are small and isolated, such that most of the overall genetic diversity is likely to be partitioned among sites. In managing such populations, there is no need for concern about human-mediated interruptions of gene flow because genetic contact has not existed naturally since the ancestral watercourses desiccated. On the contrary, precautions should be taken to avoid gene flow among populations that are isolated naturally and may, therefore, be in the process of adaptive radiation. The main concern should be the maintenance of high N_e within localities to alleviate both the genetic and demographic dangers of small population size.

Under an alternative "stream hierarchy model" (modeled after the endangered *Poeciliopsis occidentalis* and certain members of the *Cyprinodon* complex), populations in dendritic water systems exhibit varying degrees of connectedness and gene flow, such that a larger fraction of the overall genetic diversity occurs within colonies. Meffe and Vrijenhoek (1988) suggest that management programs in such situations should be aimed at preserving the genetic integrity of each species while at the same time maintaining its genetic variability. Thus, movement of conspecific individuals among colonies within a river basin should pose no special difficulties, because this probably occurs naturally. However, precautions should be taken to avoid artificial movements between separated drainages, particularly when this involves mixing of different species that might compete or hybridize.

Molecular studies of population genetic structure also have been conducted on endangered plants. For example, the meadowfoam *Limnanthes floccosa californica* is a geographically restricted annual endemic to vernal pools in Butte County, California. The taxon is of special agronomic interest as a source for a sperm whale oil substitute. Among the total of 11 known populations, protein electrophoretic analyses revealed that nearly all (> 95%) of the total genetic diversity was distributed among (rather than within) populations, such that estimates of interpopulational genetic exchange were extremely low ($Nm < 0.02$; Dole and Sun, 1992). Based on these findings, the authors recommended a conservation plan that emphasizes preservation of as many populations as possible, even at the possible expense of fewer total individuals.

STOCK IDENTIFICATION

Particularly in the field of fisheries management, much discussion has centered on questions of how best to recognize and characterize population "stocks." To develop quotas and other management guidelines for a sustainable fishery, biologists must be able to identify and demographically characterize the exploited populations. The need for such population information has become more critical each year, as growing numbers of commercial and sport fisheries approach economic collapse through overharvesting. Genetic stock assessment [GSA (Utter, 1991)] represents a practical application of the principles and procedures of population structure analysis (Chapter 6). A large literature devoted explicitly to GSA in fishery biology has been reviewed elsewhere (e.g., Ovenden, 1990; Ryman and Utter, 1987; Shaklee, 1983), so only selected examples and general conclusions will be presented here.

Mixed-Stock Fisheries   Many commercial or sport fisheries entail mixtures of native and introduced (hatchery-produced or otherwise transplanted) stocks. In several instances, genetic markers that distinguish the introduced from native populations have been employed to monitor the fate of the introductions and to assess whether hybridization and introgression subsequently have taken place. Results of such genetic appraisals have varied. Among trout (*Salmo trutta*) in the Conwy River of North Wales, the introduction of fry from anadromous populations into landlocked populations resulted in considerable hybridization between the two trout forms, as gauged by allozyme markers (Hauser et al., 1991). Thus, a stocking program that had been designed to bolster catches of trout in landlocked bodies of water may have come at the risk of introgressive loss of the unique genetic character of the native landlocked forms. On the other hand, for this same species in Spain, hatchery supplementation appears to have been a ''failure'': Allozymic analyses showed that genetically marked fry introduced from hatcheries into indigenous populations failed to reach sexual maturity and apparently did not contribute to the pool of catchable and reproductive fish (Moran et al., 1991). In the western United States, a widespread practice of stocking and transplanting non-indigenous trout has resulted in extensive introgressive hybridization with native forms, as judged by allozyme and mtDNA evidence (see section on hybridization and introgression later in this chapter). On the other hand, similar molecular studies of the Japanese ayu (*Plecoglossus altivelis*) demonstrated that introduced stocks contributed little to reproduction in a native river population of this fish species (Pastene et al., 1991).

Several fisheries involve exploitation of mixed native stocks, an example being the taking of anadromous salmon at sea. A long-sought goal in salmon management has been to find diagnostic genetic or other markers that would cleanly distinguish fish stemming from different rivers, such that the proportionate contribution of various breeding populations to the mixed oceanic fishery could be determined. With such knowledge in hand, harvesting strategies might then be geared to the varying strengths (population sizes) of the respective breeding stocks. Unfortunately, native salmon populations within a management jurisdiction seldom have proved to exhibit fixed allelic differences between rivers that would make such stock assignments unambiguous. Rather, observed genetic differences normally involve shifts (often statistically significant) in frequencies of alleles at polymorphic loci, such that overall relationships among the source populations must be summarized merely in a quantitative phenetic sense (Fig. 9.2). Several statistical approaches have been developed for estimating the proportionate contributions to a mixed fishery from breeding stocks that differ in frequency distributions of alleles or other measured characters (e.g., Millar, 1987; Pella and Milner, 1987).

Inherited Versus Acquired Markers   An important challenge, faced particularly in the field of fisheries management, has been to distinguish carefully between the kinds of information provided by different classes of populational markers (Ihssen et al., 1981). One useful start at organizing such thought is to make a distinction between inherited (genetic) versus "acquired" or "environmentally-induced" characters (Booke, 1981). The latter include a variety of attributes commonly employed in stock discrimination; for example, parasite loads, heavy metal concentrations, and perhaps many morphological characteristics such as vertebral counts that may be developmentally or phenotypically plastic (T.P. Quinn et al., 1987). They also include the physical tags and radio transmitters applied to individuals to monitor their movements. Acquired markers unquestionably serve an important role in population analysis because they can reveal where individuals have spent various portions of their lives. Furthermore, acquired and genetic markers often should disclose concordant population partitions. For example, Bermingham et al. (1991) used molecular markers to assay physically tagged salmon caught in the West Greenland fishery—among 68 tagged fish examined, 67 were correctly assigned by mtDNA genotype to the true continent of origin (North America versus Europe). Nonetheless, because physical tags or other acquired characteristics are not transmitted across generations, they do not ineluctably delineate the reproductive units that also are relevant to population management programs, nor for this reason can they reveal the principal sources of phylogeographic diversity within a species.

Figure 9.3 illustrates four possible relationships between the apparent stock structures registered by genetic versus environmentally induced markers. These two classes of marker may reveal high and concordant structures, as for example when populations have been geographically isolated (leading to genetic differences) and the disjunct habitats induce different acquired phenotypes. Such an outcome might be anticipated for isolated or sedentary species (e.g., freshwater fishes in different drainages). Or, both genetic and acquired markers might reveal relative homogeneity over broad areas, as in many vagile marine species. However, discordant outcomes also are conceivable, as when significant population structure is evidenced by genetic markers despite the absence of populational differences in acquired characteristics. This outcome might arise in a natally-homing anadromous species assayed at the freshwater adult life stage, provided that the acquired characters were incorporated during the oceanic portion of the life cycle. Conversely, significant geographic structure in acquired characters

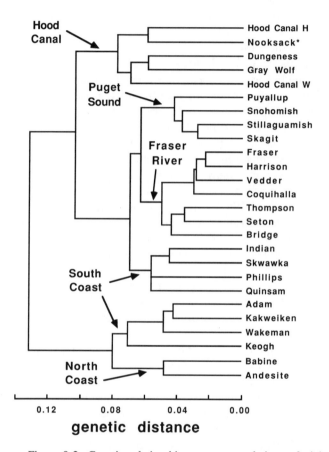

Figure 9.2. Genetic relationships among populations of pink salmon (*Oncorhynchus gorbuscha*) from 26 streams in Washington and British Columbia (after Shaklee et al., 1991). Shown is a cluster phenogram based on allelic frequencies at more than 50 allozyme loci. Indicated by arrows are five major geographic regions for which reasonably strong genetic clustering is evident. The Nooksack River (indicated by asterisk), however, enters Puget Sound.

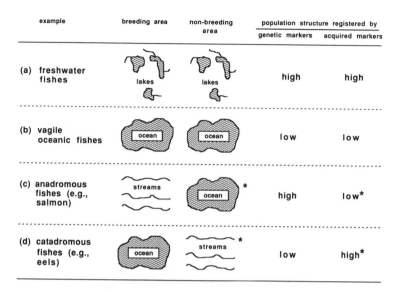

| example | breeding area | non-breeding area | population structure registered by | |
			genetic markers	acquired markers
(a) freshwater fishes	lakes	lakes	high	high
(b) vagile oceanic fishes	ocean	ocean	low	low
(c) anadromous fishes (e.g., salmon)	streams	ocean *	high	low*
(d) catadromous fishes (e.g., eels)	ocean	streams *	low	high*

Figure 9.3. Four possible categories of relationship between magnitudes of apparent population structure as registered by the distributions of inherited versus environmentally acquired markers (see text). For categories (c) and (d) in the figure, it is assumed that the environmental markers were acquired during the oceanic and freshwater life stages, respectively, as indicated by the asterisks.

might be evidenced despite an absence of significant genetic differentiation. This outcome might arise in a random-mating catadromous species sampled during the freshwater portion of its life cycle, provided that the acquired characters were incorporated during the freshwater phase. In other words, because acquired phenotypic markers by definition are appropriated through contact with the environment, they may not reflect true population genetic partitions when the appropriations took place at a locality or life stage other than where reproduction occurs.

These distinctions between genetic and acquired markers could in some situations assume management and conservation relevance. For example, regarding anadromous salmon originating in different rivers, diagnostic populational markers that happened to coincide with political (state or federal) boundaries could be ideal for equitably apportioning oceanic catches to fishermen from the relevant jurisdictions. For these purposes, any marker (genetic or otherwise) would do. Thus, naturally-acquired stream-specific parasite loads, and/or dyes or tags artificially applied to smolts (juvenile salmon in streams), would be perfectly suitable for these management purposes. Yet, the "stocks" so identified could have little or no evolutionary genetic significance if, for example, sufficient gene flow between stream populations occurs via "mistakes" in natal homing. Thus, in principle, an anadromous species with dramatic spatial structure in acquired characteristics nonetheless might remain panmictic throughout its range. Both inherited and acquired markers can find gainful employment for various management objectives, but only genetic markers are of immediate relevance to population genetic and evolutionary issues.

Shallow Versus Deep Stock Structures   Another general point about population stocks relevant to conservation biology involves the distinction emphasized in Chapter 6 between shallow (contemporary) versus deep (long-term historical) population genetic structures. Populations of many species may be isolated strongly from one another at the present time (and at most points of time in the past), but nonetheless remain tightly connected in a phylogenetic sense through recent or pulsed episodes of gene flow. Both shallow and deep evolutionary separations are relevant to management strategies. The shallow genetic separations are important because they indicate that populations not strongly connected by gene flow at the present time are unlikely to recover from overexploitation by significant natural recruitment from other populational sources. Deeper genetic separations within a species are significant also, because they register the major sources of evolutionary genetic diversity that ultimately may be of greatest importance to long-term conservation efforts. These new phylogeographic perspectives on population genetic stocks are just beginning to make an appearance in the literature of fisheries management (Avise, 1987; Dizon et al., 1992).

CONCLUSIONS ABOUT INTRASPECIFIC PHYLOGEOGRAPHY

Ehrlich (1992) recently lamented that time is running out for saving biological diversity and that "The sort of intensive, species-focused research that I and my colleagues have carried out . . . [on checkerspot butterflies] . . . appears to have a very limited future in conservation biology. Instead, if a substantial portion of remaining biodiversity is to be conserved, detailed studies of single species must be replaced with "quick and dirty" methods of evaluating entire ecosystems, designing reserves to protect them, and determining whether those reserves are working." Willers (1992) goes further by cautioning that "To dwell endlessly on the tasks of obtaining more and ever more data for the expressed purpose of managing a biological reserve is to suggest that enough knowledge is just around the corner. This is not so." In a sense, these authors are quite right. The severe and broad problems faced by conservation biology cannot be solved solely by detailed genetic and ecological studies of particular taxa or regions. Indeed, the lack of political and social will to implement existing scientific understanding (including the need for human population control) is certainly a far greater impediment to the preservation of global biodiversity than is a lack of detailed scientific information per se.

On the other hand, there are some important and general lessons to be gained from detailed case histories of the sort described in this chapter. To emphasize this point, consider again the genetic studies conducted over the past decade on the phylogeography of the freshwater and maritime faunas of the southeastern United States (Chapter 6). Several previously underappreciated perspectives of general conservation relevance have flowed from these and related studies based on molecular markers (Avise, 1992). Thus, it has become abundantly clear that most species should not be viewed as undifferentiated monotypic entities, but rather as consisting of a series of geographically varying populations with a hierarchical and sometimes deep genetic structure. With recognition of this fact come several general management guidelines for conservation of this phylogeographic diversity:

Limit Unnecessary Transplantations   Although most biologists recognize that introductions of exotic species can cause irreparable harm to regional biodiversity by forcing extinctions of native species, they have been slower to appreciate the problems that can stem from transplantation and mixing of genetic stocks within species. In fact, many public and private management agencies actively promote geographic transplantations from one area to another within a species' range for purposes such as bolstering local population sizes, introducing "desirable" genetic traits into a region, or increasing local heterozygosity. Unfortunately, several undesirable consequences also may arise from such geographic transplantations, including the possibility of disease or parasite spread, the irretrievable loss of the rich historical genetic records of populations, and inevitable erosion of the overall genetic diversity within a species (much of which we now recognize to be generated and maintained through historical geographic separations). Transplantations sometimes may be justified, as in the reintroduction of a native species to its former range where it had been extirpated by humans. However, a developing perspective is that the burden of proof in any proposed transplantation program normally should rest on advocates of this strategy rather than on those who would question the desirability of transplantations on the grounds cited above.

Design Regional Reserves   Molecular data on the fauna of the southeastern United States indicate that particular areas are geographic centers for a substantial fraction of regional, intraspecific biogenetic diversity. These biotic provinces also have been recognized previously by a different type of biogeographic data—concentrations of species' distributional limits (see Avise et al., 1987a). Such concordant lines of evidence for historical centers of biodiversity already give ample support for special efforts to preserve the integrity of these regional biotas. Such efforts might include the design of regional biological preserves, as well as the implementation of strict guidelines to discourage unnecessary transplantations between phylogeographic areas. Although such regional perspectives on genetic diversity cannot hope to capture the idiosyncratic population structures and subdivisions of all species, they can provide useful general guidelines for management strategies, particularly as natural environments come under increased pressure and decisions of conservation triage become inevitable.

Speciation and Conservation Biology

Just as issues of close kinship grade into those of population structure and intraspecific phylogeny, so too do the latter grade into issues regarding the speciation process and associated taxonomic judgements. In the field of conservation biology, these taxonomic issues often come into especially sharp focus in discussions of "endangered species" (Box 9.2).

RECOGNITION OF ENDANGERED SPECIES

Many taxonomic assignments in use today (including the formal designations of species and subspecies) were first proposed in the last century, often from limited phenotypic information and preliminary assessments of geographic variation. How adequately do these traditional taxonomies summarize true biogenetic diversity? This question remains to be answered for most groups, through continued systematic reappraisals to which molecular approaches can contribute significantly. The problems and challenges are far more important than a mere concern with nomenclature might at first imply. For better or for worse, nomenclatural assignments inevitably shape our most fundamental perceptions of how the biological world is organized. Ineluctably, formal names summarize the biotic units that are perceived and, therefore, discussed. In a conservation context, these perceived entities provide the pool of candidates from which are chosen the populations toward which special management efforts may be directed. A "dusky seaside sparrow" by any other name is just as melanistic, but without a taxonomic name this population would not likely have been recognized as a biotic unit worthy of particular conservation attention.

In the case of the dusky seaside sparrow, this dark-plumaged population endemic to Brevard County, Florida first was described in the late 1800s as a species (*Ammodramus nigrescens*) distinct from other seaside sparrows (*A. maritimus*) common along the Atlantic and Gulf coasts of North America. Although the dusky later was demoted to formal subspecific status (*A.m. nigrescens*), the nomenclatural legacy stemming from the original taxonomic description prompted continued special focus on this federally "endangered species" when, during the 1960s, the Brevard County population declined severely due to deterioration of its salt marsh habitat. In 1987, the last known dusky died in captivity, as last-ditch efforts to save the population through captive breeding failed.

Following extinction of the dusky seaside sparrow, a retrospective molecular study of nearly the entire seaside sparrow complex (which includes nine con-

Box 9.2. The United States Endangered Species Act of 1973

In 1966, passage of the Endangered Species Preservation Act (P.L. 89-669, Sections 1–3) initiated federal efforts in the United States to protect rare species from extinction. Three years later, attempts to remedy perceived deficiencies in this act (e.g., lack of habitat protection and applicability only to native wildlife) resulted in the Endangered Species Conservation Act of 1969 (P.L. 91-135, 83 Stat. 275). Again, shortcomings of the new act were recognized and, in 1973, Congress enacted a more comprehensive "Endangered Species Act" (henceforth ESA) that today remains the country's strongest conservation strategy for rare species.

The ESA was intended to "provide a means whereby the ecosystems upon which endangered species and threatened species depend may be conserved, [and] to provide a program for the conservation of such . . . species. . . ." (16 U.S.C. Section 1531[b]). To qualify for ESA protection, a species must appear on an official list, which is prepared and updated under auspices of the U.S. Fish and Wildlife Service (Interior Department) and the National Marine Fisheries Service (Commerce Department). Regulations and official interpretations of the ESA are made by these agencies as well, who receive legal advice from the Solicitor's Office and the Office of General Council within the Departments of Interior and Commerce, respectively.

Listings under the ESA may be made for species, subspecies, or "distinct population segments." For example, grizzly bears in the lower 48 states are listed as threatened, whereas the large Alaska population receives no protection under the ESA. Criteria for listing are that the plant or animal in question occurs in numbers or habitats sufficiently depleted to critically threaten survival. The current lists include over 600 species within U.S. boundaries and 500 species that occur elsewhere in the world. Another 600 + species are acknowledged by federal agencies to be just as vulnerable but have yet to be listed, and at least 3000 more remain on a waiting list pending additional investigation (Gibbons, 1992).

The ESA has been criticized on numerous grounds (e.g., Geist, 1992; O'Brien and Mayr, 1991; Rohlf, 1991), including (a) its focus on the preservation of particular species rather than on ecosystem health or biodiversity, (b) its emphasis on a few high-profile species such as birds and large mammals, (c) the lack of clear guidelines as to what constitutes a distinct species or population, or precisely what constitutes being threatened or endangered, and (d) the considerable timelag involved in the formal listing process. On the other hand, some of these problems appear to be those of implementation (including funding, political will, and incomplete scientific knowledge) rather than the legislation itself (O'Connell, 1992). In the future, the species-based approach of the ESA may well be replaced or supplemented with legislation designed more explicitly to recognize and encompass the broader goals of ecosystem maintenance and biodiversity protection. But at least as an interim measure, the ESA has represented a revolutionary and enlightened piece of governmental legislation.

ventionally recognized subspecies) produced a surprise (Avise and Nelson, 1989). In terms of mtDNA sequence, the dusky proved essentially indistinguishable from other Atlantic coast genotypes, whereas all Atlantic coast genotypes clearly were distinct phylogenetically from those observed along the Gulf of Mexico coastline (Figs. 6.8 and 9.4). These results likely are attributable to an ancient (Pleistocene) population separation, as evidenced further by a striking similarity between the phylogeographic pattern of the seaside sparrow and those of several other estuarine species complexes in the southeastern United States (Chapter 6). Thus, the traditional taxonomy for the seaside sparrow complex (upon which the endangered species status and the management efforts were based) apparently had failed to capture the true phylogenetic partitions within the group, in two respects: (a) by giving special emphasis to a presumed biotic partition that was shallow or nonexistent and (b) by failing to recognize a deep phylogeographic subdivision (between Atlantic and Gulf coast birds) that should prove most important in any future conservation plans for remaining members of the seaside sparrow assemblage.

Because of the special political and emotional sensitivities surrounding the dusky seaside sparrow (or other well-publicized endangered species), an additional note should be provided here. The molecular findings in no way excuse or justify any poor land-management practices [chronicled by Walters (1992)] that may have led to the dusky's extinction, nor should publication of the molecular results be interpreted as heartlessness over this population's loss. The extinction of any natural population is regrettable, particularly in this age of rapid deteri-

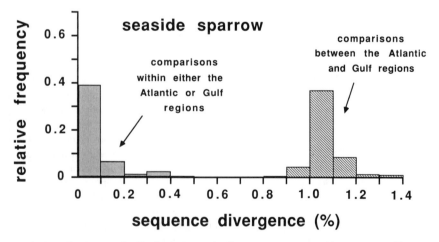

Figure 9.4. Frequency distribution of genetic distances among seaside sparrows. Shown are estimates of mtDNA divergence between pairs of birds from locales along the Atlantic and Gulf of Mexico coasts of the southeastern United States.

oration of natural habitats and biodiversity. However, the field of conservation biology finds itself in a difficult situation, with hard choices to be made about which taxa can be saved, and how best to allocate limited resources among competing conservation needs. Especially in such circumstances, management decisions should be based on the best available scientific information.

Another example of a misleading taxonomy for an "endangered species" involved the pocket gopher *Geomys colonus* endemic to Camden County, Georgia. This taxon was first described in 1898 on the basis of rather cursory descriptions of pelage and cranial characteristics. The population remained essentially unnoticed and unstudied until the 1960s, when gophers in Camden County were "rediscovered." The population referable to *G. colonus* then consisted of less than 100 individuals and was listed as an endangered species by the State of Georgia. Subsequently, a molecular genetic survey was conducted using allozyme assays, karyotypic examination, and RFLP analyses of mtDNA. None of these genetic methods detected any consistent genetic distinction between "*G. colonus*" and geographically adjacent populations of a common congener, *G. pinetis* (Laerm et al., 1982). Results were not attributable to a lack of sensitivity in the techniques employed because each molecular method revealed dramatic genetic differences among a broader geographic array of *G. pinetis* populations (particularly those in eastern versus western portions of the species' range; e.g., Fig. 6.6). The conclusion in this case was clear: "*G. colonus*" did not warrant recognition as a distinct species. Either its description in 1898 was inappropriate, or an original valid *G. colonus* species had gone extinct early in this century and been replaced by recent *G. pinetis* immigrants into Camden County.

Of course, in principle as well as practice, no study can prove the null hypothesis that genetic differences between putative taxa are absent. One example of a "valid" species that appeared nearly indistinguishable in genetic composition from a close relative involved the endangered pallid sturgeon (*Scaphirhynchus albus*). In allozyme assays of 37 monomorphic and polymorphic loci, *S. albus* could not be separated from its more common congener, the shovelnose sturgeon *S. platorynchus* (Phelps and Allendorf, 1983). Nevertheless, on the basis of pronounced morphological differences between these taxa and a sympatric distribution, the pallid and shovelnose sturgeons appear to qualify as "good" biological species [the possibility of a phenotypic polymorphism was deemed unlikely (Phelps and Allendorf, 1983)]. The molecular results suggest either some degree of introgressive hybridization between these sturgeons or that a complete separation was accomplished only recently in evolution. In general, molecular data should not be interpreted in isolation in making taxonomic judgments, but rather should be integrated with other available lines of evidence.

Molecular reappraisals of taxonomically suspect species may, of course, also bolster the rationale for special conservation efforts. One case in point involves the nearly extinct silvery minnow (*Hybognathus amarus*) endemic to the Rio

Grande River in the southwestern United States. This species has had a troubled taxonomic history, with some researchers viewing it as a distinct species and others placing it in synonymy with *H. nuchalis* and *H. placitus*. However, based on an allozyme survey of 22 loci, Cook et al. (1992) observed several fixed allelic differences between these taxa, and overall levels of genetic distance (*D* > 0.10) somewhat greater than those normally distinguishing conspecific populations in other fish groups. The authors concluded that there is little justification for considering *H. amarus* conspecific with other species with which it previously had been placed in synonymy.

Another case in point involves the highly endangered Kemp's ridley turtle (*Lepidochelys kempi*), which nests almost exclusively at a single locale in the western Gulf of Mexico, and currently is the subject of the largest international preservation effort for any marine turtle. Morphological similarities between *L. kempi* and the related *L. olivacea,* and a geographic distribution of these supposed sister taxa that makes little sense under modern conditions of climate and geography (*L. olivacea* is distributed globally in warm waters), have raised questions about the true evolutionary distinctiveness of *L. kempi*. Nonetheless, a reappraisal of the complex based on mtDNA assays indicated that *L. kempi* is more distinct from *L. olivacea* than are assayed Atlantic versus Pacific populations of the latter from one another (Bowen et al., 1991). Furthermore, the differentiation between the Kemp's and olive ridleys surpassed (slightly) levels of genetic divergence observed among any conspecific populations of the globally distributed green or loggerhead turtles (Fig. 9.5).

Neglected taxonomies for endangered forms also can kill, as exemplified by studies of the tuataras of New Zealand. These impressive lizards have been treated by government and management authorities as belonging to a single species, despite molecular (and morphological) evidence for three distinctive and taxonomically described groups (Daugherty et al., 1990). Official neglect of this genetic diversity may unwittingly have consigned one form of tuatara (*"Sphenodon punctatus reischeki"*) to extinction, whereas another form (*"S. guntheri"*) appears to have survived to this point only by sheer good fortune. As noted by Daugherty et al. (1990), good taxonomies "are not irrelevant abstractions, but the essential foundations of conservation practice."

MOLECULAR FORENSICS AND LAW ENFORCEMENT

A common problem in field enforcement of wildlife laws involves identifying the species of origin of blood, carcasses, or meat when more obvious morphological characters such as feathers or fur are not available. Because of the species-diagnostic power of several molecular methods (Chapter 7), assays of proteins or DNA recovered from problematic tissues should find wide application in wildlife forensics. For example, Cronin et al. (1991c) compiled a list of

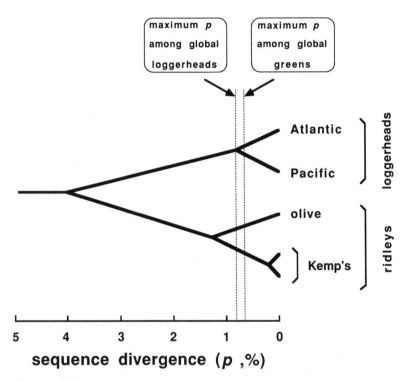

Figure 9.5. Phylogeny for ridley and loggerhead marine turtles estimated from mtDNA data (after Bowen et al., 1991). Assayed olive ridleys from the Atlantic and Pacific Oceans proved indistinguishable and, hence, are plotted here as a single OTU. Also shown are the *maximum* levels of mtDNA genetic distance observed among loggerhead turtles (*Caretta caretta*) and green turtles (*Chelonia mydas*), respectively, from around the world.

diagnostic molecular characters (mtDNA digestion profiles in this case) for 22 species of large mammals that are the frequent objects of illegal poaching. These molecular markers provide a potential means for law enforcement agencies to identify meat or other tissues confiscated from illegal sources.

What follows are a few examples of law enforcement applications in molecular wildlife forensics. In 1978, a Japanese trawler in U.S. coastal waters was suspected of illegal harvest of a rockfish species *Sebastes alutus*. Tissues confiscated by U.S. enforcement officers indeed proved upon protein-electrophoretic examination to have come from *S. alutus,* thereby contradicting claims in the trawler's log that no such specimens had been taken (Utter, 1991). A similar case in Texas involved a suspected illegal sale of flathead catfish (*Pylodictis olivaris*) and related species. Protein-electrophoretic analyses verified the species' identity of the frozen fish fillets and led to a fine levied against the seller (Harvey, 1990). Findings in molecular forensics also can exonerate the falsely accused. In another case in Texas, electrophoretic analyses of confiscated fillets revealed that fishermen were innocent of the charge of illegal possession of red drum (*Sciaenops ocellatus*) and spotted sea trout (*Cynoscion nebulosus*) (Harvey, 1990).

Some forensic applications in law enforcement involve distinguishing conspecific populations from different areas. For example, a commercial catch of king crab (*Paralithodes camtschatica*) that was claimed to have been harvested in an area of Alaska open to fishing proved upon protein-electrophoretic examination to have come from a closed area northwest in the Bering Sea (Seeb et al., 1990). A particularly interesting example of a molecular forensic case in a geographic context involved the largemouth bass (*Micropterus salmoides*). Bass-fishing tournaments in the southern United States are "big business," with first-place cash awards often in excess of $50,000. In one recent tournament in Texas, the apparent winner was suspected of having smuggled in a large bass from outside the tournament site. Tissue samples of the prize-winning fish were examined electrophoretically and shown to have come from the genetically distinctive Florida subspecies of the largemouth bass (see Philipp et al., 1981), thereby confirming that the trophy specimen had been imported illegally. Based on these genetic findings, the fisherman was charged and subsequently found guilty of fraud (Harvey, 1990).

In 1989, an organized and official approach to wildlife forensics became a reality with the opening of a U.S. Fish and Wildlife Service laboratory in Ashford, Oregon. The first of its kind in the world, this "Scotland Yard for animals" conducts forensic investigations on wildlife products that have been confiscated from professional poachers, overzealous sportsmen, illegal traders in wildlife products, or even uninformed tourists merely wanting a souvenir from some endangered species. The forensics facility is divided into three laboratory sections, Morphology, Criminalistics, and Serology, the latter devoted to the identification of samples by DNA and protein evidence. The task is daunting: Unlike the 360 police crime labs in the United States which deal with a single species (*Homo sapiens*), the Ashland facility must cope with the remainder of the biological world.

Hybridization and Introgression

In the context of rare and endangered species, both biological and legal issues (Box 9.3) have arisen concerning instances of introgressive hybridization that have been documented using molecular markers.

Biological Issues

One evolutionary threat to a rare species is genetic swamping through extensive hybridization with related taxa. An empirical example involves the nearly

Box 9.3. The Hybrid Policy of the Endangered Species Act

In a series of opinions issued by the Solicitor's office beginning in 1977, it was concluded that natural or artificial hybrids occurring between endangered species, subspecies, or populations should not receive protection under the Endangered Species Act (ESA). The rationale appears to be that even if such hybrids were themselves to breed, they would not produce purebreds of either parental form and, hence, would not promote the purposes of the ESA. These decisions have become known as the "Hybrid Policy."

The Hybrid Policy has several serious ramifications. For example, it has prompted formal petitions from various land-use constituencies to remove certain protected taxa from the endangered species list, on the grounds that hybridization and introgression have compromised the genetic integrity of the otherwise endangered forms. But as noted by Grant and Grant (1992), (a) many if not most species of plants and animals hybridize at least occasionally in nature, so that if hybridizing species fall outside the limits of protection afforded by the ESA, few candidate taxa will ever qualify for protection and (b) "if rarity increases the chances of interbreeding with a related species, presumably because conspecific mates are scarce, then the species most in need of protection, by virtue of their rarity, are the ones most likely to lose it under current practice, by hybridizing."

O'Brien and Mayr (1991) have attacked the Hybrid Policy on additional grounds, claiming that definitional and related operational difficulties "have led to confusion, conflict, and, we believe, certain misinterpretations of the [Endangered Species] Act by well-intentioned government officials." They conclude that whereas the Hybrid Policy should discourage management programs based on hybridization between species, it should not be applied to subspecies or populations because the latter retain the potential to interbreed freely as part of ongoing natural processes. With regard to three case histories involving hybridization of endangered taxa (described in the text), O'Brien and Mayr (1991) conclude:

(a) the status of "endangered" should be retained for extant red wolves because they are the only available descendants of a historic wolf subspecies;

(b) there is no justification for eliminating protection of the gray wolf as a species, because hybridization with the coyote is limited to a narrow zone that developed recently;

(c) the Florida panther should receive continued protection because it clearly qualifies as a subspecies.

extinct mahogany tree *Cercocarpus traskiae,* an endemic to Santa Catalina Is-land in Los Angeles County, California. Allozymic (and morphological) ap-praisals of the seven remaining adults of this species revealed that some were, in fact, the products of hybridization between *C. traskiae* and other more abundant *Cercocarpus* species on the island (Rieseberg et al., 1989). However, several seedlings were found that did exhibit "pure" *C. traskiae* genotypes. These genetic discoveries led to two management suggestions intended to lower the probability of further hybridization (Rieseberg et al., 1989): (a) eliminate indi-viduals of other *Cercocarpus* species from near the remaining pure *C. traskiae* specimens and (b) transplant seedlings and established cuttings from nonhybrid individuals to more remote areas on Santa Catalina Island.

The cutthroat trout native to the western United States and Canada is a com-plex of approximately 15 recognized subspecies, many of which are threatened by human alteration of habitat and by artificial introductions of non-native trout species. Eleven subspecies currently have protected legal status, and two already may be extinct. Surveys of allozymes and mtDNA in scores of cutthroat popu-lations have revealed a complex phylogeographic pattern, with some subspecies virtually identical genetically and others as distinct as normal congeneric species (Allendorf and Leary, 1988, and references therein). With regard to the current discussion, these molecular markers also have documented extensive hybridiza-tion between the native and introduced trout species, as well as between trans-planted cutthroat forms. For example, introgression from rainbow trout was observed in 7 of 39 assayed populations in Utah (Martin et al., 1985); in Mon-tana, 40% of the 80 populations formerly thought to be pure "westslope" cut-throats proved upon molecular examination to include products of hybridization with either rainbow trout or the "Yellowstone" cutthroat form. In the Flathead River drainage in Montana (considered one of the last remaining strongholds of native westslope cutthroat trout), only 2 of 19 headwater lakes sampled con-tained pure populations and detailed genetic analyses further revealed that the hybridized headwater populations were "leaking" foreign genes into down-stream areas (Allendorf and Leary, 1988). Such findings on the introgressed structure of western cutthroats have led to two conservation-related concerns (Allendorf and Leary, 1988). First, hybrids between genetically differentiated trout forms often appear to exhibit reduced fitness due to developmental abnor-malities. Second, extensive introgressive hybridization carries the danger of genetic swamping and loss of locally adapted populations. As stated by Allen-dorf and Leary (1988), "The eventual outcome of widespread introgression and continued introduction of hatchery rainbow trout is the homogenization of west-ern North American trout into a single taxon (*Salmo ubiquiti?*). Thus, we would exchange all of the diversity within and between many separate lineages, pro-duced by millions of years of evolution . . . for a single new mongrel species."

Some endangered species currently recognized may themselves be the products of past hybridization and introgression. One example perhaps involves the red wolf (*Canis rufus*), which formerly ranged throughout the southeastern United States but declined precipitously in numbers after 1900 and became extinct in the wild in about 1975. Molecular analyses of remaining captive animals (as well as museum-preserved skins and blood samples from extinct wild populations) revealed that extant red wolves likely contain some genetic material derived through hybridization with coyotes (*C. latrans*), but are otherwise extremely similar to gray wolves (*C. lupis*) [Wayne, 1992; Wayne and Jenks, 1991; but see also Ferrell et al. (1980), Nowak (1992), and Dowling et al. (1992) for alternative interpretations of the data]. It is uncertain whether this hypothesized introgression of coyote genes into red wolf populations would have occurred naturally in ancient times and/or more recently following a range expansion of coyote populations (facilitated by the human clearing of eastern forests). In any event, on the basis of these genetic findings (as well as questions concerning the distinctiveness of red wolves from gray wolves), much controversy has arisen over taxonomy of the red wolf (Phillips and Henry, 1992) as well as the appropriateness of the intensive management program (Gittleman and Pimm, 1991), which has included a call for reintroduction of this species into the wild.

LEGAL ISSUES

The "Hybrid Policy" of the Endangered Species Act [ESA—Box 9.3] has been the focal point of several legal and ethical controversies surrounding the topic of hybridization and introgression in conservation biology. The sentiment of this policy, which denies formal protection for organisms of hybrid ancestry, has served as a basis for calling into question several existing endangered species designations and associated management programs. The red wolf situation described above provides one example. So too does the gray wolf, which also appears to have hybridized with the coyote. Across a small portion of the gray wolf's range in the northern United States and southern Canada, the presence of "coyote-type" mtDNA genotypes in populations that otherwise appear wolflike has led to the conclusion that male wolves occasionally hybridized with female coyotes in this area (Lehman et al., 1991) (Fig. 9.6). This genetics-based conclusion, when interpreted against the philosophical platform of the Hybrid Policy, in turn has prompted at least one petition to the Interior Department to have the gray wolf removed from its status as an endangered species in the northern United States. [This petition was, however, turned down by the U.S. Fish and Wildlife Service (Fergus, 1991).]

Another example of a political controversy in conservation genetics involves the Florida panther (*Felis concolor coryi*), a severely threatened relict population of puma or mountain lion whose historic range included much of the southeastern United States. The small remaining population occurs only in the Big Cypress

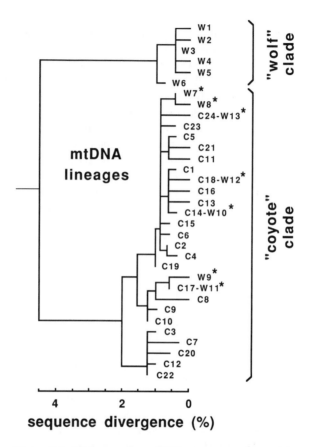

Figure 9.6. Phylogeny for mtDNA genotypes observed in gray wolves and coyotes from North America (after Lehman et al., 1991). Note that whereas several genotypes observed in wolves (W) group separately from those of coyotes (C), others (indicated by asterisks) do not, these exceptions supposedly being attributable to recent hybridizations between wolf males and coyote females that led to introgression of coyote mtDNA across the species boundary. This scenario recently was bolstered by the discovery that mtDNA genotypes in gray wolves from the Old World belong clearly to the "wolf lineage" (Wayne et al., 1992).

Swamp and the Everglades of extreme southern Florida. A molecular genetic evaluation of this and other puma subspecies based on nuclear and mtDNA markers indicated the existence of two distinct genetic stocks in the area (O'Brien et al., 1990)—one confined to Big Cypress that appears to be a bona fide descendant of the *F.c. coryi* lineage, and a second found largely in the Everglades that appears to be descended from pumas that evolved in South or Central America but probably were released (by man) into south Florida in this century. Importantly, some individuals also appeared to be the products of hybridization between these distinct genetic stocks. Thus, a strict interpretation of the Hybrid Policy conceivably could result in removal of the Florida panther from federal protection under the ESA.

Clearly, for several reasons including those mentioned in Box 9.3, the mere documentation of hybridization involving an endangered population should not alone be sufficient grounds for removal from the endangered or threatened species lists. An additional consideration that may be especially relevant to the Florida panther is the following. Most of the panthers in southern Florida appear to be inbred severely, presumably as a result of the pronounced population reduction. The evidence includes increased frequencies of several abnormal traits that often are associated with known inbreeding, e.g., deformed tail vertebrae (causing a 90° kink near the tip of the puma's tail), a "cowlick" whorl on the animal's back, cryptorchism in males (a condition in which one testicle fails to descend into the scrotum), and 90% defective sperm. Perhaps the infusion of puma genes from South America into *F.c. coryi,* inadvertent though it was, actually may increase the fitness and survival probabilities of the southern Florida population (O'Brien et al., 1990).

Species Phylogenies and Macroevolution

Beyond the taxonomic level of species, phylogenetic considerations can also be of relevance to conservation biology. Such interest may be partly academic. For example, what is the ancestry and geographic origin of the endangered Hawaiian goose (*Nesochen sandvicensis*)? Conventionally, this morphologically distinctive species has been placed in a separate genus, with curiosity centering on its phylogenetic and geographic roots. Recent comparisons based on mtDNA restriction sites and sequences revealed that the Hawaiian goose is allied more closely to the Canada goose (*Branta canadensis*) than it is to other extant candidate species—the black brant (*Branta bernicla*) or the emperor goose (*Chen canagica*) (Quinn et al., 1991). From this molecular evidence on maternal lineages, it was concluded that the Hawaiian goose's ancestors colonized the islands from North America about 0.9 mya.

The black-footed ferret (*Mustela nigripes*) of the North American plains is a highly endangered member of Mustelidae (weasel and skunk family). An abundant species with broad distribution in the last century, the black-footed ferret was decimated primarily through human eradication of its principal prey base and associate, the prairie dog (*Cynonys* sp.). In 1981, a few remaining specimens of *M. nigripes* were discovered in Wyoming, and molecular analyses showed that this small population was characterized by extremely low genetic variability (Table 9.1), probably as a result of the severe population bottlenecks. The black-footed ferret has more common relatives elsewhere, however, including the presumed sister taxon *M. eversmanni* (steppe polecat) of Siberia. O'Brien et al. (1989) employed allozyme assays to assess the phylogenetic position of *M. nigripes* within the Mustelidae (Fig. 9.7). The molecular data confirmed that *M. eversmanni* is the closest living relative of the black-footed ferret and suggested that these sister taxa are only about as differentiated genetically ($D \cong 0.08$) as are closely related congeners in other mammalian groups (Fig. 1.2). These two species likely separated about 0.5–2.0 mya (O'Brien et al., 1989).

Most of the seven to eight species of marine turtles conventionally recognized are listed as threatened or endangered in the official "Red Data Book" of the IUCN (International Union for the Conservation of Nature and Natural Resources). Several phylogenetic issues ranging from population-level distinctions to deeper evolutionary alliances among species, genera, and families have

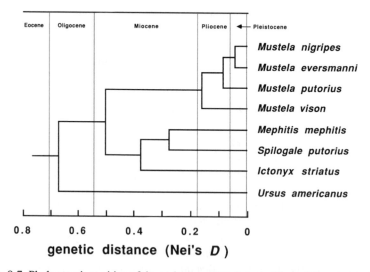

Figure 9.7. Phylogenetic position of the endangered black-footed ferret (*Mustela nigripes*) within the Mustelidae, based on allozymic comparisons (after O'Brien et al., 1989). Other species assayed were the Siberian polecat (*M. eversmanni*), European polecat (*M. putorius*), mink (*M. vison*), striped skunk (*Mephitis mephitis*), spotted skunk (*Spilogale putorius*), African striped skunk (*Ictonyx striatus*), and as outgroup the American black bear (Ursidae: *Ursus americanus*).

plagued the systematics of these reptilian mariners, with consequences some-times extending to conservation priorities. Case histories involving the green and ridley turtles already have been described (Figs. 9.1 and 9.5). Other troublesome questions have included the following: Is the east Pacific black turtle (*Chelonia agassizi*) specifically distinct from the green turtle (*C. mydas*)? Is the spongiv-orous hawksbill turtle (*Eretmochelys imbricata*) allied phylogenetically to the herbivorous green turtle, or alternatively to the carnivorous loggerhead turtle (*Caretta caretta*)? Is the Australian flatback turtle (*Natator depressa*) allied closely to greens (as its earlier placement within *Chelonia* suggested), or might it be a relative of the loggerhead turtle? Are all marine turtles including the distinctive leatherback (*Dermochelys coriacea*) of monophyletic origin? Not all of these questions have as yet been answered definitively, but a molecular start has been made using mtDNA restriction sites and nucleotide sequences from the cytochrome b gene (Fig. 9.8). Based on these matrilineal markers, preliminary answers to the above questions are as follows: The black turtle falls well within the range of genetic differentiation exhibited among green turtle rookeries world-wide; the hawksbill turtle is related more closely to the loggerhead complex than to green turtles and, hence, probably evolved from a carnivorous rather than herbivorous ancestor; the flatback is highly distinct from both the loggerhead and green complexes and is roughly equidistant to both; and, finally, both the leath-erback and the freshwater snapping turtle (*Chelydra serpentina*) exhibit large genetic distances from all other marine turtles, such that altogether an unresolved phylogenetic trichotomy exists under the available molecular evidence (Bowen et al., 1992, 1993a).

Such phylogenetic appraisals of endangered (or other) assemblages are of relevance to conservation biology in the general sense that they provide an understanding of the evolutionary relationships among species and higher taxa. Should the ramifications go farther and perhaps include phylogenetic position as a guide to prioritizing taxa with regard to conservation value? For example, does the sole living representative of an ancient and phylogenetically unique lineage (e.g., the leatherback turtle) warrant greater conservation concern than another species (e.g., the Kemp's ridley turtle) that belongs to a recent and more speciose clade? Some authors have answered this humbling question in the affirmative, on the grounds that species which are most distinctive in a phylogenetic sense make a disproportionately large contribution to the world's extant evolutionary diver-sity (Vane-Wright et al., 1991). On the other hand, each biological species is unique, and currently independent from all others genetically. Furthermore, every species alive today traces back through an unbroken chain of ancestry over thousands of millions of years, irrespective of how many speciation events have intervened along its phylogenetic journey. How can societies deign to play God with any creatures of such evolutionary fortitude? The mere fact that such issues of preservation priority must be raised represents a sad and shameful commen-

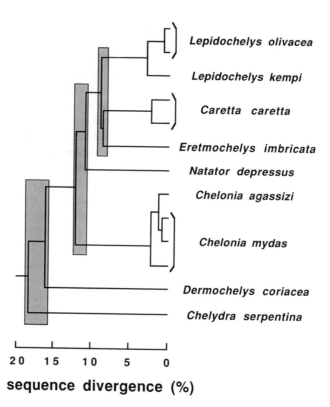

Figure 9.8. Phylogeny for all recognized species of marine turtles (plus the freshwater snapping turtle *C. serpentina*) estimated using sequence data from the mtDNA cytochrome b gene (after Bowen et al., 1993a). For species represented by more than one sample, the individuals came from different oceanic basins. Exact orders of the nodes within the shaded boxes are uncertain, differing slightly with alternative methods of data analysis.

tary on the degree to which the planet's natural heritage has succumbed to the overwhelming impact of one species, humankind.

SUMMARY

1.   Most discussions of genetics in conservation biology have centered on the topic of heterozygosity or related measures of the within-population component of genetic variation. Molecular heterozygosity indeed is exceptionally low in many rare or endangered species, presumably because of genetic drift and inbreeding accompanying severe population reductions.

2. Although it is tempting to manage endangered populations for enhancement of genetic variation, in most cases causal links between heterozygosity (as estimated by molecular markers) and fitness have proven difficult to establish firmly. For these and other reasons, some authors have argued that behavioral and demographic concerns should take priority over heterozygosity issues in management programs for endangered species. In truth, both classes of concern are important.

3. Molecular markers also can serve the field of conservation biology by revealing phylogenetic relationships within and among rare or endangered species. Such phylogenetic assessments can range from parentage evaluations in captive breeding programs to the identification of the major sources of regional phylogeographic diversity around which management guidelines and natural reserves might be established.

4. Some programs for endangered species have been directed toward taxa whose evolutionary distinctiveness has been questioned by molecular genetic reappraisals. In other cases, endangered "species" that were taxonomically suspect have proved upon molecular reexamination to be highly distinct genetically, thus adding to the rationale for special preservation efforts.

5. Particularly in the field of fisheries biology, molecular markers are employed widely to identify and characterize populational stocks under exploitation. Contributions of genetic stocks to mixed fisheries can be quantified. Molecular approaches to stock assessment also have stimulated thought about the nature of information provided by inherited versus acquired markers (such as physical tags) and of the distinction between evolutionarily deep versus shallow population genetic structures.

6. Molecular markers provide powerful forensic tools for government agencies charged with enforcement of wildlife legislation.

7. Hybridization and genetic introgression documented by molecular markers have raised both biological and legal questions concerning the status of several endangered species. Molecular appraisals of phylogenetic relationships among reproductively isolated taxa also have contributed to management discussions.

CONCLUSION

I want to close this chapter, and the book, by reiterating a sentiment expressed by E.O. Wilson in the quote that opened this chapter. Modern biology indeed has produced a genuinely new way of looking at the world. The molecular perspectives emphasized in this text do not supplant traditional approaches to the study of natural history and evolution, but rather they enrich our understanding of life. Herein lies the greatest value of molecular methods in conservation biology or elsewhere. To the degree that we come to understand and appreciate other organisms, we will increasingly cherish the earth's biological heritage, and our own. "We cannot win this battle to save species and environments without forging an emotional bond between ourselves and nature as well—for we will not fight to save what we do not love . . . We really must make room for nature in our hearts" (Gould, 1991). Think back to even a few of the fascinating organisms whose natural histories and evolutionary patterns have been elucidated using molecular markers—the honey mushrooms and their giant clones on the floors of northern forests; the hybridogenetic livebearing fishes in the arroyos of northwestern Mexico and the substantial evolutionary ages that some of these unisexual biotypes have achieved; the bluebirds in the pasturelands of the eastern United States, and their unsuspected and sometimes devious means of achieving parentage; the naked mole rats in the deserts of Africa, with their eusocial behaviors and tight fabrics of kinship; and the female green turtles, who after periods measured in decades return faithfully to natal sites across thousands of kilometers of open ocean. If this book has accomplished nothing else, I hope that it may have engendered an increased awareness, respect, and love, for the marvelous biological diversity on this fragile planet.

Literature Cited

Abbott, R.J. 1992. Plant invasions, interspecific hybridization and the evolution of new plant taxa. Trends Ecol. Evol. 7:401–405.

Abramoff, P., R.M. Darnell, and J.S. Balsano. 1968. Electrophoretic demonstration of the hybrid origin of the gynogenetic teleost *Poecilia formosa*. Amer. Natur. 102:555–558.

Achmann, R., K.-G. Heller, and J.T. Epplen. 1992. Last-male sperm precedence in the bushcricket *Poecilimon veluchianus* (Orthoptera, Tettigonioidea) demonstrated by DNA fingerprinting. Mol. Ecol. 1:47–54.

Adams, W.T., A.R. Griffin, and G.F. Moran. 1992. Using paternity analysis to measure effective pollen dispersal in plant populations. Amer. Natur. 140:762–780.

Adelman, R., R.L. Saul, and B.N. Ames. 1988. Oxidative damage to DNA: relation to species metabolic rate and life span. Proc. Natl. Acad. Sci. USA 85:2706–2708.

Adkins, R.M. and R.L. Honeycutt. 1991. Molecular phylogeny of the superorder Archonta. Proc. Natl. Acad. Sci. USA 88:10317–10321.

Aguadé, M., N. Miyashita, and C.H. Langley. 1989a. Restriction-map variation at the *Zeste-tko* region in natural populations of *Drosophila melanogaster*. Mol. Biol. Evol. 6:123–130.

Aguadé, M., N. Miyashita, and C.H. Langley. 1989b. Reduced variation in the *yellow-achaete-scute* region in natural populations of *Drosophila melanogaster*. Genetics 122:607–615.

Aide, T.M. 1986. The influence of wind and animal pollination on variation in outcrossing rates. Evolution 40:434–435.

Alatalo, R.V., L. Gustafsson, and A. Lundberg. 1984. High frequency of cuckoldry in pied and collared flycatchers. Oikos 42:41–47.

Albert, V.A., S.E. Williams, and M.W. Chase. 1992. Carnivorous plants: phylogeny and structural evolution. Science 257:1491–1495.

Allard, M.W. and R.L. Honeycutt. 1992. Nucleotide sequence variation in the mitochondrial 12S rDNA gene and the phylogeny of African mole-rats (Rodentia: Bathyergidae). Mol. Biol. Evol. 9:27–40.

Allard, R.W. 1975. The mating system and microevolution. Genetics 79:115–126.

Allard, R.W., G.R. Babbel, M.T. Clegg, and A.L. Kahler. 1972. Evidence for coadaptation in *Avena barbata*. Proc. Natl. Acad. Sci. USA 69:3043–3048.

399

Allendorf, F.W. 1983. Isolation, gene flow, and genetic differentiation among populations. Pp. 51–65 *in:* Genetics and Conservation, C.M. Schonewald-Cox, S.M. Chambers, B. MacBryde, and L. Thomas (eds.), Benjamin/Cummings, London.

Allendorf, F.W. and R.F. Leary. 1986. Heterozygosity and fitness in natural populations of animals. Pp. 57–76 *in:* Conservation Biology—The Science of Scarcity and Diversity, M. Soulé (ed.), Sinauer, Sunderland, MA.

Allendorf, F.W. and R.F. Leary. 1988. Conservation and distribution of genetic variation in a polytypic species, the cutthroat trout. Conserv. Biol. 2:170–184.

Allendorf, F.W. and F.M. Utter. 1979. Population genetics. Pp. 407–454 *in:* Fish Physiology Vol. 8, W.S. Hoar, D.J. Randall, and J.R. Brett (eds.), Academic Press, New York.

Amos, B., J. Barrett, and G.A. Dover. 1991a. Breeding system and social structure in the Faroese pilot whale as revealed by DNA fingerprinting. Pp. 255–268 *in:* Genetic Ecology of Whales and Dolphins, A.R. Hoelzel (ed.), Black Bear Press, Cambridge, England.

Amos, B., J. Barrett, and G.A. Dover. 1991b. Breeding behavior of pilot whales revealed by DNA fingerprinting. Heredity 67:49–55.

Anderson, E. 1949. Introgressive Hybridization. John Wiley and Sons, New York.

Andersson, M. 1984. The evolution of eusociality. Annu. Rev. Ecol. Syst. 15:165–189.

Andrewartha, H.G. and L.C. Birch. 1954. The Distribution and Abundance of Animals. Univ. Chicago Press, Chicago, IL.

Andrews, P. 1987. Aspects of hominoid phylogeny. Pp. 23–52 *in:* Molecules and Morphology in Evolution: Conflict or Compromise? C. Patterson (ed.), Cambridge Univ. Press, Cambridge, England.

Anonymous, 1992. The great fungus. Nature 357:179.

Antonovics, J. and A.D. Bradshaw. 1970. Evolution in closely adjacent plant populations. VIII. Clinal patterns at a mine boundary. Heredity 25:349–362.

Apfelbaum, L. and O.A. Reig. 1989. Allozyme genetic distances and evolutionary relationships in species of akodontine rodents (Cricetidae: Sigmodontinae). Biol. J. Linn. Soc. 38:257–280.

Appels, R. and J. Dvorak. 1982. The wheat ribosomal DNA spacer region: its structure and variation in populations and among species. Theoret. Appl. Genet. 63:337–348.

Appels, R. and R.L. Honeycutt. 1986. rDNA: evolution over a billion years. Pp. 81–135 *in:* DNA Systematics, Vol. II, S.K. Dutta (ed.), CRC Press, Boca Raton, FL.

Apuya, N.R., B.L. Frazier, P. Keim, E. Jill Roth, and K.G. Lark. 1988. Restriction fragment length polymorphisms as genetic markers in soybean, *Glycine max* (L.) merrill. Theor. Appl. Genet. 75:889–901.

Aquadro, C.F. and J.C. Avise. 1981. Genetic divergence between rodent species assessed by using two-dimensional electrophoresis. Proc. Natl. Acad. Sci. USA 78:3784–3788.

Aquadro, C.F. and J.C. Avise. 1982a. Evolutionary genetics of birds. VI. A reexamination of protein divergence using varied electrophoretic conditions. Evolution 36:1003–1019.

Aquadro, C.F. and J.C. Avise. 1982b. An assessment of "hidden" heterogeneity within electromorphs at three enzyme loci in deer mice. Genetics 102:269–284.

Aquadro, C.F., S.F. Desse, M.M. Bland, C.H. Langley and C.C. Laurie-Ahlberg. 1986. Molecular population genetics of the alcohol dehydrogenase gene region of *Drosophila melanogaster*. Genetics 114:1165–1190.

Aquadro, C.F. and B.D. Greenberg. 1983. Human mitochondrial DNA variation and evolution: analysis of nucleotide sequences from seven individuals. Genetics 103:287–312.

Aquadro, C.F., A.L. Weaver, S.W. Schaeffer, and W.W. Anderson. 1991. Molecular evolution of inversions in *Drosophila pseudoobscura:* the amylase gene region. Proc. Natl. Acad. Sci. USA 88:305–309.

Árnason, E., S. Pálsson, and A. Arason. 1992. Gene flow and lack of population differentiation in Atlantic cod, *Gadus morhua* L., from Iceland, and comparison of cod from Norway and Newfoundland. J. Fish Biol. 40:751–770.

Árnason, Ú., R. Spilliaert, Á. Pálsdóttir, and A. Árnason. 1991. Molecular identification of hybrids between the two largest whale species, the blue whale (*Balaenoptera musculus*) and the fin whale (*B. physalus*). Hereditas 115:183–189.

Arnheim, N. 1983. Concerted evolution of multigene families. Pp. 38–61 *in:* Evolution of Genes and Proteins, M. Nei (ed.), Sinauer, Sunderland, MA.

Arnheim, N., T. White, and W.E. Rainey. 1990. Application of PCR: organismal and population biology. BioScience 40:174–182.

Arnold, M.L. 1992. Natural hybridization as an evolutionary process. Annu. Rev. Ecol. Syst. 23:237–261.

Arnold, M.L., B.D. Bennett, and E.A. Zimmer. 1990a. Natural hybridization between *Iris fulva* and *I. hexagona:* pattern of ribosomal DNA variation. Evolution 44:1512–1521.

Arnold, M.L., C.M. Buckner, and J.J. Robinson. 1991. Pollen-mediated introgression and hybrid speciation in Louisiana irises. Proc. Natl. Acad. Sci. USA 88:1398–1402.

Arnold, M.L., N. Contreras, and D.D. Shaw. 1988. Biased gene conversion and asymmetrical introgression between subspecies. Chromosoma 96:368–371.

Arnold, M.L., J.L. Hamrick, and B.D. Bennett. 1990b. Allozyme variation in Louisiana irises: a test for introgression and hybrid speciation. Heredity 65:297–306.

Arnold, M.L., J.J. Robinson, C.M. Buckner, and B.D. Bennett. 1992. Pollen dispersal and interspecific gene flow in Louisiana irises. Heredity 68:399–404.

Arnold, M.L., D.D. Shaw, and N. Contreras. 1987. Ribosomal RNA-encoding DNA introgression across a narrow hybrid zone between two subspecies of grasshopper. Proc. Natl. Acad. Sci. USA 84:3946–3950.

Arntzen, J.W. and G.P. Wallis. 1991. Restricted gene flow in a moving hybrid zone of the newts *Triturus cristatus* and *T. marmoratus* in western France. Evolution 45:805–826.

Ashburner, M., R. Camfield, B. Clarke, D. Thatcher, and R. Woodruff. 1979. A genetic analysis of the locus coding for alcohol dehydrogenase, and its adjacent chromosome region, in *Drosophila melanogaster.* Pp. 95–106 *in:* Eukaryotic Gene Regulation, R. Axel, T. Maniatis, and C.F. Fox (eds.), Academic Press, New York.

Ashley, M.V., D.J. Melnick, and D. Western. 1990. Conservation genetics of the black rhinoceros (*Diceros bicornis*), I: evidence from the mitochondrial DNA of three populations. Cons. Biol. 4:71–77.

Ashley, M. and C. Wills. 1987. Analysis of mitochondrial DNA polymorphisms among channel island deer mice. Evolution 41:854–863.

Ashley, M. and C. Wills. 1989. Mitochondrial-DNA and allozyme divergence patterns are correlated among island deer mice. Evolution 43:646–650.

Ashton, P.A. and R.J. Abbott. 1992. Multiple origins and genetic diversity in the newly arisen allopolyploid species, *Senecio cambrensis* Rosser (Compositae). Heredity 68:25–32.

Asmussen, M.A., J. Arnold, and J.C. Avise. 1987. Definition and properties of disequilibrium statistics for associations between nuclear and cytoplasmic genotypes. Genetics 115:755–768.

Asmussen, M.A., J. Arnold, and J.C. Avise. 1989. The effects of assortative mating and migration on cytonuclear associations in hybrid zones. Genetics 122:923–934.

Asmussen, M.A. and A. Schnabel. 1991. Comparative effects of pollen and seed migration on the cytonuclear structure of plant populations. I. maternal cytoplasmic inheritance. Genetics 128:639–654.

Atchley, W.R. and W.M. Fitch. 1991. Gene trees and the origins of inbred strains of mice. Science 254:554–558.

Attardi, G. 1985. Animal mitochondrial DNA: an extreme example of genetic economy. Int. Rev. Cytol. 93:93–145.

Aubert, J. and M. Solignac. 1990. Experimental evidence for mitochondrial DNA introgression between *Drosophila* species. Evolution 44:1272–1282.

Avery, O.T., C.M. MacLeod, and M. McCarty. 1944. Studies on the chemical nature of the substance inducing transformation of Pneumococcal types. I. Induction of transformation by a DNA fraction isolated from *Pneumococcus* type III. J. Exp. Med. 79:137–158.

Avigad, S., B.E. Cohen, S. Bauer, G. Schwartz, M. Frydman, S.L.C. Woo, Y. Niny, and Y. Shiloh. 1990. A single origin of phenylketonuria in Yemenite Jews. Nature 344:168–170.

Avise, J.C. 1974. Systematic value of electrophoretic data. Syst. Zool. 23:465–481.

Avise, J.C. 1976. Genetic differentiation during speciation. Pp. 106–122 *in:* Molecular Evolution, F.J. Ayala (ed.), Sinauer, Sunderland, MA.

Avise, J.C. 1977a. Molecular variability and hypothesis testing. Sci. Prog. 64:85–94.

Avise, J.C. 1977b. Is evolution gradual or rectangular? Evidence from living fishes. Proc. Natl. Acad. Sci. USA 74:5083–5087.

Avise, J.C. 1977c. Genic heterozygosity and rate of speciation. Paleobiology 3:422–432.

Avise, J.C. 1983a. Commentary. Pp. 262–270 *in:* Perspectives in Ornithology, A.H. Brush and G.A. Clark, Jr. (eds.), Cambridge Univ. Press, Cambridge, England.

Avise, J.C. 1983b. Protein variation and phylogenetic reconstruction. Pp. 103–130 *in:* Protein Variation: Adaptive and Taxonomic Significance, G. Oxford and D. Rollinson (eds.), Systematics Association, British Museum Natural History, London.

Avise, J.C. 1986. Mitochondrial DNA and the evolutionary genetics of higher animals. Phil. Trans. Roy. Soc. London B312:325–342.

Avise, J.C. 1987. Identification and interpretation of mitochondrial DNA stocks in marine species. Pp. 105–136 *in:* Proc. Stock Identification Workshop, H. Kumpf and E.L. Nakamura (eds.), Publication of the National Oceanographic and Atmospheric Administration, Panama City, FL.

Avise, J.C. 1989a. Gene trees and organismal histories: a phylogenetic approach to population biology. Evolution 43:1192–1208.

Avise, J.C. 1989b. Nature's family archives. Natur. Hist. 3:24–27.

Avise, J.C. 1989c. A role for molecular genetics in the recognition and conservation of endangered species. Trends Ecol. Evol. 4:279–281.

Avise, J.C. 1990. Flocks of African Fishes. Nature 347:512–513.

Avise, J.C. 1991a. Ten unorthodox perspectives on evolution prompted by comparative population genetic findings on mitochondrial DNA. Annu. Rev. Genet. 25:45–69.

Avise, J.C. 1991b. Matriarchal liberation. Nature 352:192.

Avise, J.C. 1992. Molecular population structure and the biogeographic history of a regional fauna: a case history with lessons for conservation biology. Oikos 63:62–76.

Avise, J.C., R.T. Alisauskas, W.S. Nelson, and C.D. Ankney. 1992b. Matriarchal population genetic structure in an avian species with female natal philopatry. Evolution 46:1084–1096.

Avise, J.C., C.D. Ankney, and W.S. Nelson. 1990a. Mitochondrial gene trees and the evolutionary relationship of mallard and black ducks. Evolution 44:1109–1119.

Avise, J.C. and C.F. Aquadro. 1982. A comparative summary of genetic distances in the vertebrates. Evol. Biol. 15:151–185.

Avise, J.C. and C.F. Aquadro. 1987. Malate dehydrogenase isozymes provide a phylogenetic marker for the Piciformes (woodpeckers and allies). Auk 104:324–328.

Avise, J.C., J. Arnold, R.M. Ball, E. Bermingham, T. Lamb, J.E. Neigel, C.A. Reeb, and N.C. Saunders. 1987a. Intraspecific phylogeography: the mitochondrial DNA bridge between population genetics and systematics. Annu. Rev. Ecol. Syst. 18:489–522.

Avise, J.C. and F.J. Ayala. 1975. Genetic change and rates of cladogenesis. Genetics 81:757–773.

Avise, J.C. and F.J. Ayala. 1976. Genetic differentiation in speciose versus depauperate phylads: evidence from the California minnows. Evolution 30:46–58.

Avise, J.C. and R.M. Ball, Jr. 1990. Principles of genealogical concordance in species concepts and biological taxonomy. Oxford Surv. Evol. Biol. 7:45–67.

Avise, J.C. and R.M. Ball, Jr. 1991. Mitochondrial DNA and avian microevolution. Acta XX Congressus Internationalis Ornithologici 1:514–524.

Avise, J.C., R.M. Ball, and J. Arnold. 1988. Current versus historical population sizes in vertebrate species with high gene flow: a comparison based on mitochondrial DNA lineages and inbreeding theory for neutral mutations. Mol. Biol. Evol. 5:331–344.

Avise, J.C., E. Bermingham, L.G. Kessler, and N.C. Saunders. 1984b. Characterization of mitochondrial DNA variability in a hybrid swarm between subspecies of bluegill sunfish (*Lepomis macrochirus*). Evolution 38:931–941.

Avise, J.C., B.W. Bowen, and T. Lamb. 1989. DNA fingerprints from hypervariable mitochondrial genotypes. Mol. Biol. Evol. 6:258–269.

Avise, J.C., B.W. Bowen, T. Lamb, A.B. Meylan, and E. Bermingham. 1992a. Mitochondrial DNA evolution at a turtle's pace: evidence for low genetic variability and reduced microevolutionary rate in the Testudines. Mol. Biol. Evol. 9:457–473.

Avise, J.C. and J. Felley. 1979. Population structure of freshwater fishes I. Genetic variation of bluegill (*Lepomis macrochirus*) populations in man-made reservoirs. Evolution 33:15–26.

Avise, J.C., C. Giblin-Davidson, J. Laerm, J.C. Patton, and R.A. Lansman. 1979b. Mitochondrial DNA clones and matriarchal phylogeny within and among geographic populations of the pocket gopher, *Geomys pinetis*. Proc. Natl. Acad. Sci. USA 76:6694–6698.

Avise, J.C., G.S. Helfman, N.C. Saunders, and L.S. Hales. 1986. Mitochondrial DNA differentiation in North Atlantic eels: population genetic consequences of an unusual life history pattern. Proc. Natl. Acad. Sci. USA 83:4350–4354.

Avise, J.C. and G.B. Kitto. 1973. Phosphoglucose isomerase gene duplication in the bony fishes: an evolutionary history. Biochem. Genet. 8:113–132.

Avise, J.C. and R.A. Lansman. 1983. Polymorphism of mitochondrial DNA in populations of higher animals. Pp. 147–164 *in:* Evolution of Genes and Proteins, M. Nei and R.K. Koehn (eds.), Sinauer, Sunderland, MA.

Avise, J.C., R.A. Lansman, and R.O. Shade. 1979a. The use of restriction endonucleases to measure mitochondrial DNA sequence relatedness in natural populations. I. Population structure and evolution in the genus *Peromyscus*. Genetics 92:279–295.

Avise, J.C. and J.E. Neigel. 1984. Population biology aspects of histocompatibility polymorphisms in marine invertebrates. Pp. 223–234 *in:* Genetics: New Frontiers, V.L. Copra, B.C. Joshi, R.P. Sharma, and H.C. Banoal (eds.), Proc. XV Int. Congr. Genet., Vol. 4, Oxford Publ. Co., New Delhi.

Avise, J.C., J.E. Neigel, and J. Arnold. 1984a. Demographic influences on mitochondrial DNA lineage survivorship in animal populations. J. Mol. Evol. 20:99–105.

Avise, J.C. and W.S. Nelson. 1989. Molecular genetic relationships of the extinct Dusky Seaside Sparrow. Science 243:646–648.

Avise, J.C., W.S. Nelson, J. Arnold, R.K. Koehn, G.C. Williams, and V. Thorsteinsson. 1990b. The evolutionary genetic status of Icelandic eels. Evolution 44:1254–1262.

Avise, J.C., J.C. Patton, and C.F. Aquadro. 1980. Evolutionary genetics of birds II. Conservative protein evolution in North American sparrows and relatives. Syst. Zool. 29:323–334.

Avise, J.C., J.M. Quattro, and R.C. Vrijenhoek. 1992c. Molecular clones within organismal clones: mitochondrial DNA phylogenies and the evolutionary histories of unisexual vertebrates. Evol. Biol. 26:225–246.

Avise, J.C., C.A. Reeb, and N.C. Saunders. 1987b. Geographic population structure and species differences in mitochondrial DNA of mouthbrooding marine catfishes (Ariidae) and demersal spawning toadfishes (Batrachoididae). Evolution 41:991–1002.

Avise, J.C. and N.C. Saunders. 1984. Hybridization and introgression among species of sunfish (*Lepomis*): analysis by mitochondrial DNA and allozyme markers. Genetics 108:237–255.

Avise, J.C. and R.K. Selander. 1972. Evolutionary genetics of cave-dwelling fishes of the genus *Astyanax*. Evolution 26:1–19.

Avise, J.C., J.F. Shapira, S.W. Daniel, C.F. Aquadro, and R.A. Lansman. 1983. Mitochondrial DNA differentiation during the speciation process in *Peromyscus*. Mol. Biol. Evol. 1:38–56.

Avise, J.C. and D.Y. Shapiro. 1986. Evaluating kinship of newly settled juveniles within social groups of the coral reef fish *Anthias squamipinnis*. Evolution 40:1051–1059.

Avise, J.C., J.J. Smith, and F.J. Ayala. 1975. Adaptive differentiation with little genic change between two native California minnows. Evolution 29:411–426.

Avise, J.C. and M.H. Smith. 1974. Biochemical genetics of sunfish. I. Geographic variation and subspecific intergradation in the bluegill, *Lepomis macrochirus*. Evolution 28:42–56.

Avise, J.C. and M.H. Smith. 1977. Gene frequency comparisons between sunfish (Centrarchidae) populations at various stages of evolutionary divergence. Syst. Zool. 26:319–325.

Avise, J.C., M.H. Smith, and R.K. Selander. 1979c. Biochemical polymorphism and systematics in the genus *Peromyscus* VII. Geographic differentiation in members of the *truei* and *maniculatus* species groups. J. Mammal. 60:177–192.

Avise, J.C., M.H. Smith, R.K. Selander, T.E. Lawlor, and P.R. Ramsey. 1974. Biochemical polymorphism and systematics in the genus *Peromyscus*. V. Insular and mainland species of the subgenus *Haplomylomys*. Syst. Zool. 23:226–238.

Avise, J.C., J.C. Trexler, J. Travis, and W.S. Nelson. 1991. *Poecilia mexicana* is the recent female parent of the unisexual fish *P. formosa*. Evolution 45:1530–1533.

Avise, J.C. and R.C. Vrijenhoek. 1987. Mode of inheritance and variation of mitochondrial DNA in hybridogenetic fishes of the genus *Poeciliopsis*. Mol. Biol. Evol. 4:514–525.

Avise, J.C. and R.M. Zink. 1988. Molecular genetic divergence between avian sibling species: king and clapper rails, long-billed and short-billed dowitchers, boat-tailed and great-tailed grackles, and tufted and black-crested titmice. Auk 105:516–528.

Ayala, F.J. 1975. Genetic differentiation during the speciation process. Evol. Biol. 8:1–78.

Ayala, F.J. 1976a. Molecular Evolution. Sinauer, Sunderland, MA.

Ayala, F.J. 1976b. Molecular genetics and evolution. Pp. 1–20 *in:* Molecular Evolution, F.J. Ayala (ed.), Sinauer, Sunderland, MA.

Ayala, F.J. 1982a. The genetic structure of species. Pp. 60–82 *in:* Perspectives on Evolution, R. Milkman (ed.), Sinauer, Sunderland, MA.

Ayala, F.J. 1982b. Genetic variation in natural populations: problem of electrophoretically cryptic alleles. Proc. Natl. Acad. Sci. USA 79:550–554.

Ayala, F.J. 1982c. Of clocks and clades, or a story of old told by genes of now. Pp. 257–301 *in:* Biochemical Aspects of Evolutionary Biology, M.H. Nitecki (ed.), Univ. Chicago Press, Chicago, IL.

Ayala, F.J. 1986. On the virtues and pitfalls of the molecular evolutionary clock. J. Heredity 77:226–235.

Ayala, F.J. and C.A. Campbell. 1974. Frequency-dependent selection. Annu. Rev. Ecol. Syst. 5:115–138.

Ayala, F.J., C.A. Mourão, S. Pérez-Salas, R. Richmond, and T. Dobzhansky. 1970. Enzyme variability in the *Drosophila willistoni* group, I. Genetic differentiation among sibling species. Proc. Natl. Acad. Sci. USA 67:225–232.

Ayala, F.J. and J.R. Powell. 1972a. Enzyme variability in the *Drosophila willistoni* group. VI. Levels of polymorphism and the physiological function of enzymes. Biochem. Genet. 7:331–345.

Ayala, F.J. and J.R. Powell. 1972b. Allozymes as diagnostic characters of sibling species of *Drosophila.* Proc. Natl. Acad. Sci. USA 69:1094–1096.

Ayala, F.J., M.L. Tracey, L.G. Barr, J.F. McDonald, and S. Pérez-Salas. 1974. Genetic variation in natural populations of five *Drosophila* species and the hypothesis of the selective neutrality of protein polymorphisms. Genetics 77:343–384.

Ayala, F.J., M.L. Tracey, D. Hedgecock, and R.C. Richmond. 1975b. Genetic differentiation during the speciation process in *Drosophila.* Evolution 28:576–592.

Ayala, F.J., J.W. Valentine, T.E. Delaca, and G.S. Zumwalt. 1975a. Genetic variability of the antarctic brachiopod *Liothyrella notorcadensis* and its bearing on mass extinction hypotheses. J. Paleontol. 49:1–9.

Ayre, D.J. 1982. Inter-genotype aggression in the solitary sea anemone *Actinia tenebrosa.* Mar. Biol. 68:199–205.

Ayre, D.J. 1984. The effects of sexual and asexual reproduction on geographic variation in the sea anemone *Actinia tenebrosa.* Oecologia 62:222–229.

Ayre, D.J. and J.M. Resing. 1986. Sexual and asexual production of planulae in reef corals. Mar. Biol. 90:187–190.

Ayre, D.J. and B.L. Willis. 1988. Population structure in the coral *Pavona cactus:* clonal genotypes show little phenotypic plasticity. Mar. Biol. 99:495–505.

Baby, M.-C., L. Bernatchez, and J.J. Dodson. 1991. Genetic structure and relationships among anadromous and landlocked populations of rainbow smelt, *Osmerus mordax,* Mitchell, as revealed by mtDNA restriction analysis. J. Fish Biol. 39A:61–68.

Bailey, W.J., J.L. Slightom, and M. Goodman. 1992. Rejection of the "flying primate" hypothesis by phylogenetic evidence from the ε-globin gene. Science 256:86–89.

Baker, C.S., R.H. Lambertsen, M.T. Weinrich, J. Calambokidis, G. Early, and S.J. O'Brien. 1991.

Molecular genetic identification of the sex of humpback whales (*Megaptera novaeangliae*). Pp. 105–111 *in:* Genetic Ecology of Whales and Dolphins, A.R. Hoelzel (ed.), Black Bear Press, Cambridge, England.

Baker, C.S., M. MacCarthy, P.J. Smith, A.P. Perry, and G.K. Chambers. 1992. DNA fingerprints of orange roughy, *Hoplostethus atlanticus:* a population comparison. Mar. Biol. 113:561–567.

Baker, C.S., S.R. Palumbi, R.H. Lambertsen, M.T. Weinrich, J. Calambokidis, and S.J. O'Brien. 1990. Influence of seasonal migration on geographic distribution of mitochondrial DNA haplotypes in humpback whales. Nature 34:238–240.

Baker, R.H. 1968. Habitats and distribution. Pp. 98–126 *in:* Biology of *Peromyscus* (Rodentia), J.A. King (ed.), Special Publication American Society Mammalogy No. 2.

Baker, R.J., S.K. Davis, R.D. Bradley, M.J. Hamilton, and R.A. Van Den Bussche. 1989. Ribosomal-DNA, mitochondrial-DNA, chromosomal, and allozymic studies on a contact zone in the pocket gopher, *Geomys*. Evolution 43:63–75.

Baker, R.J., M.J. Novacek, and N.B. Simmons. 1991. On the monophyly of bats. Syst. Zool. 40:216–231.

Balazs, I., M. Baird, M. Clyne, and E. Meade. 1989. Human population genetic studies of five hypervariable DNA loci. Amer. J. Human Genet. 44:182–190.

Balazs, I., J. Neuweiler, P. Gunn, J. Kidd, K.K. Kidd, J. Kuhl, and L. Mingjun. 1992. Human population genetic studies using hypervariable loci. I. Analysis of Assamese, Australian, Cambodian, Caucasian, Chinese and Melanesian populations. Genetics 131:191–198.

Baldauf, S.L., J.R. Manhart, and J.D. Palmer. 1990. Different fates of the chloroplast *tufA* gene following its transfer to the nucleus in green algae. Proc. Natl. Acad. Sci. USA 87:5317–5321.

Baldauf, S.L. and J.D. Palmer. 1990. Evolutionary transfer of the chloroplast *tufA* gene to the nucleus. Nature 344:262–265.

Baldwin, B.G., D.W. Kyhos, and J. Dvorak. 1990. Chloroplast DNA evolution and adaptive radiation in the Hawaiian silversword alliance (Madiinae, Asteraceae). Annals Missouri Bot. Garden 77:96–109.

Ball, R.M., Jr., and J.C. Avise. 1992. Mitochondrial DNA phylogeographic differentiation among avian populations and the evolutionary significance of subspecies. Auk 109:626–636.

Ball, R.M., Jr., S. Freeman, F.C. James, E. Bermingham, and J.C. Avise. 1988. Phylogeographic population structure of red-winged blackbirds assessed by mitochondrial DNA. Proc. Natl. Acad. Sci. USA 85:1558–1562.

Ball, R.M., J.E. Neigel, and J.C. Avise. 1990. Gene genealogies within the organismal pedigrees of random-mating populations. Evolution 44:360–370.

Ballinger, S.W., T.G. Schurr, A. Torroni, Y.Y. Gan, J.A. Hodge, K. Hassan, K.-H. Chen, and D.C. Wallace. 1992. Southeast Asian mitochondrial DNA analysis reveals genetic continuity of ancient Mongoloid migrations. Genetics 130:139–152.

Ballou, J.D. 1983. Calculating inbreeding coefficients from pedigrees. Pp. 509–520 *in:* Genetics and Conservation, C.M. Schonewald-Cox, S.M. Chambers, F. MacBryde, and L. Thomas (eds.), Benjamin/Cummings, Menlo Park, CA.

Balsano, J.S., R.M. Darnell, and P. Abramoff. 1972. Electrophoretic evidence of triploidy associated with populations of the gynogenetic teleost *Poecilia formosa*. Copeia 1972:292–297.

Bannister, J.V. and M.W. Parker. 1985. The presence of a copper/zinc superoxide dismutase in the bacterium *Photobacterium leiognathi:* a likely case of gene transfer from eukaryotes to prokaryotes. Proc. Natl. Acad. Sci. USA 82:149–152.

Barlow, G.W. 1961. Causes and significance of morphological variation in fishes. Syst. Zool. 10:105–117.

Barnes, B.V. 1966. The clonal growth habit of American aspens. Ecology 47:439–447.

Barnes, P.T. and C.C. Laurie-Ahlberg. 1986. Genetic variability of flight metabolism in *Drosophila melanogaster*. III. Effects of G_{PDH} allozymes and environmental temperature on power output. Genetics 112:267–294.

Barrell, B.G., A.T. Bankier, and J. Drouin. 1979. A different genetic code in human mitochondria. Nature 282:189–194.

Barrett, S.C.H. 1989. Mating system evolution and speciation in heterostylous plants. Pp. 257–283 *in:* Speciation and Its Consequences, D. Otte and J.A. Endler (eds.), Sinauer, Sunderland, MA.

Barrowclough, G.F. 1983. Biochemical studies of microevolutionary processes. Pp. 223–261 *in:* Perspectives in Ornithology, A.H. Brush and G.A. Clark, Jr., (eds.), Cambridge Univ. Press, New York.

Barrowclough, G.F. and R.J. Gutiérrez. 1990. Genetic variation and differentiation in the spotted owl (*Strix occidentalis*). Auk 107:737–744.

Bartlett, S.E. and W.S. Davidson. 1991. Identification of *Thunnus* species by the polymerase chain reaction and direct sequence analysis of their mitochondrial cytochrome *b* genes. Can. J. Fish. Aquat. Sci. 48:309–317.

Barton, N.H. 1989. Founder effect speciation. Pp. 229–256 *in:* Speciation and Its Consequences, D. Otte and J.A. Endler (eds.), Sinauer, Sunderland, MA.

Barton, N.H. and B. Charlesworth. 1984. Genetic revolutions, founder effects, and speciation. Annu. Rev. Ecol. Syst. 15:133–164.

Barton, N.H. and G.M. Hewitt. 1985. Analysis of hybrid zones. Annu. Rev. Ecol. Syst. 16:113–148.

Barton, N.H. and G.M. Hewitt. 1989. Adaptation, speciation and hybrid zones. Nature 341:497–502.

Barton, N.H. and J.S. Jones. 1983. Mitochondrial DNA: new clues about evolution. Nature 306:317–318.

Barton, N.H. and M. Slatkin. 1986. A quasi-equilibrium theory of the distribution of rare alleles in a subdivided population. Heredity 56:409–415.

Bartz, S.H. 1979. Evolution of eusociality in termites. Proc. Natl. Acad. Sci. USA 76:5764–5768.

Bautz, E.K. and F.A. Bautz. 1964. The influence of noncomplementary bases on the stability of ordered polynucleotides. Proc. Natl. Acad. Sci. USA 52:1476–1481.

Baverstock, P.R., M. Adams, and I. Beveridge. 1985. Biochemical differentiation in bile duct cestodes and their marsupial hosts. Mol. Biol. Evol. 2:321–337.

Baverstock, P.R., J. Birrell, and M. Krieg. 1987. Albumin immunologic relationships of the Diprodonta. Pp. 229–234 *in:* Possums and Opossums: Studies in Evolution, M. Archer (ed.), Surrey Beatty and Sons, Chipping Norton, Australia.

Baverstock, P.R., S.R. Cole, B.J. Richardson, and C.H. Watts. 1979. Electrophoresis and cladistics. Syst. Zool. 28:214–219.

Bayer, R.J. 1989. Patterns of isozyme variation in the *Antennaria rosea* (Asteraceae: Inuleae) polyploid agamic complex. Syst. Bot. 14:389–397.

Beckmann, J.S., Y. Kashi, E.M. Hallerman, A. Nave, and M. Soller. 1986. Restriction fragment length polymorphism among Israeli Holstein-Friesian dairy bulls. Anim. Genet. 17:25–38.

Begun, D.J. and C.F. Aquadro. 1992. Levels of naturally occurring DNA polymorphism correlate with recombination rates in *D. melanogaster*. Nature 356:519–520.

Benjamin, D.C., J.A. Berzofsky, I.J. East, F.R.N. Gurd, C. Hannum, S.J. Leach, E. Margoliash, J.G. Michael, A. Milner, E.M. Prager, M. Reichlin, E.E. Sercarz, S.J. Smith-Gill, P.E. Todd, and A.C. Wilson. 1984. The antigenic structure of proteins: A reappraisal. Annu. Rev. Immunol. 2:67–101.

Bennett, S., L.J. Alexander, R.H. Crozier, and A.G. Mackinley. 1988. Are megabats flying primates? Contrary evidence from a mitochondrial DNA sequence. Aust. J. Biol. Sci. 41:327–332.

Berger, E. 1973. Gene-enzyme variation in three sympatric species of *Littorina*. Biol. Bull. 145: 83–90.

Berlocher, S.H. 1981. A comparison of molecular and morphological data, and phenetic and cladistic methods, in the estimation of phylogeny in *Rhagoletis* (Diptera: Tephritidae). Pp. 1–31 *in:* Applications of Genetics and Cytology in Insect Systematics and Evolution, W.M. Stock (ed.), Univ. Idaho Press, Moscow, ID.

Berlocher, S.H. and G.L. Bush. 1982. An electrophoretic analysis of *Rhagoletis* (Diptera: Tephritidae) phylogeny. Syst. Zool. 31:136–155.

Bermingham, E. and J.C. Avise. 1986. Molecular zoogeography of freshwater fishes in the southeastern United States. Genetics 113:939–965.

Bermingham, E., S.H. Forbes, K. Friedland, and C. Pla. 1991. Discrimination between Atlantic salmon (*Salmo salar*) of North American and European origin using restriction analyses of mitochondrial DNA. Can. J. Fish. Aquat. Sci. 48:884–893.

Bermingham, E., T. Lamb, and J.C. Avise. 1986. Size polymorphism and heteroplasmy in the mitochondrial DNA of lower vertebrates. J. Heredity 77:249–252.

Bermingham, E. and H.A. Lessios. 1993. Rate variation of protein and mitochondrial DNA evolution as revealed by sea urchins separated by the Isthmus of Panama. Proc. Natl. Acad. Sci. USA 90:2734–2738.

Bermingham, E., S. Rohwer, S. Freeman, and C. Wood. 1992. Vicariance biogeography in the Pleistocene and speciation in North American wood warblers: a test of Mengel's model. Proc. Natl. Acad. Sci. USA 89:6624–6628.

Bernatchez, L. and J.J. Dodson. 1990. Allopatric origin of sympatric populations of lake whitefish (*Coregonus clupeaformis*) as revealed by mitochondrial-DNA restriction analysis. Evolution 44: 1263–1271.

Bernatchez, L. and J.J. Dodson. 1991. Phylogeographic structure in mitochondrial DNA of the lake whitefish (*Coregonus clupeaformis*) and its relation to Pleistocene glaciations. Evolution 45:1016–1035.

Bernstein, I.S. 1976. Dominance, aggression and reproduction in primate societies. J. Theoret. Biol. 60:459–472.

Bernstein, S., L. Throckmorton, and J. Hubby. 1973. Still more genetic variability in natural populations. Proc. Natl. Acad. Sci. USA 70:3928–3931.

Beverley, S.M. and A.C. Wilson. 1985. Ancient origin for Hawaiian Drosophilinae inferred from protein comparisons. Proc. Natl. Acad. Sci. USA 82:4753–4757.

Billington, N. and P.D.N. Hebert. 1991. Mitochondrial DNA diversity in fishes and its implications for introductions. Can. J. Fish. Aquat. Sci. 48 (suppl. 1):80–94.

Birdsall, D.A. and D. Nash. 1973. Occurrence of successful multiple insemination of females in natural populations of deer mice (*Peromyscus maniculatus*). Evolution 27:106–110.

Birkhead, T.R., T. Burke, R. Zann, F.M. Hunter, and A.P. Krupa. 1990. Extrapair paternity and intraspecific brood parasitism in wild zebra finches *Taeniopygia guttata*, revealed by DNA fingerprinting. Behav. Ecol. Sociobiol. 27:315–324.

Birkhead, T.R. and A.P. Møller. 1992. Sperm Competition in Birds. Academic Press, New York.

Birkhead, T.R. and A.P. Møller. 1993. Female control of paternity. Trends Ecol. Evol. 8:100–104.

Birky, C.W., Jr. 1978. Transmission genetics of mitochondria and chloroplasts. Annu. Rev. Genet. 12:471–512.

Birky, C.W., Jr. 1988. Evolution and variation in plant chloroplast and mitochondrial genomes. Pp. 23–53 *in:* Plant Evolutionary Biology, L.D. Gottlieb and S.K. Jain (eds.), Chapman and Hall, New York.

Birky, C.W., Jr., P. Fuerst, and T. Maruyama. 1989. Organelle gene diversity under migration, mutation, and drift: equilibrium expectations, approach to equilibrium, effects of heteroplasmic cells, and comparison to nuclear genes. Genetics 121:613–627.

Birley, A.J. and J.H. Croft. 1986. Mitochondrial DNAs and phylogenetic relationships. Pp. 107–137 *in:* DNA Systematics, S.K. Dutta (ed.), CRC Press, Boca Raton, FL.

Birt, T.P., J.M. Green, and W.S. Davidson. 1991. Mitochondrial DNA variation reveals genetically distinct sympatric populations of anadromous and nonanadromous Atlantic salmon, *Salmo salar*. Can. J. Fish. Aquat. Sci. 48:577–582.

Bishop, C.E., P. Boursot, B. Baron, F. Bonhomme, and D. Hatat. 1985. Most classical *Mus musculus domesticus* laboratory mouse strains carry a *Mus musculus musculus* Y chromosome. Nature 315:70–72.

Black, R. and M.S. Johnson. 1979. Asexual viviparity and population genetics of *Actinia tenebrosa*. Mar. Biol. 53:27–31.

Blackwelder, R.E. and B.A. Shepherd. 1981. The Diversity of Animal Reproduction. CRC Press, Boca Raton, FL.

Blair, W.F. 1950. Ecological factors in speciation of *Peromyscus*. Evolution 4:253–275.

Blair, W.F. 1955. Mating call and stage of speciation in the *Microhyla olivacea-M. carolinensis* complex. Evolution 9:469–480.

Blanco, A. and W.H. Zinkham. 1963. Lactate dehydrogenases in human testes. Science 139:601–602.

Block, B.A., J.R. Finnerty, A.F.R. Stewart, and J. Kidd. 1993. Evolution of endothermy in fish: mapping physiological traits on a molecular phylogeny. Science 260:210–214.

Blouin, M.S., J.B. Dame, C.A. Tarrant, and C.H. Courtney. 1992. Unusual population genetics of a parasitic nematode: mtDNA variation within and among populations. Evolution 46:470–476.

Boag, P.T. 1987. Effects of nestling diet on growth and adult size of zebra finches (*Poephila guttata*). Auk 104:155–166.

Boehnke, M., N. Arnheim, H. Li, and F.S. Collins. 1989. Fine-structure mapping of human chromosomes using the polymerase chain reaction on single sperm: experimental design considerations. Amer. J. Human Genet. 45:21–32.

Bohlmeyer, D.A. and J.R. Gold. 1991. Genetic studies in marine fishes. II. A protein electrophoretic analysis of population structure in the red drum *Sciaenops ocellatus*. Mar. Biol. 108:197–206.

Bollinger, E.K. and T.A. Gavin. 1991. Patterns of extra-pair fertilizations in bobolinks. Behav. Ecol. Sociobiol. 29:1–7.

Bonhomme, F. and R.K. Selander. 1978. The extent of allelic diversity underlying electrophoretic protein variation in the house mouse. Pp. 569–589 *in:* Origins of Inbred Mice, H.C. Morse, III (ed.), Academic Press, New York.

Bonnell, M.L. and R.K. Selander. 1974. Elephant seals: genetic variation and near extinction. Science 184:908–909.

Booke, H.E. 1981. The conundrum of the stock concept—are nature and nurture definable in fishery science? Can. J. Fish. Aquat. Sci. 38:1479–1480.

Bork, P. and R.F. Doolittle. 1992. Proposed acquisition of an animal protein domain by bacteria. Proc. Natl. Acad. Sci. USA 89:8990–8994.

Bos, M., H. Harmens, and K. Vrieling. 1986. Gene flow in *Plantago* I. Gene flow and neighborhood size in *P. lanceolata*. Heredity 56:43–54.

Bosquet, J., S.H. Strauss, A.H. Doerksen, and R.A. Price. 1992. Extensive variation in evolutionary rate of *rbcL* gene sequences among seed plants. Proc. Natl. Acad. Sci. USA 89:7844–7848.

Botstein, D., R.L. White, M. Skolnick, and R.W. Davis. 1980. Construction of a genetic linkage map in man using restriction fragment length polymorphisms. Amer. J. Human Genet. 32:314–331.

Bowen, B.W. and J.C. Avise. 1990. Genetic structure of Atlantic and Gulf of Mexico populations of sea bass, menhaden, and sturgeon: influence of zoogeographic factors and life history patterns. Mar. Biol. 107:371–381.

Bowen, B.W., A.B. Meylan, and J.C. Avise. 1989. An odyssey of the green sea turtle: Ascension Island revisited. Proc. Natl. Acad. Sci. USA 86:573–576.

Bowen, B.W., A.B. Meylan, and J.C. Avise. 1991. Evolutionary distinctiveness of the endangered Kemp's ridley sea turtle. Nature 352:709–711.

Bowen, B.W., A.B. Meylan, J.P. Ross, C.J. Limpus, G.H. Balazs, and J.C. Avise. 1992. Global population structure and natural history of the green turtle (*Chelonia mydas*) in terms of matriarchal phylogeny. Evolution 46:865–881.

Bowen, B.W., W.S. Nelson, and J.C. Avise. 1993a. A molecular phylogeny for marine turtles: trait mapping, rate assessment, and conservation relevance. Proc. Natl. Acad. Sci. USA, 90:5574–5577.

Bowen, B.W., J.I. Richardson, A.B. Meylan, D. Margaritoulis, R. Hopkins Murphy, and J.C. Avise. 1993b. Population structure of loggerhead turtles (*Caretta caretta*) in the West Atlantic Ocean and Mediterranean Sea. Cons. Biol., *in press*.

Bowers, J.H., R.J. Baker, and M.H. Smith. 1973. Chromosomal, electrophoretic, and breeding studies of selected populations of deer mice (*Peromyscus maniculatus*) and black-eared mice (*P. melanotis*). Evolution 27:378–386.

Bowling, A.T. and O.A. Ryder. 1987. Genetic studies of blood markers in Przewalski's horses. J. Heredity 78:75–80.

Bowman, B.H., J.W. Taylor, A.G. Brownlee, J. Lee, S.-D. Lu, and T.J. White. 1992. Molecular evolution of the fungi: relationships of the Basidiomycetes, Ascomycetes, and Chytridiomycetes. Mol. Biol. Evol. 9:285–296.

Boyce, A.J. 1983. Computation of inbreeding and kinship coefficients on extended pedigrees. J. Heredity 74:400–404.

Boyer, J.F. 1974. Clinal and size-dependent variation at the LAP locus in *Mytilus edulis*. Biol. Bull. 147:535–549.

Breder, C.M., Jr. 1936. The reproductive habits of North American sunfishes (family Centrarchidae). Zoologica 21:1–47.

Bremer, B. and K. Bremer. 1989. Cladistic analysis of blue-green procaryote interrelationships and chloroplast origin based on 16S rRNA oligonucleotide catalogues. J. Evol. Biol. 2:13–30.

Brewer, G.J. 1970. An Introduction to Isozyme techniques. Academic Press, New York.

Bridge, D., C.W. Cunningham, B. Schierwater, R. DeSalle, and L.W. Buss. 1992. Class-level relationships in the phylum Cnidaria: evidence from mitochondrial genome structure. Proc. Natl. Acad. Sci. USA 89:8750–8753.

Briggs, J.C. 1958. A list of Florida fishes and their distribution. Bull. Fla. State Mus. Biol. Sci. 2:223–318.

Briggs, J.C. 1974. Marine Zoogeography. McGraw-Hill, New York.

Britten, R.J. 1986. Rates of DNA sequence evolution differ between taxonomic groups. Science 231:1393–1398.

Britten, R.J. and E.H. Davidson. 1969. Gene regulation for higher cells: a theory. Science 165: 349–357.

Britten, R.J. and E.H. Davidson. 1971. Repetitive and non-repetitive DNA sequences and a speculation on the origins of evolutionary novelty. Quart. Rev. Biol. 46:111–138.

Britten, R.J., D.E. Graham, and B.R. Neufeld. 1974. Analysis of repeating DNA sequences by reassociation. Methods Enzymol. 29:363–418.

Britten, R.J. and D.E. Kohne. 1968. Repeated sequences in DNA. Science 161:529–540.

Brock, M.K. and B.N. White. 1991. Multifragment alleles in DNA fingerprints of the parrot, *Amazona ventralis*. J. Heredity 82:209–212.

Brookfield, J.F.Y. 1992. DNA fingerprinting in clonal organisms. Mol. Ecol. 1:21–26.

Brooks, D.R. and D.A. McLennan. 1991. Phylogeny, Ecology, and Behavior. Univ. of Chicago Press, Chicago, IL.

Brower, A.V.Z. and T.M. Boyce. 1991. Mitochondrial DNA variation in monarch butterflies. Evolution 45:1281–1286.

Brown, A.H.D. 1989. Genetic characterization of plant mating systems. Pp. 145–162 *in:* Plant Population Genetics, Breeding and Genetic Resources, A.H.D. Brown, M.T. Clegg, A.L. Kahler, and B.S. Weir (eds.), Sinauer, Sunderland, MA.

Brown, A.H.D. and R.W. Allard. 1970. Estimation of the mating system in open-pollinated maize populations using isozyme polymorphisms. Genetics 66:133–145.

Brown, C.R. and M.B. Brown. 1988. Genetic evidence of multiple parentage in broods of cliff swallows. Behav. Ecol. Sociobiol. 23:379–387.

Brown, D.D., P.C. Wensink, and E. Jordan. 1972. A comparison of the ribosomal DNAs of *Xenopus laevis* and *Xenopus mulleri:* Evolution of tandem genes. J. Mol. Biol. 63:57–73.

Brown, G.G. and M.V. Simpson. 1982. Novel features of animal mtDNA evolution as shown by sequences of two rat cytochrome oxidase subunit II genes. Proc. Natl. Acad. Sci. USA 79:3246–3250.

Brown, J.L. 1987. Helping and Communal Breeding in Birds. Princeton Univ. Press, Princeton, NJ.

Brown, T. 1992. Ancient DNA reference list. Ancient DNA Newsl. 1(1):36–38.

Brown, W.M. 1980. Polymorphism in mitochondrial DNA of humans as revealed by restriction endonuclease analysis. Proc. Natl. Acad. Sci. USA 77:3605–3609.

Brown, W.M. 1983. Evolution of animal mitochondrial DNA. Pp. 62–88 *in:* Evolution of Genes and Proteins, M. Nei and R.K. Koehn (eds.), Sinauer, Sunderland, MA.

Brown, W.M. 1985. The mitochondrial genome of animals. Pp. 95–130 *in:* Evolutionary Genetics, R.J. MacIntyre (ed.), Plenum, New York.

Brown, W.M., M. George, Jr., and A.C. Wilson. 1979. Rapid evolution of animal mitochondrial DNA. Proc. Natl. Acad. Sci. USA 76:1967–1971.

Brown, W.M., E.M. Prager, A. Wang, and A.C. Wilson. 1982. Mitochondrial DNA sequences of primates: Tempo and mode of evolution. J. Mol. Evol. 18:225–239.

Brown, W.M. and J. Wright. 1979. Mitochondrial DNA analyses and the origin and relative age of parthenogenetic lizards (genus *Cnemidophorus*). Science 203:1247–1249.

Bruce E.J. and F.J. Ayala. 1979. Phylogenetic relationships between man and the apes: electrophoretic evidence. Evolution 33:1040–1056.

Bruford, M.W., O. Hanotte, J.F.Y. Brookfield, and T. Burke. 1992. Single-locus and multilocus DNA fingerprinting. Pp. 225–269 *in:* Molecular Genetic Analysis of Populations. A Practical Approach, A.R. Hoelzel (ed.), IRL Press, Oxford.

Brunk, C.F. and E.C. Olson (eds.). 1990. Special issue on DNA–DNA hybridization and evolution. J. Mol. Evol. 30:191–311.

Bruns, T.D., R. Fogel, T.J. White, and J.D. Palmer. 1989. Accelerated evolution of a false-truffle from a mushroom ancestor. Nature 339:140–142.

Bruns, T.D. and J.D. Palmer. 1989. Evolution of mushroom mitochondrial DNA: *Suillus* and related genera. J. Mol. Evol. 28:349–362.

Bruns, T.D., T.J. White, and J.W. Taylor. 1991. Fungal molecular systematics. Annu. Rev. Ecol. Syst. 22:525–564.

Brunsfeld, S.J., D.E. Soltis, and P.S. Soltis. 1992. Evolutionary patterns and processes in *Salix* sect. *Longifoliae:* evidence from chloroplast DNA. Syst. Bot. 17:239–256.

Bucklin, A., D. Hedgecock, and C. Hand. 1984. Genetic evidence of self-fertilization in the sea anemone *Epiactis prolifera*. Mar. Biol. 84:175–182.

Budowle, B., A.M. Giusti, J.S. Waye, F.S. Baechtel, R.M. Fourney, D.E. Adams, L.A. Presley, H.A. Deadman, and K.L. Monson. 1991. Fixed-bin analysis for statistical evaluation of continuous distributions of allelic data from VNTR loci, for use in forensic comparisons. Amer. J. Human Genet. 48:841–855.

Bull, J.J. 1980. Sex determination in reptiles. Quart. Rev. Biol. 55:3–21.

Burgoyne, P.S. 1986. Mammalian X and Y crossover. Nature 319:258–259.

Burke, T. 1989. DNA fingerprinting and other methods for the study of mating success. Trends Ecol. Evol. 4:139–144.

Burke, T. and M.W. Bruford. 1987. DNA fingerprinting in birds. Nature 327:149–152.

Burke, T., N.B. Davies, M.W. Bruford, and B.J. Hatchwell. 1989. Parental care and mating behavior of polyandrous dunnocks *Prunella modularis* related to paternity by DNA fingerprinting. Nature 338:249–251.

Burke, T., O. Hanotte, M.W. Bruford, and E. Cairns. 1991. Multilocus and single locus minisatellite analysis in population biological studies. Pp. 154–168 *in:* DNA Fingerprinting Approaches and Applications, T. Burke, G. Dolf, A.J. Jeffreys, and R. Wolff (eds.), Birkhauser Verlag, Basel, Switzerland.

Burnell, K.L. and S.B. Hedges. 1990. Relationships of West Indian *Anolis* (Sauria: Iguanidae): an approach using slow-evolving protein loci. Caribbean J. Sci. 26:7–30.

Buroker, N.E. 1983. Population genetics of the American oyster *Crassostrea virginica* along the Atlantic coast and the Gulf of Mexico. Mar. Biol. 75:99–112.

Burton, R.S. 1983. Protein polymorphisms and genetic differentiation of marine invertebrate populations. Mar. Biol. Letters 4:193–206.

Burton, R.S. 1985. Mating system of the intertidal copepod *Tigriopus californicus*. Mar. Biol. 86:247–252.

Burton, R.S. 1986. Evolutionary consequences of restricted gene flow among natural populations of the copepod, *Tigriopus californicus*. Bull. Mar. Sci. 39:526–535.

Burton, R.S. and M.W. Feldman. 1981. Population genetics of *Tigriopus californicus*. II. Differentiation among neighboring populations. Evolution 35:1192–1205.

Burton, R.S. and M.W. Feldman. 1983. Physiological effects of an allozyme polymorphism: glutamate-pyruvate transaminase and response to hyperosmotic stress in the copepod *Tigriopus californicus*. Biochem. Genet. 21:238–251.

Burton, R.S., M.W. Feldman, and J.W. Curtsinger. 1979. Population genetics of *Tigriopus californicus* (Copepoda: Harpacticoida). I. Population structure along the central California coast. Mar. Ecol. Progr. Ser. 1:29–39.

Burton, R.S. and S.G. Swisher. 1984. Population structure in the intertidal copepod *Tigriopus californicus* as revealed by field manipulation of allele frequencies. Oecologia 65:108–111.

Busack, C.A. and G.A.E. Gall. 1981. Introgressive hybridization in populations of Paiute cutthroat trout (*Salmo clarki seleniris*). Can. J. Fish. Aquat. Sci. 38:939–951.

Bush, G.L. 1969. Sympatric host race formation and speciation in frugivorous flies of the genus *Rhagoletis* (Diptera, Tephritidae). Evolution 23:237–251.

Bush, G.L. 1975. Modes of animal speciation. Annu. Rev. Ecol. Syst. 6:339–364.

Buss, L.W. 1983. Evolution, development, and the units of selection. Proc. Natl. Acad. Sci. USA 80:1387–1391.

Buss, L.W. 1985. Uniqueness of the individual revisited. Pp. 467–505 *in:* Population Biology and Evolution of Clonal Organisms, J.B.C. Jackson, L.W. Buss, and R.E. Cook (eds.), Yale Univ. Press, New Haven, CT.

Buth, D.G. 1979. Duplicate gene expression in tetraploid fishes of the tribe Moxostomatini (Cypriniformes, Catostomidae). Comp. Biochem. Physiol. 63B:7–12.

Buth, D.G. 1982. Glucosephosphate-isomerase expression in a tetraploid fish, *Moxostoma lachneri* (Cypriniformes, Catostomidae): Evidence for "retetraploidization"? Genetica 57:171–175.

Buth, D.G. 1983. Duplicate isozyme loci in fishes: origins, distribution, phyletic consequences, and locus nomenclature. Isozymes X:381–400.

Buth, D.G. 1984. The application of electrophoretic data in systematic studies. Annu. Rev. Ecol. Syst. 15:501–522.

Buth, D.G., R.W. Murphy, M.M. Miyamoto, and C.S. Lieb. 1985. Creatine kinases of amphibians and reptiles: Evolutionary and systematic aspects of gene expression. Copeia 1985:279–284.

Butlin, R. 1989. Reinforcement of premating isolation. Pp. 158–179 *in:* Speciation and Its Consequences, D. Otte and J.A. Endler (eds.), Sinauer, Sunderland, MA.

Caccone, A., G.D. Amato, and J.R. Powell. 1987. Intraspecific DNA divergence in *Drosophila:* a study on parthenogenetic *D. mercatorum*. Mol. Biol. Evol. 4:343–350.

Caccone, A., G.D. Amato, and J.R. Powell. 1988a. Rates and patterns of scnDNA and mtDNA divergence within the *Drosophila melanogaster* subgroup. Genetics 118:671–683.

Caccone, A., R. DeSalle, and J.R. Powell. 1988b. Calibration of the change in thermal stability of DNA duplexes and degree of base-pair mismatch. J. Mol. Evol. 27:212–216.

Caccone, A. and J.R. Powell. 1987. Molecular evolutionary divergence among North American cave crickets. II. DNA–DNA hybridization. Evolution 41:1215–1238.

Caccone, A. and J.R. Powell. 1989. DNA divergence among hominoids. Evolution 43:925–942.

Caccone, A. and V. Sbordoni. 1987. Molecular evolutionary divergence among North American cave crickets. I. Allozyme variation. Evolution 41:1198–1214.

Calahan, C.M. and C. Gliddon. 1985. Genetic neighborhood sizes in *Primula vulgaris*. Heredity 54:65–70.

Calie, P.J. and K.W. Hughes. 1987. The consensus land plant chloroplast gene order is present, with two alterations, in the moss *Physcomitrella patens*. Mol. Gen. Genet. 208:335–341.

Calvi, B.R., et al. 1991. Evidence for a common evolutionary origin of inverted repeat transposons in *Drosophila* and plants: hobo, Activator, and Tam3. Cell 66:465–471.

Camin, J.H. and R.R. Sokal. 1965. A method for deducing branching sequences in phylogeny. Evolution 19:311–326.

Cammarano, P., et al. 1992. Early evolutionary relationships among known life forms inferred from elongation factor EF-2/EF-G sequences: phylogenetic coherence and structure of the Archael domain. J. Mol. Evol. 34:396–405.

Campton, D.E. 1987. Natural hybridization and introgression in fishes. Pp. 161–192 *in:* Population Genetics and Fisheries Management, N. Ryman and F. Utter (eds.), Univ. Washington Press, Seattle.

Campton, D.E. and F.M. Utter. 1987. Genetic structure of anadromous cutthroat trout (*Salmo clarki clarki*) populations in the Puget Sound area: evidence for restricted gene flow. Can. J. Fish. Aquat. Sci. 44:573–582.

Cann, R.L., et al. 1984. Polymorphic sites and the mechanism of evolution in human mitochondrial DNA. Genetics 106:479–499.

Cann, R.L., et al. 1987. Mitochondrial DNA and human evolution. Nature 325:31–36.

Cannings, C. and E.A. Thompson. 1982. Genealogical and Genetic Structure. Cambridge Univ. Press, Cambridge, England.

Cano, R.J., et al. 1992. Enzymatic amplification and nucleotide sequencing of portions of the 18s rRNA gene of the bee *Proplebeia dominicana* (Apidae: Hymenoptera) isolated from 25–40 million year old Dominican amber. Med. Sci. Res. 20:619–622.

Cano, R.J., et al. 1993. Amplification and sequencing of DNA from a 120–135-million-year-old weevil. Nature 363:536–538.

Cantatore, P., M.N. Gadaleta, M. Roberti, C. Saccone, and A.C. Wilson. 1987. Duplication and remoulding of tRNA genes during the evolutionary rearrangement of mitochondrial genomes. Nature 329:853–855.

Cantatore, P. and C. Saccone. 1987. Organization, structure, and evolution of mammalian mitochondrial genes. Int. Rev. Cytol. 108:149–208.

Carlson, T.A. and B.K. Chelm. 1986. Apparent eukaryotic origin of glutamine synthetase II from the bacterium *Bradyrhizobium japonicum*. Nature 322:568–570.

Carr, A. and P.J. Coleman. 1974. Seafloor spreading theory and the odyssey of the green turtle from Brazil to Ascension Island, Central Atlantic. Nature 249:128–130.

Carr, S.M., S.W. Ballinger, J.N. Derr, L.H. Blankenship, and J.W. Bickham. 1986. Mitochondrial DNA analysis of hybridization between sympatric white-tailed deer and mule deer in west Texas. Proc. Natl. Acad. Sci. USA 83:9576–9580.

Carr, S.M. and O.M. Griffith. 1987. Rapid isolation of animal mitochondrial DNA in a small fixed-angle rotor at ultrahigh speed. Biochem. Genet. 25:385–390.

Carson, H.L. 1968. The population flush and its genetic consequences. Pp. 123–137 *in:* Population Biology and Evolution, R.C. Lewontin (ed.), Syracuse Univ. Press, Syracuse, NY.

Carson, H.L. 1976. Inference of the time of origin of some *Drosophila* species. Nature 259:395–396.

Carson, H.L. 1990. Increased genetic variance after a population bottleneck. Trends Ecol. Evol. 5:228–230.

Carson, H.L. 1992. The Galapagos that were. Nature 355:202–203.

Carson, H.L. and K.Y. Kaneshiro. 1976. *Drosophila* of Hawaii: systematics and ecological genetics. Annu. Rev. Ecol. Syst. 7:311–345.

Carson, H.L. and A.R. Templeton. 1984. Genetic revolutions in relation to speciation phenomena: the founding of new populations. Annu. Rev. Ecol. Syst. 15:97–131.

Carvalho, G.R., N. Maclean, S.D. Wratten, R.E. Carter, and J.P. Thurston. 1991. Differentiation of aphid clones using DNA fingerprints from individual aphids. Proc. Roy. Soc. London B243: 109–114.

Casanova, M., P. Leroy, C. Boucekkine, J. Weissenbach, C. Bishop, M. Fellous, M. Purrello, G. Fiori, and M. Siniscalco. 1985. A human Y-linked DNA polymorphism and its potential for estimating genetic and evolutionary distance. Science 230:1403–1406.

Castle, P.H.J. 1984. Notacanthiformes and Anguilliformes: development. Pp. 62–93 *in:* Ontogeny and Systematics of Fishes, Spec. Publ. No. 1, Amer. Soc. Ichthyologists and Herpetologists, Allen Press, Lawrence, KS.

Catzeflis, F.M., F.H. Sheldon, J.E. Ahlquist, and C.G. Sibley. 1987. DNA–DNA hybridization evidence of the rapid rate of muroid rodent DNA evolution. Mol. Biol. Evol. 4:242–253.

Caugant, D.A., L.O. Froholm, K. Bovre, E. Holten, C.F. Frasch, L.F. Mocca, W.D. Zollinger, and R.K. Selander. 1986. Intercontinental spread of a genetically distinctive complex of clones of *Neisseria meningitidis* causing epidemic disease. Proc. Natl. Acad. Sci. USA 83:4927–4931.

Caugant, D.A., B.R. Levin, and R.K. Selander. 1981. Genetic diversity and temporal variation in the *E. coli* population of a human host. Genetics 98:467–490.

Cavalier-Smith, T. 1975. The origin of nuclei and of eukaryotic cells. Nature 256:463–468.

Cavalier-Smith, T. (ed.). 1985. The Evolution of Genome Size. Wiley, New York.

Cavalier-Smith, T. 1991. Archamoebae: the ancestral eukaryotes? BioSystems 25:25–38.

Cavalli-Sforza, L. 1966. Population structure and human evolution. Proc. Royal Soc. London B164:362–379.

Cavener, D.R. and M.T. Clegg. 1981. Evidence for biochemical and physiological differences between enzyme genotypes in *Drosophila melanogaster*. Proc. Natl. Acad. Sci. USA 78:4444–4447.

Cedergren, R., M.W. Gray, Y. Abel and D. Sankoff. 1988. The evolutionary relationships among known life forms. J. Mol. Evol. 28:98–112.

Chakraborty, R. 1981. The distribution of the number of heterozygous loci in an individual in natural populations. Genetics 98:461–466.

Chakraborty, R. and K.K. Kidd. 1991. The utility of DNA typing in forensic work. Science 254:1735–1739.

Champion, A.B., E.M. Prager, D. Wachter, and A.C. Wilson. 1974. Microcomplement fixation. Pp. 397–416 *in:* C.A. Wright (ed.), Biochemical and Immunological Taxonomy of Animals, Academic Press, New York.

Chapman, R.W. and D.A. Powers. 1984. A method for rapid isolation of mtDNA from fishes. Maryland Sea Grant Technical Rep. UM-SG-TS-84-05 (11 pp.).

Chapman, R.W., J.C. Stephens, R.A. Lansman, and J.C. Avise. 1982. Models of mitochondrial DNA transmission genetics and evolution in higher eucaryotes. Genet. Res. 40:41–57.

Charlesworth, B., J.A. Coyne, and N.H. Barton. 1987. The relative rates of evolution of sex chromosomes and autosomes. Amer. Natur. 130:113–146.

Charlesworth, B., R. Lande, and M. Slatkin. 1982. A neo-Darwinian commentary on macroevolution. Evolution 36:474–498.

Charlesworth, D. and B. Charlesworth. 1987. Inbreeding depression and its evolutionary consequences. Annu. Rev. Ecol. Syst. 18:237–268.

Chase, M.W. and 37 others. 1993. Phylogenetics of seed plants: an analysis of nucleotide sequences from the plastid gene *rbcL*. Annals Missouri Bot. Gardens, *in press*.

Cheliak, W.M. and J.A. Patel. 1984. Electrophoretic identification of clones in trembling aspen. Can. J. Forest Res. 14:740–743.

Cherry, L.M., S.M. Case, and A.C. Wilson. 1978. Frog perspective on the morphological divergence between humans and chimpanzees. Science 200:209–211.

Chesser, R.K. 1983. Genetic variability within and among populations of the black-tailed prairie dog. Evolution 37:320–331.

Chesser, R.K., M.W. Smith, and M.H. Smith. 1984. Biochemical genetics of mosquitofish populations. I. Incidence and significance of multiple insemination. Genetica 64:77–81.

Childers, W.F. 1967. Hybridization of four species of sunfishes (Centrarchidae). Bull. Ill. Nat. Hist. Surv. 29:159–214.

Christie, D.M., R.A. Duncan, A.R. McBirney, M.A. Richards, W.M. White, K.S. Harpp, and C.G. Fox. 1992. Drowned islands downstream from the Galapagos hotspot imply extended speciation times. Nature 355:246–248.

Cimino, M.C. 1972. Egg production, polyploidization and evolution in a diploid all-female fish of the genus *Poeciliopsis*. Evolution 26:294–306.

Clark, A.G. 1988. Deterministic theory of heteroplasmy. Evolution 42:621–626.

Clark, A.G. 1990. Inference of haplotypes from PCR-amplified samples of diploid populations. Mol. Biol. Evol. 7:111–122.

Clarke, B. 1975. The contribution of ecological genetics to evolutionary theory: detecting the direct effects of natural selection on particular polymorphic loci. Genetics 79s:101–113.

Clausen, J., D.D. Keck, and W.M. Hiesey. 1940. Experimental studies on the nature of species. I. Effect of varied environments on western North American plants. Carnegie Inst. Wash. Publ. No. 520.

Clegg, M.T. 1980. Measuring plant mating systems. BioScience 30:814–818.

Clegg, M.T. and R.W. Allard. 1972. Patterns of genetic differentiation in the slender wild oat species *Avena barbata*. Proc. Natl. Acad. Sci. USA 69:1820–1824.

Clegg, M.T. and R.W. Allard. 1973. Viability versus fecundity selection in the slender wild oat, *Avena barbata* L. Science 181:667–668.

Clegg, M.T., R.W. Allard, and A.L. Kahler. 1972. Is the gene the unit of selection? Evidence from two experimental plant populations. Proc. Natl. Acad. Sci. USA 69:2474–2478.

Clegg, M.T., K. Ritland, and G. Zurawski. 1986. Processes of chloroplast DNA evolution. Pp. 275–294 *in:* Evolutionary Processes and Theory, S. Karlin and E. Nevo (eds.), Academic Press, New York.

Clegg, M.T. and G. Zurawski. 1992. Chloroplast DNA and the study of plant phylogeny: present status and future prospects. Pp. 1–13 *in:* Molecular Systematics of Plants, P.S. Soltis, D.E. Soltis, and J.J. Doyle (eds.), Chapman and Hall, New York.

Cobbs, G. 1977. Multiple insemination and male sexual selection in natural populations of *Drosophila pseudoobscura.* Amer. Natur. 111:641–656.

Cocks, G.T. and A.C. Wilson. 1972. Enzyme evolution in the Enterobacteriaceae. J. Bacteriol. 110:793–802.

Coffin, J.M., A. Hasse, J.A. Levy, L. Montagnier, S. Oroszlan, N. Teich, H. Temin, K. Toyoshima, H. Varmus, P. Vogt, and R. Weiss. 1986. Human immunodeficiency viruses. Science 232:697.

Coffroth, M.A., H.R. Lasker, M.E. Diamond, J.A. Bruenn, and E. Bermingham. 1992. DNA fingerprints of a gorgonian coral: a method for detecting clonal structure in a vegetative species. Mar. Biol. 114:317–325.

Cole, B.J. 1983. Multiple mating and the evolution of social behavior in the Hymenoptera. Behav. Ecol. Sociobiol. 12:191–201.

Collier, G.E. and R.J. MacIntyre. 1977. Microcomplement fixation studies on the evolution of α-glycerophosphate dehydrogenase within the genus *Drosophila.* Proc. Natl. Acad. Sci. USA 74:684–688.

Cook, J.A., K.R. Bestgen, D.L. Propst, and T.L. Yates. 1992. Allozymic divergence and systematics of the Rio Grande silvery minnow, *Hybognathus amarus* (Teleostei: Cyprinidae). Copeia 1992:36–44.

Cook, R.E. 1980. Reproduction by duplication. Natur. Hist. 89:88–93.

Cook, R.E. 1983. Clonal plant populations. Amer. Sci. 71:244–253.

Cook, R.E. 1985. Growth and development in clonal plant populations. Pp. 259–296 *in:* Population Biology and Evolution of Clonal Organisms, J.B.C. Jackson, L.W. Buss, and R.E. Cook (eds.), Yale Univ. Press, New Haven, CT.

Cooke, F., C.D. MacInnes, and J.P. Prevett. 1975. Gene flow between breeding populations of lesser snow geese. Auk 92:493–510.

Cooke, F., D.T. Parkin, and R.F. Rockwell. 1988. Evidence of former allopatry of the two color phases of lesser snow geese (*Chen caerulescens caerulescens*). Auk 105:467–479.

Cooper, A., C. Mourer-Chauvire, G.K. Chambers, A. von Haeseler, A.C. Wilson, and S. Pääbo. 1992. Independent origins of New Zealand moas and kiwis. Proc. Natl. Acad. Sci. USA 89: 8741–8744.

Costa, J.T., III and K.G. Ross. 1993. Seasonal decline in intracolony genetic relatedness in eastern tent caterpillars: implications for social evolution. Behavior. Ecol. Sociobiol 32:47–54.

Coyne, J.A. 1982. Gel electrophoresis and cryptic protein variation. Isozymes V:1–32.

Coyne, J.A. 1992. Genetics and speciation. Nature 355:511–515.

Coyne, J.A. and H.A. Orr. 1989a. Two rules of speciation. Pp. 180–207 *in:* Speciation and Its Consequences, D. Otte and J.A. Endler (eds.), Sinauer, Sunderland, MA.

Coyne, J.A. and H.A. Orr. 1989b. Patterns of speciation in *Drosophila.* Evolution 43:362–381.

Crabtree, C.B. and D.G. Buth. 1987. Biochemical systematics of the catostomid genus *Catostomus:* assessment of *C. clarki, C. plebeius* and *C. discobolus* including the Zuni sucker, *C.d. yarrowi.* Copeia 1987:843–854.

Cracraft, J. 1983. Species concepts and speciation analysis. Pp. 159–187 *in:* Current Ornithology, R.F. Johnston (ed.), Plenum Press, New York.

Cracraft, J. 1986. Origin and evolution of continental biotas: speciation and historical congruence within the Australian avifauna. Evolution 40:977–996.

Cracraft, J. 1992. Book Review—Phylogeny and Classification of Birds. Mol. Biol. Evol. 9:182–186.

Cracraft, J. and D.P. Mindell. 1989. The early history of modern birds: a comparison of molecular and morphological evidence. Pp. 389–403 *in:* The Hierarchy of Life, B. Fernholm, K. Bremer, and H. Jornvall (eds.), Elsevier, Amsterdam.

Craig, R. and R.H. Crozier. 1979. Relatedness in the polygynous ant *Myrmecia pilosula.* Evolution 33:335–341.

Crane, P.R. 1985. Phylogenetic analysis of seed plants and the origin of angiosperms. Annals Missouri Bot. Gardens 72:716–793.

Crawford, D.J. and R. Ornduff. 1989. Enzyme electrophoresis and evolutionary relationships among three species of *Lasthenia* (Asteraceae: Heliantheae). Amer. J. Bot. 76:289–296.

Crease, T.J., D.J. Stanton, and P.D.N. Hebert. 1989. Polyphyletic origins of asexuality in *Daphnia pulex.* II. Mitochondrial-DNA variation. Evolution 43:1016–1026.

Crisp, D.J. 1976. Settlement responses in marine organisms. Pp. 83–124 *in:* Adaptation to the Marine Environment, R.C. Newell (ed.), Butterworths, London.

Crisp, D.J. 1978. Genetic consequences of different reproductive strategies in marine invertebrates. Pp. 257–273 *in:* Marine Organisms: Genetics, Ecology, and Evolution, B. Battaglia and J.A. Beardmore (eds.), Plenum Press, New York.

Cronin, M.A. 1992. Intraspecific variation in mitochondrial DNA of North American cervids. J. Mammal. 73:70–82.

Cronin, M.A., S.C. Amstrup, G.W. Garner, and E.R. Vyse. 1991b. Interspecific and intraspecific mitochondrial DNA variation in North American bears (*Ursus*). Can. J. Zool. 69:2985–2992.

Cronin, M.A., M.E. Nelson, and D.F. Pac. 1991a. Spatial heterogeneity of mitochondrial DNA and allozymes among populations of white-tailed deer and mule deer. J. Heredity 82:118–127.

Cronin, M.A., D.A. Palmisciano, E.R. Vyse, and D.G. Cameron. 1991c. Mitochondrial DNA in wildlife forensic science: species identification of tissues. Wildl. Soc. Bull. 19:94–105.

Crosetti, D., W.S. Nelson, and J.C. Avise. 1993. Pronounced genetic structure of mitochondrial DNA among populations of the circum-globally distributed grey mullet (*Mugil cephalus* Linnaeus). J. Fish Biol., *in press.*

Crosland, M.W.J. 1988. Inability to discriminate between related and unrelated larvae in the ant *Rhytidoponera confusa* (Hymenoptera: Formicidae). Ann. Entomol. Soc. Amer. 81:844–850.

Crow, J.F. 1954. Breeding structure of populations. II. Effective population number. Pp. 543–556 *in:* Statistics and Mathematics in Biology, T.A. Bancroft, J.W. Gowen, and J.L. Lush (eds.), Iowa State College Press, Ames.

Crozier, R.H., Y.C. Crozier, and A.G. Mackinley. 1989. The CO-I and CO-II region of honeybee mitochondrial DNA: evidence for variation in insect mitochondrial evolutionary rates. Mol. Biol. Evol. 6:399–411.

Crozier, R.H. and P. Luykx. 1985. The evolution of termite eusociality is unlikely to have been based on a male-haploid analogy. Amer. Natur. 126:867–869.

Crozier, R.H. and R.E. Page. 1985. On being the right size: male contributions and multiple mating in social Hymenoptera. Behav. Ecol. Sociobiol. 18:105–115.

Crozier, R.H., P. Pamilo, and Y.C. Crozier. 1984. Relatedness and microgeographic genetic variation in *Rhytidoponera mayri,* an Australian arid zone ant. Behav. Ecol. Sociobiol. 15:143–150.

Crozier, R.H., B.H. Smith, and Y.C. Crozier. 1987. Relatedness and population structure of the primitively eusocial bee *Lasioglossum zephyrum* (Hymenoptera: Halictidae) in Kansas. Evolution 41:902–910.

Cuellar, O. 1974. On the origin of parthenogenesis in vertebrates: the cytogenetic factors. Amer. Natur. 108:625–648.

Cuellar, O. 1977. Animal parthenogenesis. Science 197:837–843.

Cuellar, O. 1984. Histocompatibility in Hawaiian and Polynesian populations of the parthenogenetic gecko *Lepidodactylus lugubris*. Evolution 38:176–185.

Cunningham, C.W., N.W. Blackstone, and L.W. Buss. 1992. Evolution of king crabs from hermit crab ancestors. Nature 355:539–542.

Curie-Cohen, M., D. Yoshihara, L. Luttrell, K. Benforado, J.W. MacCluer, and W.H. Stone. 1983. The effects of dominance on mating behavior and paternity in a captive troop of rhesus monkeys (*Macaca mulatta*). Amer. J. Primatol. 5:127–138.

Curtis, A.S.G., J. Kerr, and N. Knowlton. 1982. Graft rejection in sponges. Genetic structure of accepting and rejecting populations. Transplantation 30:127–133.

Curtis, S.E. and M.T. Clegg. 1984. Molecular evolution of chloroplast DNA sequences. Mol. Biol. Evol. 1:291–301.

Cutler, M.G., S.E. Bartlett, S.E. Hartley, and W.S. Davidson. 1991. A polymorphism in the ribosomal RNA genes distinguishes Atlantic Salmon (*Salmo salar*) from North America and Europe. Can. J. Fish. Aquat. Sci. 48:1655–1661.

Dancik, B.P., and F.C. Yeh. 1983. Allozyme variability and evolution of lodgepole pine (*Pinus contorta* var. *latifolia*) and jack pine (*Pinus banksiana*) in Alberta. Can. J. Genet. Cytol. 25:57–64.

Daniels, S.B., K.R. Peterson, L.D. Strausbaugh, M.G. Kidwell, and A. Chovnick. 1990. Evidence for horizontal transmission of the P transposable element between *Drosophila* species. Genetics 124:339–355.

Darlington, C.D. 1939. The Evolution of Genetic Systems. Cambridge, Univ. Press, Cambridge, England.

Darlington, P.J., Jr. 1957. Zoogeography: The Geological Distributions of Animals. John Wiley and Sons, New York.

Darlington, P.J., Jr. 1965. Biogeography of the Southern End of the World. Harvard Univ. Press, Cambridge, MA.

Darwin, C. 1859. On the Origin of Species by Means of Natural Selection, or the Preservation of Favored Races in the Struggle for Life. John Murray, London.

Daugherty, C.H., A. Cree, J.M. Hay, and M.B. Thompson. 1990. Neglected taxonomy and continuing extinctions of tuatara (*Sphenodon*). Nature 347:177–179.

Davidson, W.S., T.P. Birt, and J.M. Green. 1989. A review of genetic variation in Atlantic salmon, *Salmo salar* L., and its importance for stock identification, enhancement programmes and aquaculture. J. Fish Biol. 34:547–560.

Davis, R.F., C.F. Herreid, II, and H.L. Short. 1962. Mexican free-tailed bats in Texas. Ecol. Monogr. 32:311–346.

Davis, S.K., J.E. Strassmann, C. Hughes, L.S. Pletscher, and A.R. Templeton. 1990. Population structure and kinship in *Polistes* (Hymenoptera, Vespidae): an analysis using ribosomal DNA and protein electrophoresis. Evolution 44:1242–1253.

Dawid, I.B. and A.W. Blackler 1972. Maternal and cytoplasmic inheritance of mitochondrial DNA in *Xenopus*. Dev. Biol. 29:152–161.

Dawkins, R. 1989. The Selfish Gene (2nd Ed.). Oxford Univ. Press, Oxford, England.

Dawley, R.M. 1992. Clonal hybrids of the common laboratory fish *Fundulus heteroclitus*. Proc. Natl. Acad. Sci. USA 89:2485–2488.

Dawley, R.M. and J.P. Bogart (eds.). 1989. Evolution and Ecology of Unisexual Vertebrates. New York State Museum, Albany, NY.

Day, L. 1991. Redefining the tree of life. Mosaic 22(4):47–57.

Dayhoff, M.O. 1972. Atlas of Protein Sequence and Structure, Vol. 5. National Biomedical Research Foundation, Silver Springs, MD.

Dayhoff, M.O. and R.V. Eck. 1968. Atlas of Protein Sequence and Structure 1967–68. National Biomedical Research Foundation, Silver Springs, MD.

DeBry, R.W., and N.A. Slade. 1985. Cladistic analysis of restriction endonuclease cleavage maps within a maximum-likelihood framework. Syst. Zool. 34:21–34.

DeMarais, B.D., T.E. Dowling, M.E. Douglas, M.L. Minckley, and P.C. Marsh. 1992. Origin of *Gila seminuda* (Teleostei: Cyprinidae) through introgressive hybridization: implications for evolution and conservation. Proc. Natl. Acad. Sci. USA 89:2747–2751.

Denaro, M., H. Blanc, M.J. Johnson, K.H. Chen, E. Wilmsen, L.L. Cavalli-Sforza, and D.C. Wallace. 1981. Ethnic variation in *Hpa*I endonuclease cleavage patterns of human mitochondrial DNA. Proc. Natl. Acad. Sci. USA 78:5768–5772.

Densmore, L.D. III, C.C. Moritz, J.W. Wright, and W.M. Brown. 1989a. Mitochondrial-DNA analyses and the origin and relative age of parthenogenetic lizards (genus *Cnemidophorus*). IV. Nine *sexlineatus*-group unisexuals. Evolution 43:969–983.

Densmore, L.D., III., J.W. Wright, and W.M. Brown. 1989b. Mitochondrial-DNA analyses and the origin and relative age of parthenogenetic lizards (genus *Cnemidophorus*). II. *C. neomexicanus* and the *C. tesselatus* complex. Evolution 43:943–957.

dePamphilis, C.W. and J.D. Palmer. 1989. Evolution and function of plastid DNA: a review with special reference to non-photosynthetic plants. Pp. 182–202 *in:* Physiology, Biochemistry, and Genetics of Nongreen Plastids, C.D. Boyer, J.C. Shannon, and R.C. Hardison (eds.), American Society of Plant Physiology, Rockville, MD.

dePamphilis, C.W. and R. Wyatt. 1989. Hybridization and introgression in buckeyes (*Aesculus:* Hippocastanaceae): a review of the evidence and a hypothesis to explain long-distance gene flow. Syst. Bot. 14:593–611.

dePamphilis, C.W. and R. Wyatt. 1990. Electrophoretic confirmation of interspecific hybridization in *Aesculus* (Hippocastanaceae) and the genetic structure of a broad hybrid zone. Evolution 44:1295–1317.

de Queiroz, K. and M.J. Donoghue. 1988. Phylogenetic systematics and the species problem. Cladistics 4:317–338.

deRuiter, J.R. 1992. Monastic meeting questions paternity. Trends Ecol. Evol. 7:72–73.

Desai, S.M., V.S. Kalyanaraman, J.M. Casey, A. Srinivasan, P.R. Andersen, and S.G. Devare. 1986. Molecular cloning and primary nucleotide sequence analysis of a distinct human immunodeficiency virus isolate reveal significant divergence in its genomic sequences. Proc. Natl. Acad. Sci. USA 83:8380–8384.

DeSalle, R. 1992a. The origin and possible time of divergence of the Hawaiian Drosophilidae: evidence from DNA sequences. Mol. Biol. Evol. 9:905–916.

DeSalle, R. 1992b. The phylogenetic relationships of flies in the family Drosophilidae deduced from mtDNA sequences. Mol. Phylogenet. Evol. 1:31–40.

DeSalle, R., T. Freedman, E.M. Prager, and A.C. Wilson. 1987. Tempo and mode of sequence evolution in mitochondrial DNA of Hawaiian *Drosophila*. J. Mol. Evol. 26:157–164.

DeSalle, R., J. Gatesy, W. Wheeler, and D. Grimaldi. 1992. DNA sequences from a fossil termite in Oligo-Miocene amber and their phylogenetic implications. Science 257:1933–1936.

DeSalle, R. and L.V. Giddings. 1986. Discordance of nuclear and mitochondrial DNA phylogenies in Hawaiian *Drosophila*. Proc. Natl. Acad. Sci. USA 83:6902–6906.

DeSalle, R., L.V. Giddings, and K.Y. Kaneshiro. 1986b. Mitochondrial DNA variability in natural populations of Hawaiian Drosophila. II. Genetic and phylogenetic relationships of natural populations of *D. silvestris* and *D. heteroneura*. Heredity 56:87–92.

DeSalle, R., L.V. Giddings, and A.R. Templeton. 1986a. Mitochondrial DNA variability in natural populations of Hawaiian Drosophila. I. Methods and levels of variability in *D. silvestris* and *D. heteroneura* populations. Heredity 56:75–86.

DeSalle, R. and A.R. Templeton. 1988. Founder effects and the rate of mitochondrial DNA evolution in Hawaiian *Drosophila*. Evolution 42:1076–1084.

Desjardins, P. and R. Morais. 1990. Sequence and gene organization of the chicken mitochondrial genome. J. Mol. Biol. 212:599–634.

Desjardins, P. and R. Morais. 1991. Nucleotide sequence and evolution of coding and noncoding regions of a quail mitochondrial genome. J. Mol. Evol. 32:153–161.

Dessauer, H.C. and C.J. Cole. 1986. Clonal inheritance in parthenogenetic whiptail lizards: biochemical evidence. J. Heredity 77:8–12.

Dessauer, H.C. and C.J. Cole. 1989. Diversity between and within nominal forms of unisexual teiid lizards. Pp. 49–71 *in:* Evolution and Ecology of Unisexual Vertebrates, R.M. Dayley and J.P. Bogart (eds.), New York State Museum, Albany, N.Y.

Devine, M.C. 1984. Potential for sperm competition in reptiles: behavioral and physiological consequences. Pp. 509–521 *in:* Sperm Competition and the Evolution of Animal Mating Systems, R.L. Smith (ed.), Academic Press, New York.

Devlin, B. and N.C. Ellstrand. 1990. The development and application of a refined method for estimating gene flow from angiosperm paternity analysis. Evolution 44:248–259.

Devlin, B., N. Risch, and K. Roeder. 1991. Estimation of allele frequencies for VNTR loci. Amer. J. Human Genet. 48:662–676.

Devlin, B., N. Risch, and K. Roeder. 1992. Forensic inference from DNA fingerprints. J. Amer. Stat. Assn. 87:337–350.

deVries, H. 1910. The Mutation Theory. Translated by J.B. Farmer and A.D. Darbishire. Open Court, Chicago, IL.

DeWinter, A.J. 1992. The genetic basis and evolution of acoustic mate recognition signals in a *Ribautodelphax* planthopper (Homoptera, Delphacidae) I. The female call. J. Evol. Biol. 5:249–265.

Dewsbury, D.A. 1982. Dominance rank, copulatory behavior, and differential reproduction. Quart. Rev. Biol. 57:135–159.

Diamond, J. 1992. The mysterious origin of AIDS. Natur. Hist. 101(9):24–29.

Diamond, J.M. and J.I. Rotter. 1987. Observing the founder effect in human evolution. Nature 329:105–106.

Diehl, S.R. and G.L. Bush. 1989. The role of habitat preference in adaptation and speciation. Pp. 345–365 *in:* Speciation and Its Consequences, D. Otte and J.A. Endler (eds.), Sinauer, Sunderland, MA.

Dillon, R.T., Jr. 1988. Evolution from transplants between genetically distinct populations of freshwater snails. Genetica 76:111–119.

DiMichele, L., J.A. DiMichele, and D.A. Powers. 1986. Developmental and physiological consequences of genetic variation at enzyme synthesizing loci in *Fundulus heteroclitus*. Amer. Zool. 26:201–210.

DiMichele, L., K. Paynter, and D.A. Powers. 1991. Lactate dehydrogenase-B allozymes directly effect development of *Fundulus heteroclitus*. Science 253:898–900.

Dinerstein, E. and G.F. McCracken. 1990. Endangered greater one-horned rhinoceros carry high levels of genetic variation. Cons. Biol. 4:417–422.

DiRienzo, A. and A.C. Wilson. 1991. Branching pattern in the evolutionary tree for human mitochondrial DNA. Proc. Natl. Acad. Sci. USA 88:1597–1601.

Dizon, A.E., C. Lockyer, W.F. Perrin, D.P. Demaster, and J. Sisson. 1992. Rethinking the stock concept: a phylogeographic approach. Cons. Biol. 6:24–36.

Dizon, A.E., S.O. Southern, and W.F. Perrin. 1991. Molecular analysis of mtDNA types in exploited populations of spinner dolphins (*Stenella longirostris*). Pp. 183–202 *in:* Genetic Ecology of Whales and Dolphins, A.R. Hoelzel (ed.), Black Bear Press, Cambridge, England.

Dobzhansky, T. 1937. Genetics and the Origin of Species. Columbia Univ. Press, New York.

Dobzhansky, T. 1940. Speciation as a stage in evolutionary divergence. Amer. Natur. 74:312–321.

Dobzhansky, T. 1951. Genetics and the Origin of Species, 3rd ed. Columbia Univ. Press, New York.

Dobzhansky, T. 1955. A review of some fundamental concepts and problems of population genetics. Cold Spring Harbor Symp. Quant. Biol. 20:1–15.

Dobzhansky, T. 1970. Genetics of the Evolutionary Process. Columbia Univ. Press, New York.

Dobzhansky, T. 1973. Nothing in biology makes sense except in the light of evolution. Amer. Biol. Teacher 35:125–129.

Dobzhansky, T. 1974. Genetic analysis of hybrid sterility within the species *Drosophila pseudoobscura*. Hereditas 77:81–88.

Dobzhansky, T. 1976. Organismic and molecular aspects of species formation. Pp. 95–105 *in:* Molecular Evolution, F.J. Ayala (ed.), Sinauer, Sunderland, MA.

Dodd, B.E. 1985. DNA fingerprinting in matters of family and crime. Nature 318:506–507.

Doebley, J.F. 1989. Molecular evidence for a missing wild relative of maize and the introgression of its chloroplast genome into *Zea perennis*. Evolution 43:1555–1558.

Dole, J.A. and M. Sun. 1992. Field and genetic survey of the endangered Butte County meadowfoam—*Limnanthes floccosa* subsp. *californica* (Limnanthaceae). Cons. Biol. 6:549–558.

Dong, J., D.B. Wagner, A.D. Yanchuk, M.R. Carlson, S. Magnussen, X.-R. Wang, and A.E. Szmidt. 1992. Paternal chloroplast DNA inheritance in *Pinus contorta* and *Pinus banksiana*: independence of parental species or cross direction. J. Heredity 83:419–422.

Donoghue, M.J. 1985. A critique of the biological species concept and recommendations for a phylogenetic alternative. Bryologist 88:172–181.

Doolittle, R.F. 1987. The evolution of the vertebrate plasma proteins. Biol. Bull. 172:269–283.

Doolittle, R.F., D.-F. Feng, K.L. Anderson, and M.R. Alberro. 1990. A naturally occurring horizontal gene transfer from a eukaryote to a prokaryote. J. Mol. Evol. 31:383–388.

Doolittle, R.F., D.-F. Feng, M.S. Johnson, and M.A. McClure. 1989. Origins and evolutionary relationships of retroviruses. Quart. Rev. Biol. 64:1–30.

Doran, G.H., D.N. Dickel, W.E. Ballinger, Jr., O.F. Agee, P.J. Laipis, and W.W. Hauswirth. 1986. Anatomical, cellular and molecular analysis of 8,000-yr-old human brain tissue from the Windover archaeological site. Nature 323:803–806.

Dotson, R.C. and J.E. Graves. 1984. Biochemical identification of a bluefin tuna establishes a new California size record. Calif. Fish Game 70:62–64.

Doty, P., J. Marmur, J. Eigner, and C. Schildkraut. 1960. Strand separation and specific recombination in deoxyribonucleic acids: physical chemical studies. Proc. Natl. Acad. Sci. USA 46:461–476.

Douglas, D.A. 1989. Clonal growth of *Salix setchelliana* on glacial river gravel bars in Alaska. J. Ecol. 77:112–126.

Douglas, M.E. and J.C. Avise. 1982. Speciation rates and morphological divergence in fishes: tests of gradual versus rectangular modes of evolutionary change. Evolution 36:224–232.

Dover, G.A. 1982. Molecular drive: A cohesive mode of species evolution. Nature 299:111–117.

Dowling, T.E. and W.R. Hoeh. 1991. The extent of introgression outside the contact zone between *Notropis cornutus* and *Notropis chrysocephalus* (Teleostei: Cyprinidae). Evolution 45:944–956.

Dowling, T.E., W.L. Minckley, M.E. Douglas, P.C. Marsh, and B.D. Demarais. 1992. Response to Wayne, Nowak, and Phillips and Henry: use of molecular characters in conservation biology. Cons. Biol. 6:600–603.

Dowling, T.E., C. Moritz, and J.D. Palmer. 1990. Nucleic acids II: Restriction site analysis. Pp. 250–317 *in:* Molecular Systematics, D.M. Hillis and C. Moritz (eds.), Sinauer, Sunderland, MA.

Dowling, T.E., G.R. Smith, and W.M. Brown. 1989. Reproductive isolation and introgression between *Notropis cornutus* and *Notropis chrysocephalus* (family Cyprinidae): comparison of morphology, allozymes, and mitochondrial DNA. Evolution 43:620–634.

Downie, S.R., R.G. Olmstead, G. Zurawski, D.E. Soltis, P.S. Soltis, J.C. Watson, and J.D. Palmer. 1991. Six independent losses of the chloroplast DNA *rpl2* intron in dicotyledons: molecular and phylogenetic implications. Evolution 45:1245–1259.

Downie, S.R. and J.D. Palmer. 1992. Use of chloroplast DNA rearrangements in reconstructing plant phylogeny. Pp. 14–35 *in:* Molecular Systematics of Plants, P.S. Soltis, D.E. Soltis, and J.J. Doyle (eds.), Chapman and Hall, New York.

Doyle, J.J. 1992. Gene trees and species trees: molecular systematics as one-character taxonomy. Syst. Bot. 17:144–163.

Doyle, J.J., J.L. Doyle, A.H.D. Brown, and J.P. Grace. 1990. Multiple origins of polyploids in the *Glycine tabacina* complex inferred from chloroplast DNA polymorphism. Proc. Natl. Acad. Sci. USA 87:714–717.

Doyle, J.J., M. Lavin, and A. Bruneau. 1992. Contributions of molecular data to papilionoid legume systematics. Pp. 223–251 *in:* Molecular Systematics of Plants, P.S. Soltis, D.E. Soltis, and J.J. Doyle (eds.), Chapman and Hall, New York.

DuBose, R.F., D.E. Dykhuisen, and D.L. Hartl. 1988. Genetic exchange among natural isolates of bacteria: Recombination within the *phoA* gene of *Escherichia coli*. Proc. Natl. Acad. Sci. USA 85:7036–7040.

Duellman, W.E. and D.M. Hillis. 1987. Marsupial frogs (Anura: Hylidae: *Gastrotheca*) of the Ecuadorian Andes: Resolution of taxonomic problems and phylogenetic relationships. Herpetologica 43:135–167.

Dunning, A.M., P. Talmud, and S.E. Humphries. 1988. Errors in the polymerase chain reaction. Nucleic Acids Res. 16:10393.

Dutta, S.K. 1986. DNA Systematics, Vols. I and II. CRC Press, Boca Raton, FL.

Duvall, M.R., M.T. Clegg, M.W. Chase, W.D. Clark, W.J. Kress, H.G. Hills, L.E. Equiarte, J.F. Smith, B.S. Gaut, E.A. Zimmer, and G.H. Learn. 1993. Phylogenetic hypotheses for the monocotyledons constructed from *rbcL* sequence data. Annals Missouri Bot. Gard., *in press.*

Duvall, S.W., I.S. Bernstein, and T.P. Gordon. 1976. Paternity and status in a rhesus monkey group. J. Reprod. Fert. 47:25–31.

Eanes, W.F. and R.K. Koehn. 1978a. Relationship between subunit size and number of rare electrophoretic alleles in human enzymes. Biochem. Genet. 16:971–985.

Eanes, W.F. and R.K. Koehn. 1978b. An analysis of genetic structure in the monarch butterfly, *Danaus plexippus* L. Evolution 32:784–797.

Easteal, S. 1991. The relative rate of DNA evolution in primates. Mol. Biol. Evol. 8:115–127.

Echelle, A.A. and P.J. Connor. 1989. Rapid, geographically extensive genetic introgression after secondary contact between two pupfish species (Cyprinodon, Cyprinodontidae). Evolution 43: 717–727.

Echelle, A.A., T.E. Dowling, C.C. Moritz, and W.M. Brown. 1989. Mitochondrial-DNA diversity and the origin of the *Menidia clarkhubbsi* complex of unisexual fishes (Atherinidae). Evolution 43:984–993.

Echelle, A.A. and A.F. Echelle. 1984. Evolutionary genetics of a "species flock:" atherinid fishes on the Mesa Central of Mexico. Pp. 93–110 *in:* Evolution of Fish Species Flocks, A.A. Echelle and I. Kornfield (eds.), Univ. Maine Press, Orono.

Echelle, A.A., A.F. Echelle, and D.R. Edds. 1987. Population structure of four pupfish species (Cyprinodontidae: *Cyprinodon*) from the Chihuahuan desert region of New Mexico and Texas: allozymic variation. Copeia 1987:668–681.

Echelle, A.A. and I. Kornfield (eds.). 1984. Evolution of Fish Species Flocks. Univ. Maine Press, Orono.

Echelle, A.A. and D.T. Mosier. 1981. All-female fish: a cryptic species of *Menidia* (Atherinidae). Science 212:1411–1413.

Echelle, A.F., A.A. Echelle, and D.R. Edds. 1989. Conservation genetics of a spring-dwelling desert fish, the Pecos gambusia (*Gambusia nobilis*, Poeciliidae). Cons. Biol. 3:159–169.

Edman, J.C., J.A. Kovacs, H. Masur, D.V. Santi, H.J. Elwood, and M.L. Sogin. 1988. Ribosomal RNA sequence shows *Pneumocystic carinii* to be a member of the fungi. Nature 334:519–522.

Edwardson, J.R. 1970. Cytoplasmic male sterility. Bot. Rev. 36:341–420.

Ehrlich, P.R. 1975. The population biology of coral reef fishes. Annu. Rev. Ecol. Syst. 6:211–248.

Ehrlich, P.R. 1992. Population biology of checkerspot butterflies and the preservation of global biodiversity. Oikos 63:6–12.

Ehrlich, P.R. and A.H. Ehrlich. 1991. Healing the Planet. Addison-Wesley, Reading, MA.

Ehrlich, P.R. and P.H. Raven. 1969. Differentiation of populations. Science 165:1228–1232.

Eldredge, N. and J. Cracraft. 1980. Phylogenetic Patterns and the Evolutionary Process. Columbia Univ. Press, New York.

Eldredge, N. and S.J. Gould. 1972. Punctuated equilibria: an alternative to phyletic gradualism. Pp. 82–115 *in:* Models in Paleobiology, T.J.M. Schopf (ed.), Freeman, Cooper and Co., San Francisco, CA.

Ellegren, H. 1991. Fingerprinting birds' DNA with a synthetic polynucleotide probe $(TG)_n$. Auk 108:956–958.

Ellis, N., A. Taylor, B.O. Bengtsson, J. Kidd, J. Rogers, and P. Goodfellow. 1990. Population structure of the human pseudoautosomal boundary. Nature 34:663–665.

Ellstrand, N.C. 1984. Multiple paternity within the fruits of the wild radish, *Raphanus sativus*. Amer. Natur. 123:819–828.

Ellstrand, N.C. and D.L. Marshall. 1985. Interpopulation gene flow by pollen in wild radish, *Raphanus sativus*. Amer. Natur. 126:606–616.

Ellstrand, N.C. and M.L. Roose. 1987. Patterns of genotypic diversity in clonal plant species. Amer. J. Bot. 74:123–131.

Endler, J.A. 1977. Geographic Variation, Speciation, and Clines. Princeton Univ. Press, Princeton, NJ.

Ennis, P.D., J. Zemmour, R.D. Salter, and P. Parham. 1990. Rapid cloning of HLA-A,B cDNA by using the polymerase chain reaction: frequency and nature of errors produced in amplification. Proc. Natl. Acad. Sci. USA 87:2833–2837.

Ennos, R.A. and M.T. Clegg. 1982. Effect of population substructuring on estimates of outcrossing rate in plant populations. Heredity 48:283–292.

Ennos, R.A. and K.W. Swales. 1987. Estimation of the mating system in a fungal pathogen *Crumenulopsis sororia* (Karst.) Groves using isozyme markers. Heredity 59:423–430.

Epperson, B.K. and M.T. Clegg. 1987. First-pollination primacy and pollen selection in the morning glory, *Ipomoea purpurea*. Heredity 58:5–14.

Erlich, H.A. and N. Arnheim. 1992. Genetic analysis using the polymerase chain reaction. Annu. Rev. Genet. 26:479–506.

Erlich, H.A., D. Gelfand, and J.J. Sninsky. 1991. Recent advances in the polymerase chain reaction. Science 252:1643–1651.

Evarts, S. and C.J. Williams. 1987. Multiple paternity in a wild population of mallards. Auk 104:597–602.

Ewens, W.J., R.S. Speilman, and H. Harris. 1981. Estimation of genetic variation at the DNA level from restriction endonuclease data. Proc. Natl. Acad. Sci. USA 78:3748–3750.

Excoffier, L. 1990. Evolution of human mitochondrial DNA: evidence for departure from a pure neutral model of populations at equilibrium. J. Mol. Evol. 30:125–139.

Farris, J.S. 1970. Methods for computing Wagner trees. Syst. Zool. 34:21–34.

Farris, J.S. 1971. The hypothesis of nonspecificity and taxonomic congruence. Annu. Rev. Ecol. Syst. 2:227–302.

Farris, J.S. 1972. Estimating phylogenetic trees from distance matrices. Amer. Natur. 106:645–668.

Farris, J.S. 1973. On comparing the shapes of evolutionary trees. Syst. Zool. 27:27–33.

Farris, J.S. 1977. Phylogenetic analysis under Dollo's Law. Syst. Zool. 26:77–88.

Feder, J.L., C.A. Chilcote, and G.L. Bush. 1988. Genetic differentiation between sympatric host races of the apple maggot fly *Rhagoletis pomonella*. Nature 336:61–64.

Feder, J.L., C.A. Chilcote, and G.L. Bush. 1990a. The geographic pattern of genetic differentiation between host associated populations of *Rhagoletis pomonella* (Diptera: Tephritidae) in the eastern United States and Canada. Evolution 44:570–594.

Feder, J.L., C.A. Chilcote, and G.L. Bush. 1990b. Regional, local and microgeographic allele frequency variation between apple and hawthorne populations of *Rhagoletis pomonella* in western Michigan. Evolution 44:595–608.

Fell, J.W., A. Statzell-Tallman, M.J. Lutz, and C.P. Kurtzman. 1992. Partial rRNA sequences in marine yeasts: a model for identification of marine eukaryotes. Mol. Mar. Biol. Biotech. 1:175–186.

Felsenstein, J. 1974. The evolutionary advantage of recombination. Genetics 78:737–756.

Felsenstein, J. 1978. The number of evolutionary trees. Syst. Zool. 27:27–33.

Felsenstein, J. 1981. Skepticism towards Santa Rosalia, or why are there so few kinds of animals. Evolution 35:124–138.

Felsenstein, J. 1982. Numerical methods for inferring evolutionary trees. Quart. Rev. Biol. 57:379–404.

Felsenstein, J. 1983. Parsimony in systematics: biological and statistical issues. Annu. Rev. Ecol. Syst. 14:313–333.

Felsenstein, J. 1985a. Confidence limits on phylogenies: an approach using the bootstrap. Evolution 39:783–791.

Felsenstein, J. 1985b. Phylogenies and the comparative method. Amer. Natur. 125:1–15.

Felsenstein, J. 1988. Phylogenies from molecular sequences: Inference and reliability. Annu. Rev. Genet. 22:521–565.

Felsenstein, J. 1992. Estimating effective population size from samples of sequences: inefficiency of pairwise and segregating sites as compared to phylogenetic estimates. Genet. Res. Camb. 59:139–147.

Fenster, C.B. 1991. Gene flow in *Chamaecrista fasciculata* (Leguminosae). I. Gene dispersal. Evolution 45:398–409.

Fergus, C. 1991. The Florida panther verges on extinction. Science 251:1178–1180.

Ferguson, A. 1989. Genetic differences among brown trout, *Salmo trutta,* stocks and their importance for the conservation and management of the species. Freshwat. Biol. 21:35–46.

Ferguson, M.M. 1992. Enzyme heterozygosity and growth in rainbow trout: genetic and physiological explanations. Hereditary 68:115–122.

Ferguson, M.M. and L.R. Drahushchak. 1990. Disease resistance and enzyme heterozygosity in rainbow trout. Heredity 64:413–418.

Fernholm, B., K. Bremer, and H. Jornvall (eds.). 1989. The Hierarchy of Life, Elsevier, Amsterdam.

Ferrell, R.E., D.C. Morizot, J. Horn, and C.J. Carley. 1980. Biochemical markers in a species endangered by introgression: the Red Wolf. Biochem. Genet. 18:39–49.

Ferris, S.D. and W.J. Berg. 1986. The utility of mitochondrial DNA in fish genetics and fishery management. Pp. 277–299 *in:* Population Genetics and Fishery Management, N. Ryman and F. Utter (eds.), Univ. Washington Press, Seattle, WA.

Ferris, S.D., R.D. Sage, C.-M. Huang, J.T. Nielsen, U. Ritte, and A.C. Wilson. 1983a. Flow of mitochondrial DNA across a species boundary. Proc. Natl. Acad. Sci. USA 80:2290–2294.

Ferris, S.D., R.D. Sage, E.M. Prager, U. Ritte, and A.C. Wilson. 1983b. Mitochondrial DNA evolution in mice. Genetics 105:681–721.

Ferris, S.D. and G.S. Whitt. 1977. Loss of duplicate gene expression after polyploidization. Nature 265:258–260.

Ferris, S.D. and G.S. Whitt. 1978. Phylogeny of tetraploid catastomid fishes based on the loss of duplicate gene expression. Syst. Zool. 27:189–206.

Ferris, S.D. and G.S. Whitt. 1979. Evolution of the differential regulation of duplicate genes after polyploidization. J. Mol. Evol. 12:267–317.

Fiala, K.L. and R.R. Sokal. 1985. Factors determining the accuracy of cladogram estimation: Evaluation using computer simulation. Evolution 39:609–622.

Field, K.G., G.J. Olsen, D.J. Lane, S.J. Giovannoni, M.T. Ghiselin, E.C. Raff, N.R. Pace, and R.A. Raff. 1988. Molecular phylogeny of the animal kingdom. Science 239:748–753.

Figueroa, F., E. Gunther, and J. Klein. 1988. MHC polymorphisms pre-dating speciation. Nature 335:265–271.

Finnegan, D.J. 1983. Retroviruses and transposable elements—which came first? Nature 302:105–106.

Finnegan, D.J. 1989. Eukaryotic transposable elements and genome evolution. Trends Genet. 5:103–107.

Finnerty, V. and F.H. Collins. 1988. Ribosomal DNA probes for identification of member species of the *Anopheles gambiae* complex. Fla. Entomol. 71:288–294.

Fisher, R.A. 1930. The Genetical Theory of Natural Selection. Clarendon Press, Oxford.

Fisher, S.E., J.B. Shaklee, S.D. Ferris, and G.S. Whitt. 1980. Evolution of five multilocus isozyme systems in the chordates. Genetica 52:73–85.

Fisher, S.E. and G.S. Whitt. 1978. Evolution of isozyme loci and their differential tissue expression. Creatine kinase as a model system. J. Mol. Evol. 12:25–55.

Fitch, W.M. 1970. Distinguishing homologous from analogous proteins. Syst. Zool. 19:99–113.

Fitch, W.M. 1971. Toward defining the course of evolution: minimal change for a specific tree topology. Syst. Zool. 20:406–416.

Fitch, W.M. 1976. Molecular evolutionary clocks. Pp. 160–178 *in:* Molecular Evolution, F.J. Ayala (ed.), Sinauer, Sunderland, MA.

Fitch, W.M. and E. Margoliash. 1967. Construction of phylogenetic trees. Science 155:279–284.

Flavell, A.J. 1992. *Ty1-copia* group retrotransposons and the evolution of retroelements in the eukaryotes. Genetica 86:203–214.

Fletcher, D.J.C. and C.D. Michener (eds.). 1987. Kin recognition in animals. John Wiley and Sons, New York.

Foltz, D.W. 1981. Genetic evidence for long-term monogamy in a small rodent, *Peromyscus polionotus*. Amer. Natur. 117:665–675.

Foltz, D.W. and J.L. Hoogland. 1981. Analysis of the mating system in the black-tailed prairie dog. (*Cynomys ludovicianus*) by likelihood of paternity. J. Mamm. 62:706–712.

Foltz, D.W. and J.L. Hoogland. 1983. Genetic evidence of outbreeding in the black-tailed prairie dog (*Cynomys ludovicianus*). Evolution 37:273–281.

Foltz, D.W., H. Ochman, J.S. Jones, S.M. Evangelisti, and R.K. Selander. 1982. Genetic population structure and breeding systems in arionid slugs (Mollusca: Pulmonata). Biol. J. Linn. Soc. 17:225–241.

Foltz, D.W., H. Ochman, and R.K. Selander. 1984. Genetic diversity and breeding systems in terrestrial slugs of the families Limacidae and Arionidae. Malacologia 25:593–605.

Foltz, D.W. and P.L. Schwagmeyer. 1989. Sperm competition in the thirteen-lined ground squirrel: differential fertilization success under field conditions. Amer. Natur. 133:257–265.

Foote, C.J., C.C. Wood, and R.E. Withler. 1989. Biochemical genetic composition of sockeye salmon and kokanee, the anadromous and nonanadromous forms of *Oncorhynchus nerka*. Can. J. Fish. Aquat. Sci. 46:149–158.

Forbes, S.H. and F.W. Allendorf. 1991. Associations between mitochondrial and nuclear genotypes in cutthroat trout hybrid swarms. Evolution 45:1332–1349.

Ford, E.B. 1964. Ecological Genetics. Metheun, London.

Ford, H. 1985. Life history strategies in two coexisting agamospecies of dandelion. Biol. J. Linn. Soc. 25:169–186.

Ford, H. and A.J. Richards. 1985. Isozyme variation within and between *Taraxacum* agamospecies in a single locality. Heredity 55:289–291.

Forsthoefel, N.R., H.J. Bohnert, and S.E. Smith. 1992. Discordant inheritance of mitochondrial and plastid DNA in diverse alfalfa genotypes. J. Heredity 83:342–345.

Fos, M., M.A. Dominguez, A. Latorre, and A. Moya. 1990. Mitochondrial DNA evolution in experimental populations of *Drosophila subobscura*. Proc. Natl. Acad. Sci. USA 87:4198–4201.

Fox, G.E., L.J. Magrum, W.E. Balch, R.S. Wolfe, and C.R. Woese. 1977. Classification of methanogenic bacteria by 16S ribosomal RNA characterization. Proc. Natl. Acad. Sci. USA 74:4537–4541.

Fox, G.E. and 18 others. 1980. The phylogeny of prokaryotes. Science 209:457–463.

Fox, T.D. 1983. Mitochondrial genes in the nucleus. Nature 301:371–372.

Frankel, O.H. and J.G. Hawkes (eds.). 1975. Crop Genetic Resources for Today and Tomorrow. Cambridge Univ. Press, Cambridge, England.

Franklin, O.H. 1980. Evolutionary change in small populations. Pp. 135–150 *in:* Conservation Biology: An Evolutionary-Ecological Perspective, M.E. Soulé and B.A. Wilcox (eds.), Sinauer, Sunderland, MA.

Franklin, O.H. and M.E. Soulé. 1981. Conservation and Evolution. Cambridge Univ. Press, Cambridge, England.

Friedman, S.T. and W.T. Adams. 1985. Estimation of gene flow into two seed orchards of loblolly pine (*Pinus taeda*). Theoret. Appl. Genet. 69:609–615.

Fries, R., A. Eggen, and G. Stranzinger. 1990. The bovine genome contains polymorphic microsatellites. Genomics 8:403–406.

Fritsch, P. and L.H. Rieseberg. 1992. High outcrossing rates maintain male and hermaphroditic individuals in populations of the flowering plant *Datisca glomerata*. Nature 359:633–636.

Fryer, G. and T.D. Iles. 1972. The Cichlid Fishes of the Great Lakes of Africa. TFH, Neptune City, NJ.

Fuhrman, J.A., K. McCallum, and A.A. Davis. 1992. Novel major archaebacterial group from marine plankton. Nature 356:148–149.

Furnier, G.R., M.P. Cummings, and M.T. Clegg. 1990. Evolution of the avacados as revealed by DNA restriction fragment variation. J. Heredity 81:183–188.

Furnier, G.R., P. Knowles, M.A. Clyde, and B.P. Dancik. 1987. Effects of avian seed dispersal on the genetic structure of whitebark pine populations. Evolution 41:607–612.

Futuyma, D.J. and G.C. Mayer. 1980. Non-allopatric speciation in animals. Syst. Zool. 29:254–271.

Gaffney, P.M. and B. McGee. 1992. Multiple paternity in *Crepidula fornicata* (Linnaeus). Veliger 35:12–15.

Gallez, G.P. and L.D. Gottlieb. 1982. Genetic evidence for the hybrid origin of the diploid plant *Stephanomeria diegensis*. Evolution 36:1158–1167.

Gallo, R.C. 1987. The AIDS virus. Sci. Amer. 256(1):46–56.

Gamow, G. 1954. Possible relation between deoxyribonucleic acid and protein structures. Nature 173:318.

Ganders, F.R. 1989. Adaptive radiation in Hawaiian *Bidens.* Pp. 99–112 *in:* Genetics, Speciation and the Founder Principle, L.V. Giddings, K.Y. Kaneshiro, and W.W. Anderson (eds.), Oxford Univ. Press, New York.

Gantt, J.S., S.L. Baldauf, P.J. Calie, N.F. Weeden, and J.D. Palmer. 1991. Transfer of *rpl22* to the nucleus greatly preceded its loss from the chloroplast and involved the gain of an intron. The EMBO J. 10:3073–3078.

Garten, C.T., Jr. 1977. Relationships between exploratory behaviour and genic heterozygosity in the oldfield mouse. Anim. Behav. 25:328–332.

Garton, D.W., R.K. Koehn, and T.M. Scott. 1984. Multiple-locus heterozygosity and the physiological energetics of growth in the coot clam, *Mulinia lateralis,* from a natural population. Genetics 108:445–455.

Gartside, D.F., J.S. Rogers, and H.C. Dessauer. 1977. Speciation with little genic and morphological differentiation in the ribbon snakes *Thamnophis proximus* and *T. sauritus* (Colubridae). Copeia 1977:697–705.

Gastony, G.J. 1986. Electrophoretic evidence for the origin of fern species by unreduced spores. Amer. J. Bot. 73:1563–1569.

Gaudet, J., J. Julien, J.F. Lafay, and Y. Brygoo. 1989. Phylogeny of some *Fusarium* species, as determined by large-subunit rRNA sequence comparisons. Mol. Biol. Evol. 6:227–242.

Gaut, B.S., S.V. Muse, W.D. Clark, and M.T. Clegg. 1992. Relative rates of nucleotide substitution at the *rbcL* locus of monocotyledonous plants. J. Mol. Evol. 35:292–303.

Gavin, T.A. and E.K. Bollinger. 1985. Multiple paternity in a territorial passerine: the bobolink. Auk 102:550–555.

Geist, V. 1992. Endangered species and the law. Nature 357:274–276.

Gellissen, G., J.Y. Bradfield, B.N. White, and G.R. Wyatt. 1983. Mitochondrial DNA sequences in the nuclear genome of a locust. Nature 301:631–634.

Gelter, H.P. and H. Tegelström. 1992. High frequency of extra-pair paternity in Swedish pied flycatchers revealed by allozyme electrophoresis and DNA fingerprinting. Behav. Ecol. Sociobiol. 31:1–7.

Gelter, H.P., H. Tegelström, and L. Gustafsson. 1992. Evidence from hatching success and DNA fingerprinting for the fertility of hybrid pied X collared flycatchers *Ficedula hypoleuca* X *albicollis.* Ibis 134:62–68.

George, M., Jr. and O.A. Ryder. 1986. Mitochondrial DNA evolution in the genus *Equus.* Mol. Biol. Evol. 3:535–546.

Georges, M., P. Cochaux, A.S. Lequarre, M.W. Young, and G. Vassart. 1987. DNA fingerprinting in man using a mouse probe related to part of the *Drosophila* "Per" gene. Nucleic Acids Res. 15:7193.

Gerbi, S.A. 1985. Evolution of ribosomal RNA. Pp. 419–518 *in:* Molecular Evolutionary Genetics, R.J. MacIntyre (ed.), Plenum Press, New York.

Geyer, C.J., E.A. Thompson, and O.A. Ryder. 1989. Gene survival in the Asian wild horse (*Equus przewalskii*): II. Gene survival in the whole population, in subgroups, and through history. Zoo Biol. 8:313–329.

Gibbons, A. 1991. Looking for the father of us all. Science 251:378–380.

Gibbons, A. 1992. Mission impossible: saving all endangered species. Science 256:1386.

Gibbs, H.L., P.T. Boag, B.N. White, P.J. Weatherhead, and L.M. Tabak. 1991. Detection of a hypervariable DNA locus in birds by hybridization with a mouse MHC probe. Mol. Biol. Evol. 8:433–446.

Gibbs, H.L., P.J. Weatherhead, P.T. Boag, B.N. White, L.M. Tabak, and D.J. Hoysak. 1990. Realized reproductive success of polygynous red-winged blackbirds revealed by DNA markers. Science 250:1394–1397.

Gibson, A.R. and J.B. Falls. 1975. Evidence for multiple insemination in the common garter snake, *Thamnophis sirtalis*. Can. J. Zool. 53:1362–1368.

Giddings, L.V., K.Y. Kaneshiro, and W.W. Anderson (eds.). 1989. Genetics, Speciation and the Founder Principle. Oxford Univ. Press, New York.

Gilbert, D.A., N. Lehman, S.J. O'Brien, and R.K. Wayne. 1990. Genetic fingerprinting reflects population differentiation in the California Channel Island fox. Nature 344:764–767.

Gilbert, D.A., C. Packer, A.E. Pusey, J.C. Stephens, and S.J. O'Brien. 1991. Analytical DNA fingerprinting in lions: parentage, genetic diversity, and kinship. J. Heredity 82:378–386.

Gilbert, D.G. and R.C. Richmond. 1982. Studies of esterase 6 in *Drosophila melanogaster* XII. Evidence for temperature selection of *Est 6* and *Adh* alleles. Genetica 58:109–119.

Giles, E. and S.H. Ambrose. 1986. Are we all out of Africa? Nature 322:21–22.

Giles, R.E., H. Blanc, H.M. Cann, and D.C. Wallace. 1980. Maternal inheritance of human mitochondrial DNA. Proc. Natl. Acad. Sci. USA 77:6715–6719.

Gill, F.B. 1990. Ornithology. W.H. Freeman and Co., New York.

Gill, P., A.J. Jeffreys, and D.J. Werrett. 1985. Forensic application of DNA "fingerprints." Nature 318:577–579.

Gillespie, J.H. 1986. Variability of evolutionary rates of DNA. Genetics 113:1077–1091.

Gillespie, J.H. 1987. Molecular evolution and the neutral allele theory. Oxford Surv. Evol. Biol. 4:10–37.

Gillespie, J.H. 1988. More on the overdispersed molecular clock. Genetics 118:385–386.

Gillespie, J.H. and C.H. Langley. 1974. A general model to account for enzyme variation in natural populations. Genetics 76:837–884.

Gillham, N.W. 1978. Organelle Heredity. Raven Press, New York.

Gilpin, M.E. and M.E. Soulé. 1986. Minimum viable populations: processes of species extinction. Pp. 19–34 *in:* Conservation Biology: the Science of Scarcity and Diversity, M.E. Soulé (ed.), Sinauer, Sunderland, MA.

Gilpin, M.E. and C. Wills. 1991. MHC and captive breeding: a rebuttal. Cons. Biol. 5:554–555.

Giorgi, P.P. 1992. Sex and the male stick insect. Nature 357:444–445.

Giovannoni, S.J., T.B. Britschgi, C.L. Moyer, and K.G. Field. 1990. Genetic diversity in Sargasso Sea bacterioplankton. Nature 345:60–63.

Giovannoni, S.J., S. Turner, G.J. Olsen, S. Barns, D.J. Lane, and N.R. Pace. 1988. Evolutionary relationships among cyanobacteria and green chloroplasts. J. Bacteriol. 170:3584–3592.

Gittleman, J.L. and S.L. Pimm. 1991. Crying wolf in North America. Nature 351:524–525.

Goddard, K.A., R.M. Dawley, and T.E. Dowling. 1989. Origin and genetic relationships of diploid, triploid, and diploid-triploid mosaic biotypes in the *Phoxinus eos-neogaeus* unisexual complex. Pp. 268–280 *in:* Evolution and Ecology of Unisexual Vertebrates, R. Dawley and J. Bogart (eds.), New York State Museum, Albany, NY.

Gogarten, J.P. and 12 others. 1989. Evolution of the vacuolar H^+-ATPase: implications for the origin of eukaryotes. Proc. Natl. Acad. Sci. USA 86:6661–6665.

Gold, J.R. and L.R. Richardson. 1991. Genetic studies in marine fishes. IV. An analysis of population structure in the red drum (*Sciaenops ocellatus*) using mitochondrial DNA. Fish. Res. 12:213–241.

Goldman, D., P.R. Giri, and S.J. O'Brien. 1989. Molecular genetic-distance estimates among the Ursidae as indicated by one- and two-dimensional protein electrophoresis. Evolution 43:282–295.

Goldman, N. and N.H. Barton. 1992. Genetics and geography. Nature 357:440–441.

Goldschmidt, R. 1940. The Material Basis of Evolution. Yale Univ. Press, New Haven, CT.

Golenberg, E.M. 1989. Migration patterns and the development of multilocus associations in a selfing annual, *Triticum dicoccoides*. Evolution 43:595–606.

Golenberg, E.M., D.E. Giannasi, M.T. Clegg, C.J. Smiley, M. Durbin, D. Henderson, and G. Zurawski. 1990. Chloroplast DNA sequence from a Miocene *Magnolia* species. Nature 344:656–658.

Gomendio, M. and E.R.S. Roldan. 1993. Mechanisms of sperm competition: linking physiology and behavioural ecology. Trends Ecol. Evol. 8:95–100.

González-Villaseñor, L.I. and D.A. Powers. 1990. Mitochondrial-DNA restriction-site polymorphisms in the teleost *Fundulus heteroclitus* support secondary intergradation. Evolution 44:27–37.

Gooch, J.L. 1975. Mechanisms of evolution and population genetics. Pp. 349–409 *in:* Marine Ecology, Vol. 2, Part 1, O. Kinne (ed.), John Wiley and Sons, London.

Good, D.A., and J.W. Wright. 1984. Allozymes and the hybrid origin of the parthenogenetic lizard *Cnemidophorus exsanguis*. Experientia 40:1012–1014.

Goodman, M. 1962. Immunochemistry of the primates and primate evolution. Ann. N.Y. Acad. Sci. 102:219–234.

Goodman, M. 1963. Serological analysis of the systematics of recent hominoids. Human Biol. 35:377–424.

Goodman, M., J. Barnabas, G. Matsuda, and G.W. Moor. 1971. Molecular evolution in the descent of man. Nature 233:604–613.

Goodman, M., J. Czelusniak, and J.E. Beeber. 1985. Phylogeny of primates and other eutherian orders: a cladistic analysis using amino acid and nucleotide sequence data. Cladistics 1:171–185.

Goodman, M., D.A. Tagle, D.H.A. Fitch, W. Bailey, J. Czelusniak, B.F. Koop, P. Benson, and J.L. Slightom. 1990. Primate evolution at the DNA level and a classification of hominoids. J. Mol. Evol. 30:260–266.

Gore, P.L., B.M. Potts, P.W. Volker, and J. Megalos. 1990. Unilateral cross-incompatibility in *Eucalyptus:* the case of hybridization between *E. globulus* and *E. nitens*. Aust. J. Bot. 38:383–394.

Gorr, T., T. Kleinschmidt, and H. Fricke. 1991. Close tetrapod relationship of the coelacanth *Latimeria* indicated by haemoglobin sequences. Nature 351:394–397.

Gottlieb, L.D. 1973a. Enzyme differentiation and phylogeny in *Clarkia franciscana*, *C. rubicunda*, and *C. amoena*. Evolution 27:205–214.

Gottlieb, L.D. 1973b. Genetic differentiation, sympatric speciation and the origin of a diploid species of *Stephanomeria*. Amer. J. Bot. 60:545–553.

Gottlieb, L.D. 1974. Genetic confirmation of the origin of *Clarkia lingulata*. Evolution 28:244–250.

Gottlieb, L.D. 1977. Electrophoretic evidence and plant systematics. Annals Missouri Bot. Gardens 64:161–180.

Gottlieb, L.D. 1981. Electrophoretic evidence and plant populations. Progr. Phytochem. 7:2–46.

Gottlieb, L.D. 1988. Towards molecular genetics in *Clarkia:* gene duplications and molecular characterization of PGI genes. Annals Missouri Bot. Gardens 75:1169–1179.

Gottlieb, L.D. and G. Pilz. 1976. Genetic similarity between *Gaura longiflora* and its obligately outcrossing derivative *G. demareei.* Syst. Bot. 1:181–187.

Gould, S.J. 1977. Ontogeny and Phylogeny. Belknap Press, Harvard University, Cambridge, MA.

Gould, S.J. 1980. Is a new and general theory of evolution emerging? Paleobiology 6:119–130.

Gould, S.J. 1983. Hen's Teeth and Horses Toes. W.W. Norton, New York.

Gould, S.J. 1985. A clock of evolution. Natur. Hist. 94(4):12–25.

Gould, S.J. 1991. Unenchanted evening. Natur. Hist. 100(5):7–14.

Gould, S.J. 1992. We are all monkeys' uncles. Natur. Hist. 101(6):14–21.

Gould, S.J. and N. Eldredge. 1977. Punctuated equilibria: the tempo and mode of evolution reconsidered. Paleobiology 3:115–151.

Gould, S.J. and R.C. Lewontin. 1979. The spandrels of San Marco and the Panglossian paradigm: A critique of the adaptationist programme. Proc. Roy. Soc. London B205:581–598.

Gouy, M. and W.-H. Li. 1989. Phylogenetic analysis based on rRNA sequences supports the archaebacterial rather than the eocyte tree. Nature 339:145–147.

Gowaty, P.A. and W.C. Bridges. 1991a. Nestbox availability affects extra-pair fertilizations and conspecific nest parasitism in eastern bluebirds, *Sialia sialis.* Anim. Behav. 41:661–675.

Gowaty, P.A. and W.C. Bridges. 1991b. Behavioral, demographic, and environmental correlates of extrapair fertilizations in eastern bluebirds, *Sialia sialis.* Behav. Ecol. 2:339–350.

Gowaty, P.A. and A.A. Karlin. 1984. Multiple maternity and paternity in single broods of apparently monogamous eastern bluebirds (*Sialia sialis*). Behav. Ecol. Sociobiol. 15:91–95.

Grachev, M.A., S. Ja. Slobodyanyuk, N.G. Kholodilov, S.P. Fyodorov, S.I. Belikov, D. Yu. Sherbakov, V.G. Sideleva, A.A. Zubin, and V.V. Kharchenko. 1992. Comparative study of two protein-coding regions of mitochondrial DNA from three endemic sculpins (Cottoidei) of Lake Baikal. J. Mol. Evol. 34:85–90.

Grant, P.R. and B.R. Grant. 1992. Hybridization of bird species. Science 256:193–197.

Grant, V. 1963. The Origin of Adaptations. Columbia Univ. Press, New York.

Grant, V. 1981. Plant Speciation, 2nd Ed. Columbia Univ. Press, New York.

Grant, W.S. 1987. Genetic divergence between congeneric Atlantic and Pacific Ocean fishes. Pp. 225–246 *in:* Population Genetics and Fisheries Management, N. Ryman and F. Utter (eds.), Univ. of Washington Press, Seattle.

Grassle, J.P. and J.F. Grassle. 1976. Sibling species in the marine pollution indicator *Capitella* (Polychaeta). Science 192:567–569.

Graves, J.E., M.J. Curtis, P.A. Oeth, and R.S. Waples. 1989. Biochemical genetics of Southern California basses of the genus *Paralabrax:* specific identification of fresh and ethanol-preserved individual eggs and early larvae. Fish. Bull. 88:59–66.

Graves, J.E. and A.E. Dizon. 1989. Mitochondrial DNA sequence similarity of Atlantic and Pacific albacore tuna (*Thunnus alalunga*). Can. J. Fish. Aquat. Sci. 46:870–873.

Graves, J.E., S.D. Ferris, and A.E. Dizon. 1984. Close genetic similarity of Atlantic and Pacific skipjack tuna (*Katsuwonus pelamis*) demonstrated with restriction endonuclease analysis of mitochondrial DNA. Mar. Biol. 79:315–319.

Graves, J.E., M.A. Simovich, and K.M. Schaefer. 1988. Electrophoretic identification of early juvenile yellowfin tuna, *Thunnus albacares.* Fish. Bull. 86:835–838.

Gray, M.W. 1989. Origin and evolution of mitochondrial DNA. Annu. Rev. Cell Biol. 5:25–50.

Gray, M.W. 1992. The endosymbiont hypothesis revisited. Int. Rev. Cytol. 141:233–357.

Gray, M.W., R. Cedergren, Y. Abel, and D. Sankoff. 1989. On the evolutionary origin of the plant mitochondrion and its genome. Proc. Natl. Acad. Sci. USA 86:2267–2271.

Greenberg, B.D., J.E. Newbold, and A. Sugino. 1983. Intraspecific nucleotide sequence variability surrounding the origin of replication in human mitochondrial DNA. Gene 21:33–49.

Greenlaw, J.S. 1993. Behavioral and morphological diversification in Atlantic coastal Sharp-tailed Sparrows (*Ammodramus caudacutus*). Auk, *in press.*

Greenwood, P.J. 1980. Mating systems, philopatry and dispersal in birds and mammals. Anim. Behav. 28:1140–1162.

Greenwood, P.J. and P.H. Harvey. 1982. The natal and breeding dispersal of birds. Annu. Rev. Ecol. Syst. 13:1–21.

Greenwood, P.H. 1980. Towards a phyletic classification of the "genus" *Haplochromis* (Pisces Cichlidae) and related taxa. Part 2. The species from Lake Victoria, Nabugabo, Edward, George and Kivu. Bull. Brit. Mus. Nat. Hist. (Zool.) 39:1–101.

Greenwood, P.H. 1981. The Haplochromine Fishes of the East African Lakes. Cornell Univ. Press, Ithaca, N.Y.

Griffin, A.R., I.P. Burgess, and L. Wolf. 1988. Patterns of natural and manipulated hybridization in the genus *Eucalyptus* L'Herit.—a review. Aust. J. Bot. 36:41–66.

Griffiths, R. and P. Holland. 1990. A novel avian W chromosome DNA repeat sequence in the lesser black-backed gull (*Larus fuscus*). Chromosoma 99:243–250.

Gromko, M.H., D.G. Gilbert, and R.C. Richmond. 1984. Sperm transfer and use in the multiple mating system of *Drosophila*. Pp. 371–426 *in:* Sperm Competition and the Evolution of Animal Mating Systems, R.L. Smith (ed.), Academic Press, New York.

Grosberg, R.K. 1988. The evolution of allorecognition specificity in clonal invertebrates. Quart. Rev. Biol. 63:377–412.

Grosberg, R.K. 1991. Sperm-mediated gene flow and the genetic structure of a population of the colonial ascidian *Botryllus schlosseri*. Evolution 45:130–142.

Grosberg, R.K. and J.F. Quinn. 1986. The genetic control and consequences of kin recognition by the larvae of a colonial marine invertebrate. Nature 322:456–459.

Gross, M.R. 1979. Cuckoldry in sunfishes (*Lepomis:* Centrarchidae). Can. J. Zool. 57:1507–1509.

Gross, M.R. and E.L. Charnov. 1980. Alternative male life histories in bluegill sunfish. Proc. Natl. Acad. Sci. USA 77:6937–6940.

Grudzien, T.A. and B.J. Turner. 1984. Genic identity and geographic differentiation of trophically dichotomous *Ilyodon* (Teleostei: Goodeidae). Copeia 1984:102–107.

Guillemette, J.G. and P.N. Lewis. 1983. Detection of subnanogram quantities of DNA and RNA on native and denaturing polyacrylamide and agarose gels by silver staining. Electrophoresis 4:92–94.

Guries, R.P. and F.T. Ledig. 1982. Genetic diversity and population structure in pitch pine (*Pinus rigida* Mill.). Evolution 36:387–402.

Gyllensten, U.B. 1985. The genetic structure of fish: differences in the intraspecific distribution of biochemical genetic variation between marine, anadromous, and freshwater species. J. Fish Biol. 26:691–699.

Gyllensten, U.B., S. Jakobsson, and H. Temrin. 1990. No evidence for illegitimate young in monogamous and polygynous warblers. Nature 343:168–170.

Gyllensten, U.B., R.F. Leary, F.W. Allendorf, and A.C. Wilson. 1985b. Introgression between two cutthroat trout subspecies with substantial karyotypic, nuclear and mitochondrial genomic divergence. Genetics 111:905–915.

Gyllensten, U.B., D. Wharton, A. Josefsson, and A.C. Wilson. 1991. Paternal inheritance of mitochondrial DNA in mice. Nature 352:255–257.

Gyllensten, U.B., D. Wharton, and A.C. Wilson. 1985a. Maternal inheritance of mitochondrial DNA during backcrossing of two species of mice. J. Heredity 76:321–324.

Gyllensten, U.B. and A.C. Wilson. 1987a. Mitochondrial DNA of salmonids. Pp. 301–317 *in:* Population Genetics and Fisheries Management, N. Ryman and F. Utter (eds.), Univ. Washington Press, Seattle.

Gyllensten, U.B. and A.C. Wilson. 1987b. Interspecific mitochondrial DNA transfer and the colonization of Scandinavia by mice. Genet. Res. Camb. 49:25–29.

Hachtel, W. 1980. Maternal inheritance of chloroplast DNA in some *Oenothera* species. J. Heredity 71:191–194.

Hadrys, H., M. Balick, and B. Schierwater. 1992. Applications of random amplified polymorphic DNA (RAPD) in molecular ecology. Mol. Ecol. 1:55–63.

Hafner, M.S., J.C. Hafner, J.L. Patton, and M.F. Smith. 1987. Macrogeographic patterns of genetic differentiation in the pocket gopher (*Thomomys umbrinus*). Syst. Zool. 36:18–34.

Hagelberg, E., I.C. Gray, and A.J. Jeffreys. 1991. Identification of the skeletal remains of a murder victim by DNA analysis. Nature 352:427–429.

Haig, S.M., J.D. Ballou, and S.R. Derrickson. 1990. Management options for preserving genetic diversity: reintroduction of Guam rails to the wild. Cons. Biol. 4:290–300.

Haig, S.M. and L.W. Oring. 1988. Genetic differentiation of piping plovers across North America. Auk 105:260–267.

Haldane, J.B.S. 1922. Sex ratio and unisexual sterility of hybrid animals. J. Genet. 12:101–109.

Haldane, J.B.S. 1932. The Causes of Evolution. Longmans and Green, London.

Hale, L.R. and R.S. Singh. 1986. Extensive variation and heteroplasmy in size of mitochondrial DNA among geographic populations of *Drosophila melanogaster*. Proc. Natl. Acad. Sci. USA 83:8813–8817.

Hale, L.R. and R.S. Singh. 1987. Mitochondrial DNA variation and genetic structure in populations of *Drosophila melanogaster*. Mol. Biol. Evol. 4:622–637.

Hale, L.R. and R.S. Singh. 1991. A comprehensive study of genic variation in natural populations of *Drosophila melanogaster*. IV. Mitochondrial DNA variation and the role of history *vs.* selection in the genetic structure of geographic populations. Genetics 129:102–117.

Hall, H.G. 1990. Parental analysis of introgressive hybridization between African and European honeybees using nuclear DNA RFLP's. Genetics 125:611–621.

Hall, H.G. and K. Muralidharan. 1989. Evidence from mitochondrial DNA that African honey bees spread as continuous maternal lineages. Nature 339:211–213.

Hall, H.G. and D.R. Smith. 1991. Distinguishing African and European honeybee matrilines using amplified mitochondrial DNA. Proc. Natl. Acad. Sci. USA 88:4548–4552.

Hamada, H., M.G. Petrino, T. Kakunaga, M. Seidman, and B.D. Stollar. 1984. Characterization of genomic poly (dT-dG) poly (dC-dA) sequences: structure, organization, and conformation. Mol. Cell. Biol. 4:2610–2621.

Hamby, R.K. and E.A. Zimmer. 1992. Ribosomal RNA as a phylogenetic tool in plant systematics. Pp. 50–91 *in:* Molecular Systematics of Plants, P.S. Soltis, D.E. Soltis, and J.J. Doyle (eds.), Chapman and Hall, New York.

Hames, B.D. and S.J. Higgins (eds.). 1985. Nucleic Acid Hybridization: A Practical Approach. IRL Press, Oxford, England.

Hamilton, W.D. 1964. The genetical evolution of social behavior. J. Theoret. Biol. 7:1–52.

Hamrick, J.L. and R.W. Allard. 1972. Microgeographical variation in allozyme frequencies in *Avena barbata*. Proc. Natl. Acad. Sci. USA 69:2100–2104.

Hamrick, J.L., H.M. Blanton, and K.J. Hamrick. 1989. Genetic structure of geographically marginal populations of ponderosa pine. Amer. J. Bot. 76:1559–1568.

Hamrick, J.L. and M.J.W. Godt. 1989. Allozyme diversity in plants. Pp. 43–63 *in:* Plant Population Genetics, Breeding and Genetic Resources, A.H.D. Brown, M.T. Clegg, A.L. Kahler, and B.S. Weir (eds.), Sinauer, Sunderland, MA.

Hamrick, J.L., M.J.W. Godt, and S.L. Sherman-Broyles. 1992. Factors influencing levels of genetic diversity in woody plant species. New Forests 6:95–124.

Hamrick, J.L., Y.B. Linhart, and J.B. Mitton. 1979. Relationships between life history characteristics and electrophoretically detectable genetic variation in plants. Annu. Rev. Ecol. Syst. 10: 173–200.

Hamrick, J.L. and M.D. Loveless. 1989. The genetic structure of tropical tree populations: associations with reproductive biology. Pp. 129–146 *in:* Plant Evolutionary Ecology, J.H. Bock and Y.B. Linhart (eds.), Westview Press, Boulder, CO.

Hamrick, J.L. and D.A. Murawski. 1990. The breeding structure of tropical tree populations. Plant Species Biol. 5:157–165.

Hanken, J. and P.W. Sherman. 1981. Multiple paternity in Belding's ground squirrel litters. Science 212:351–353.

Hanotte, O., M.W. Bruford, and T. Burke. 1992a. Multilocus DNA fingerprints in gallinaceous birds: general approach and problems. Heredity 68:481–494.

Hanotte, O., T. Burke, J.A.L. Armour, and A.J. Jeffreys. 1991. Cloning, characterization and evolution of Indian peafowl *Pavo cristatus* minisatellite loci. Pp. 193–216 *in:* DNA Fingerprinting Approaches and Applications, T. Burke, G. Dolf, A.J. Jeffreys, and R. Wolff (eds.), Birkhauser Verlag, Basel, Switzerland.

Hanotte, O., E. Cairns, T. Robson, M.C. Double, and T. Burke. 1992b. Cross-species hybridization of a single-locus minisatellite probe in passerine birds. Mol. Ecol. 1:127–130.

Hanson, M.R. 1991. Plant mitochondrial mutations and male sterility. Annu. Rev. Genet. 25:461–486.

Hanson, M.R. and O. Folkerts. 1992. Structure and function of the higher plant mitochondrial genome. Int. Rev. Cytol. 141:129–172.

Hare, M.P. and G.F. Shields. 1992. Mitochondrial-DNA variation in the polytypic Alaskan song sparrow. Auk 109:126–132.

Hare, P.E. (ed.). 1980. The Biogeochemistry of Amino Acids. John Wiley and Sons, New York.

Harper, J.L. 1977. The Population Biology of Plants. Academic Press, New York.

Harper, J.L. 1985. Modules, branches, and the capture of resources. Pp. 1–33 *in:* Population Biology and Evolution of Clonal Organisms, J.B.C. Jackson, L.W. Buss, and R.E. Cook, Yale Univ. Press, New Haven, CT.

Harrington, R.W., Jr. 1961. Oviparous hermaphroditic fish with internal self-fertilization. Science 134:1749–1750.

Harrington, R.W., Jr. and K.D. Kallman. 1968. The homozygosity of clones of the self-fertilizing hermaphroditic fish *Rivulus marmoratus* Poey (Cyprinodontidae, Atheriniformes). Amer. Natur. 102:337–343.

Harris, H. 1966. Enzyme polymorphisms in man. Proc. Roy. Soc. Lond. B164:298–310.

Harris, H. and D.A. Hopkinson. 1976 et seq. Handbook of Enzyme Electrophoresis in Human Genetics. North-Holland, Amsterdam.

Harrison, R.G. 1979. Speciation in North American field crickets: evidence from electrophoretic comparisons. Evolution 33:1009–1023.

Harrison, R.G. 1989. Animal mitochondrial DNA as a genetic marker in population and evolutionary biology. Trends Ecol. Evol. 4:6–11.

Harrison, R.G. 1990. Hybrid zones: windows on evolutionary process. Oxford Surv. Evol. Biol. 7:69–128.

Harrison, R.G. 1991. Molecular changes at speciation. Annu. Rev. Ecol. Syst. 22:281–308.

Harrison, R.G. (ed.). 1993. Hybrid Zones and the Evolutionary Process. Oxford Univ. Press, Oxford, England.

Harrison, R.G. and D.M. Rand. 1989. Mosaic hybrid zones and the nature of species boundaries. Pp. 111–133 *in:* Speciation and Its Consequences, D. Otte and J.A. Endler (eds.), Sinauer, Sunderland, MA.

Harrison, R.G., D.M. Rand, and W.C. Wheeler. 1985. Mitochondrial DNA size variation within individual crickets. Science 228:1446–1448.

Harrison, R.G., D.M. Rand, and W.C. Wheeler. 1987. Mitochondrial DNA variation in field crickets across a narrow hybrid zone. Mol. Biol. Evol. 4:144–158.

Harry, J.L. and D.A. Briscoe. 1988. Multiple paternity in the loggerhead turtle (*Caretta caretta*). J. Heredity 79:96–99.

Hartl, D.L. and A.G. Clark. 1989. Principles of Population Genetics, 2nd Ed. Sinauer, Sunderland, MA.

Hartl, D.L. and D.E. Dykhuizen. 1984. The population genetics of *Escherichia coli*. Annu. Rev. Genet. 18:31–68.

Harvey, P.H. and M.D. Pagel. 1991. The Comparative Method in Evolutionary Biology. Oxford Univ. Press, Oxford, England.

Harvey, P.H. and A. Purvis. 1991. Comparative methods for explaining adaptations. Nature 351: 619–624.

Harvey, W.D. 1990. Electrophoretic techniques in forensics and law enforcement. Pp. 313–321 *in*. Electrophoretic and Isoelectric Focusing Techniques in Fisheries Management, D.H. Whitmore (ed.), CRC Press, Boca Raton, FL.

Hasegawa, M. 1990. Phylogeny and molecular evolution in primates. Japan J. Genet. 65:243–266.

Hasegawa, M. and S. Horai. 1991. Time of the deepest root for polymorphism in human mitochondrial DNA. J. Mol. Evol. 32:37–42.

Hasegawa, M. and H. Kishino. 1989. Heterogeneity of tempo and mode of mitochondrial DNA evolution among mammalian orders. Japan J. Genet. 64:243–258.

Hauser, L., A.R. Beaumont, G.T.H. Marshall, and R.J. Wyatt. 1991. Effects of sea trout stocking on the population genetics of landlocked brown trout, *Salmo trutta* L., in the Conwy River system, North Wales, U.K. J. Fish Biol. 39A:109–116.

Hayden, M.R., H.C. Hopkins, M. MacRae, and P.H. Brighton. 1980. The origin of Huntington's chorea in the Afrikaner population of South Africa. S. African Med. J. 58:197–200.

Heath, D.J., J.R. Radford, B.J. Riddoch, and D. Childs. 1990. Multiple mating in a natural population of the isopod *Sphaeroma rugicauda;* evidence from distorted ratios in offspring. Heredity 64:81–85.

Hebert, P.D.N. 1974a. Enzyme variability in natural populations of *Daphnia magna* III. Genotypic frequencies in intermittent populations. Genetics 77:335–341.

Hebert, P.D.N. 1974b. Ecological differences between genotypes in a natural population of *Daphnia magna*. Heredity 33:327–337.

Hebert, P.D.N. 1974c. Enzyme variability in natural populations of *Daphnia magna* II. Genotypic frequencies in permanent populations. Genetics 77:323–334.

Hebert, P.D.N., M.J. Beaton, S.S. Schwartz, and D.J. Stanton. 1989. Polyphyletic origin of asexuality in *Daphnia pulex*. I. Breeding-system variation and levels of clonal diversity. Evolution 43:1004–1015.

Hebert, P.D.N. and T.J. Crease. 1980. Clonal existence in *Daphnia pulex* (Leydig): another planktonic paradox. Science 207:1363–1365.

Hebert, P.D.N. and R.D. Ward. 1972. Inheritance during parthenogenesis in *Daphnia magna*. Genetics 71:639–642.

Hebert, P.D.N. and R.D. Ward. 1976. Enzyme variability in natural populations of *Daphnia magna*. Heredity 36:331–341.

Hedgecock, D. 1979. Biochemical genetic variation and evidence of speciation in *Chthamalus* barnacles of the tropical eastern Pacific Ocean. Mar. Biol. 54:207–214.

Hedgecock, D. 1986. Is gene flow from pelagic larval dispersal important in the adaptation and evolution of marine invertebrates? Bull. Mar. Sci. 39:550–564.

Hedgecock, D., V. Chow, and R.S. Waples. 1992. Effective population numbers of shellfish broodstocks estimated from temporal variance in allelic frequencies. Aquaculture 108:215–232.

Hedgecock, D. and F. Sly. 1990. Genetic drift and effective population sizes of hatchery-propagated stocks of the Pacific oyster, *Crassostrea gigas*. Aquaculture 88:21–38.

Hedges, R.W. 1972. The pattern of evolutionary change in bacteria. Heredity 28:39–48.

Hedges, S.B. 1989. An island radiation: allozyme evolution in Jamaican frogs of the genus *Eleutherodactylus* (Leptodactylidae). Caribbean J. Sci. 25:123–147.

Hedges, S.B. 1992. The number of replications needed for accurate estimation of the bootstrap *P* value in phylogenetic studies. Mol. Biol. Evol. 9:366–369.

Hedges, S.B., R.L. Bezy, and L.R. Maxson. 1991. Phylogenetic relationships and biogeography of Xantusiid lizards, inferred from mitochondrial DNA sequences. Mol. Biol. Evol. 8:767–780.

Hedges, S.B., J.P. Bogart, and L.R. Maxson. 1992a. Ancestry of unisexual salamanders. Nature 356:708–710.

Hedges, S.B. and K.L. Burnell. 1990. The Jamaican radiation of *Anolis* (Sauria: Iguanidae): an analysis of relationships and biogeography using sequential electrophoresis. Caribbean J. Sci. 26:31–44.

Hedges, S.B., C.A. Hass, and L.R. Maxson. 1992b. Caribbean biogeography: molecular evidence for dispersal in West Indian terrestrial vertebrates. Proc. Natl. Acad. Sci. USA 89:1909–1913.

Hedges, S.B., S. Kumar, and K. Tamura. 1992c. Human origins and analysis of mitochondrial DNA sequences. Science 255:737–739.

Hedges, S.B., K.D. Moberg, and L.R. Maxson. 1990. Tetrapod phylogeny inferred from 18S and 28S ribosomal RNA sequences and a review of the evidence for amniote relationships. Mol. Biol. Evol. 7:607–633.

Hedrick, P.W. 1986. Genetic polymorphisms in heterogeneous environments: a decade later. Annu. Rev. Ecol. Syst. 17:535–566.

Hedrick, P.W. 1992. Shooting the RAPDs. Nature 355:679–680.

Hedrick, P.W., P.F. Brussard, F.W. Allendorf, J.A. Beardmore, and S. Orzack. 1986. Protein variation, fitness, and captive propagation. Zoo Biol. 5:91–99.

Hedrick, P.W. and P.S. Miller. 1992. Conservation genetics: techniques and fundamentals. Ecol. Applic. 2:30–46.

Hedrick, P.W., T.S. Whittam, and P. Parham. 1991. Heterozygosity at individual amino acid sites: extremely high levels for *HLA-A* and *-B* genes. Proc. Natl. Acad. Sci. USA 88:5897–5901.

Heinemann, J.A. 1991. Genetics of gene transfer between species. Trends Genet. 7:181–185.

Heinemann, J.A. and G.F. Sprague, Jr. 1989. Bacterial conjugative plasmids mobilize DNA transfer between bacteria and yeast. Nature 340:205–209.

Hennig, W. 1966. Phylogenetic Systematics. Univ. of Illinois Press, Urbana.

Hepper, P.G. (ed.). 1991. Kin Recognition. Cambridge Univ. Press, New York.

Herbers, J.M. 1986. Nest site limitation and facultative polygyny in the ant *Leptothorax longispinosus*. Behav. Ecol. Sociobiol. 19:115–122.

Hermanutz, L.A., D.J. Innes, and I.M. Weis. 1989. Clonal structure of arctic dwarf birch (*Betula glandulosa*) at its northern limit. Amer. J. Bot. 76:755–761.

Hewitt, G.M. 1988. Hybrid zones—natural laboratories for evolutionary studies. Trends Ecol. Evol. 3:158–166.

Hewitt, G.M. 1989. The subdivision of species by hybrid zones. Pp. 85–110 *in:* Speciation and Its Consequences, D. Otte and J.A. Endler (eds.), Sinauer, Sunderland, MA.

Heyward, A.J. and J.A. Stoddart. 1985. Genetic structure of two species of *Montipora* on a patch reef: conflicting results from electrophoresis and histocompatibility. Mar. Biol. 85:117–121.

Heywood, J.S. 1991. Spatial analysis of genetic variation in plant populations. Annu. Rev. Ecol. Syst. 22:335–355.

Hickey, D.A. 1982. Selfish DNA: a sexually-transmitted nuclear parasite. Genetics 101:519–531.

Hickey, D.A. and M.D. McLean. 1980. Selection for ethanol tolerance and Adh allozymes in natural populations of *Drosophila melanogaster*. Genet. Res. 36:11–15.

Hickey, R.J., M.A. Vincent, and S.I. Guttman. 1991. Genetic variation in running buffalo clover (*Trifolium stoloniferum,* Fabaceae). Cons. Biol. 5:309–316.

Hickman, C.S. and J.H. Lipps. 1985. Geologic youth of Galapagos Islands confirmed by marine stratigraphy and paleontology. Science 227:1578–1580.

Higgs, D.R., J.S. Wainscoat, J. Flint, A.V.S. Hill, S.L. Thein, R.D. Nichols, H. Teal, H. Ayyub, T.E.A. Peto, A.G. Falusi, A.P. Jarman, J.B. Clegg, and D.J. Weatherall. 1986. Analysis of the human α-globin gene cluster reveals a highly informative genetic locus. Proc. Natl. Acad. Sci. USA 83:5165–5169.

Highton, R. 1984. A new species of woodland salamander of the *Plethodon glutinosus* group from the southern Appalachian Mountains. Brimleyana 9:1–20.

Highton, R., G.C. Maha, and L.R. Maxson. 1989. Biochemical evolution in the slimy salamanders of the *Plethodon glutinosus* complex in the eastern United States. Illinois Biol. Monogr. 57:1–153.

Higuchi, R.G., B. Bowman, M. Freiberger, O.A. Ryder, and A.C. Wilson. 1984. DNA sequence from the quagga, an extinct member of the horse family. Nature 312:282–284.

Higuchi, R.G., L.A. Wrischnik, E. Oakes, M. George, B. Tong, and A.C. Wilson. 1987. Mitochondrial DNA of the extinct quagga: relatedness and extent of postmortem change. J. Mol. Evol. 25:283–287.

Hilbish, T.J., L.E. Deaton, and R.K. Koehn. 1982. Effect of an allozyme polymorphism on regulation of cell volume. Nature 298:688–689.

Hilbish, T.J. and R.K. Koehn. 1985. The physiological basis of natural selection at the *LAP* locus. Evolution 39:1302–1317.

Hildemann, W.H., R.L. Raison, G. Cheung, C.J. Hull, L. Akaka, and J. Okamoto. 1977. Immunological specificity and memory in a scleractinian coral. Nature 270:219–223.

Hill, W.G. 1987. DNA fingerprints applied to animal and bird populations. Nature 327:98–99.

Hillis, D.M. 1987. Molecular versus morphological approaches to systematics. Annu. Rev. Ecol. Syst. 18:23–42.

Hillis, D.M. 1989. Genetic consequences of partial self-fertilization on populations of *Liguus fasciatus* (Mollusca: Pulmonata: Bulimulidae). Amer. Malacolog. Bull 7:7–12.

Hillis, D.M. and J.J. Bull. 1991. Of genes and genomes. Science 254:528–558.

Hillis, D.M., J.J. Bull, M.E. White, M.R. Badgett, and I.J. Molineux. 1992. Experimental phylogenetics: generation of a known phylogeny. Science 255:589–592.

Hillis, D.M. and M.T. Dixon. 1989. Vertebrate phylogeny: evidence from 28S ribosomal DNA sequences. Pp. 355–367 *in:* The Hierarchy of Life, B. Fernholm, K. Bremer, and H. Jornvall (eds.), Elsevier, Amsterdam.

Hillis, D.M. and M.T. Dixon. 1991. Ribosomal DNA: molecular evolution and phylogenetic inference. Quart. Rev. Biol. 66:411–453.

Hillis, D.M., M.T. Dixon, and A.L. Jones. 1991. Minimal genetic variation in a morphologically diverse species (Florida tree snail, *Liguus fasciatus*). J. Heredity 82:282–286.

Hillis, D.M., A. Larson, S.K. Davis, and E.A. Zimmer. 1990. Nucleic acids III: Sequencing. Pp. 318–370 *in:* Molecular Systematics, D.M. Hillis and C. Moritz (eds.), Sinauer, Sunderland, MA.

Hillis, D.M. and C. Moritz. 1990. Molecular Systematics. Sinauer, Sunderland, MA.

Hindar, K., B. Jonsson, N. Ryman, and G. Ståhl. 1991. Genetic relationships among landlocked, resident, and anadromous brown trout, *Salmo trutta* L. Heredity 66:83–91.

Hobbs, R.J. 1992. The role of corridors in conservation: solution or bandwagon? Trends Ecol. Evol. 7:389–392.

Hoeh, W.R., K.H. Blakley, and W.M. Brown. 1991. Heteroplasmy suggests limited biparental inheritance of *Mytilus* mitochondrial DNA. Science 251:1488–1490.

Hoelzel, A.R. (ed.). 1991a. Genetic Ecology of Whales and Dolphins. Black Bear Press, Cambridge, England.

Hoelzel, A.R. 1991b. Analysis of regional mitochondrial DNA variation in the killer whale; implications for cetacean conservation. Pp. 225–233 *in:* Genetic Ecology of Whales and Dolphins, A.R. Hoelzel (ed.), Black Bear Press, Cambridge, England.

Hoelzel, A.R. (ed.). 1992. Molecular Genetic Analysis of Populations: A Practical Approach. IRL Press, Oxford, England.

Hoelzel, A.R. and G.A. Dover. 1991a. Molecular Genetic Ecology. Oxford Univ. Press, Oxford, England.

Hoelzel, A.R. and G.A. Dover. 1991b. Genetic differentiation between sympatric killer whale populations. Heredity 66:191–196.

Hoelzel, A.R., J. Halley, C. Campagna, T. Arnbom, B. Le Boeuf, S.J. O'Brien, K. Ralls, and G.A. Dover. 1993. Elephant seal genetic variation and the use of simulation models to investigate historical population bottlenecks. J. Heredity, *in press*.

Hoffman, R.J. 1983. The mating system of the terrestrial slug *Deroceras laeve*. Evolution 37:423–425.

Hoffman, R.J. 1986. Variation in contributions of asexual reproduction to the genetic structure of populations of the sea anemone *Metridium senile*. Evolution 40:357–365.

Hoffmann, R.J., J.L. Boore, and W.M. Brown. 1992. A novel mitochondrial genome organization for the blue mussel, *Mytilus edulis*. Genetics 131:397–412.

Hölldobler, B. and E.O. Wilson. 1990. The Ants. Belknap Press, Cambridge, MA.

Holmes, E.C., L.Q. Zhang, P. Simmonds, C.A. Ludlam, and A.J.L. Brown. 1992. Convergent and divergent sequence evolution in the surface envelope glycoprotein of human immunodeficiency virus type 1 within a single infected patient. Proc. Natl. Acad. Sci. USA 89:4835–4839.

Holmes, W.G. and P.W. Sherman. 1982. The ontogeny of kin recognition in two species of ground squirrels. Amer. Zool. 22:491–517.

Holmquist, G. 1989. Evolution of chromosome bands: molecular ecology of noncoding DNA. J. Mol. Evol. 28:469–486.

Honeycutt, R.L. 1992. Naked mole-rats. Amer. Sci. 80:43–53.

Hoogland, J.L. 1982. Prairie dogs avoid extreme inbreeding. Science 215:1639–1641.

Hoogland, J.L. and D.W. Foltz. 1982. Variance in male and female reproductive success in a harem-polygynous mammal, the black-tailed prairie dog (Sciuridae: *Cynomys ludovicianus*). Behav. Ecol. Sociobiol. 11:155–163.

Horai, S., Y. Satta, K. Hayasaka, R. Kondo, T. Inoue, T. Ishida, S. Hayashi, and N. Takahata. 1992. Man's place in Hominoidea revealed by mitochondrial DNA genealogy. J. Mol. Evol. 35:32–43.

Hori, H., B.-L. Lim, and S. Osawa. 1985. Evolution of green plants as deduced from 5S rRNA sequences. Proc. Natl. Acad. Sci. USA 82:820–823.

Hori, H. and S. Osawa. 1987. Origin and evolution of organisms as deduced from 5S ribosomal RNA sequences. Mol. Biol. Evol. 4:445–472.

Horn, G.T., B. Richards, and K.W. Klinger. 1989. Amplification of a highly polymorphic VNTR segment by the polymerase chain reaction. Nucleic Acids Res. 17:2140.

Horowitz, J.J. and G.S. Whitt. 1972. Evolution of a nervous system specific lactate dehydrogenase isozyme in fish. J. Exp. Zool. 180:13–31.

Höss, M., M. Kohn, S. Pääbo, F. Knauer, and W. Schröder. 1992. Excrement analysis by PCR. Nature 359:199.

Houck, M.A., J.B. Clark, K.R. Peterson, and M.G. Kidwell. 1991. Possible horizontal transfer of *Drosophila* genes by the mite *Proctolaelaps regalis*. Science 253:1125–1129.

Howe, C.J., T.J. Beanland, A.W.D. Larkum, and P.J. Lockhart. 1992. Plastid origins. Trends Ecol. Evol. 7:378–383.

Hu, Y.-P., R.A. Lutz, and R.C. Vrijenhoek. 1992. Electrophoretic identification and genetic analysis of bivalve larvae. Mar. Biol. 113:227–230.

Hubbs, C.L. 1955. Hybridization between fish species in nature. Syst. Zool. 4:1–20.

Hubbs, C.L. and L.C. Hubbs. 1933. The increased growth, predominant maleness, and apparent infertility of hybrid sunfishes. Pap. Mich. Acad. Sci. Arts Lett. 17:613–641.

Hubby, J.L. and L.H. Throckmorton. 1968. Protein differences in *Drosophila*. IV. A study of sibling species. Amer. Natur. 102:193–205.

Hudson, R.R. 1990. Gene genealogies and the coalescent process. Oxford Surv. Evol. Biol. 7:1–44.

Hudson, R.R., M. Slatkin, and W.P. Maddison. 1992. Estimation of levels of gene flow from DNA sequence data. Genetics 132:583–589.

Hughes, A.L. 1991. MHC polymorphism and the design of captive breeding programs. Cons. Biol. 5:249–251.

Hughes, A.L. 1992. Avian species described on the basis of DNA only. Trends Ecol. Evol. 7:2–3.

Hughes, A.L. and M. Nei. 1988. Pattern of nucleotide substitution at major histocompatibility complex class I loci reveals overdominant selection. Nature 335:167–170.

Hughes, A.L. and M. Nei. 1989. Nucleotide substitution at major histocompatibility complex class II loci: evidence for overdominant selection. Proc. Natl. Acad. Sci. USA 86:958–962.

Hughes, J. and A.J. Richards. 1988. The genetic structure of populations of sexual and asexual *Taraxacum* (dandelions). Heredity 60:161–171.

Hughes, J. and A.J. Richards. 1989. Isozymes, and the status of *Taraxacum* (Asteraceae) agamospecies. Bot. J. Linn. Soc. 99:365–376.

Hughes, M.B. and J.C. Lucchesi. 1977. Genetic rescue of a lethal "null" activity allele of 6-phosphogluconate dehydrogenase in *Drosophila melanogaster*. Science 196:1114–1115.

Hull, D.L. 1988. Science as a Process. Univ. Chicago Press, Chicago, IL.

Humphries, C.J. and L. Parenti. 1986. Cladistics-Biogeography. Oxford Monographs on Biogeography, Clarendon Press, Oxford, England.

Humphries, J.M. 1984. Genetics of speciation in pupfishes from Laguna Chichancanab, Mexico. Pp. 129–139 *in:* Evolution of Fish Species Flocks, A.A. Echelle and I. Kornfield (eds.), Univ. Maine Press, Orono.

Hunt, A. and D.J. Ayre. 1989. Population structure in the sexually reproducing sea anemone *Oulactis muscosa*. Mar. Biol. 102:537–544.

Hunt, W.G. and R.K. Selander. 1973. Biochemical genetics of hybridisation in European house mice. Heredity 31:11–33.

Hunter, C.L. 1985. Assessment of clonal diversity and population structure of *Porites compressa* (Cnideria, Scleractinia). Proc. 5th Int. Coral Reef Symp. 6:69–74.

Hunter, R.L. and C.L. Markert. 1957. Histochemical demonstration of enzymes separated by zone electrophoresis in starch gels. Science 125:1294–1295.

Hutchison, C.A., III, J.E. Newbold, S.S. Potter, and M.H. Edgell. 1974. Maternal inheritance of mammalian mitochondrial DNA. Nature 251:536–538.

Hutton, J.R. and J.G. Wetmur. 1973. Effect of chemical modification on the rate of renaturation of deoxyribonucleic acid. Deaminated and glyoxalated deoxyribonucleic acid. Biochemistry 12:558–563.

Hyman, B.C., J.L. Beck, and K.C. Weiss. 1988. Sequence amplification and gene arrangement in parasitic nematode mitochondrial DNA. Genetics 120:707–712.

Ihssen, P.E., H.E. Booke, J.M. Casselman, J.M. McGlade, N.R. Payne, and F.M. Utter. 1981. Stock identification: materials and methods. Can. J. Fish. Aquat. Sci. 38:1838–1855.

Innes, D.J. 1987. Genetic structure of asexually reproducing *Enteromorpha linza* (Ulvales: Chlorophyta) in Long Island Sound. Mar. Biol. 94:459–467.

Innes, D.J. and C. Yarish. 1984. Genetic evidence for the occurrence of asexual reproduction in populations of *Enteromorpha linza* (L.) J. Ag. (Chlorophyta, Ulvales) from Long Island Sound. Phycologia 23:311–320.

Innis, M.A., D.H. Gelfand, J.J. Sninsky, and T.J. White. 1990. PCR Protocols: A Guide to Methods and Applications. Academic Press, New York.

Innis, M.A., K.B. Myambo, D.H. Gelfand, and M.A.D. Brow. 1988. DNA sequencing with *Thermus aquaticus* DNA polymerase and direct sequencing of polymerase chain reaction-amplified DNA. Proc. Natl. Acad. Sci. USA 85:9436–9440.

Ioerger, T.R., A.G. Clark, and T.-H. Kao. 1990. Polymorphism at the self-incompatibility locus in Solanaceae predates speciation. Proc. Natl. Acad. Sci. USA 87:9732–9735.

Irwin, D.M., T.D. Kocher, and A.C. Wilson. 1991. Evolution of the cytochrome *b* gene of mammals. J. Mol. Evol. 32:128–144.

Istock, C.A., K.E. Duncan, N. Ferguson, and X. Zhou. 1992. Sexuality in a natural population of bacteria—*Bacillus subtilis* challenges the clonal paradigm. Mol. Ecol. 1:95–103.

Iwabe, N., K.-I. Kuma, M. Hasegawa, S. Osawa, and T. Miyata. 1989. Evolutionary relationship of archaebacteria, eubacteria, and eukaryotes inferred from phylogenetic trees of duplicated genes. Proc. Natl. Acad. Sci. USA 86:9355–9359.

Jablonski, D. 1986. Larval ecology and macroevolution in marine invertebrates. Bull. Mar. Sci. 39:565–587.

Jackson, J.B.C. 1985. Distribution and ecology of clonal and aclonal benthic invertebrates. Pp. 297–355 *in:* Population Biology and Evolution of Clonal Organisms, J.B.C. Jackson, L.W. Buss, and R.E. Cook (eds.), Yale Univ. Press, New Haven, CT.

Jackson, J.B.C. 1986. Modes of dispersal of clonal benthic invertebrates: consequences for species' distributions and genetic structure of local populations. Bull. Mar. Sci. 39:588–606.

Jackson, J.B.C., L.W. Buss, and R.E. Cook (eds.). 1985. Population Biology and Evolution of Clonal Organisms. Yale Univ. Press, New Haven, CT.

Jacobs, H., S. Asakawa, T. Araki, K. Mikura, M.J. Smith, and K. Watanabe. 1989. Conserved tRNA gene cluster in starfish mitochondrial DNA. Curr. Genet. 15:193–206.

Jacobs, H., D. Elliot, V. Math, and A. Farguharson. 1988. Nucleotide sequence and gene organization of sea urchin mitochondrial DNA. J. Mol. Biol. 201:185–217.

Jacobson, D.J. and T.R. Gordon. 1990. Variability of mitochondrial DNA as an indicator of relationships between populations of *Fusarium oxysporum f. sp. melonis*. Mycol. Res. 94:734–744.

Jaenike, J., E.D. Parker, Jr., and R.K. Selander. 1980. Clonal niche structure in the parthenogenetic earthworm *Octolasion tyrtaeum*. Amer. Natur. 116:196–205.

Jaenike, J. and R.K. Selander. 1979. Evolution and ecology of parthenogenesis in earthworms. Amer. Zool. 19:729–737.

Jain, S.K. and R.W. Allard. 1966. The effects of linkage, epistasis and inbreeding on population changes under selection. Genetics 53:633–659.

James, F.C. 1983. Environmental component of morphological differentiation in birds. Science 221:184–186.

Janczewski, D.N., N. Yuhki, D.A. Gilbert, G.T. Jefferson, and S.J. O'Brien. 1992. Molecular phylogenetic inference from saber-toothed cat fossils of Rancho La Brea. Proc. Natl. Acad. Sci. USA 89:9769–9773.

Jansen, R.K., H.J. Michaels, R.S. Wallace, K.-J. Kim, S.C. Keeley, L.E. Watson, and J.D.

Palmer. 1992. Chloroplast DNA variation in the Asteraceae: phylogenetic and evolutionary implications. Pp. 252–279 *in:* Molecular Systematics of Plants, P.S. Soltis, D.E. Soltis, and J.J. Doyle (eds.), Chapman and Hall, New York.

Jansen, R.K. and J.D. Palmer. 1987. A chloroplast DNA inversion marks an ancient evolutionary split in the sunflower family (Asteraceae). Proc. Natl. Acad. Sci. USA 84:5818–5822.

Jansen, R.K. and J.D. Palmer. 1988. Phylogenetic implications of chloroplast DNA restriction site variation in the Mutisieae (Asteraceae). Amer. J. Bot. 75:753–766.

Janson, K. 1987. Allozyme and shell variation in two marine snails (*Littorina,* Prosobranchia) with different dispersal abilities. Biol. J. Linn. Soc. 30:245–256.

Jarman, A.P., R.D. Nichols, D.J. Weatherall, J.B. Clegg, and D.R. Higgs. 1986. Molecular characterization of a hypervariable region downstream of the human α-globin gene cluster. The EMBO J. 5:1857–1863.

Jarman, A.P. and R.A. Wells. 1989. Hypervariable minisatellites: recombinators or innocent bystanders? Trends Genet. 5:367–371.

Jarne, P. and B. Delay. 1991. Population genetics of freshwater snails. Trends Ecol. Evol. 6:383–386.

Jarne, P., B. Delay, C. Bellec, G. Roizes, and G. Cuny. 1990. DNA fingerprinting in schistosome-vector snails. Biochem. Genet. 28:577–583.

Jarne, P., B. Delay, C. Bellec, G. Roizes, and G. Cuny. 1992. Analysis of mating systems in the schistosome-vector hermaphrodite snail *Bulinus globosus* by DNA fingerprinting. Heredity 68: 141–146.

Jarvis, J.U.M. 1981. Eusociality in a mammal: cooperative breeding in naked mole-rat colonies. Science 212:571–573.

Jeffreys, A.J. 1987. Highly variable minisatellites and DNA fingerprints. Biochem. Soc. Trans. 15:309–317.

Jeffreys, A.J., J.F.Y. Brookfield, and R. Semenoff. 1985c. Positive identification of an immigration test-case using human DNA fingerprints. Nature 317:818–819.

Jeffreys, A.J., A. MacLeod, K. Tamaki, D.L. Neil, and D.G. Monckton. 1991. Minisatellite repeat coding as a digital approach to DNA typing. Nature 354:204–209.

Jeffreys, A.J. and D.B. Morton. 1987. DNA fingerprints of dogs and cats. Anim. Genet. 18:1–15.

Jeffreys, A.J., R. Neumann, and V. Wilson. 1990. Repeat unit sequence variation in minisatellites: a novel source of DNA polymorphism for studying variation and mutation by single molecule analysis. Cell 60:473–485.

Jeffreys, A.J., N.J. Royle, V. Wilson, and Z. Wong. 1988b. Spontaneous mutation rates to new length alleles at tandem-repetitive hypervariable loci in human DNA. Nature 332:278–281.

Jeffreys, A.J., V. Wilson, R. Kelly, B.A. Taylor, and G. Bulfield. 1987. Mouse DNA "fingerprints": analysis of chromosome localization and germline stability of hypervariable loci in recombinant inbred strains. Nucleic Acids Res. 15:2823–2836.

Jeffreys, A.J., V. Wilson, R. Neumann, and J. Keyte. 1988a. Amplification of human minisatellites by the polymerase chain reaction: towards DNA fingerprinting of single cells. Nucleic Acids Res. 16:10953–10971.

Jeffreys, A.J., V. Wilson, and S.L. Thein. 1985a. Hypervariable "minisatellite" regions in human DNA. Nature 314:67–73.

Jeffreys, A.J., V. Wilson, and S.L. Thein. 1985b. Individual-specific "fingerprints" of human DNA. Nature 316:76–79.

Jermiin, L.S., V. Loeschcke, V. Simonsen, and V. Mahler. 1991. Electrophoretic and morphometric analyses of two sibling species pairs in *Trachyphloeus* (Coleoptera: Curculionidae). Entomol. Scandinavica 22:159–170.

Jin, L. and M. Nei. 1991. Relative efficiencies of the maximum-parsimony and distance-matrix methods of phylogeny construction for restriction data. Mol. Biol. Evol. 8:356–365.

Johnson, F.M., C.G. Kanapi, R.H. Richardson, M.R. Wheeler, and W.S. Stone. 1966. An analysis of polymorphisms among isozyme loci in dark and light *Drosophila ananassae* strains from American and Western Samoa. Proc. Natl. Acad. Sci. USA 56:119–125.

Johnson, G. 1976a. Hidden alleles at the α-glycerophosphate dehydrogenase locus in *Colias* butterflies. Genetics 83:149–167.

Johnson, G. 1976b. Genetic polymorphism and enzyme function. Pp. 46–59 *in:* Molecular Evolution, F.J. Ayala (ed.), Sinauer, Sunderland, MA.

Johnson, G. 1977. Evaluation of the charge state model of electrophoretic mobility: comparison of the gel sieving behavior of alleles at the esterase-5 locus of *Drosophila pseudoobscura*. Genetics 87:139–157.

Johnson, M.S. and R. Black. 1982. Chaotic genetic patchiness in an intertidal limpet, *Siphonaria* sp. Mar. Biol. 70:157–164.

Johnson, M.S. and T.J. Threlfall. 1987. Fissiparity and population genetics of *Coscinasterias calamaria*. Mar. Biol. 93:517–525.

Johnson, M.S., D.C. Wallace, S.D. Ferris, M.C. Rattazzi, and L.L. Cavalli-Sforza. 1983. Radiation of human mitochondrial DNA types analyzed by restriction endonuclease cleavage patterns. J. Mol. Evol. 19:255–271.

Jones, C.S., H. Tegelström, D.S. Latchman, and R.J. Berry. 1988. An improved rapid method for mitochondrial DNA isolation suitable for use in the study of closely related populations. Biochem. Genet. 26:83–88.

Jones, J.S., S.H. Bryant, R.C. Lewontin, J.A. Moore, and T. Prout. 1981. Gene flow and the geographical distribution of a molecular polymorphism in *Drosophila pseudoobscura*. Genetics 98:157–178.

Jorgensen, R.A. and P.D. Cluster. 1988. Modes and tempos in the evolution of nuclear ribosomal DNA: new characters for evolutionary studies and new markers for genetic and population studies. Annals Missouri Bot. Gardens 75:1238–1247.

Joste, N., J.D. Ligon, and P.B. Stacey. 1985. Shared paternity in the acorn woodpecker (*Melanerpes formicivorus*). Behav. Ecol. Sociobiol. 17:39–41.

Jukes, T.H. and C.R. Cantor. 1969. Evolution of protein molecules. Pp. 21–132 *in:* Mammalian Protein Metabolism, H.N. Munro (ed.), Academic Press, New York.

Kallman, K.D. and R.W. Harrington, Jr. 1964. Evidence for the existence of homozygous clones in the self-fertilizing hermaphroditic teleost *Rivulus marmoratus* (Poey). Biol. Bull. 126:101–114.

Karl, S.A. and J.C. Avise. 1992. Balancing selection at allozyme loci in oysters: implications from nuclear RFLPs. Science 256:100–102.

Karl, S.A. and J.C. Avise. 1993. PCR-based assays of Mendelian polymorphisms from anonymous single-copy nuclear DNA: Techniques and applications for population genetics. Mol. Biol. Evol. 10:342–361.

Karl, S.A., B.W. Bowen, and J.C. Avise. 1992. Global population genetic structure and male-mediated gene flow in the green turtle (*Chelonia mydas*): RFLP analyses of anonymous nuclear loci. Genetics 131:163–173.

Kawamura, S., H. Tanabe, Y. Watanabe, K. Kurosaki, N. Saitou, and S. Ueda. 1991. Evolutionary rate of immunoglobulin alpha noncoding region is greater in hominoids than in Old World monkeys. Mol. Biol. Evol. 8:743–752.

Kempenaers, B., G.R. Verheyen, M. Van den Broeck, T. Burke, C. Van Broeckhoven, and A.A. Dhondt. 1992. Extra-pair paternity results from female preference for high-quality males in the blue tit. Nature 357:494–496.

Kemperman, J.A. and B.V. Barnes. 1976. Clone size in American aspens. Can. J. Bot. 54:2603–2607.

Kennedy, P.K., M.L. Kennedy, and M.H. Smith. 1985. Microgeographic genetic organization of populations of largemouth bass and two other species in a reservoir. Copeia 1985:118–125.

Kennedy, P.K., M.L. Kennedy, E.G. Zimmerman, R.K. Chesser, and M.H. Smith. 1986. Biochemical genetics of mosquitofish. V. Perturbation effects on genetic organization of populations. Copeia 1986:937–945.

Kessler, C. 1987. Class II restriction endonucleases. Pp. 225–279 *in:* Cytogenetics, G. Obe and A. Basler (eds.), Springer-Verlag, Berlin.

Kessler, L.G. and J.C. Avise. 1985. Microgeographic lineage analysis by mitochondrial genotype: variation in the cotton rat (*Sigmodon hispidus*). Evolution 39:831–837.

Kidwell, M.G. 1992. Horizontal transfer of *P* elements and other short inverted repeat transposons. Genetica 86:275–286.

Kim, K.-J., R.K. Jansen, R.S. Wallace, H.J. Michaels, and J.D. Palmer. 1992. Phylogenetic implications of *rbcL* sequence variation in the Asteraceae. Annals Missouri Bot. Gardens 79:428–445.

Kim, S.S. and S.K. Narang. 1990. Restriction site polymorphism of mtDNA for differentiating *Anopheles quadrimaculatus* (Say) sibling species. Korean J. Appl. Entomol. 29:132–135.

Kimura, M. 1968a. Evolutionary rate at the molecular level. Nature 217:624–626.

Kimura, M. 1968b. Genetic variability maintained in a finite population due to mutational production of neutral and nearly neutral isoalleles. Genet. Res. 11:247–269.

Kimura, M. 1980. A simple method for estimating evolutionary rate of base substitutions through comparative studies of nucleotide sequences. J. Mol. Evol. 16:111–120.

Kimura, M. 1983. The Neutral Theory of Molecular Evolution. Cambridge Univ. Press, Cambridge, England.

Kimura, M. 1990. The present status of the neutral theory. Pp. 1–16 *in:* Population Biology of Genes and Molecules, N. Takahata and J.F. Crow (eds.), Baifukan, Tokyo.

Kimura, M. 1991. Recent developments of the neutral theory viewed from the Wrightian tradition of theoretical population genetics. Proc. Natl. Acad. Sci. USA 88:5969–5973.

Kimura, M. and T. Ohta. 1971. Theoretical Aspects of Population Genetics. Princeton Univ. Press, Princeton, NJ.

King, J.L. and T.H. Jukes. 1969. Non-Darwinian evolution: random fixation of selectively neutral mutations. Science 164:788–798.

King, M.-C. and A.C. Wilson. 1975. Evolution at two levels in humans and chimpanzees. Science 188:107–116.

King, P.S. 1987. Macro- and microgeographic structure of a spatially subdivided beetle species in nature. Evolution 41:401–416.

King, R.C. and W.D. Stansfield. 1990. A Dictionary of Genetics. Oxford Univ. Press, New York.

Kirby, L.T. 1990. DNA Fingerprinting. Stockton Press, New York.

Kirkpatrick, K.J. and H.D. Wilson. 1988. Interspecific gene flow in Cucurbita: *C. texana* vs. *C. pepo*. Amer. J. Bot. 75:519–527.

Kirkpatrick, M. and R.K. Selander. 1979. Genetics of speciation in lake whitefishes in the Allegash basin. Evolution 33:478–485.

Kirsch, J.A.W. 1977. The comparative serology of Marsupialia, and a classification of marsupials. Aust. J. Zool. Suppl. Ser. 52:1–152.

Kirsch, J.A.W., M.S. Springer, C. Krajewski, M. Archer, K. Aplin, and A.W. Dickerman. 1990. DNA/DNA hybridization studies of the carnivorous marsupials. I: The intergeneric relationships of bandicoots (Marsupialia: Perameloidea). J. Mol. Evol. 30:434–448.

Kishino, H. and M. Hasegawa. 1989. Evaluation of the maximum likelihood estimate of the evolutionary tree topologies from DNA sequence data, and the branching order in Hominoidea. J. Mol. Evol. 29:170–179.

Kishino, H., T. Miyata, and M. Hasegawa. 1990. Maximum likelihood inference of protein phylogeny and the origin of chloroplasts. J. Mol. Evol. 31:151–160.

Kleiman, D.G. 1977. Monogamy in mammals. Quart. Rev. Biol. 52:39–69.

Klein, J. 1986. Natural History of the Major Histocompatibility Complex. Wiley-Interscience, New York.

Kluge, A.G. and J.S. Farris. 1969. Quantitative phyletics and the evolution of anurans. Syst. Zool. 18:1–32.

Knoblach, I.W. 1972. Intergeneric hybridization in flowering plants. Taxon 21:97–103.

Knoll, A.H. 1992. The early evolution of eukaryotes: a geological perspective. Science 256:622–627.

Knowlton, N. and S.R. Greenwell. 1984. Male sperm competition avoidance mechanisms: the influence of female interests. Pp. 61–84 *in:* Sperm Competition and the Evolution of Animal Mating Systems, R.L. Smith (ed.), Academic Press, New York.

Knowlton, N. and B.D. Keller. 1986. Larvae which fall short of their potential: highly localized recruitment in an alpheid shrimp with extended larval development. Bull. Mar. Sci. 39:213–223.

Knowlton, N., E. Weil, L.A. Weigt, and H.M. Guzman. 1992. Sibling species in *Montastraea annularis,* coral bleaching, and the coral climate record. Science 255:330–333.

Knox, E.B., S.R. Downie, and J.D. Palmer. 1993. Chloroplast genome rearrangements and the evolution of giant lobelias from herbaceous ancestors. Mol. Biol. Evol. 10:414–430.

Kocher, T.D., W.K. Thomas, A. Meyer, S.V. Edwards, S. Pääbo, F.X. Villablanca, and A.C. Wilson. 1989. Dynamics of mitochondrial DNA evolution in animals: amplification and sequencing with conserved primers. Proc. Natl. Acad. Sci. USA 86:6196–6200.

Kochert, G. 1989. Introduction to RFLP mapping and plant breeding applications. Special Publication Rockefeller Foundation, New York.

Koehn, R.K. 1978. Physiology and biochemistry of enzyme variation: the interface of ecology and population genetics. Pp. 51–72 *in:* Ecological Genetics: The Interface, P. Brussard (ed.), Springer-Verlag, New York.

Koehn, R.K., W.J. Diehl, and T.M. Scott. 1988. The differential contribution by individual enzymes of glycolysis and protein catabolism to the relationship between heterozygosity and growth rate in the coot clam, *Mulinia lateralis*. Genetics 118:121–130.

Koehn, R.K. and W.F. Eanes. 1976. An analysis of allelic diversity in natural populations of

Drosophila: The correlation of rare alleles with heterozygosity. Pp. 377–390 in: Population Genetics and Ecology, S. Karlin and E. Nevo (eds.), Academic Press, New York.

Koehn, R.K. and W.F. Eanes. 1978. Molecular structure and protein variation within and among populations. Evol. Biol. 11:39–100.

Koehn, R.K. and T.J. Hilbish. 1987. The adaptive importance of genetic variation. Amer. Sci. 75:134–141.

Koehn, R.K. and F.W. Immerman. 1981. Biochemical studies of animopeptidase in *Mytilus edulis.* I. Dependence of enzyme activity on season, tissue and genotype. Biochem. Genet. 19:1115–1142.

Koehn, R.K., R. Milkman, and J.B. Mitton. 1976. Population genetics of marine pelecypods. IV. Selection, migration and genetic differentiation in the blue mussel, *Mytilus edulis.* Evolution 30:2–32.

Koehn, R.K. and G.C. Williams. 1978. Genetic differentiation without isolation in the American eel, *Anguilla rostrata.* II. Temporal stability of geographic variation. Evolution 32:624–637.

Koehn, R.K., A.J. Zera, and J.G. Hall. 1983. Enzyme polymorphism and natural selection. Pp. 115–136 *in:* Evolution of Genes and Proteins, M. Nei and R.K. Koehn (eds.), Sinauer, Sunderland, MA.

Kohne, D.E. 1970. Evolution of higher organism DNA. Quart. Rev. Biophys. 33:327–375.

Kojima, K., J. Gillespie, and Y.N. Tobari. 1970. A profile of *Drosophila* species' enzymes assayed by electrophoresis. I. Number of alleles, heterozygosities, and linkage disequilibrium in glucose-metabolizing systems and some other enzymes. Biochem. Genet. 4:627–637.

Kondo, R., Y. Satta, E.T. Matsuura, H. Ishiwa, N. Takahata, and S.I. Chigusa. 1990. Incomplete maternal transmission of mitochondrial DNA in *Drosophila.* Genetics 126:657–663.

Kondrashov, A.S. and M.V. Mina. 1986. Sympatric speciation: When is it possible? Biol. J. Linn. Soc. 27:201–223.

Konkle, B.R. and D.P. Philipp. 1992. Asymmetric hybridization between two species of sunfishes (*Lepomis:* Centrarchidae). Mol. Ecol., 1:215–222.

Koop, B.F., M. Goodman, P. Zu, J.L. Chan, and J.L. Slighton. 1986. Primate η-globin DNA sequences and man's place among the great apes. Nature 319:234–238.

Koop, B.F., D.A. Tagle, M. Goodman, and J.L. Slightom. 1989. A molecular view of primate phylogeny and important systematic and evolutionary questions. Mol. Biol. Evol. 6:580–612.

Koopman, K.F. 1950. Natural selection for reproductive isolation between *Drosophila pseudoobscura* and *Drosophila persimilis.* Evolution 4:135–148.

Kornfield, I.L. and K.E. Carpenter. 1984. Cyprinids of Lake Lanao, Philippines: taxonomic validity, evolutionary rates and speciation scenarios. Pp. 69–84 *in:* Evolution of Fish Species Flocks, A.A. Echelle and I. Kornfield (eds.), Univ. Maine Press, Orono.

Kornfield, I.L. and R.K. Koehn. 1975. Genetic variation and speciation in New World cichlids. Evolution 29:427–437.

Kornfield, I.L., D.C. Smith, P.S. Gagnon, and J.N. Taylor. 1982. The cichlid fish of Cuatro Ciénegas, Mexico: direct evidence of conspecificity among distinct trophic morphs. Evolution 36:658–664.

Krane, D.E., R.W. Allen, S.A. Sawyer, D.A. Petrov, and D.L. Hartl. 1992. Genetic differences at four DNA typing loci in Finnish, Italian, and mixed Caucasian populations. Proc. Natl. Acad. Sci. USA 89:10583–10587.

Kraus, F. and M.M. Miyamoto. 1990. Mitochondrial genotype of a unisexual salamander of hybrid origin is unrelated to either of its nuclear haplotypes. Proc. Natl. Acad. Sci. USA 87:2235–2238.

Kreiswirth, B., J. Kornblum, R.D. Arbeit, W. Eisner, J.N. Maslow, A. McGeer, D.E. Low, and R.P. Novick. 1993. Evidence for a clonal origin of methicillin resistance in *Staphylococcus aureus*. Science 259:227–230.

Kreitman, M. 1987. Molecular population genetics. Oxford Surv. Evol. Biol. 4:38–60.

Kreitman, M. 1991. Detecting selection at the level of DNA. Pp. 204–221 *in:* Evolution at the Molecular Level, R.K. Selander, A.G. Clark, and T.S. Whittam (eds.), Sinauer, Sunderland, MA.

Krieber, M. and M.R. Rose. 1986. Molecular aspects of the species barrier. Annu. Rev. Ecol. Syst. 17:465–485.

Krueger, C.C. and B. May. 1987. Stock identification of naturalized brown trout in Lake Superior tributaries: differentiation based on allozyme data. Trans. Amer. Fish. Soc. 116:785–794.

Kuhnlein, U., D. Zadworny, Y. Dawe, R.W. Fairfull and J.S. Gavora. 1990. Assessment of inbreeding by DNA fingerprinting: development of a calibration curve using defined strains of chickens. Genetics 125:161–165.

Kumazaki, T., H. Hori, and S. Osawa. 1983. Phylogeny of protozoa deduced from 5S rRNA sequences. J. Mol. Evol. 19:411–419.

Kusakabe, T., K.W. Makabe, and N. Satoh. 1992. Tunicate muscle actin genes: Structure and organization as a gene cluster. J. Mol. Biol. 227:955–960.

Kwiatowski, J., D. Skarecky, S. Hernandez, D. Pham, F. Quijas, and F.J. Ayala. 1991. High fidelity of the polymerase chain reaction. Mol. Biol. Evol. 8:884–887.

Lacson, J.M. 1992. Minimal genetic variation among samples of six species of coral reef fishes collected at La Parguera, Puerto Rico, and Discovery Bay, Jamaica. Mar. Biol. 112:327–331.

Lacy, R.C. 1980. The evolution of eusociality in termites: a haplodiploid analogy? Amer. Natur. 116:449–451.

Lacy, R.C. 1989. Analysis of founder representation in pedigrees: founder equivalents and founder genome equivalents. Zoo Biol. 8:111–123.

Lacy, R.C. 1992. The effects of inbreeding on isolated populations: are minimum viable population sizes predictable? Pp. 277–296 *in:* Conservation Biology, P.L. Fiedler and S.K. Jain (eds.), Chapman and Hall, New York.

Laerm, J., J.C. Avise, J.C. Patton, and R.A. Lansman. 1982. Genetic determination of the status of an endangered species of pocket gopher in Georgia. J. Wildlife Mgt. 46:513–518.

Laidlaw, H.H., Jr. and R.E. Page, Jr. 1984. Polyandry in honey bees (*Apis mellifera* L.): sperm utilization and intracolony genetic relationships. Genetics 108:985–997.

Laikre, L. and N. Ryman. 1991. Inbreeding depression in a captive wolf (*Canis lupus*) population. Cons. Biol. 5:33–40.

Laipis, P.J., M.J. Van de Walle, and W.W. Hauswirth. 1988. Unequal partitioning of bovine mitochondrial genotypes among siblings. Proc. Natl. Acad. Sci. USA 85:8107–8110.

Laird, C.D., B.L. McConaughy, and B.J. McCarthy. 1969. Rate of fixation of nucleotide substitutions in evolution. Nature 224:149–154.

Lake, J.A. 1988. Origin of the eukaryotic nucleus determined by rate-invariant analysis of rRNA sequences. Nature 331:184–186.

Lake, J.A. 1989. Origin of the eukaryotic nucleus determined by rate-invariant analyses of ribosomal RNA genes. Pp. 87–101 *in:* The Hierarchy of Life, B. Fernholm, K. Bremer, and H. Jornvall (eds.), Elsevier, Amsterdam.

Lake, J.A. 1991. Tracing origins with molecular sequences: metazoan and eukaryotic beginnings. Trends Biochem. Sci. 16:46–50.

Lamb, T. and J.C. Avise. 1986. Directional introgression of mitochondrial DNA in a hybrid population of tree frogs: the influence of mating behavior. Proc. Natl. Acad. Sci. USA 83:2526–2530.

Lamb, T. and J.C. Avise. 1987. Morphological variability in genetically defined categories of anuran hybrids. Evolution 41:157–165.

Lamb, T. and J.C. Avise. 1992. Molecular and population genetics aspects of mitochondrial DNA variability in the diamondback terrapin, *Malaclemys terrapin*. J. Heredity 83:262–269.

Lamb, T., J.C. Avise, and J.W. Gibbons. 1989. Phylogeographic patterns in mitochondrial DNA of the desert tortoise (*Xerobates agassizi*), and evolutionary relationships among the North American gopher tortoises. Evolution 43:76–87.

Lamb, T., T.R. Jones, and J.C. Avise. 1992. Phylogeographic histories of representative herpetofauna of the desert southwest: mitochondrial DNA variation in the chuckwalla (*Sauromalus obesus*) and desert iguana (*Dipsosaurus dorsalis*). J. Evol. Biol. 5:465–480.

Lambert, M.E., J.F. McDonald, and I.B. Weinstein. 1988. Eukaryotic Transposable Elements As Mutagenic Agents. Cold Spring Harbor Press, Cold Spring Harbor, NY.

Lande, R. 1981. The minimum number of genes contributing to quantitative variation between and within populations. Genetics 99:541–553.

Lande, R. 1988. Genetics and demography in biological conservation. Science 241:1455–1460.

Lande, R. and D.W. Schemske. 1985. The evolution of self-fertilization and inbreeding depression in plants. I. Genetic models. Evolution 39:24–40.

Lander, E.S. 1989. DNA fingerprinting on trial. Nature 339:501–505.

Lander, E.S. 1991. Research on DNA typing catching up with courtroom application. Amer. J. Human Genet. 48:819–823.

Lander, E.S. and D. Botstein. 1989. Mapping Mendelian factors underlying quantitative traits using RFLP linkage maps. Genetics 121:185–199.

Langley, C.H. and C.F. Aquadro. 1987. Restriction-map variation in natural populations of *Drosophila melanogaster: White*-locus region. Mol. Biol. Evol. 4:651–663.

Langley, C.H. and W.M. Fitch. 1974. An examination of the constancy of the rate of molecular evolution. J. Mol. Evol. 3:161–177.

Langley, C.H., A.E. Shrimpton, T. Yamazaki, N. Miyashita, Y. Matsuo, and C.F. Aquadro. 1988. Naturally-occurring variation in the restriction map of the *Amy* region of *Drosophila melanogaster*. Genetics 119:619–629.

Lank, D.B., P. Mineau, R.F. Rockwell, and F. Cooke. 1989. Intraspecific nest parasitism and extra-pair copulation in lesser snow geese. Anim. Behav. 37:74–89.

Lansman, R.A., J.C. Avise, C.F. Aquadro, J.F. Shapira, and S.W. Daniel. 1983. Extensive genetic variation in mitochondrial DNA's among geographic populations of the deer mouse, *Peromyscus maniculatus*. Evolution 37:1–16.

Lansman, R.A., R.O. Shade, J.F. Shapira, and J.C. Avise. 1981. The use of restriction endonucleases to measure mitochondrial DNA sequence relatedness in natural populations. III. Techniques and potential applications. J. Mol. Evol. 17:214–226.

Lanyon, S.M. 1992. Interspecific brood parasitism in blackbirds (Icterinae): a phylogenetic perspective. Nature 255:77–79.

Lanyon, S.M. and R.M. Zink. 1987. Genetic variability in piciform birds: Monophyly and generic and familial relationships. Auk 104:724–732.

Larson, A. 1984. Neontological inferences of evolutionary pattern and process in the salamander family Plethodontidae. Evol. Biol. 17:119–127.

Larson, A. 1989. The relationship between speciation and morphological evolution. Pp. 579–598 *in:* Speciation and Its Consequences, D. Otte and J.A. Endler (eds.), Sinauer, Sunderland, MA.

Larson, A., D.B. Wake, and K.P. Yanev. 1984. Measuring gene flow among populations having high levels of genetic fragmentation. Genetics 106:293–308.

Lasker, G.W. 1985. Surnames and Genetic Structure. Cambridge Univ., Press, New York.

Latorre, A., A. Moya, and F.J. Ayala. 1986. Evolution of mitochondrial DNA in *Drosophila suboobscura*. Proc. Natl. Acad. Sci. USA 83:8649–8653.

Lavin, M., J.J. Doyle, and J.D. Palmer. 1990. Evolutionary significance of the loss of chloroplast-DNA inverted repeat in the Leguminosae subfamily Papilionoideae. Evolution 44:390–402.

Lawlor, D.A., F.E. Ward, P.D. Ennis, A.P. Jackson, and P. Parham. 1988. HLA-A and B polymorphisms predate the divergence of humans and chimpanzees. Nature 335:268–271.

Lawrence, J.G. and D.L. Hartl. 1992. Inference of horizontal genetic transfer from molecular data: an approach using the bootstrap. Genetics 131:753–760.

Layton, C.R. and F.R. Ganders. 1984. The genetic consequences of contrasting breeding systems in *Plectritis* (Valerianaceae). Evolution 38:1308–1325.

Learn, G.L. and B.A. Schaal. 1987. Population subdivision for rDNA repeat variants in *Clematis fremontii*. Evolution 41:433–437.

Leary, R.B., F.W. Allendorf, K.L. Knudsen, and G.H. Thorgaard. 1985. Heterozygosity and developmental stability in gynogenetic diploid and triploid rainbow trout. Heredity 54:219–225.

Leary, R.F., F.W. Allendorf, S.R. Phelps, and K.L. Knudsen. 1984. Introgression between westslope cutthroat and rainbow trout in Clark Fork River drainage, Montana. Proc. Mont. Acad. Sci. 43:1–18.

Ledig, F.T. and M.T. Conkle. 1983. Gene diversity and genetic structure in a narrow endemic, Torrey pine (*Pinus torreyana* Parry ex Carr). Evolution 37:79–85.

Ledig, F.T., R.P. Guries, and B.A. Bonefield. 1983. The relation of growth to heterozygosity in pitch pine. Evolution 37:1227–1238.

Lehman, N., A. Eisenhawer, K. Hansen, D.L. Mech, R.O. Peterson, J.P. Gogan, and R.K. Wayne. 1991. Introgression of coyote mitochondrial DNA into sympatric North American gray wolf populations. Evolution 45:104–119.

Lehman, N. and R.K. Wayne. 1991. Analysis of coyote mitochondrial DNA genotype frequencies: estimation of the effective number of alleles. Genetics 128:405–416.

Leigh Brown, A.J. and C.H. Langley. 1979. Re-evaluation of level of genic heterozygosity in natural populations of *Drosophila melanogaster* by two-dimensional electrophoresis. Proc. Natl. Acad. Sci. USA 76:2381–2384.

Leinaas, H.P. 1983. A haplodiploid analogy in the evolution of termite eusociality? Reply to Lacy. Amer. Natur. 121:302–304.

Lemen, C.A. and P.W. Freeman. 1981. A test of macroevolutionary problems with neontological data. Paleobiology 7:311–315.

Lemen, C.A. and P.W. Freeman. 1989. Testing macroevolutionary hypotheses with cladistic analysis: evidence against rectangular evolution. Evolution 43:1538–1554.

Leone, C.A. 1964. Taxonomic Biochemistry and Serology. Ronald Press, New York.

Lerner, I.M. 1954. Genetic Homeostasis. Oliver and Boyd, Edinburgh, Scotland.

Lesica, P., R.F. Leary, F.W. Allendorf, and D.E. Bilderback. 1988. Lack of genic diversity within and among populations of an endangered plant, *Howellia aquatilis*. Cons. Biol. 2:275–282.

Leslie, J.F. and H. Dingle. 1983. Interspecific hybridization and genetic divergence in milkweed bugs (*Oncopeltus:* Hemiptera: Lygaeidae). Evolution 37:583–591.

Lessios, H.A. 1979. Use of Panamanian sea urchins to test the molecular clock. Nature 280:599–601.

Lessios, H.A. 1981. Divergence in allopatry: molecular and morphological differentiation between sea urchins separated by the Isthmus of Panama. Evolution 35:618–634.

Leunissen, J.A.M. and W.W. de Jong. 1986. Copper/zinc superoxide dismutase: how likely is gene transfer from ponyfish to *Photobacterium leiognathi?* J. Mol. Evol. 23:250–258.

Levene, H. 1953. Genetic equilibrium when more than one ecological niche is available. Amer. Natur. 87:331–333.

Levin, B.R. 1981. Periodic selection, infectious gene exchange and the genetic structure of *E. coli* populations. Genetics 99:1–23.

Levin, D.A. 1979. The nature of plant species. Science 204:381–384.

Levin, D.A. and H.W. Kerster. 1971. Neighborhood structure in plants under diverse reproductive methods. Amer. Natur. 105:345–354.

Levin, D.A. and H.W. Kerster. 1974. Gene flow in seed plants. Evol. Biol. 7:139–220.

Levin, L.A. 1990. A review of methods for labeling and tracking marine invertebrate larvae. Ophelia 32:115–144.

Levine, H. 1953. Genetic equilibrium when more than one ecological niche is available. Amer. Natur. 87:331–333.

Levins, R. 1968. Evolution in Changing Environments. Princeton Univ. Press, Princeton, NJ.

Lewin, B. 1990. Genes IV. Oxford Univ. Press, New York.

Lewis, R.E., Jr. and J.M. Cruse. 1992. DNA typing in human parentage using multilocus and single-locus probes. Pp. 3–17 *in:* Paternity in Primates: Genetic Tests and Theories, R.D. Martin, A.F. Dixson, and E.J. Wickings (eds.), Karger, Basel, Switzerland.

Lewis, W. 1980. Polyploidy: Biological Relevance. Plenum Press, New York.

Lewontin, R.C. 1972. The apportionment of human diversity. Evol. Biol. 6:381–398.

Lewontin, R.C. 1974. The Genetic Basis of Evolutionary Change. Columbia Univ. Press, New York.

Lewontin, R.C. 1985. Population genetics. Annu. Rev. Genet. 19:81–102.

Lewontin, R.C. 1988. On measures of gametic disequilibrium. Genetics 120:849–852.

Lewontin, R.C. 1991. Twenty-five years ago in GENETICS: Electrophoresis in the development of evolutionary genetics: milestone or millstone? Genetics 128:657–662.

Lewontin, R.C. and D.L. Hartl. 1991. Population genetics in forensic DNA typing. Science 254:1745–1750.

Lewontin, R.C. and J.L. Hubby. 1966. A molecular approach to the study of genic heterozygosity in natural populations. II. Amount of variation and degree of heterozygosity in natural populations of *Drosophila pseudoobscura*. Genetics 54:595–609.

Lewontin, R.C. and J. Krakauer. 1973. Distribution of gene frequency as a test of the theory of the selective neutrality of polymorphisms. Genetics 74:175–195.

Lewontin, R.C. and J. Krakauer. 1975. Testing the heterogeneity of F values. Genetics 80:397.

Li, H., U.B. Gyllensten, X. Cui, R.K. Saiki, H.A. Erlich, and N. Arnheim. 1988. Amplification and analysis of DNA sequences in single human sperm and diploid cells. Nature 335:414–417.

Li, W.-H. 1978. Maintenance of genetic variability under the joint effect of mutation, selection, and random genetic drift. Genetics 90:349–382.

Li, W.-H. and D. Graur. 1991. Fundamentals of Molecular Evolution. Sinauer, Sunderland, MA.

Li, W.-H. and L.A. Sadler. 1991. Low nucleotide diversity in man. Genetics 129:513–523.

Li, W.-H. and L.A. Sadler. 1992. DNA variation in humans and its implications for human evolution. Oxford Surv. Evol. Biol. 8:111–134.

Li, W.-H. and M. Tanimura. 1987. The molecular clock runs more slowly in man than in apes and monkeys. Nature 326:93–96.

Li, W.-H., M. Tanimura, and P.M. Sharp. 1987. An evaluation of the molecular clock hypothesis using mammalian DNA sequences. J. Mol. Evol. 25:330–342.

Li, W.-H., M. Tanimura, and P.M. Sharp. 1988. Rates and dates of divergence between AIDS virus nucleotide sequences. Mol. Biol. Evol. 5:313–330.

Lidholm, J., A.E. Szmidt, J.E. Hallgren, and P. Gustafsson. 1988. The chloroplast genomes of conifers lack one of the rRNA-encoding inverted repeats. Mol. Gen. Genet. 212:6–10.

Liebherr, J.K. 1988. Gene flow in ground beetles (Coleoptera: Carabidae) of differing habitat preference and flight-wing development. Evolution 42:129–137.

Lifjeld, J.T., T. Slagsvold, and H.M. Lampe. 1991. Low frequency of extra-pair paternity in pied flycatchers revealed by DNA fingerprinting. Behav. Ecol. Sociobiol. 29:95–101.

Linn, S. and W. Arber. 1968. Host specificity of DNA produced by *Escherichia coli*. X. In vitro restriction of phage fd replicative form. Proc. Natl. Acad. Sci. USA 59:1300–1306.

Linnaeus, C. 1759. *Systema Naturae*. Reprinted 1964 by Wheldon and Wesley Ltd., New York.

Liston, A., L.H. Rieseberg, and T.S. Elias. 1989. Genetic similarity is high between intercontinental disjunct species of *Senecio* (Asteraceae). Amer. J. Bot. 76:383–388.

Litt, M. and J.A. Luty. 1989. A hypervariable microsatellite revealed by in vitro amplification of a dinucleotide repeat within the cardiac muscle actin gene. Amer. J. Human Genet. 44:387–401.

Liu, L.L., D.W. Foltz, and W.B. Stickle. 1991. Genetic population structure of the southern oyster drill *Stramonita* (= *Thais*) *haemostoma*. Mar. Biol. 111:71–79.

Lloyd, M., G. Kritsky, and C. Simon. 1983. A simple Mendelian model for 13- and 17-year life cycles of periodical cicadas, with historical evidence of hybridization between them. Evolution 37:1162–1180.

Lloyd, M. and J.A. White. 1976. Sympatry of periodical cicada broods and the hypothetical four-year acceleration. Evolution 30:786–801.

Lobo, J.A., M.A. Del Lama, and M.A. Mestriner. 1989. Population differentiation and racial admixture in the Africanized honeybee (*Apis mellifera* L.). Evolution 43:794–802.

Lockhart, P.J., T.J. Beanland, C.J. Howe, and A.W.D. Larkum. 1992. Sequence of *Prochloron didemni atpBE* and the inference of chloroplast origins. Proc. Natl. Acad. Sci. USA 89:2742–2746.

Long, E.O. and I.B. Dawid. 1980. Repeated sequences in eukaryotes. Annu. Rev. Biochem. 49:727–764.

Longmire, J.L., G.F. Gee, C.L. Hardekopf, and G.A. Mark. 1992. Establishing paternity in whooping cranes (*Grus americana*) by DNA analysis. Auk 109:522–529.

Longmire, J.L., P.M. Kraemer, N.C. Brown, L.C. Hardekopf, and L.L. Deaven. 1990. A new multi-locus DNA fingerprinting probe: pV47-2. Nucleic Acids Res. 18:1658.

Lopez, T.J., E.D. Hauselman, L.R. Maxson, and J.W. Wright. 1992. Preliminary analysis of phylogenetic relationships among Galapagos Island lizards of the genus *Tropidurus*. Amphibia-Reptilia 13:327–339.

Lotka, A.J. 1931. Population analysis—the extinction of families. I. J. Washington Acad. Sci. 21:377–380.

Loveless, M.D. and J.L. Hamrick. 1984. Ecological determinants of genetic structure in plant populations. Annu. Rev. Ecol. Syst. 15:65–95.

Lowenstein, J.M., V.M. Sarich, and B.J. Richardson. 1981. Albumin systematics of the extinct mammoth and Tasmanian wolf. Nature 291:409–411.

Lynch, M. 1988. Estimation of relatedness by DNA fingerprinting. Mol. Biol. Evol. 5:584–589.

Lynch, M. and T.J. Crease. 1990. The analysis of population survey data on DNA sequence variation. Mol. Biol. Evol. 7:377–394.

MacIntyre, R.J. 1976. Evolution and ecological value of duplicate genes. Annu. Rev. Ecol. Syst. 7:421–468.

MacIntyre, R.J., M.R. Dean, and G. Batt. 1978. Evolution of acid phosphatase-1 in the genus *Drosophila*. Immunological studies. J. Mol. Evol. 12:121–142.

MacNeil, D. and C. Strobeck. 1987. Evolutionary relationships among colonies of Columbian ground squirrels as shown by mitochondrial DNA. Evolution 41:873–881.

MacRae, A.F. and W.W. Anderson. 1988. Evidence for non-neutrality of mitochondrial DNA haplotypes in *Drosophila pseudoobscura*. Genetics 120:485–494.

Maddison, D.R. 1991. African origins of human mitochondrial DNA reexamined. Syst. Zool. 40:355–363.

Maddox, G.D., R.E. Cook, P.H. Wimberger, and S. Gardescu. 1989. Clone structure in four *Solidago altissima* (Asteraceae) populations: rhizome connections within genotypes. Amer. J. Bot. 76:318–326.

Maeda, N., C.-I. Wu, J. Bliska, and J. Reneke. 1988. Molecular evolution of intergenic DNA in higher primates: pattern of DNA changes, molecular clock, and evolution of repetitive sequences. Mol. Biol. Evol. 5:1–20.

Maha, G.C., R. Highton, and L.R. Maxson. 1989. Biochemical evolution in the slimy salamanders of the *Plethodon glutinosus* complex in the eastern United States. Illinois Biol. Monogr. 57:81–150.

Manhart, J.R. and J.D. Palmer. 1990. The gain of two chloroplast tRNA introns marks the green algal ancestors of land plants. Nature 345:268–270.

Maniatis, T., R.C. Hardison, E. Lacy, J. Lauer, C. O'Connell, D. Quon, D.K. Sim, and A. Efstratiadis. 1978. The isolation of structural genes from libraries of eucaryotic DNA. Cell 15:687–701.

Mantovani, B. and V. Scali. 1992. Hybridogenesis and androgenesis in the stick-insect *Bacillus rossius-grandii benazzii* (Insecta, Phasmatodea). Evolution 46:783–796.

Mantovani, B., V. Scali, and F. Tinti. 1991. Allozyme analysis and phyletic relationships of two new stick-insects from north-west Sicily: *Bacillus grandii benazzii* and *B. rossius-grandii benazzii* (Insecta Phasmatodea). J. Evol. Biol. 4:279–290.

Marchant, A.D. 1988. Apparent introgression of mitochondrial DNA across a narrow hybrid zone in the *Caledia captiva* species-complex. Heredity 60:39–46.

Margoliash, E. 1963. Primary structure and evolution of cytochrome c. Proc. Natl. Acad. Sci. USA 50:672–679.

Margulis, L. 1970. Origin of Eukaryotic Cells. Yale Univ. Press, New Haven, CT.

Margulis, L. 1981. Symbiosis in Cell Evolution: Life and Its Environment in the Early Earth. Freeman, San Francisco.

Markert, C.L. and I. Faulhaber. 1965. Lactate dehydrogenase isozyme patterns of fish. J. Exp. Zool. 159:319–332.

Markert, C.L., J.B. Shaklee, and G.S. Whitt. 1975. Evolution of a gene. Science 189:105–114.

Marshall, D.E. and N.C. Ellstrand. 1985. Proximal causes of multiple paternity in wild radish, *Raphanus sativus*. Amer. Natur. 126:596–605.

Marshall, D.E. and N.C. Ellstrand. 1986. Sexual selection in *Raphanus sativus:* experimental data on nonrandom fertilization, maternal choice, and consequences of multiple paternity. Amer. Natur. 127:446–461.

Martin, A.P., G.J.P. Naylor, and S.R. Palumbi. 1992. Rates of mitochondrial DNA evolution in sharks are slow compared with mammals. Nature 357:153–155.

Martin, A.P. and S.R. Palumbi. 1993. Body size, metabolic rate, generation time and the molecular clock. Proc. Natl. Acad. Sci. USA 90:4087–4091.

Martin, A.P. and C. Simon. 1988. Anomalous distribution of nuclear and mitochondrial DNA markers in periodical cicadas. Nature 336:237–239.

Martin, A.P. and C. Simon. 1990. Differing levels of among-population divergence in the mitochondrial DNA of periodical cicadas related to historical biogeography. Evolution 44:1066–1080.

Martin, B., J. Nienhuis, G. King, and A. Schaefer. 1989. Restriction fragment length polymorphisms associated with water use efficiency in tomato. Science 243:1725–1728.

Martin, J.P. and I. Fridovich. 1981. Evidence for a natural gene transfer from the ponyfish to its bioluminescent bacterial symbiont *Photobacter leiognathi*. The close relationship between bacteriocuprein and the copper-zinc superoxide dismutase of teleost fishes. J. Biol. Chem. 256:6080–6089.

Martin, M.A., D.K. Shiozawa, E.J. Loudenslager, and J.N. Jensen. 1985. Electrophoretic study of cutthroat trout populations in Utah. Great Basin Natur. 45:677–687.

Martin, R.D., A.F. Dixson, and E.J. Wickings (eds.). 1992. Paternity in Primates: Genetic Tests and Theories. Karger, Basel, Switzerland.

Martin, W., C.C. Somerville, and S. Loiseaux-de Goer. 1992. Molecular phylogenies of plastid origins and algal evolution. J. Mol. Evol. 35:385–404.

Martyniuk, J. and J. Jaenike. 1982. Multiple mating and sperm usage patterns in natural populations of *Prolinyphia marginata* (Aranae: Linyphiidae). Ann. Entomol. Soc. Amer. 75:516–518.

Maruyama, T. and D.L. Hartl. 1991. Evidence for interspecific transfer of the transposable element mariner between *Drosophila* and *Zaprionus*. J. Mol. Evol. 33:514–524.

Maruyama, T. and M. Kimura. 1980. Genetic variability and effective population size when local extinction and recolonization of subpopulations are frequent. Proc. Natl. Acad. Sci. USA 77:6710–6714.

Matson, R.H. 1989. Distribution of the testis-specific LDH-X among avian taxa with comments on the evolution of the LDH gene family. Syst. Zool. 38:106–115.

Maxam, A.M. and W. Gilbert. 1977. A new method for sequencing DNA. Proc. Natl. Acad. Sci. USA 74:560–564.

Maxam, A.M. and W. Gilbert. 1980. Sequencing end-labeled DNA with base-specific chemical cleavages. Meth. Enzymol. 65:499–559.

Maxson, L.R. and R.D. Maxson. 1986. Micro-complement fixation: A quantitative estimator of protein evolution. Mol. Biol. Evol. 3:375–388.

Maxson, L.R. and R.D. Maxson. 1990. Proteins II: Immunological techniques. Pp. 127–155 *in:* Molecular Systematics, D.M. Hillis and C. Moritz (eds.), Sinauer, Sunderland, MA.

Maxson, L.R., E. Pepper, and R.D. Maxson. 1977. Immunological resolution of a diploid-tetraploid species complex of tree frogs. Science 197:1012–1013.

Maxson, L.R. and J.D. Roberts. 1984. Albumin and Australian frogs: molecular data a challenge to speciation model. Science 225:957–958.

Maxson, L.R. and A.C. Wilson. 1975. Albumin evolution and organismal evolution in tree frogs (Hylidae). Syst. Zool. 24:1–15.

Mayden, R.L. 1986. Speciose and depauperate phylads and tests of punctuated and gradual evolution: fact or artifact? Syst. Zool. 35:147–152.

Maynard Smith, J. 1966. Sympatric speciation. Amer. Natur. 100:637–650.

Maynard Smith, J. 1978. The Evolution of Sex. Cambridge Univ. Press, Cambridge, England.

Maynard Smith, J. 1990. The Y of human relationships. Nature 344:591–592.

Maynard Smith, J. 1992. Age and the unisexual lineage. Nature 356:661–662.

Mayr, E. 1954. Change of genetic environment and evolution. Pp. 157–180 *in:* Evolution as a Process, J. Huxley, A.C. Hardy, and E.B. Ford (eds.), Allen and Unwin, London.

Mayr, E. 1963. Animal Species and Evolution. Harvard Univ. Press, Cambridge, MA.

Mayr, E. 1969. Principles of Systematic Zoology. McGraw-Hill, New York.

Mayr, E. 1990. A natural system of organisms. Nature 348:491.

Mazodier, P. and J. Davies. 1991. Gene transfer between distantly related bacteria. Annu. Rev. Genet. 25:147–171.

McCauley, D.E. and R. O'Donnell. 1984. The effect of multiple mating on genetic relatedness in larval aggregations of the imported willow leaf beetle (*Plagiodera versicolora,* Coleoptera: Chrysomelidae). Behav. Ecol. Sociobiol. 15:287–291.

McClenaghan, L.R., Jr., J. Berger, and H.D. Truesdale. 1990. Founding lineages and genic variability in plains bison (*Bison bison*) from Badlands National Park, South Dakota. Cons. Biol. 4:285–289.

McClenaghan, L.R., Jr., and T.J. O'Shea. 1988. Genetic variability in the Florida manatee (*Trichechus manatus*). J. Mammal. 69:481–488.

McClenaghan, L.R., Jr., M.H. Smith, and M.W. Smith. 1985. Biochemical genetics of mosquitofish. IV. Changes in allele frequencies through time and space. Evolution 39:451–459.

McConnell, T.J., W.S. Talbot, R.A. McIndoe, and E.K. Wakeland. 1988. The origin of MHC class II gene polymorphism within the genus *Mus.* Nature 332:651–654.

McCorquodale, D.B. 1988. Relatedness among nestmates in a primitively social wasp, *Cerceris antipodes* (Hymenoptera: Sphecidae). Behav. Ecol. Sociobiol. 23:401–406.

McCoy, E.D. and K.L. Heck, Jr. 1976. Biogeography of corals, sea grasses, and mangroves: an alternative to the center of origin concept. Syst. Zool. 25:201–210.

McCracken, G.F. 1984. Communal nursing in Mexican free-tailed bat maternal colonies. Science 223:1090–1091.

McCracken, G.F. and J.W. Bradbury. 1977. Paternity and genetic heterogeneity in the polygynous bat, *Phyllostomus hastatus.* Science 198:303–306.

McCracken, G.F. and J.W. Bradbury. 1981. Social organization and kinship in the polygynous bat *Phyllostomus hastatus*. Behav. Ecol. Sociobiol. 8:11–34.

McDermott, J.M., B.A. McDonald, R.W. Allard, and R.K. Webster. 1989. Genetic variability for pathogenicity, isozyme, ribosomal DNA and colony color variants in populations of *Rhynchosporium secalis*. Genetics 122:561–565.

McDonald, J.F. 1983. The molecular basis of adaptation: a critical review of relevant ideas and observations. Annu. Rev. Ecol. Syst. 14:77–102.

McDonald, J.F. 1989. The potential evolutionary significance of retroviral-like transposable elements in peripheral populations. Pp. 190–205 *in:* Evolutionary Biology of Transient Unstable Populations, A. Fontdevila (ed.), Springer-Verlag, New York.

McDonald, J.F. 1990. Macroevolution and retroviral elements. BioScience 40:183–191.

McDonald, J.F. and F.J. Ayala. 1974. Genetic response to environmental heterogeneity. Nature 250:572–574.

McDowell, R. and S. Prakash. 1976. Allelic heterogeneity within allozymes separated by electrophoresis in *Drosophila pseudoobscura*. Proc. Natl. Acad. Sci. USA 73:4150–4153.

McKinney, F., K.M. Cheng, and D.J. Bruggers. 1984. Sperm competition in apparently monogamous birds. Pp. 523–545 *in:* Sperm competition and the Evolution of Animal Mating Systems, R.L. Smith (ed.), Academic Press, New York.

McKitrick, M.C. 1990. Genetic evidence for multiple parentage in eastern kingbirds (*Tyrannus tyrannus*). Behav. Ecol. Sociobiol. 26:149–155.

McKitrick, M.C. and R.M. Zink. 1988. Species concepts in ornithology. Condor 90:1–14.

McMillan, W.O., R.A. Raff, and S.R. Palumbi. 1992. Population genetic consequences of developmental evolution in sea urchins (genus *Heliocidaris*). Evolution 46:1299–1312.

McNeilly, T. and J. Antonovics. 1968. Evolution in closely adjacent plant populations. IV. Barriers to gene flow. Heredity 23:205–218.

McPheron, B.A., D.C. Smith and S.H. Berlocher. 1988. Genetic differences between host races of *Rhagoletis pomonella*. Nature 336:64–67.

Meagher, S. and T.E. Dowling. 1991. Hybridization between the cyprinid fishes *Luxilus albeolus, L. cornutus,* and *L. cerasinus* with comments on the proposed hybrid origin of *L. albeolus*. Copeia 1991:979–991.

Meagher, T.R. 1986. Analysis of paternity within a natural population of *Chamaelirium luteum*. I. Identification of most-likely male parents. Amer. Natur. 128:199–215.

Meagher, T.R. 1991. Analysis of paternity within a natural population of *Chamaelirium luteum*. II. Patterns of male reproductive success. Amer. Natur. 137:738–752.

Meagher, T.R. and E. Thompson. 1987. Analysis of parentage for naturally established seedlings of *Chamaelirium luteum* (Liliaceae). Ecology 68:803–812.

Meffe, G.K. and R.C. Vrijenhoek. 1988. Conservation genetics in the management of desert fishes. Cons. Biol. 2:157–169.

Melnick, D.J. and G.A. Hoelzer. 1992. Differences in male and female macaque dispersal lead to contrasting distributions of nuclear and mitochondrial DNA variation. Int. J. Primatol. 13:379–393.

Melson, K.E. 1990. Legal and ethical considerations. Pp. 189–215 *in:* DNA Fingerprinting, L.T. Kirby (ed.), Stockton Press, New York.

Meng, A., R.E. Carter, and D.T. Parkin. 1990. The variability of DNA fingerprints in three species of swan. Heredity 64:73–80.

Mengel, R.N. 1964. The probable history of species formation in some northern wood warblers (Parulidae). Living Bird 3:9–43.

Menotti-Raymond, M. and S.J. O'Brien. 1993. Dating the genetic bottleneck of the African cheetah. Proc. Natl. Acad. Sci. USA 90:3172–3176.

Merenlender, A.M., D.S. Woodruff, O.A. Ryder, R. Kock, and J. Váhala. 1989. Allozyme variation and differentiation in African and Indian rhinoceroses. J. Heredity 80:377–382.

Merriwether, D.A., A.G. Clark, S.W. Ballinger, T.G. Schurr, H. Soodyall, T. Jenkins, S.T. Sherry, and D.C. Wallace. 1991. The structure of human mitochondrial DNA variation. J. Mol. Evol. 33:543–555.

Meselson, M. and R. Yuan. 1968. DNA restriction enzyme from *E. coli*. Nature 217:1110–1114.

Metcalf, R.A. and G.S. Whitt. 1977. Intra-nest relatedness in the social wasp *Polistes metricus*. A genetic analysis. Behav. Ecol. Sociobiol. 2:339–351.

Metzlaff, M., T. Borner, and R. Hagemann. 1981. Variations of chloroplast DNAs in the genus *Pelargonium* and their biparental inheritance. Theoret. Appl. Genet. 60:37–41.

Meyer, A. 1987. Phenotypic plasticity and heterochrony in *Cichlasoma managuense* (Pisces, Cichlidae) and their implications for speciation in cichlid fishes. Evolution 41:1357–1369.

Meyer, A. 1993. Molecular phylogenetic studies of fishes. In: Evolution and Genetics of Aquatic Organisms, A.R. Beaumont (ed.), Chapman and Hall, N.Y.

Meyer, A. and S.I. Dolven. 1992. Molecules, fossils, and the origin of tetrapods. J. Mol. Evol. 35:102–113.

Meyer, A., T.D. Kocher, P. Basasibwaki, and A.C. Wilson. 1990. Monophyletic origin of Lake Victoria cichlid fishes suggested by mitochondrial DNA sequences. Nature 347:550–553.

Meyer, A. and A.C. Wilson. 1990. Origin of tetrapods inferred from their mitochondrial DNA affiliation to lungfish. J. Mol. Evol. 31:359–364.

Meylan, A.B., B.W. Bowen, and J.C. Avise. 1990. A genetic test of the natal homing versus social facilitation models for green turtle migration. Science 248:724–727.

Michener, G.R. 1983. Kin identification, matriarchies, and the evolution of sociality in ground-dwelling sciurids. Pp. 528–572 *in*: Advances in the Study of Mammalian Behavior, J.F. Eisenberg and D.G. Kleiman (eds.), American Society Mammalogy, Special Publ. No. 7.

Michod, R.E. and W.W. Anderson. 1979. Measures of genetic relationship and the concept of inclusive fitness. Amer. Natur. 114:637–647.

Mickevich, M.F. 1978. Taxonomic congruence. Syst. Zool. 27:143–158.

Miles, S.J. 1978. Enzyme variations in the *Anopheles gambiae* group of species (Diptera, Culicidae). Bull. Entomol. Res. 68:85–96.

Milinkovitch, M.C. 1992. DNA-DNA hybridization support ungulate ancestry of Cetacea. J. Evol. Biol. 5:149–160.

Milinkovitch, M.C., G. Ortí, and A. Meyer. 1993. Revised phylogeny of whales suggested by mitochondrial ribosomal DNA sequences. Nature 361:346–348.

Milkman, R. 1973. Electrophoretic variation in *Escherichia coli* from natural sources. Science 182:1024–1026.

Milkman, R. 1975. Allozyme variation in *E. coli* of diverse natural origins. Isozymes IV:273–285.

Milkman, R. 1976. Further studies on thermostability variation within electrophoretic mobility classes of enzymes. Biochem. Genet. 14:383–387.

Milkman, R. and M.M. Bridges. 1990. Molecular evolution of the *E. coli* chromosome. III. Clonal frames. Genetics 126:505–517.

Milkman, R. and A. Stoltzfus. 1988. Molecular evolution of the *E. coli* chromosome. II. Clonal segments. Genetics 120:359–366.

Milkman, R. and R.R. Zeitler. 1974. Concurrent multiple paternity in natural and laboratory populations of *Drosophila melanogaster*. Genetics 78:1191–1193.

Millar, C.D., D.M. Lambert, A.R. Bellamy, P.M. Stapleton, and E.C. Young. 1992. Sex-specific restriction fragments and sex ratios revealed by DNA fingerprinting in the brown skua. J. Heredity 83:350–355.

Millar, C.I. 1983. A steep cline in *Pinus muricata*. Evolution 37:311–319.

Millar, R.B. 1987. Maximum likelihood estimation of mixed stock fishery composition. Can. J. Fish. Aquat. Sci. 44:583–590.

Miller, P.S. and P.W. Hedrick. 1991. MHC polymorphism and the design of captive breeding programs: simple solutions are not the answer. Cons. Biol. 5:556–558.

Miller, S., R.W. Pearcy, and E. Berger. 1975. Polymorphism at the alpha-glycerophosphate dehydrogenase locus in *Drosophila melanogaster*. I. Properties of adult allozymes. Biochem. Genet. 13:175–188.

Milligan, B.G., J.N. Hampton, and J.D. Palmer. 1989. Dispersed repeats and structural reorganization in subclover chloroplast DNA. Mol. Biol. Evol. 6:355–368.

Mindell, D.P., C.W. Dick, and R.J. Baker. 1991. Phylogenetic relationships among megabats, microbats, and primates. Proc. Natl. Acad. Sci. USA 88:10322–10326.

Mindell, D.P. and R.L. Honeycutt. 1990. Ribosomal RNA in vertebrates: evolution and phylogenetic applications. Annu. Rev. Ecol. Syst. 21:541–566.

Mindell, D.P. and J.W. Sites, Jr. 1987. Tissue expression patterns of avian isozymes: a preliminary study of phylogenetic applications. Syst. Zool. 36:137–152.

Mindell, D.P., J.W. Sites, Jr., and D. Graur. 1990. Mode of allozyme evolution: increased genetic distance associated with speciation events. J. Evol. Biol. 3:125–131.

Mishler, B.D. and M.J. Donoghue. 1982. Species concepts: A case for pluralism. Syst. Zool. 31:491–503.

Mitchell, S.E., S.K. Narang, A.F. Cockburn, J.A. Seawright, and M. Goldenthal. 1993. Mitochondrial and ribosomal DNA variation among members of the *Anopheles quadrimaculatus* (Diptera: Culicidae) species complex. Genome, in press.

Mitton, J.B. 1993. Theory and data pertinent to the relationship between heterozygosity and fitness. Pp. 17–41 *in:* The Natural History of Inbreeding and Outbreeding, N. Thornhill (ed.), Univ. Chicago Press, Chicago.

Mitton, J.B. and M.C. Grant. 1984. Associations among protein heterozygosity, growth rate, and developmental homeostasis. Annu. Rev. Ecol. Syst. 15:479–499.

Mitton, J.B. and R.K. Koehn. 1975. Genetic organization and adaptive response of allozymes to ecological variables in *Fundulus heteroclitus*. Genetics 79:97–111.

Mitton, J.B. and R.K. Koehn. 1985. Shell shape variation in the blue mussel, *Mytilus edulis* L., and its association with enzyme heterozygosity. J. Exp. Mar. Biol. Ecol. 90:73–90.

Mitton, J.B. and Lewis, W.M., Jr. 1989. Relationships between genetic variability and life-history features of bony fishes. Evolution 43:1712–1723.

Mitton, J.B. and B.A. Pierce. 1980. The distribution of individual heterozygosity in natural populations. Genetics 95:1043–1054.

Miyamoto, M.M. and J. Cracraft. 1991. Phylogenetic Analysis of DNA Sequences. Oxford Univ. Press, New York.

Miyamoto, M.M. and M. Goodman. 1986. Biomolecular systematics of Eutherian mammals: phylogenetic patterns and classification. Syst. Zool. 35:230–240.

Miyamoto, M.M. and M. Goodman. 1990. DNA systematics and evolution of primates. Annu. Rev. Ecol. Syst. 21:197–220.

Mizrokhi, L.J. and A.M. Mazo. 1990. Evidence for horizontal transmission of the mobile element *jockey* between distant *Drosophila* species. Proc. Natl. Acad. Sci. USA 87:9216–9220.

Mock, D.W. 1983. On the study of avian mating systems. Pp. 55–84 *in:* Perspectives in Ornithology, A.H. Brush and G.A. Clark (eds.), Cambridge Univ. Press, London.

Møller, A.P. 1992. Female swallow preference for symmetrical male sexual ornaments. Nature 357:238–240.

Montagna, W. 1942. The Sharp-tailed sparrows of the Atlantic coast. Wilson Bull. 54:107–120.

Monteiro, W., J.M.G. Almeida., Jr., and B.S. Dias. 1984. Sperm sharing in *Biomphalaria* snails: a new behavioural strategy. Nature 308:727–729.

Moody, M.D. 1989. DNA analysis in forensic science. BioScience 39:31–36.

Moore, W.S. and D.B. Buchanan. 1985. Stability of the northern flicker hybrid zone in historical times: implications for adaptive speciation theory. Evolution 39:135–151.

Mopper, S., J.B. Mitton, T.G. Whitham, N.S. Cobb, and K.M. Christensen. 1991. Genetic differentiation and heterozygosity in pinyon pine associated with resistance to herbivory and environmental stress. Evolution 45:989–999.

Moran, P., A.M. Pendas, E. Garcia-Vazquez, and J. Izquierdo. 1991. Failure of a stocking policy, of hatchery reared brown trout, *Salmo trutta* L., in Asturias, Spain, detected using *LDH-5** as a genetic marker. J. Fish Biol. 39A:117–121.

Morden, C.W., C.F. Delwiche, M. Kuhsel, and J.D. Palmer. 1992. Gene phylogenies and the endosymbiotic origin of plastids. BioSystems 28:75–90.

Morgan, R.P. 1975. Distinguishing larval white perch and striped bass by electrophoresis. Chesapeake Sci. 16:68–70.

Morgan, T.H. 1919. The Physical Basis of Heredity. Lippincott, Philadelphia, PA.

Morin, P.A., J.J. Moore, and D.S. Woodruff. 1992. Identification of chimpanzee subspecies with DNA from hair and allele-specific probes. Proc. Royal Soc. London B249:293–297.

Morita, T., H. Kubota, K. Murata, M. Nozaki, C. Delarbre, K. Willison, Y. Satta, M. Sakaizumi, N. Takahata, G. Gachelin, and Z. Matsushiro. 1992. Evolution of the mouse *t* haplotype: recent and worldwide introgression to *Mus musculus*. Proc. Natl. Acad. Sci. USA 89:6851–6855.

Moritz, C. 1991. The origin and evolution of parthenogenesis in *Heteronotia binoei* (Gekkonidae): evidence for recent and localized origins of widespread clones. Genetics 129:211–219.

Moritz, C. and W.M. Brown. 1986. Tandem duplication of D-loop and ribosomal RNA sequences in lizard mitochondrial DNA. Science 233:1425–1427.

Moritz, C. and W.M. Brown. 1987. Tandem duplications in animal mitochondrial DNAs: Variation in incidence and gene content among lizards. Proc. Natl. Acad. Sci. USA 84:7183–7187.

Moritz, C., T.E. Dowling, and W.M. Brown. 1987. Evolution of animal mitochondrial DNA: relevance for population biology and systematics. Annu. Rev. Ecol. Syst. 18:269–292.

Moritz, C.C., C.J. Schneider, and D.B. Wake. 1992. Evolutionary relationships within the *Ensatina eschscholtzii* complex confirm the ring species interpretation. Syst. Biol. 41:273–291.

Moritz, C.C., J.W. Wright, and W.M. Brown. 1989. Mitochondrial-DNA analyses and the origin and relative age of parthenogenetic lizards (genus *Cnemidophorus*). III. *C. velox* and *C. exsanguis*. Evolution 43:958–968.

Morton, E.S., L. Forman, and M. Braun. 1990. Extrapair fertilizations and the evolution of colonial breeding in purple martins. Auk 107:275–283.

Morton, N.E. 1992. Genetic structure of forensic populations. Proc. Natl. Acad. Sci. USA 89: 2556–2560.

Muller, H.J. 1950. Our load of mutations. Amer. J. Human Genet. 2:111–176.

Muller, H.J. 1964. The relevance of mutation to mutational advance. Mutat. Res. 1:2–9.

Mullis, K.B. 1990. The unusual origin of the polymerase chain reaction. Sci. Amer. 262(4):56–65.

Mullis, K. and F. Faloona. 1987. Specific synthesis of DNA in vitro via a polymerase catalyzed chain reaction. Methods Enzymol. 155:335–350.

Mullis, K., F. Faloona, S. Scharf, R. Saiki, G. Horn, and H. Erlich. 1986. Specific enzymatic amplification of DNA in vitro: the polymerase chain reaction. Cold Spring Harb. Symp. Quant. Biol. 51:263–273.

Mulvey, J. and R.C. Vrijenhoek. 1981. Multiple paternity in the hermaphroditic snail, *Biomphalaria obstructa*. J. Heredity 72:308–312.

Mumme, R.L., W.D. Koenig, R.M. Zink, and J.A. Martin. 1985. Genetic variation and parentage in a California population of acorn woodpeckers. Auk 102:305–312.

Munstermann, L.E. 1988. Biochemical systematics of nine nearctic *Aedes* mosquitoes (subgenus *Ochlerotatus, annulipes* group B). Pp. 135–147 *in:* Biosystematics of Haematophagous Insects, M.W. Service (ed.), Clarendon, Oxford, England.

Murawski, D.A. and J.L. Hamrick. 1990. Local genetic and clonal structure in the tropical terrestrial bromeliad, *Aechmea magdalenae*. Amer. J. Bot. 77:1201–1208.

Murphy, R.W. 1988. The problematic phylogenetic analysis of interlocus heteropolymer isozyme characters: A case study from sea snakes and cobras. Can. J. Zool. 66:2628–2633.

Murphy, R.W., J.W. Sites, Jr., D.G. Buth, and C.H. Haufler. 1990. Proteins I: isozyme electrophoresis. Pp. 45–126 *in:* Molecular Systematics, D.M. Hillis and C. Moritz (eds.), Sinauer, Sunderland, MA.

Murray, J., O.C. Stine, and M.S. Johnson. 1991. The evolution of mitochondrial DNA in *Partula*. Heredity 66:93–104.

Musser, J.M., S.J. Barenkamp, D.M. Granoff, and R.K. Selander. 1986. Genetic relationships of serologically nontypable and serotype b strains of *Haemophilus influenzae*. Infect. Immunol. 52:183–191.

Musser, J.M., D.A. Bemis, H. Ishikawa, and R.K. Selander. 1987. Clonal diversity and host distribution in *Bordetella bronchiseptica*. J. Bacteriol. 169:2793–2803.

Musser, J.M., D.M. Granoff, P.E. Pattison, and R.K. Selander. 1985. A population genetic framework for the study of invasive diseases caused by serotype b strains of *Haemophilus influenzae*. Proc. Natl. Acad. Sci. USA 82:5078–5082.

Musser, J.M., P.M. Schlievert, A.W. Chow, P. Ewan, B.N. Kreiswirth, V.T. Rosdahl, A.S. Naidu, W. Witte, and R.K. Selander. 1990. A single clone of *Staphyloccus aureus* causes the majority of cases of toxic shock syndrome. Proc. Natl. Acad. Sci. USA 87:225–229.

Nairn, C.J. and R.J. Ferl. 1988. The complete nucleotide sequence of the small-subunit ribosomal RNA coding region for the cycad *Zamia pumila:* phylogenetic implications. J. Mol. Evol. 27: 133–141.

Nakamura, Y., M. Leppert, P. O'Connell, R. Wolff, T. Holm, M. Culver, C. Martin, E. Fujimoto, M. Hoff, E. Kumlin, and R. White. 1987. Variable number of tandem repeat (VNTR) markers for human gene mapping. Science 235:1616–1622.

Narang, S.K., P.E. Kaiser, and J.A. Seawright. 1989a. Dichotomous electrophoretic key for identification of sibling species A, B, and C of the *Anopheles quadrimaculatus* complex (Diptera: Culicidae). J. Med. Entomol. 26:94–99.

Narang, S.K., P.E. Kaiser, and J.A. Seawright. 1989b. Identification of species D, a new member of the *Anopheles quadrimaculatus* species complex: a biochemical key. J. Amer. Mosquito Control Assn. 5:317–324.

Narang, S.K., S.R. Toniolo, J.A. Seawright, and P.E. Kaiser. 1989c. Genetic differentiation among sibling species A, B, and C of the *Anopheles quadrimaculatus* complex (Diptera: Culicidae). Ann. Entomol. Soc. Amer. 82:508–515.

National Research Council. 1992. DNA Technology in Forensic Science. National Academy Press, Washington, D.C.

Nei, M. 1972. Genetic distance between populations. Amer. Natur. 106:283–292.

Nei, M. 1973. Analysis of gene diversity in subdivided populations. Proc. Natl. Acad. Sci. USA 70:3321–3323.

Nei, M. 1975. Molecular Population Genetics and Evolution. North-Holland, Amsterdam.

Nei, M. 1977. *F*-statistics and analysis of gene diversity in subdivided populations. Ann. Human Genet. London 41:225–233.

Nei, M. 1978. Estimation of average heterozygosity and genetic distance from a small number of individuals. Genetics 23:341–369.

Nei, M. 1983. Genetic polymorphism and the role of mutation in evolution. Pp. 165–190 *in:* Evolution of Genes and Proteins, M. Nei and R.K. Koehn (eds.), Sinauer, Sunderland, MA.

Nei, M. 1985. Human evolution at the molecular level. Pp. 41–64 *in:* Population Genetics and Molecular Evolution, T. Ohta and K. Aoki (eds.), Japan Science Society Press and Springer-Verlag, Tokyo.

Nei, M. 1987. Molecular Evolutionary Genetics. Columbia Univ. Press, New York.

Nei, M. and D. Graur. 1984. Extent of protein polymorphism and the neutral mutation theory. Evol. Biol. 17:73–118.

Nei, M. and A.L. Hughes. 1991. Polymorphism and evolution of the major histocompatibility complex loci in mammals. Pp. 222–247 *in:* Evolution at the Molecular Level, R.K. Selander, A.G. Clark, and T.S. Whittam (eds.), Sinauer, Sunderland, MA.

Nei, M. and R.K. Koehn. 1983. Evolution of Genes and Proteins. Sinauer, Sunderland, MA.

Nei, M. and W.-H. Li. 1979. Mathematical model for studying genetic variation in terms of restriction endonucleases. Proc. Natl. Acad. Sci. USA 76:5269–5273.

Nei, M. and G. Livshits. 1990. Evolutionary relationships of Europeans, Asians, and Africans at the molecular level. Pp. 251–265 *in:* Population Biology of Genes and Molecules, N. Takahata and J.F. Crow (eds.), Baifukan, Tokyo.

Nei, M. and T. Maruyama. 1975. Lewontin-Krakauer test for neutral genes. Genetics 80:395.

Nei, M., T. Maruyama, and R. Chakraborty. 1975. The bottleneck effect and genetic variability in populations. Evolution 29:1–10.

Nei, M. and A.K. Roychoudhury. 1982. Genetic relationship and evolution of human races. Evol. Biol. 14:1–59.

Nei, M., J.C. Stephens, and N. Saitou. 1985. Methods for computing the standard errors of branching points in an evolutionary tree and their application to molecular data from humans and apes. Mol. Biol. Evol. 2:66–85.

462 *Literature Cited*

Nei, M. and F. Tajima. 1981. DNA polymorphism detectable by restriction endonucleases. Genetics 97:145–163.

Nei, M. and F. Tajima. 1985. Evolutionary change of restriction cleavage sites and phylogenetic inference for man and apes. Mol. Biol. Evol. 2:189–205.

Nei, M., F. Tajima, and Y. Tateno. 1983. Accuracy of estimated phylogenetic trees from molecular data. II. Gene frequency data. J. Mol. Evol. 19:153–170.

Neigel, J.E. and J.C. Avise. 1983a. Clonal diversity and population structure in a reef-building coral, *Acropora cervicornis:* self-recognition analysis and demographic interpretation. Evolution 37:437–453.

Neigel, J.E. and J.C. Avise. 1983b. Histocompatibility bioassays of population structure in marine sponges. J. Heredity 74:134–140.

Neigel, J.E. and J.C. Avise. 1985. The precision of histocompatibility response in clonal recognition in tropical marine sponges. Evolution 39:724–732.

Neigel, J.E. and J.C. Avise. 1986. Phylogenetic relationships of mitochondrial DNA under various demographic models of speciation. Pp. 515–534 *in:* Evolutionary Processes and Theory, E. Nevo and S. Karlin (eds.), Academic Press, New York.

Neigel, J.E., R.M. Ball, Jr., and J.C. Avise. 1991. Estimation of single generation migration distances from geographic variation in animal mitochondrial DNA. Evolution 45:423–432.

Nelson, G. and N.I. Platnick. 1981. Systematics and Biogeography: Cladistics and Vicariance. Columbia Univ. Press, New York.

Nelson, G. and D.E. Rosen (eds.). 1981. Vicariance Biogeography, A Critique. Columbia Univ. Press, New York.

Nelson, K., R.J. Baker, and R.L. Honeycutt. 1987. Mitochondrial DNA and protein differentiation between hybridizing cytotypes of the white-footed mouse, *Peromyscus leucopus.* Evolution 41:864–872.

Nelson, K. and D. Hedgecock. 1977. Electrophoretic evidence of multiple paternity in the lobster *Homarus americanus* (Milne-Edwards). Amer. Natur. 111:361–365.

Nelson, K., T.S. Whittam, and R.K. Selander. 1991. Nucleotide polymorphism and evolution in the glyceraldehyde-3-phosphate dehydrogenase gene (*gapA*) in natural populations of *Salmonella* and *Escherichia coli.* Proc. Natl. Acad. Sci. USA 88:6667–6671.

Nevo, E. 1978. Genetic variation in natural populations: patterns and theory. Theoret. Pop. Biol. 13:121–177.

Nevo, E., Y.J. Kim, C.R. Shaw, and C.S. Thaeler. 1974. Genetic variation, selection, and speciation in *Thomomys talpoides* pocket gophers. Evolution 28:1–23.

Nevo, E. and C.R. Shaw. 1972. Genetic variation in a subterranean mammal, *Spalax ehrenbergi.* Biochem. Genet. 7:235–241.

Newton, A.C., C.E. Caten, and R. Johnson. 1985. Variation for isozymes and double-stranded RNA among isolates of *Puccinia striiformis* and two other cereal rusts. Plant Pathol. 34:235–247.

Nichols, R.A. and D.J. Balding. 1991. Effects of population structure on DNA fingerprint analysis in forensic science. Heredity 66:297–302.

Normark, B.B., A.R. McCune, and R.G. Harrison. 1991. Phylogenetic relationships of neopterygian fishes, inferred from mitochondrial DNA sequences. Mol. Biol. Evol. 8:819–834.

Novacek, M.J. 1992. Mammalian phylogeny: shaking the tree. Nature 356:121–125.

Nowak, R.M. 1992. The red wolf is not a hybrid. Cons. Biol. 6:593–595.

Nugent, J.M. and J.D. Palmer. 1991. RNA-mediated transfer of the gene *coxII* from the mitochondrion to the nucleus during flowering plant evolution. Cell 66:473–481.

Nuttall, G.H.F. 1904. Blood Immunity and Blood Relationship. Cambridge Univ. Press, Cambridge, England.

Nybom, H., J. Ramser, D. Kaemmer, G. Kahl, and K. Weising. 1992. Oligonucleotide DNA fingerprinting detects a multiallelic locus in box elder (*Acer negundo*). Mol. Ecol. 1:65–67.

Nybom, H. and B.A. Schaal. 1990. DNA "fingerprints" reveal genotypic distributions in natural populations of blackberries and raspberries (*Rubus*, Rosaceae). Amer. J. Bot. 77:883–888.

Oakeshott, J.G., J.B. Gibson, P.R. Anderson, W.R. Knibb, D.G. Anderson, and G.K. Chambers. 1982. Alcohol dehydrogenase and glycerol-3-phosphate dehydrogenase clines in *Drosophila melanogaster* on different continents. Evolution 36:86–96.

O'Brien, S.J. 1987. The ancestry of the giant panda. Sci. Amer. 257(5):102–107.

O'Brien, S.J. and J.F. Evermann. 1988. Interactive influence of infectious disease and genetic diversity in natural populations. Trends Ecol. Evol. 3:254–259.

O'Brien, S.J. and R.J. MacIntyre. 1972. The alpha-glycerophosphate cycle in *Drosophila melanogaster*. I. Biochemical and developmental aspects. Biochem. Genet. 7:141–161.

O'Brien, S.J., J.S. Martenson, M.A. Eichelberger, E.T. Thorne, and F. Wright. 1989. Genetic variation and molecular systematics of the black-footed ferret. Pp. 21–33 *in:* Conservation Biology and the Black-Footed Ferret, U.S. Seal, E.T. Thorne, M.A. Bogan, and S.H. Anderson (eds.), Yale Univ. Press, New Haven, CT.

O'Brien, S.J. and E. Mayr. 1991. Bureaucratic mischief: recognizing endangered species and subspecies. Science 251:1187–1188.

O'Brien, S.J., W.G. Nash, D.E. Wildt, M.E. Bush, and R.E. Benveniste. 1985a. A molecular solution to the riddle of the giant panda's phylogeny. Nature 317:140–144.

O'Brien, S.J., M.E. Roelke, L. Marker, A. Newman, C.A. Winkler, D. Meltzer, L. Colly, J.F. Evermann, M. Bush, and D.E. Wildt. 1985b. Genetic basis for species vulnerability in the cheetah. Science 227:1428–1434.

O'Brien, S.J., M.E. Roelke, N. Yuhki, K.W. Richards, W.E. Johnson, W.L. Franklin, A.E. Anderson, O.L. Bass, Jr., R.C. Belden, and J.S. Martenson. 1990. Genetic introgression within the Florida panther *Felis concolor coryi*. Natl. Geog. Res. 6:485–494.

O'Brien, S.J., D.E. Wildt, M. Bush, T.M. Caro, C. FitzGibbon, I. Aggundey, and R.E. Leakey. 1987. East African cheetahs: evidence for two population bottlenecks? Proc. Natl. Acad. Sci. USA 84:508–511.

O'Brien, S.J., D.E. Wildt, D. Goldman, C.R. Merril, and M. Bush. 1983. The cheetah is depauperate in genetic variation. Science 221:459–462.

Ochman, H. and R.K. Selander. 1984. Evidence for clonal population structure in *Escherichia coli*. Proc. Natl. Acad. Sci. USA 81:198–201.

Ochman, H. and A.C. Wilson. 1987. Evolution in bacteria: evidence for a universal substitution rate in cellular genomes. J. Mol. Evol. 26:74–86.

O'Connell, M. 1992. Response to: "Six biological reasons why the endangered species act doesn't work and what to do about it." Cons. Biol. 6:140–143.

O'Farrell, P.H. 1975. High resolution two-dimensional electrophoresis of proteins. J. Biol. Chem. 250:4007–4021.

Ohno, S. 1967. Sex Chromosomes and Sex-Linked Genes. Springer-Verlag, New York.

Ohno, S. 1970. Evolution by Gene Duplication. Springer-Verlag, New York.

Ohta, T. 1980. Evolution and Variation of Multigene Families. Springer-Verlag, Berlin.

Ohta, T. 1984. Some models of gene conversion for treating the evolution of multigene families. Genetics 106:517–528.

Ohta, T. 1992a. The nearly neutral theory of molecular evolution. Annu. Rev. Ecol. Syst. 23:263–286.

Ohta, T. 1992b. The meaning of natural selection revisited at the molecular level. Trends Ecol. Evol. 7:311–312.

Ohta, T. and G.A. Dover. 1983. Population genetics of multi-gene families that are dispersed into two or more chromosomes. Proc. Natl. Acad. Sci. USA 80:4079–4083.

Ohta, T. and M. Kimura. 1971. On the constancy of the evolutionary rate of cistrons. J. Mol. Evol. 1:18–25.

Ohta, T. and M. Kimura. 1973. A model of mutation appropriate to estimate the number of electrophoretically detectable alleles in a finite population. Genet. Res. 22:201–204.

Ohta, T. and H. Tachida. 1990. Theoretical study of near neutrality. I. Heterozygosity and rate of mutant substitution. Genetics 126:219–229.

O'hUigin, C. and W.-H. Li. 1992. The molecular clock ticks regularly in muroid rodents and hamsters. J. Mol. Evol. 35:377–384.

Ohyama, K., H. Fukuzawa, T. Kohchi, H. Shirai, T. Sano, S. Sano, K. Umesono, Y. Shiki, M. Takeuchi, Z. Chang, S. Aota, H. Inokuchi, and H. Ozeki. 1986. Chloroplast gene organization deduced from complete sequence of liverwort *Marchantia polymorpha* chloroplast DNA. Nature 322:572–574.

Okimoto, R., J.L. Macfarlane, D.O. Clary, and D.R. Wolstenholme. 1992. The mitochondrial genomes of two nematodes, *Caenorhabditis elegans* and *Ascaris suum*. Genetics 130:471–498.

Oliver, S.G. and 145 others. 1992. The complete DNA sequence of yeast Chromosome III. Nature 357:38–46.

Olmstead, R.G., R.K. Jansen, H.J. Michaels, S.R. Downie, and J.D. Palmer. 1990. Chloroplast DNA and phylogenetic studies in the Asteridae. Pp. 119–134 *in:* Biological Approaches and Evolutionary Trends in Plants, S. Kawano (ed.), Academic Press, San Diego, CA.

Olmstead, R.G., H.J. Michaels, K.M. Scott, and J.D. Palmer. 1992. Monophyly of the Asteridae and identification of their major lineages inferred from DNA sequences of *rbcL*. Annals Missouri Bot. Gardens 79:249–265.

Olson, R.R., J.A. Runstadler, and T.D. Kocher. 1991. Whose larvae? Nature 351:357–358.

Oring, L.W., R.C. Fleischer, J.M. Reed, and K.E. Marsden. 1992. Cuckoldry through stored sperm in the sequentially polyandrous spotted sandpiper. Nature 359:631–633.

Orr, H.A. 1987. Genetics of male and female sterility in hybrids of *Drosophila pseudoobscura* and *D. persimilis*. Genetics 116:555–563.

Orr, H.A. 1992. Mapping and characterization of a "speciation gene" in *Drosophila*. Genet. Res. Camb. 59:73–80.

Oste, C. 1988. Polymerase chain reaction. BioTechniques 6:162–167.

Ou, C.-Y. and 17 others. 1992. Molecular epidemiology of HIV transmission in a dental practice. Science 256:1165–1171.

Ovenden, J.R. 1990. Mitochondrial DNA and marine stock assessment: a review. Aust. J. Mar. Freshwater Res. 41:835–853.

Ovenden, J.R., D.J. Brasher, and R.W.G. White. 1992. Mitochondrial DNA analyses of the red rock lobster *Jasus edwardsii* supports an apparent absence of population subdivision throughout Australasia. Mar. Biol. 112:319–326.

Ovenden, J.R. and R.W.G. White. 1990. Mitochondrial and allozyme genetics of incipient speciation in a landlocked population of *Galaxias truttaceus* (Pisces: Galaxiidae). Genetics 124:701–716.

Pääbo, D. 1985. Molecular cloning of ancient mummy DNA. Nature 314:644–645.

Pääbo, S. 1989. Ancient DNA: extraction, characterization, molecular cloning and enzymatic amplification. Proc. Natl. Acad. Sci. USA 86:1939–1943.

Pääbo, S., R.G. Higuchi, and A.C. Wilson. 1989. Ancient DNA and the polymerase chain reaction. J. Biol. Chem. 264:9709–9712.

Pääbo, S., W.K. Thomas, K.M. Whitfield, Y. Kumazawa, and A.C. Wilson. 1991. Rearrangements of mitochondrial transfer RNA genes in marsupials. J. Mol. Evol. 33:426–430.

Pääbo, S. and A.C. Wilson. 1988. Polymerase chain reaction reveals cloning artefacts. Nature 334:387–388.

Packer, C., D.A. Gilbert, A.E. Pusey, and S.J. O'Brien. 1991. A molecular genetic analysis of kinship and cooperation in African lions. Nature 351:562–564.

Packer, L. 1991. The evolution of social behavior and nest architecture in sweat bees of the subgenus *Evylaeus* (Hymenoptera: Halictidae): a phylogenetic approach. Behav. Ecol. Sociobiol. 29:153–160.

Page, D.C., A. de la Chapelle, and J. Weissenbach. 1985. Chromosome Y-specific DNA in related human XX males. Nature 315:224–226.

Page, D.C., B. DeMartinville, D. Barker, A. Wyman, R. White, U. Francke, and D. Botstein. 1982. Single copy sequence hybridizes to polymorphic and homologous loci on human X and Y chromosomes. Proc. Natl. Acad. Sci. USA 79:5352–5356.

Page, D.C., M.E. Harper, J. Love, and D. Botstein. 1984. Occurrence of a transposition from the X-chromosome long arm to the Y-chromosome short arm during human evolution. Nature 311:119–123.

Paige, K.N., W.C. Capman, and P. Jennetten. 1991. Mitochondrial inheritance patterns across a cottonwood hybrid zone: cytonuclear disequilibria and hybrid zone dynamics. Evolution 45:1360–1369.

Palmer, A.R. and C. Strobeck. 1986. Fluctuating asymmetry: measurement, analysis, patterns. Annu. Rev. Ecol. Syst. 17:391–421.

Palmer, J.D. 1985. Evolution of chloroplast and mitochondrial DNA in plants and algae. Pp. 131–240 *in:* Molecular Evolutionary Genetics, R.J. MacIntyre (ed.), Plenum Press, New York.

Palmer, J.D. 1987. Chloroplast DNA evolution and biosystematic uses of chloroplast DNA variation. Amer. Natur. 130:S6–S29.

Palmer, J.D. 1990. Contrasting modes and tempos of genome evolution in land plant organelles. Trends. Genet. 6:115–120.

Palmer, J.D. 1992. Mitochondrial DNA in plant systematics: applications and limitations. Pp. 36–49 *in:* Molecular Systematics of Plants, P.S. Soltis, D.E. Soltis, and J.J. Doyle (eds.), Chapman and Hall, New York.

Palmer, J.D. and L.A. Herbon. 1986. Tripartite mitochondrial genomes of *Brassica* and *Raphanus:* reversal of repeat configurations by inversion. Nucleic Acids Res. 14:9755–9765.

Palmer, J.D. and L.A. Herbon. 1988. Plant mitochondrial DNA evolves rapidly in structure, but slowly in sequence. J. Mol. Evol. 28:87–97.

Palmer, J.D., R.K. Jansen, H.J. Michaels, M.W. Chase, and J.R. Manhart. 1988a. Chloroplast DNA variation and plant phylogeny. Annals Missouri Bot. Gardens 75:1180–1206.

Palmer, J.D., R.A. Jorgenson, and W.F. Thompson. 1985. Chloroplast DNA variation and evolution in *Pisum:* patterns of change and phylogenetic analysis. Genetics 109:195–213.

Palmer, J.D., B. Osorio, and W.F. Thompson. 1988b. Evolutionary significance of inversions in legume chloroplast DNAs. Curr. Genet. 14:65–74.

Palmer, J.D. and C.R. Shields. 1984. Tripartite structure of the *Brassica campestris* mitochondrial genome. Nature 307:437–440.

Palmer, J.D., C.R. Shields, D.B. Cohen, and T.J. Orten. 1983. Chloroplast DNA evolution and the origin of amphidiploid *Brassica* species. Theoret. Appl. Genet. 65:181–189.

Palmer, J.D. and W.F. Thompson. 1981. Rearrangements in the chloroplast genomes of mung bean and pea. Proc. Natl. Acad. Sci. USA 78:5533–5537.

Palumbi, S.R. and J. Benzie. 1991. Large mitochondrial DNA differences between morphologically similar Penaeid shrimp. Mol. Mar. Biol. Biotech. 1:27–34.

Palumbi, S.R., A. Martin, S. Romano, W.O. McMillan, L. Stice, and G. Grabowski. 1991. The Simple Fool's Guide to PCR, Version 2. University of Hawaii Zoology Department, Honolulu, HI.

Palumbi, S.R. and A.C. Wilson. 1990. Mitochondrial DNA diversity in the sea urchins *Strongylocentrotus purpuratus* and *S. droebachiensis*. Evolution 44:403–415.

Palva, T.K. and E.T. Palva. 1985. Rapid isolation of animal mitochondrial DNA by alkaline extraction. FEBS Letters 192:267–270.

Pamilo, P. 1981. Genetic organization of *Formica sanguinea* populations. Behav. Ecol. Sociobiol. 9:45–50.

Pamilo, P. 1982. Genetic population structure in polygynous *Formica* ants. Heredity 48:95–106.

Pamilo, P. 1984a. Genotypic correlation and regression in social groups: multiple alleles, multiple loci and subdivided populations. Genetics 107:307–320.

Pamilo, 1984b. Genetic relatedness and evolution of insect sociality. Behav. Ecol. Sociobiol. 15: 241–248.

Pamilo, P. and R.H. Crozier. 1982. Measuring genetic relatedness in natural populations: methodology. Theoret. Pop. Biol. 21:171–193.

Pamilo, P. and M. Nei. 1988. Relationships between gene trees and species trees. Mol. Biol. Evol. 5:568–583.

Pamilo, P. and S.-L. Varvio-Aho. 1979. Genetic structure of nests in the ant *Formica sanguinea*. Behav. Ecol. Sociobiol. 6:91–98.

Parker, G.A. 1970. Sperm competition and its evolutionary consequences in the insects. Biol. Rev. 45:525–567.

Parker, G.A. 1984. Sperm competition and the evolution of male mating strategies. Pp. 1–60 *in:* Sperm Competition and the Evolution of Animal Mating Systems, R.L. Smith (ed.), Academic Press, New York.

Parker, K.C. and J.L. Hamrick. 1992. Genetic diversity and clonal structure in a columnar cactus, *Lophocereus schottii*. Amer. J. Bot. 79:86–96.

Pasdar, M., D.P. Philipp, and G.S. Whitt. 1984. Linkage relationships of nine enzyme loci in sunfishes (Lepomis; Centrarchidae). Genetics 107:435–446.

Pastene, L.A., K. Numachi, and K. Tsukamoto. 1991. Examination of reproductive success of transplanted stocks in an amphidromous fish, *Plecoglossus altivelis* (Temmink et Schlegel) using mitochondrial DNA and isozyme markers. J. Fish Biol. 39A:93–100.

Paterson, A., E. Lander, J. Hewitt, S. Peterson, S. Lincoln, and S. Tanksley. 1988. Resolution of quantitative traits into Mendelian factors by using a complete linkage map of restriction length polymorphisms. Nature 335:721–726.

Paterson, H.E.H. 1985. The recognition concept of species. Pp. 21–29 *in:* Species and Speciation, E.S. Vrba (ed.), Transvaal Museum Monograph No. 4, Pretoria, South Africa.

Patterson, C. (ed.) 1987. Molecules and Morphology in Evolution: Conflict or Compromise? Cambridge Univ. Press, Cambridge, England.

Patton, J.C. and J.C. Avise. 1983. An empirical evaluation of qualitative Hennigian analyses of protein electrophoretic data. J. Mol. Evol. 19:244–254.

Patton, J.L. and M.S. Hafner. 1983. Biosystematics of the native rodents of the Galapagos archipelago. Pp. 539–568 *in:* Patterns of Evolution in Galapagos Organisms, R.I. Bowman, M. Berson, and A.E. Leviton (eds.), American Association for the Advancement of Science, San Francisco, CA.

Patton, J.L. and M.F. Smith. 1981. Molecular evolution in *Thomomys:* phyletic systematics, paraphyly, and rates of evolution. J. Mammal. 62:493–500.

Patton, J.L. and M.F. Smith. 1989. Population structure and the genetic and morphological divergence among pocket gopher species (genus *Thomomys*). Pp. 284–304 *in:* Speciation and Its Consequences, D. Otte and J.A. Endler (eds.), Sinauer, Sunderland, MA.

Pearson, B. 1983. Intra-colonial relatedness amongst workers in a population of nests of the polygynous ant, *Myrmica rubra* Latreille. Behav. Ecol. Sociobiol. 12:1–4.

Pella, J.J. and G.B. Milner. 1987. Use of genetic marks in stock composition analysis. Pp. 247–276 *in:* Population Genetics and Fishery Management, N. Ryman and F. Utter (eds.), Univ. Washington Press, Seattle.

Perkins, H.D. and A.J. Howells. 1992. Genomic sequences with homology to the *P* element of *Drosophila melanogaster* occur in the blowfly *Lucilia cuprina*. Proc. Natl. Acad. Sci. USA 89:10753–10757.

Pernin, P., A. Ataya, and M.L. Cariou. 1992. Genetic structure of natural populations of the free-living amoeba, *Naegleria lovaniensis*. Evidence for sexual reproduction. Heredity 68:173–181.

Petrie, M. and A.P. Møller. 1991. Laying eggs in others' nests: intraspecific brood parasitism in birds. Trends Ecol. Evol. 6:315–320.

Petter, S.C., D.B. Miles, and M.M. White. 1990. Genetic evidence of mixed reproductive strategy in a monogamous bird. The Condor 92:702–708.

Pettigrew, J.D. 1986. Flying primates? Megabats have the advanced pathway from eye to midbrain. Science 231:1304–1306.

Pettigrew, J.D. 1991. Wings or brain? Convergent evolution in the origins of bats. Syst. Zool. 40:199–216.

Phelps, S.R. and F.W. Allendorf. 1983. Genetic identity of pallid and shovelnose sturgeon (*Scaphirhynchus albus* and *S. platorynchus*). Copeia 1983:696–700.

Philipp, D.P., W.F. Childers, and G.S. Whitt. 1981. Management implications for different genetic stocks of largemouth bass (*Micropterus salmoides*) in the United States. Can. J. Fish. Aquat. Sci. 38:1715–1723.

Philipp, D.P., H.R. Parker, and G.S. Whitt. 1983. Evolution of gene regulation: isozymic analysis of patterns of gene expression during hybrid fish development. Isozymes X:193–237.

Phillips, M.K. and V.G. Henry. 1992. Comments on red wolf taxonomy. Cons. Biol. 6:596–599.

Pierce, B.A. and J.B. Mitton. 1982. Allozyme heterozygosity and growth in the tiger salamander, *Ambystoma tigrinum.* J. Heredity 73:250–253.

Pilbeam, D. 1984. The descent of hominoids and hominids. Sci. Amer. 250(3):84–96.

Place, A.R. and D.A. Powers. 1979. Genetic variation and relative catalytic efficiencies: Lactate dehydrogenase B allozymes of *Fundulus heteroclitus.* Proc. Natl. Acad. Sci. USA 76:2354–2358.

Place, A.R. and D.A. Powers. 1984. The LDH-B allozymes of *Fundulus heteroclitus:* II. Kinetic analyses. J. Biol. Chem. 259:1309–1318.

Plante, Y., P.T. Boag, and B.N. White. 1989. Microgeographic variation in mitochondrial DNA of meadow voles (*Microtus pennsylvanicus*) in relation to population density. Evolution 43:1522–1537.

Pleasants, J.M. and J.F. Wendel. 1989. Genetic diversity in a clonal narrow endemic, *Erythronium propullans,* and its widespread progenitor, *Erythronium albidum.* Amer. J. Bot. 76:1136–1151.

Polans, N.O. 1983. Enzyme polymorphisms in Galapagos finches. Pp. 219–236 *in:* Patterns of Evolution in Galapagos Organisms, R.I. Bowman, M. Berson, and A.E. Leviton (eds.), American Association for the Advancement of Science, San Francisco, CA.

Pope, T.R. 1990. The reproductive consequences of male cooperation in the red howler monkey: paternity exclusion in multi-male and single-male troops using genetic markers. Behav. Ecol. Sociobiol. 27:439–446.

Popper, K.R. 1968. The Logic of Scientific Discovery. Harper Torchbooks, New York.

Potts, B.M. and J.B. Reid. 1985. Analysis of a hybrid swarm between *Eucalyptus risdonii* Hook.f. and *E. amygdalina* Labill. Aust. J. Bot. 33:543–562.

Powell, J.R. 1971. Genetic polymorphisms in varied environments. Science 174:1035–1036.

Powell, J.R. 1975. Protein variation in natural populations of animals. Evol. Biol. 3:79–119.

Powell, J.R. 1983. Interspecific cytoplasmic gene flow in the absence of nuclear gene flow: evidence from *Drosophila.* Proc. Natl. Acad. Sci. USA 80:492–495.

Powell, J.R. 1991. Monophyly/paraphyly/polyphyly and gene/species trees: an example from *Drosophila.* Mol. Biol. Evol. 8:892–896.

Powell, J.R., A. Caccone, G.D. Amato, and C. Yoon. 1986. Rates of nucleotide substitutions in *Drosophila* mitochondrial DNA and nuclear DNA are similar. Proc. Natl. Acad. Sci. USA 83:9090–9093.

Powell, J.R., W.J. Tabachnick, and J. Arnold. 1980. Genetics and the origin of a vector population: *Aedes aegypti,* a case study. Science 208:1385–1387.

Powell, J.R. and C.E. Taylor. 1979. Genetic variation in ecologically diverse environments. Amer. Sci. 67:590–596.

Powell, J.R. and M.C. Zuninga. 1983. A simplified procedure for studying mtDNA polymorphisms. Biochem. Genet. 21:1051–1055.

Powers, D.A. 1991. Evolutionary genetics of fish. Adv. Genet. 29:120–228.

Powers, D.A., T. Lauerman, D. Crawford, and L. DiMichele. 1991a. Genetic mechanisms for adapting to a changing environment. Annu. Rev. Genet. 25:629–659.

Powers, D.A., T. Lauerman, D. Crawford, M. Smith, I. González-Villaseñor, and L. DiMichele. 1991b. The evolutionary significance of genetic variation at enzyme synthesizing loci in the teleost *Fundulus heteroclitus.* J. Fish Biol. 39A:169–184.

Powers, D.A., I. Ropson, D.C. Brown, R. Van Beneden, R. Cashon, et al. 1986. Genetic variation in *Fundulus heteroclitus:* geographic distribution. Amer. Zoo. 26:131–144.

Prager, E.M., A.H. Brush, R.A. Nolan, M. Nakanishi, and A.C. Wilson. 1974. Slow evolution of transferrin and albumin in birds according to micro-complement fixation analysis. J. Mol. Evol. 3:243–262.

Prager, E.M. and A.C. Wilson. 1975. Slow evolutionary loss of the potential for interspecific hybridization in birds: a manifestation of slow regulatory evolution. Proc. Natl. Acad. Sci. USA 72:200–204.

Prager, E.M. and A.C. Wilson. 1976. Congruency of phylogenies derived from different proteins. J. Mol. Evol. 9:45–57.

Prager, E.M. and A.C. Wilson. 1978. Construction of phylogenetic trees for proteins and nucleic acids: empirical evaluation of alternative matrix methods. J. Mol. Evol. 11:129–142.

Prager, E.M. and A.C. Wilson. 1993. Information content of immunological distances. Methods Enzymol. 224: in press.

Prager, E.M., A.C. Wilson, J.M. Lowenstein, and V.M. Sarich. 1980. Mammoth albumin. Science 209:287–289.

Prager, E.M., A.C. Wilson, D.T. Osuga, and R.E. Feeney. 1976. Evolution of flightless land birds on southern continents: transferrin comparison shows monophyletic origin of ratites. J. Mol. Evol. 8:283–294.

Prakash, S. 1977. Allelic variants at the xanthine dehydrogenase locus affecting enzyme activity in *Drosophila pseudoobscura*. Genetics 87:159–168.

Preziosi, R.F. and D.J. Fairbairn. 1992. Genetic population structure and levels of gene flow in the stream dwelling waterstrider, *Aquarius* (= *Gerris*) *remigis* (Hemiptera: Gerridae). Evolution 46: 430–444.

Price, D.K., G.E. Collier, and C.F. Thompson. 1989. Multiple parentage in broods of house wrens: genetic evidence. J. Heredity 80:1–5.

Price, M.V. and N.M. Waser. 1979. Pollen dispersal and optimal outcrossing in *Delphinium nelsoni*. Nature 277:294–297.

Pring, D.R. and D.M. Lonsdale. 1985. Molecular biology of higher plant mitochondrial DNA. Int. Rev. Cytol. 97:1–46.

Prinsloo, P. and T.J. Robinson. 1992. Geographic mitochondrial DNA variation in the rock hyrax, *Procavia capensis*. Mol. Biol. Evol. 9:447–456.

Prodöhl, P.A., J.B. Taggart, and A. Ferguson. 1992. Genetic variability within and among sympatric brown trout (*Salmo trutta*) populations: multi-locus DNA fingerprint analysis. Hereditas 117:45–50.

Provine, W.B. 1986. Sewell Wright and Evolutionary Biology. Univ. Chicago Press, Chicago, IL.

Provine, W.B. 1989. Founder effects and genetic revolutions in microevolution and speciation: an historical perspective. Pp. 43–76 *in:* Genetics, Speciation and the Founder Principle, L.V. Giddings, K.Y. Kaneshiro, and W.W. Anderson (eds.), Oxford Univ. Press, New York.

Pühler, G., H. Leffers, F. Gropp, P. Palm, H.-P. Klenk, F. Lottspeich, R.A. Garrett, and W. Zillig. 1989. Archaebacterial DNA-dependent RNA polymerases testify to the evolution of the eukaryotic nuclear genome. Proc. Natl. Acad. Sci. USA 86:4569–4573.

Quattro, J.M., J.C. Avise, and R.C. Vrijenhoek. 1991. Molecular evidence for multiple origins of hybridogenetic fish clones (Poeciliidae: *Poeciliopsis*). Genetics 127:391–398.

Quattro, J.M., J.C. Avise, and R.C. Vrijenhoek. 1992a. An ancient clonal lineage in the fish genus *Poeciliopsis* (Atheriniformes: Poeciliidae). Proc. Natl. Acad. Sci. USA 89:348–352.

Quattro, J.M., J.C. Avise, and R.C. Vrijenhoek. 1992b. Mode of origin and sources of genotypic diversity in triploid fish clones (*Poeciliopsis:* Poeciliidae). Genetics 130:621–628.

Quattro, J.M. and R.C. Vrijenhoek. 1989. Fitness differences among remnant populations of the endangered Sonoran topminnow. Science 245:976–978.

Quattro, J.M., H.A. Woods, and D.A. Powers. 1993. Sequence analysis of teleost retina-specific lactate dehydrogenase C: Evolutionary implications for the vertebrate lactate dehydrogenase gene family. Proc. Natl. Acad. Sci. USA 90:242–246.

Queller, D.C. and K.F. Goodnight. 1989. Estimating relatedness using genetic markers. Evolution 43:258–275.

Queller, D.C., J.E. Strassmann, and C.R. Hughes. 1988. Genetic relatedness in colonies of tropical wasps with multiple queens. Science 242:1155–1157.

Quinn, T.P., C.C. Wood, L. Margolis, B.E. Riddell, and K.D. Hyatt. 1987. Homing in wild sockeye salmon (*Oncorhynchus nerka*) populations as inferred from differences in parasite prevalance and allozyme allele frequencies. Can. J. Fish. Aquat. Sci. 44:1963–1968.

Quinn, T.W. 1988. DNA sequence variation in the Lesser Snow Goose, *Anser caerulescens caerulescens*. Ph.D. Dissertation, Queen's University, Kingston, Ontario, Canada.

Quinn, T.W. 1992. The genetic legacy of mother goose—phylogeographic patterns of lesser snow goose *Chen caerulescens caerulescens* maternal lineages. Mol. Ecol. 1:105–117.

Quinn, T.W., F. Cooke, and B.N. White. 1990. Molecular sexing of geese using a cloned Z chromosomal sequence with homology to the W chromosome. Auk 107:199–202.

Quinn, T.W., J.C. Davies, F. Cooke, and B.N. White. 1989. Genetic analysis of offspring of a female-female pair in the lesser snow goose (*Chen c. caerulescens*). Auk 106:177–184.

Quinn, T.W., J.S. Quinn, F. Cooke, and B.N. White. 1987. DNA marker analysis detects multiple maternity and paternity in single broods of the lesser snow goose. Nature 326:392–394.

Quinn, T.W., G.F. Shields, and A.C. Wilson. 1991. Affinities of the Hawaiian goose based on two types of mitochondrial DNA data. Auk 108:585–593.

Quinn, T.W. and B.N. White. 1987. Identification of restriction-fragment-length polymorphisms in genomic DNA of the lesser snow goose (*Anser caerulescens caerulescens*). Mol. Biol. Evol. 4:126–143.

Rabenold, P.P., W.H. Piper, M.D. Decker, and D.J. Minchella. 1991. Polymorphic minisatellite amplified on avian W chromosome. Genome 34:489–493.

Rabenold, P.P., K.N. Rabenold, W.H. Piper, J. Haydock, and S.W. Zack. 1990. Shared paternity revealed by genetic analysis in cooperatively breeding tropical wrens. Nature 348:538–540.

Racine, R.R. and C.H. Langley. 1980. Genetic heterozygosity in a natural population of *Mus musculus* assessed using two-dimensional electrophoresis. Nature 283:855–857.

Ralin, D.B. 1976. Behavioral and genetic differentiation in a diploid-tetraploid cryptic species complex of treefrogs. Herpetol. Rev. 7:97–98.

Ralls, K., J.D. Ballou, and A. Templeton. 1988. Estimates of lethal equivalents and the cost of inbreeding in mammals. Cons. Biol. 2:185–193.

Ramshaw, J.A.M., J.A. Coyne, and R.C. Lewontin. 1979. The sensitivity of gel electrophoresis as a detector of genetic variation. Genetics 93:1019–1037.

Rand, D.M. and R.G. Harrison. 1986. Mitochondrial DNA transmission genetics in crickets. Genetics 114:955–970.

Rand, D.M. and R.G. Harrison. 1989. Ecological genetics of a mosaic hybrid zone: mitochondrial, nuclear, and reproductive differentiation of crickets by soil type. Evolution 43:432–449.

Randall, T. 1991. Mitochondrial DNA: a new frontier in acquired and inborn gene defects. JAMA 266:1739–1740.

Rapacz, J., L. Chen, E. Butler-Brunner, M.-J. Wu, J.O. Hasler-Rapacz, R. Butler, and V.N. Schumaker. 1991. Identification of the ancestral haplotype for apolipoprotein B suggests an African origin of *Homo sapiens sapiens* and traces their subsequent migration to Europe and the Pacific. Proc. Natl. Acad. Sci. USA 88:1403–1406.

Rasheed, B.K.A., E.C. Whisenant, R. Fernandez, H. Ostrer, and Y.M. Bhatnagar. 1991. A Y-chromosomal DNA fragment is conserved in human and chimpanzee. Mol. Biol. Evol. 8:416–432.

Raubeson, L.A. and R.K. Jansen. 1992a. Chloroplast DNA evidence on the ancient evolutionary split in vascular land plants. Science 255:1697–1699.

Raubeson, L.A. and R.K. Jansen. 1992b. A rare chloroplast-DNA structural mutation is shared by all conifers. Biochem. Syst. Ecol. 20:17–24.

Raven, P.H. 1979. A survey of reproductive biology in Onagraceae. N. Zeal. J. Bot. 17:575–593.

Raymond, M., A. Callaghan, P. Fort, and N. Pasteur. 1991. Worldwide migration of amplified resistance genes in mosquitoes. Nature 350:151–153.

Recipon, H., R. Perasso, A. Adoutte, and F. Quetier. 1992. ATP synthase subunit c/III/9 gene sequences as a tool for interkingdom and metaphytes molecular phylogenies. J. Mol. Evol. 34:292–303.

Reeb, C.A. and J.C. Avise. 1990. A genetic discontinuity in a continuously distributed species: mitochondrial DNA in the American oyster, *Crassostrea virginica*. Genetics 124:397–406.

Reeve, H.K., D.F. Westneat, W.A. Noon, P.W. Sherman, and C.F. Aquadro. 1990. DNA "fingerprinting" reveals high levels of inbreeding in colonies of the eusocial naked mole-rat. Proc. Natl. Acad. Sci. USA 87:2496–2500.

Regnery, R.L., C.L. Spruill, and B.D. Plikaytis. 1991. Genotypic identification of rickettsiae and estimation of intraspecies sequence divergence for portions of two rickettsial genes. J. Bacteriol. 173:1576–1589.

Resing, J.M. and D.J. Ayre. 1985. The usefulness of the tissue grafting bioassay as an indicator of clonal identity in scleractinian corals. Proc. Fifth Int. Coral Reef Congr. 6:75–81.

Ribble, D.O. 1991. The monogamous mating system of *Peromyscus californicus* as revealed by DNA fingerprinting. Behav. Ecol. Sociobiol. 29:161–166.

Richards, R.I., K. Holman, K. Friend, E. Kremer, D. Hillen, A. Staples, W.T. Brown, P. Goonewardena, J. Tarleton, C. Schwartz, and G.R. Sutherland. 1992. Evidence of founder chromosomes in fragile X syndrome. Nature Genetics 1:257–260.

Richman, A.D. and T. Price. 1992. Evolution of ecological differences in the Old World leaf warblers. Nature 355:817–821.

Richmond, R.C., D.G. Gilbert, K.B. Sheehan, M.H. Gromko, and F.M. Butterworth. 1980. Esterase 6 and reproduction in *Drosophila melanogaster*. Science 207:1483–1485.

Rick, C.M., E. Kesicki, J.F. Forbes, and M. Holle. 1976. Genetic and biosystematic studies of two new sibling species of *Lycopersicon* from interandean Peru. Theoret. Appl. Genet. 47:55–68.

Rick, C.M. and P.G. Smith. 1953. Novel variation in tomato species hybrids. Amer. Natur. 87: 359–373.

Ricklefs, R.E. 1980. Phyletic gradualism vs. punctuated equilibrium: applicability of neontological data. Paleobiology 6:271–275.

Riddle, B.R. and R.L. Honeycutt. 1990. Historical biogeography in North American arid regions: an approach using mitochondrial-DNA phylogeny in grasshopper mice (genus *Onychomys*). Evolution 44:1–15.

Rieseberg, L.H. 1991. Homoploid reticulate evolution in *Helianthus* (Asteraceae): evidence from ribosomal genes. Amer. J. Bot. 78:1218–1237.

Rieseberg, L.H., S. Beckstrom-Sternberg, and K. Doan. 1990b. *Helianthus annuus* spp. *texanum* has chloroplast DNA and nuclear ribosomal RNA genes of *Helianthus debilis* spp *cucumerifolius*. Proc. Natl. Acad. Sci. USA 87:593–597.

Rieseberg, L.H., S.M. Beckstrom-Sternberg, A. Liston, and D.M. Arias. 1991. Phylogenetic and systematic inferences from chloroplast DNA and isozyme variation in *Helianthus* sect. *Helianthus* (Asteraceae). Syst. Bot. 16:50–76.

Rieseberg, L.H. and S.J. Brunsfeld. 1992. Molecular evidence and plant introgression. Pp. 151–176 *in:* Molecular Systematics of Plants, P.S. Soltis, D.E. Soltis, and J.J. Doyle (eds.), Chapman and Hall, New York.

Rieseberg, L.H., R. Carter, and S. Zona. 1990a. Molecular tests of the hypothesized hybrid origin of two diploid *Helianthus* species (Asteraceae). Evolution 44:1498–1511.

Rieseberg, L.H. and M.F. Doyle. 1989. Tetrasomic segregation in the naturally occurring autotetraploid *Allium nevii* (Alliaceae). Hereditas 111:31–36.

Rieseberg, L.H., M.A. Hanson, and C.T. Philbrick. 1992. Androdioecy is derived from dioecy in Datiscaceae: evidence from restriction site mapping of PCR-amplified chloroplast DNA fragments. Syst. Bot. 17:324–326.

Rieseberg, L.H. and D.E. Soltis. 1991. Phylogenetic consequences of cytoplasmic gene flow in plants. Evol. Trends Plants 5:65–84.

Rieseberg, L.H., D.E. Soltis, and J.D. Palmer. 1988. A molecular reexamination of introgression between *Helianthus annuus* and *H. bolanderi* (Compositae). Evolution 42:227–238.

Rieseberg, L.H. and J.F. Wendel. 1993. Introgression and its consequences in plants. *In:* Hybrid Zones and the Evolutionary Process, R.G. Harrison (ed.), Oxford Univ. Press, Oxford, England, pp. 70–109.

Rieseberg, L.H., S. Zona, L. Aberbom, and T.D. Martin. 1989. Hybridization in the island endemic, Catalina mahogany. Cons. Biol. 3:52–58.

Riley, M.A., M.E. Hallas, and R.C. Lewontin. 1989. Distinguishing the forces controlling genetic variation at the *Xdh* locus in *Drosophila pseudoobscura*. Genetics 123:359–369.

Rinderer, T.E., J.A. Stelzer, B.P. Oldroyd, S.M. Buco, and W.L. Rubink. 1991. Hybridization between European and Africanized honey bees in the neotropical Yucutan peninsula. Science 253:309–311.

Risch, N.J. and B. Devlin. 1992. On the probability of matching DNA fingerprints. Science 255:717–720.

Rising, J.D. and J.C. Avise. 1993. An application of genealogical concordance principles to the taxonomy and evolutionary history of the Sharp-tailed Sparrow (*Ammodramus caudacutus*). Auk, *in press.*

Ritland, K, and M.T. Clegg. 1987. Evolutionary analyses of plant DNA sequences. Amer. Natur. 130:S74–S100.

Ritland, K. and S.K. Jain. 1981. A model for the estimation of outcrossing rate and gene frequencies using m independent loci. Heredity 47:35–52.

Roberts, J.D. and L.R. Maxson. 1985. Tertiary speciation models in Australian anurans: molecular data challenge Pleistocene scenario. Evolution 39:325–334.

Roberts, J.R. 1984. Restriction and modification enzymes and their recognition sequences. Nucleic Acids Res. 12:r167–r204.

Roberts, L. 1991. Fight erupts over DNA fingerprinting. Science 254:1721–1723.

Robertson, A. 1975. Remarks on the Lewontin-Krakauer test. Genetics 80:396.

Rogers, J.S. 1972. Measures of genetic similarity and genetic distance. Studies in Genetics VII, Univ. Texas Publ. 7213:145–153.

Rogers, S.O., S. Honda, and A.J. Bendich. 1986. Variation in the ribosomal RNA genes among individuals of *Vicia faba*. Plant. Mol. Biol. 6:339–345.

Rogstad, S.H., H. Nybom, and B.A. Schaal. 1991. The tetrapod "DNA fingerprinting" M13 repeat probe reveals genetic diversity and clonal growth in quaking aspen (*Populus tremuloides*, Salicaceae). Plant Syst. Evol. 175:115–123.

Rogstad, S.H., J.C. Patton II, and B.A. Schaal. 1988. M13 repeat probe detects minisatellite-like sequences in gymnosperms and angiosperms. Proc. Natl. Acad. Sci. USA 85:9176–9178.

Rohlf, D.J. 1991. Six biological reasons why the endangered species act doesn't work—and what to do about it. Cons. Biol. 5:273–282.

Rollinson, D. 1986. Reproductive strategies of some species of *Bulinus*. Proc. 8th Malacological Congr. Budapest 1983:221–226.

Rollinson, D., R.A. Kane, and J.R.L. Lines. 1989. An analysis of fertilization in *Bulinus cernicus* (Gastropoda: Planorbidae). J. Zool. 217:295–310.

Roose, M.L. and L.D. Gottlieb. 1976. Genetic and biochemical consequences of polyploidy in *Tragopogon*. Evolution 30:818–830.

Rose, M.R. and W.F. Doolittle. 1983. Molecular biological mechanisms of speciation. Science 220:157–162.

Rosen, D.E. 1978. Vicariant patterns and historical explanation in biogeography. Syst. Zool. 27: 159–188.

Rosenblatt, R.H. and R.S. Waples. 1986. A genetic comparison of allopatric populations of shore fish species from the eastern and central Pacific Ocean: dispersal or vicariance? Copeia 1986: 275–284.

Ross, K.G. and J.M. Carpenter. 1991. Population genetic structure, relatedness, and breeding systems. Pp. 451–479 *in:* The Social Biology of Wasps, K.G. Ross and R.W. Matthews (eds.), Comstock Publ., Ithaca, NY.

Ross, K.G. and D.J.C. Fletcher. 1985. Comparative study of genetic and social structure in two forms of the fire ant *Solenopsis invicta* (Hymenoptera: Formicidae). Behav. Ecol. Sociobiol. 17:349–356.

Ross, K.G. and R.W. Matthews. 1989a. Population genetic structure and social evolution in the sphecid wasp *Microstigmus comes*. Amer. Natur. 134:574–598.

Ross, K.G. and R.W. Matthews. 1989b. New evidence for eusociality in the sphecid wasp *Microstigmus comes*. Anim. Behav. 38:613–619.

Rowan, R. and D.A. Powers. 1991. A molecular genetic classification of zooxanthellae and the evolution of animal-algal symbioses. Science 251:1348–1351.

Rowan, R. and D.A. Powers. 1992. Ribosomal RNA sequences and the diversity of symbiotic dinoflagellates (zooxanthellae). Proc. Natl. Acad. Sci. USA 89:3639–3643.

Roy, B.A. and L.H. Rieseberg. 1989. Evidence for apomixis in *Arabis*. J. Heredity 80:506–508.

Ruano, G., K.H. Kidd, and J.C. Stephens. 1990. Haplotype of multiple polymorphisms resolved by enzymatic amplification of single DNA molecules. Proc. Natl. Acad. Sci. USA 87:6296–6300.

Rudloe, A. 1979. *Limulus polyphemus:* a review of the ecologically significant literature. Pp. 27–35 *in:* Biomedical Applications of the Horseshoe Crab (Limulidae), E. Cohen (ed.), Alan R. Liss, New York.

Ryman, N. 1983. Patterns of distribution of biochemical genetic variation in salmonids: differences between species. Aquaculture 33:1–21.

Ryman, N. and F. Utter. 1987. Population Genetics and Fishery Management. Univ. Washington Press, Seattle.

Ryskov, A.P., A.G. Jincharadze, M.I. Prosnyak, P.L. Ivanov, and S.A. Limborska. 1988. M13 phage DNA as a universal marker for DNA fingerprinting of animals, plants and microorganisms. FEBS Letters 233:388–392.

Sage, R.D., D. Heyneman, K.-C. Lim, and A.C. Wilson. 1986. Wormy mice in a hybrid zone. Nature 324:60–63.

Sage, R.D., P.V. Loiselle, P. Basasibwaki, and A.C. Wilson. 1984. Molecular versus morphological change among cichlid fishes of Lake Victoria. Pp. 185–201 *in:* Evolution of Fish Species Flocks, A.A. Echelle and I. Kornfield (eds.), Univ. Maine Press, Orono.

Sage, R.D. and R.K. Selander. 1975. Trophic radiation through polymorphism in cichlid fishes. Proc. Natl. Acad. Sci. USA 72:4669–4673.

Sage, R.D. and J.O. Wolff. 1986. Pleistocene glaciations, fluctuating ranges, and low genetic variability in a large mammal (*Ovis dalli*). Evolution 40:1092–1095.

Saghai-Maroof, M.A., K.M. Soliman, R.A. Jorgensen, and R.W. Allard. 1984. Ribosomal DNA spacer-length polymorphisms in barley: Mendelian inheritance, chromosomal location, and population dynamics. Proc. Natl. Acad. Sci. USA 81:8014–8018.

Saiki, R.K., D.H. Gelfand, S. Stoffel, S.J. Scharf, R. Higuchi, G.T. Horn, K.B. Mullis, and H.A. Erlich. 1988. Primer-directed enzymatic amplification of DNA with a thermostabile DNA polymerase. Science 239:487–491.

Saiki, R.K., S. Scharf, F. Faloona, K.B. Mullis, G.T. Horn, H.A. Erlich, and N. Arnheim. 1985. Enzymatic amplification of B-globin genomic sequences and restriction site analysis for diagnosis of sickle cell anemia. Science 230:1350–1354.

Saitou, N. and M. Nei. 1986. The number of nucleotides required to determine the branching order of three species, with special reference to the human-chimpanzee-gorilla divergence. J. Mol. Evol. 24:189–204.

Saitou, N. and M. Nei. 1987. The neighbor-joining method: a new method for reconstructing phylogenetic trees. Mol. Biol. Evol. 4:406–425.

Salama, M., W. Sandine, and S. Giovannoni. 1991. Development and application of oligonucleotide probes for identification of *Lactococcus lactis* subsp. *cremoris*. Appl. Environ. Microbiol. 57: 1313–1318.

Sambrook, J., E.F. Fritsch, and T. Maniatis. 1989. Molecular Cloning. 2nd Ed., Cold Spring Harbor Lab Press, Cold Spring Harbor, NY.

Sampsell, B. and S. Sims. 1982. Effect of *adh* genotype and heat stress on alcohol tolerance in *Drosophila melanogaster*. Nature 296:853–855.

Sanger, F., S. Nicklen, and A.R. Coulson. 1977. DNA sequencing with chain-terminating inhibitors. Proc. Natl. Acad. Sci. USA 74:5463–5467.

Sankoff, D., G. Leduc, N. Antoine, B. Paquin, B.F. Lang, and R. Cedergren. 1992. Gene order comparisons for phylogenetic inference: Evolution of the mitochondrial genome. Proc. Natl. Acad. Sci. USA 89:6575–6579.

Saperstein, D.A. and J.M. Nickerson. 1991. Restriction fragment length polymorphism analysis using PCR coupled to restriction digests. BioTechniques 10:488–489.

Sarich, V.M. 1973. The giant panda is a bear. Nature 245:218–220.

Sarich, V.M., C.W. Schmid, and J. Marks. 1989. DNA hybridization as a guide to phylogenies: A critical analysis. Cladistics 5:3–32.

Sarich, V.M. and A.C. Wilson. 1966. Quantitative immunochemistry and the evolution of primate albumins: Micro-complement fixation. Science 154:1563–1566.

Sarich, V.M. and A.C. Wilson. 1967. Immunological time scale for hominid evolution. Science 158:1200–1203.

Sarich, V.M. and A.C. Wilson. 1973. Generation time and genomic evolution in primates. Science 179:1144–1147.

Sassaman, C. 1978. Mating systems in porcellionid isopods: Multiple paternity and sperm mixing in *Porcellio scaber* Latr. Heredity 41:385–397.

Saunders, N.C., L.G. Kessler, and J.C. Avise. 1986. Genetic variation and geographic differentiation in mitochondrial DNA of the horseshoe crab, *Limulus polyphemus*. Genetics 112:613–627.

Scanlan, B.E., L.R. Maxson, and W.E. Duellman. 1980. Albumin evolution in marsupial frogs (Hylidae: *Gastrotheca*). Evolution 34:222–229.

Schaal, B.A. 1980. Measurement of gene flow in *Lupinus texensis*. Nature 284:450–451.

Schaal, B.A. 1985. Genetic variation in plant populations: from demography to DNA. Pp. 321–342 *in:* Structure and Functioning of Plant Populations (J. Haeck and J. Woldendorp, eds.), North-Holland, Amsterdam.

Schaal, B.A., W.J. Leverich, and J. Nieto-Sotelo. 1987. Ribosomal DNA variation in the native plant *Phlox divaricata*. Mol. Biol. Evol. 4:611–621.

Schaal, B.A., S.L. O'Kane, Jr., and S.H. Rogstad. 1991. DNA variation in plant populations. Trends Ecol. Evol. 6:329–332.

Schaeffer, S.W., C.F. Aquadro, and W.W. Anderson. 1987. Restriction-map variation in the alcohol dehydrogenase region of *Drosophila pseudoobscura*. Mol. Biol. Evol. 4:254–265.

Schaeffer, S.W., C.F. Aquadro, and C.H. Langley. 1988. Restriction-map variation in the *Notch* region of *Drosophila melanogaster*. Mol. Biol. Evol. 5:30–40.

Schaeffer, S.W. and E.L. Miller. 1992. Molecular population genetics of an electrophoretically monomorphic protein in the alcohol dehydrogenase region of *Drosophila pseudoobscura*. Genetics 132:163–178.

Schaller, G.B. 1972. The Serengeti Lion. Univ. Chicago Press, Chicago. IL.

Scharf, S.J., G.T. Horn, and H.A. Erlich. 1986. Direct cloning and sequence analysis of enzymatically amplified genomic sequences. Science 233:1076–1078.

Scheller, G., M.T. Conkle, and L. Griswald. 1985. Local differentiation among Mediterranean populations of Aleppo pine in their isozymes. Silv. Genet. 35:11–19.

Scheltema, R.S. 1986. On dispersal and planktonic larvae of benthic invertebrates: an eclectic overview and summary of problems. Bull. Mar. Sci. 39:290–322.

Schemske, D.W. and R. Lande. 1985. The evolution of self-fertilization and inbreeding depression in plants. II. Empirical observations. Evolution 39:41–52.

Schlötterer, C., B. Amos, and D. Tautz. 1991. Conservation of polymorphic simple sequence loci in cetacean species. Nature 354:63–65.

Schmidt, T.M., E.F. DeLong, and N.R. Pace. 1991. Analysis of a marine picoplankton community by 16S rRNA gene cloning and sequencing. J. Bacteriol. 173:4371–4378.

Schnabel, A. and M.A. Asmussen. 1989. Definition and properties of disequilibria within nuclear-mitochondrial-chloroplast and other nuclear-dicytoplasmic systems. Genetics 123:199–215.

Schoen, D.J. 1988. Mating system estimation via the one pollen parent model with the progeny array as the unit of observation. Heredity 60:439–444.

Schoen, D.J. and A.H.D. Brown. 1991. Intraspecific variation in population gene diversity and effective population size correlates with the mating system in plants. Proc. Natl. Acad. Sci. USA 88:4494–4497.

Schoen, D.J. and M.T. Clegg. 1984. Estimation of mating system parameters when outcrossing events are correlated. Proc. Natl. Acad. Sci. USA 81:5258–5262.

Schultz, R.J. 1969. Hybridization, unisexuality and polyploidy in the teleost *Poeciliopsis* (Poeciliidae) and other vertebrates. Amer. Natur. 103:605–619.

Schultz, R.J. 1973. Unisexual fish: laboratory synthesis of a "species." Science 179:180–181.

Schuster, W.S. and J.B. Mitton. 1991. Relatedness within clusters of a bird-dispersed pine and the potential for kin interactions. Heredity 67:41–48.

Schwagmeyer, P.L. 1988. Scramble-competition polygyny in an asocial mammal: male mobility and mating success. Amer. Natur. 131:885–892.

Schwartz, F.J. 1972. World Literature to Fish Hybrids, with an Analysis by Family, Species, and Hybrid. Publ. Gulf Coast Res. Lab. Mus. 3:1–328.

Schwartz, F.J. 1981. World Literature to Fish Hybrids, with an Analysis by Family, Species, and Hybrid: Supplement I. NOAA Tech. Report NMFS SSRF-750, U.S. Dept. of Commerce.

Schwartz, M.P. 1987. Intra-colony relatedness and sociality in the allodapine bee *Exoneura bicolor*. Behav. Ecol. Sociobiol. 21:387–392.

Schwartz, R.M. and M.O. Dayhoff. 1978. Origins of prokaryotes, eukaryotes, mitochondria, and chloroplasts. Science 199:395–403.

Scribner, K.T. 1991. Heterozygosity as an indicator of fitness and historical population demography. Pp. 77–84 *in:* Genetics and Wildlife Conservation, E. Randi (ed.), Supplemento alle Richerche di Biologia della Selvaggina Vol. XVIII, Bologna, Italy.

Scribner, K.T. and J.C. Avise. 1993a. Cytonuclear genetic architecture in mosquitofish populations, and the possible roles of introgressive hybridization. Mol. Ecol., *in press*.

Scribner, K.T. and J.C. Avise. 1993b. "Population cage" experiments with a vertebrate: the temporal demography and cytonuclear genetics of hybridization in *Gambusia* fishes. Evolution, *in press*.

Scribner, K.T. and M.H. Smith. 1990. Genetic variability and antler development. Pp. 457–469 *in:* G.A. Bubenik and B. Bubenik (eds.), Pronghorns, Horns, and Antlers, Springer-Verlag, New York.

Sebens, K.P. 1984. Agonistic behavior in the intertidal sea anemone *Anthopleura xanthogrammica*. Biol. Bull. 166:457–472.

Seeb, J.E., G.H. Kruse, L.W. Seeb, and R.G. Weck. 1990. Genetic structure of red king crab populations in Alaska facilitates enforcement of fishing regulations. Alaska Sea Grant College Program Report 90-04:91–502.

Selander, R.K. 1970. Behavior and genetic variation in natural populations. Amer. Zool. 10:53–66.

Selander, R.K. 1975. Stochastic factors in the genetic structure of populations. Pp. 284–332 *in:* Proceedings of the Eighth International Conference on Numerical Taxonomy, G.F. Estabrook (ed.), W.H. Freeman, San Francisco.

Selander, R.K. 1976. Genic variation in natural populations. Pp. 21–45 *in:* Molecular Evolution, F.J. Ayala (ed.), Sinauer, Sunderland, MA.

Selander, R.K. 1982. Phylogeny. Pp. 32–59 in: Perspectives on Evolution, R. Milkman (ed.), Sinauer, Sunderland, MA.

Selander, R.K., P. Beltran, and N.H. Smith. 1991. Evolutionary genetics of *Salmonella*. Pp. 25–57 *in:* Evolution at the Molecular Level, R.K. Selander, A.G. Clark, and T.S. Whittam (eds.), Sinauer, Sunderland, MA.

Selander, R.K., D.A. Caugant, and T.S. Whittam. 1987a. Genetic structure and variation in natural populations of *Escherichia coli*. Pp. 1625–1648 *in: Escherichia coli* and *Salmonella typhimurium* Cellular and Molecular Biology, Vol. 2, F.C. Neidhardt et al. (eds.), American Society for Microbiology, Washington, D.C.

Selander, R.K. and R.O. Hudson. 1976. Animal population structure under close inbreeding: the land snail *Rumina* in southern France. Amer. Natur. 110:695–718.

Selander, R.K., W.G. Hunt, and S.Y. Yang. 1969. Protein polymorphism and genic heterozygosity in two European subspecies of the house mouse. Evolution 23:379–390.

Selander, R.K. and D.W. Kaufman. 1973a. Genic variability and strategies of adaptation in animals. Proc. Natl. Acad. Sci. USA 70:1875–1877.

Selander, R.K. and D.W. Kaufman. 1973b. Self-fertilization and genetic population structure in a colonizing land snail. Proc. Natl. Acad. Sci. USA 70:1186–1190.

Selander, R.K. and D.W. Kaufman. 1975a. Genetic population structure and breeding systems. Isozymes IV:27–48.

Selander, R.K. and D.W. Kaufman. 1975b. Genetic structure of populations of the brown snail (*Helix aspersa*). I. Microgeographic variation. Evolution 29:385–401.

Selander, R.K., D.W. Kaufman, and R.S. Ralin. 1974. Self-fertilization in the terrestrial snail *Rumina decollata*. Veliger 16:265–270.

Selander, R.K. and B.R. Levin. 1980. Genetic diversity and structure in *Escherichia coli* populations. Science 210:545–547.

Selander, R.K., R.M. McKinney, T.S. Whittam, W.F. Bibb, D.J. Brenner, F.S. Nolte, and P.E. Pattison. 1985. Genetic structure of populations of *Legionella pneumophila*. J. Bacteriol. 163: 1021–1037.

Selander, R.K. and J.M. Musser. 1990. The population genetics of bacterial pathogenesis. Pp. 11–36 *in:* Molecular Basis of Bacterial Pathogenesis, B.H. Iglewski and V.L. Clark (eds.), Academic Press, Orlando, FL.

Selander, R.K., J.M. Musser, D.A. Caugant, M.N. Gilmour, and T.S. Whittam. 1987b. Population genetics of pathogenic bacteria. Microbial Pathogen. 3:1–7.

Selander, R.K., M.H. Smith, S.Y. Yang, W.E. Johnson, and J.B. Gentry. 1971. Biochemical polymorphism and systematics in the genus *Peromyscus*. II. Genic heterozygosity and genetic similarity among populations of the old-field mouse (*Peromyscus polionotus*). Studies in Genetics VI, Univ. Texas Publ. 7103:49–90.

Selander, R.K. and T.S. Whittam. 1983. Protein polymorphism and the genetic structure of populations. Pp. 89–114 *in:* Evolution of Genes and Proteins, M. Nei and R.K. Koehn (eds.), Sinauer, Sunderland, MA.

Sene, F.M. and H.L. Carson. 1977. Genetic variation in Hawaiian *Drosophila*. IV. Allozymic similarity between *D. sylvestris* and *D. heteroneura* from the island of Hawaii. Genetics 86:187–198.

Seutin, G., P.T. Boag, B.N. White, and L.M. Ratcliffe. 1991. Sequential polyandry in the common redpoll (*Carduelis flammea*). Auk 108:166–198.

478 *Literature Cited*

Shaklee, J.B. 1983. The utilization of isozymes as gene markers in fisheries management and conservation. Isozymes: Curr. Top. Biol. Med. Res. 11:213–247.

Shaklee, J.B. 1984. Genetic variation and population structure in the damselfish, *Stegastes fasciolatus,* throughout the Hawaiian archipelago. Copeia 1984:629–640.

Shaklee, J.B., K.L. Kepes, and G.S. Whitt. 1973. Specialized lactate dehydrogenase isozymes: The molecular and genetic basis for the unique eye and liver LDHs of teleost fishes. J. Exp. Zool. 185:217–240.

Shaklee, J.B., D.C. Klaybor, S. Young, and B.A. White. 1991. Genetic stock structure of odd-year pink salmon, *Oncorhynchus gorbuscha* (Walbaum), from Washington and British Columbia and potential mixed-stock fisheries applications. J. Fish Biol. 39A:21–34.

Shaklee, J.B., C.S. Tamaru, and R.S. Waples. 1982. Speciation and evolution of marine fishes studied by the electrophoretic analysis of proteins. Pacific Sci. 36:141–157.

Shapiro, D.Y. 1983. On the possibility of kin groups in coral reef fishes. Pp. 39–45 *in:* Ecology of Deep and Shallow Coral Reefs, M.L. Reaka (ed.), Coral Reef Symp. Ser. Undersea Res. Vol. 1, National Oceanographic and Atmospheric Administration, Washington, D.C.

Sharp, G.D. and S.W. Pirages. 1978. The distribution of red and white swimming muscles, their biochemistry, and the biochemical pathway of selected scombrid fishes. Pp. 41–78 *in:* G.D. Sharp and A.E. Dizon (eds.), The Physiological Ecology of Tunas, Academic Press, New York.

Sharp, P.M., A.T. Lloyd, and D.G. Higgins. 1991. Coelacanth's relationships. Nature 353:218–219.

Shatters, R.G. and M.L. Kahn. 1989. Glutamine synthetase II in *Rhizobium:* reexamination of the proposed horizontal transfer of DNA from eukaryotes to prokaryotes. J. Mol. Evol. 29:422–428.

Shaw, C.R. and R. Prasad. 1970. Starch gel electrophoresis of enzymes—a compilation of recipes. Biochem. Genet. 4:297–320.

Shaw, D.V., A.L. Kahler, and R.W. Allard. 1981. A multilocus estimator of the mating system parameters in plant populations. Proc. Natl. Acad. Sci. USA 78:1298–1302.

Sheldon, F.H. and A.H. Bledsoe. 1989. Indexes to the reassociation and stability of solution DNA hybrids. J. Mol. Evol. 29:328–343.

Sheppard, W.S., T.E. Rinderer, J.A. Mazzoli, J.A. Stelzer, and H. Shimanuki. 1991. Gene flow between African- and European-derived honey bee populations in Argentina. Nature 349:782–784.

Sheridan, M. and R.H. Tamarin. 1986. Kinships in a natural meadow vole population. Behav. Ecol. Sociobiol. 19:207–211.

Sherman, P.W., J.U.M. Jarvis, and R.D. Alexander (eds.). 1991. The Biology of the Naked Mole-Rat. Princeton Univ. Press, Princeton, NJ.

Sherman, P.W. and M.L. Morton. 1988. Extra-pair fertilizations in mountain white-crowned sparrows. Behav. Ecol. Sociobiol. 22:413–421.

Sherwin, W.B., N.D. Murray, J.A.M. Graves, and P.R. Brown. 1991. Measurement of genetic variation in endangered populations: bandicoots (Marsupialia: Peramelidae) as an example. Cons. Biol. 5:103–108.

Shields, G.F. and A.C. Wilson. 1987a. Calibration of mitochondrial DNA evolution in geese. J. Mol. Evol. 24:212–217.

Shields, G.F. and A.C. Wilson. 1987b. Subspecies of the Canada goose (*Branta canadensis*) have distinct mitochondrial DNAs. Evolution 41:662–666.

Shinozaki, K. and 22 others. 1986. The complete nucleotide sequence of the tobacco chloroplast genome: its gene organization and expression. The EMBO J. 5:2043–2049.

Shively, C. and D.G. Smith. 1985. Social status and reproductive success of male *Macaca fascicularis*. Amer. J. Primatol. 9:129–135.

Shows, T.B. 1983. Human genome organization of enzyme loci and metabolic diseases. Isozymes X:323–339.

Shykoff, J.A. and P. Schmid-Hempel. 1991a. Genetic relatedness and eusociality: parasite-mediated selection on the genetic composition of groups. Behav. Ecol. Sociobiol. 28:371–376.

Shykoff, J.A. and P. Schmid-Hempel. 1991b. Parasites and the advantage of genetic variability within social insect colonies. Proc. Roy. Soc. London B243:55–58.

Sibley, C.G. 1991. Phylogeny and classification of birds from DNA comparisons. Acta XX Congressus Internationalis Ornithologici, Vol. 1:111–126.

Sibley, C.G. and J.E. Ahlquist. 1981. The phylogeny and relationships of the ratite birds as indicated by DNA-DNA hybridization. Pp. 301–335 *in:* Evolution Today, G.G.E. Scudder and J.L. Reveal (eds.), Hunt Institute Botanical Document, Pittsburgh, PA.

Sibley, C.G. and J.E. Ahlquist. 1984. The phylogeny of the hominoid primates, as indicated by DNA-DNA hybridization. J. Mol. Evol. 20:2–15.

Sibley, C.G. and J.E. Ahlquist. 1986. Reconstructing bird phylogeny by comparing DNA's. Sci. Amer. 254(2):82–93.

Sibley, C.G. and J.E. Ahlquist. 1987. DNA hybridization evidence of hominoid phylogeny: results from an expanded data set. J. Mol. Evol. 26:99–121.

Sibley, C.G. and J.E. Ahlquist. 1990. Phylogeny and Classification of Birds. Yale Univ. Press, New Haven, CT.

Sibley, C.G., J.E. Ahlquist, and B.L. Monroe, Jr. 1988. A classification of the living birds of the world based on DNA-DNA hybridization studies. Auk 105:409–423.

Sibley, C.G., J.A. Comstock, and J.E. Ahlquist. 1990. DNA hybridization evidence of hominoid phylogeny: a reanalysis of the data. J. Mol. Evol. 30:202–236.

Sibley, L.D. and J.C. Boothroyd. 1992. Virulent strains of *Toxoplasma gondii* comprise a single clonal lineage. Nature 359:82–85.

Sidell, B.D., R.G. Otto, and D.A. Powers. 1978. A biochemical method for distinction of striped bass and white perch larvae. Copeia 1978:340–343.

Silander, J.A., Jr. 1985. Microevolution in clonal plants. Pp. 107–152 *in:* Population Biology and Evolution of Clonal Organisms, J.B.C. Jackson, L.W. Buss, and R.E. Cook (eds.), Yale Univ Press, New Haven, CT.

Silberman, J.D. and P.J. Walsh. 1992. Species identification of spiny lobster phyllosome larvae via ribosomal DNA analysis. Mol. Mar. Biol. Biotech. 1:195–205.

Simberloff, D. 1988. The contribution of population and community biology to conservation science. Annu. Rev. Ecol. Syst. 19:473–511.

Simberloff, D. and J. Cox. 1987. Consequences and costs of conservation corridors. Cons. Biol. 1:63–71.

Simberloff, D., J.A. Farr, J. Cox, and D.W. Mehlman. 1992. Movement corridors: conservation bargains or poor investments. Cons. Biol. 6:493–504.

Simmons, G.M. 1992. Horizontal transfer of *hobo* transposable elements within the *Drosophila melanogaster* species complex: evidence from DNA sequencing. Mol. Biol. Evol. 9:1050–1060.

Simon, C.M. 1979. Evolution of periodical cicadas: phylogenetic inferences based on allozyme data. Syst. Zool. 28:22–39.

Simonsen, V. and O. Frydenberg. 1972. Genetics of *Zoarces* populations II. Three loci determining esterase isozymes in eye and brain tissue. Hereditas 70:235–242.

Simpson, G.G. 1940. Mammals and land bridges. J. Washington Acad. Sci. 30:137–163.

Simpson, G.G. 1945. The principles of classification and a classification of mammals. Bull. Amer. Mus. Natur. Hist. 85:1–350.

Simpson, G.G. 1951. The species concept. Evolution 5:285–298.

Sinclair, A.H., P. Berta, M.S. Palmer, J.R. Hawkins, B.L. Griffiths, M.J. Smith, J.W. Foster, A.-M. Frischauf, R. Lovell-Badge, and P.N. Goodfellow. 1990. A gene from the human sex-determining region encodes a protein with homology to a conserved DNA-binding motif. Nature 346:240–244.

Singh, R.S. and L.R. Rhomberg. 1987. A comprehensive study of genic variation in natural populations of *Drosophila melanogaster*. I. Estimates of gene flow from rare alleles. Genetics 115: 313–322.

Singh, S.M. and E. Zouros. 1978. Genetic variability associated with growth rate in the American oyster (*Crassostrea virginica*). Evolution 32:342–353.

Sites, J.W., Jr., R.L. Bezy, and P. Thompson. 1986. Nonrandom heteropolymer expression of lactate dehydrogenase isozymes in the lizard family Xantusiidae. Biochem. Syst. Ecol. 14:539–545.

Sites, J.W., Jr. and I.F. Greenbaum. 1983. Chromosome evolution in the iguanid lizard *Sceloporus grammicus*. II. Allozyme variation. Evolution 37:54–65.

Sites, J.W. and C. Moritz. 1987. Chromosomal evolution and speciation revisited. Syst. Zool. 36:153–174.

Skaala, Ø. and G. Nævdal. 1989. Genetic differentiation between freshwater resident and anadromous brown trout, *Salmo trutta*, within watercourses. J. Fish Biol. 34:597–605.

Skrochowska, S. 1969. Migrations of the sea-trout (*Salmo trutta* L.), brown trout (*Salmo trutta* m. *fario* L.), and their crosses. Pol. Arch. Hydrobiol. 16:125–192.

Slatkin, M. 1985a. Rare alleles as indicators of gene flow. Evolution 39:53–65.

Slatkin, M. 1985b. Gene flow in natural populations. Annu. Rev. Ecol. Syst. 16:393–430.

Slatkin, M. 1987. Gene flow and the geographic structure of natural populations. Science 236:787–792.

Slatkin, M. and N.H. Barton. 1989. A comparison of three indirect methods for estimating average levels of gene flow. Evolution 43:1349–1368.

Slatkin, M. and R.R. Hudson. 1991. Pairwise comparisons of mitochondrial DNA sequences in stable and exponentially growing populations. Genetics 129:555–562.

Slatkin, M. and W.P. Maddison. 1989. A cladistic measure of gene flow inferred from the phylogenies of alleles. Genetics 123:603–613.

Small, M.F. and D.G. Smith. 1982. The relationship between maternal and paternal rank in rhesus macaques (*Macaca mulatta*). Anim. Behav. 30:626–633.

Smith, A.B. 1992. Echinoderm phylogeny: morphology and molecules approach accord. Trends Ecol. Evol. 7:224–239.

Smith, D.G. 1981. The association between rank and reproductive success in male rhesus monkeys. Amer. J. Primatol. 1:83–90.

Smith, D.G. and S. Smith. 1988. Parental rank and reproductive success of natal rhesus males. Anim. Behav. 36:554–562.

Smith, D.R., O.R. Taylor, and W.M. Brown. 1989. Neotropical Africanized honey bees have African mitochondrial DNA. Nature 339:213–215.

Smith, E.F.G., P. Arctander, J. Fjeldsa, and O.G. Amir. 1991. A new species of shrike (Laniidae: *Laniarius*) from Somalia, verified by DNA sequence data from the only known individual. Ibis 133:227–235.

Smith, H.G., R. Montgomerie, T. Poldmaa, B.N. White, and P.T. Boag. 1991. DNA fingerprinting reveals relation between tail ornaments and cuckoldry in barn swallows, *Hirundo rustica*. Behav. Ecol. 2:90–98.

Smith, M.F., W.K. Thomas, and J.L. Patton. 1992. Mitochondrial DNA-like sequence in the nuclear genome of an akodontine rodent. Mol. Biol. Evol. 9:204–215.

Smith, M.H., C.T. Garten, Jr., and P.R. Ramsey. 1975. Genic heterozygosity and population dynamics in small mammals. Isozymes IV:85–102.

Smith, M.H., K.T. Scribner, J.D. Hernandez, and M.C. Wooten. 1989. Demographic, spatial, and temporal genetic variation in *Gambusia*. Pp. 235–257 *in:* Ecology and Evolution of Livebearing Fishes (Poeciliidae), G.K. Meffe and F.F. Snelson, Jr. (eds.), Prentice Hall, Englewood Cliffs, N.J.

Smith, M.J., A. Arndt, S. Gorski, and E. Fajber. 1993. The phylogeny of echinoderm classes based on mitochondrial gene arrangements. J. Mol. Evol. 36:545–554.

Smith, M.J., D.K. Banfield, K. Doteval, S. Gorski, and D.J. Kowbel. 1989. Gene arrangement in sea star mitochondrial DNA demonstrates a major inversion event during echinoderm evolution. Gene 76:181–185.

Smith, M.J., D.K. Banfield, K. Doteval, S. Gorski, and D.J. Kowbel. 1990. Nucleotide sequence of nine protein-coding genes and 22 tRNAs in the mitochondrial DNA of the sea star *Pisaster ochraceus*. J. Mol. Evol. 31:195–204.

Smith, M.L., J.N. Bruhn, and J.B. Anderson. 1992. The fungus *Armillaria bulbosa* is among the largest and oldest living organisms. Nature 356:428–431.

Smith, M.L., L.C. Duchesne, J.N. Bruhn, and J.B. Anderson. 1990. Mitochondrial genetics in a natural population of the plant pathogen *Armillaria*. Genetics 126:575–582.

Smith, M.W. and R.F. Doolittle. 1992a. A comparison of evolutionary rates of the two major kinds of superoxide dismutase. J. Mol. Evol. 34:175–184.

Smith, M.W. and R.F. Doolittle. 1992b. Anomalous phylogeny involving the enzyme glucose-6-phosphate isomerase. J. Mol. Evol. 34:544–545.

Smith, M.W., D.-F. Feng, and R.F. Doolittle. 1992. Evolution by acquisition: the case for horizontal gene transfers. Trends Biochem. Sci. 17:489–493.

Smith, P.J. 1986. Genetic similarity between samples of the orange roughy *Hoplostethus atlanticus* from the Tasman Sea, south-west Pacific Ocean and north-east Atlantic Ocean. Mar. Biol. 91: 173–180.

Smith, P.J. and P.G. Benson. 1980. Electrophoretic identification of larval and O-group flounders (*Rhombosolea* spp.) from Wellington Harbour, N.Z. J. Mar. Freshwater Res. 14:401–404.

Smith, P.J. and J. Crossland. 1977. Identification of larvae of snapper, *Chrysophrys auratus* Forster, by electrophoretic separation of tissue enzymes. N.Z. Mar. Freshwater Res. 11:795–798.

Smith, P.J. and Y. Fujio. 1982. Genetic variation in marine teleosts: high variability in habitat specialists and low variability in habitat generalists. Mar. Biol. 69:7–20.

Smith, R.L. (ed.). 1984. Sperm Competition and the Evolution of Animal Mating Systems. Academic Press, New York.

Smith, R.L. and K.J. Sytsma. 1990. Evolution of *Populus nigra* L. (sect. *Aigeiros*): introgressive hybridization and the chloroplast contribution of *Populus alba* L. (sect. *Populus*). Amer. J. Bot. 77:1176–1187.

Smith, S.C., R.R. Racine, and C.H. Langley. 1980. Lack of genic variation in the abundant proteins of human kidney. Genetics 96:967–974.

Smithies, O. 1955. Zone electrophoresis in starch gels: group variations in the serum proteins of normal individuals. Biochem. J. 61:629–641.

Smithies, O. and P.A. Powers. 1986. Gene conversions and their relation to homologous chromosome pairing. Phil. Trans. Roy. Soc. Lond. B312:291–302.

Smouse, 1986. The fitness consequences of multiple-locus heterozygosity under the inbreeding and multiplicative overdominance and inbreeding depression models. Evolution 40:946–957.

Smyth, C.A. and J.L. Hamrick. 1987. Realized gene flow via pollen in artificial populations of the musk thistle, *Carduus nutans* L. Evolution 41:613–619.

Sneath, P.H.A. and R.R. Sokal. 1973. Numerical Taxonomy. W.H. Freeman and Co., San Francisco.

Snow, A.A. 1990. Effects of pollen-load size and number of donors on sporophyte fitness in wild radish (*Raphanus raphanistrum*). Amer. Natur. 136:742–758.

Snow, A.A. and T.P. Spira. 1991. Differential pollen-tube growth rates and nonrandom fertilization in *Hibiscus moscheutos* (Malvaceae). Amer. J. Bot. 78:1419–1426.

Sober, E. 1983. Parsimony in systematics: philosophical issues. Annu. Rev. Ecol. Syst. 14:335–357.

Sogin, M.L. 1991. Early evolution and the origin of eukaryotes. Curr. Opinion Genet. Develop. 1:457–463.

Sogin, M.L., U. Edman, and H. Elwood. 1989. A single kingdom of eukaryotes. Pp. 133–143 *in:* The Hierarchy of Life, B. Fernholm, K. Bremer, and H. Jornvall (eds.), Elsevier, Amsterdam.

Sokal, R.R. and N.L. Oden. 1978a. Spatial autocorrelation in biology. 1. Methodology. Biol. J. Linn. Soc. 10:199–228.

Sokal, R.R. and N.L. Oden. 1978b. Spatial autocorrelation in biology. 2. Some biological implications and four applications of evolutionary and ecological interest. Biol. J. Linn. Soc. 10:229–249.

Sokal, R.R. and P.H.A. Sneath. 1963. Principles of Numerical Taxonomy. W.H. Freeman and Co., San Francisco.

Solé-Cava, A.M. and J.P. Thorpe. 1986. Genetic differentiation between morphotypes of the marine sponge *Suberites ficus* (Demospongiae: Hadromerida). Mar. Biol. 93:247–253.

Solé-Cava, A.M. and J.P. Thorpe. 1989. Biochemical genetics of genetic variation in marine lower invertebrates. Biochem. Genet. 27:303–312.

Solignac, M., J. Génermont, M. Monnerot, and J.-C. Mounolou. 1984. Genetics of mitochondria in *Drosophila:* mtDNA inheritance in heteroplasmic strains of *D. mauritiana*. Mol. Gen. Genet. 197:183–188.

Solignac, M., J. Génermont, M. Monnerot, and J.-C. Mounolou. 1987. *Drosophila* mitochondrial genetics: evolution of heteroplasmy through germ line cell divisions. Genetics 117:687–696.

Solignac, M. and M. Monnerot. 1986. Race formation, speciation, and introgression within *Drosophila simulans, D. mauritiana,* and *D. sechellia* inferred from mitochondrial DNA analysis. Evolution 40:531–539.

Soltis, D.E., P.S. Soltis, M.T. Clegg, and M. Durbin. 1990. *rbcL* sequence divergence and phylogenetic relationships in Saxifragaceae sensu lato. Proc. Natl. Acad. Sci. USA 87:4640–4644.

Soltis, D.E., P.S. Soltis, T.G. Collier, and M.L. Edgerton. 1991. Chloroplast DNA variation within and among genera of the *Heuchera* group: evidence for chloroplast capture and paraphyly. Amer. J. Bot. 78:1091–1112.

Soltis, D.E., P.S. Soltis, and B.G. Milligan. 1992. Intraspecific chloroplast DNA variation: systematic and phylogenetic implications. Pp. 117–150 *in:* Molecular Systematics of Plants, P.S. Soltis, D.E. Soltis, and J.J. Doyle (eds.), Chapman and Hall, New York.

Soltis, D.E., P.S. Soltis, and B.D. Ness. 1989. Chloroplast-DNA variation and multiple origins of autopolyploidy in *Heuchera micrantha* (Saxifragaceae). Evolution 43:650–656.

Soltis, P.S., D.E. Soltis, and J.J. Doyle. 1990. Molecular Systematics of Plants. Chapman and Hall, New York.

Soltis, P.S., D.E. Soltis, and L.D. Gottlieb. 1987. Phosphoglucomutase gene duplications in *Clarkia* (Onagraceae) and their phylogenetic implications. Evolution 41:667–671.

Soltis, P.S., D.E. Soltis, and C.J. Smiley. 1992a. An *rbcL* sequence from a Miocene *Taxodium* (bald cypress). Proc. Natl. Acad. Sci. USA 89:449–451.

Soltis, P.S., D.E. Soltis, T.L. Tucker, and F.A. Lang. 1992b. Allozyme variability is absent in the narrow endemic *Bensoniella oregona* (Saxifragaceae). Cons. Biol. 6:131–134.

Soulé, M. 1976. Allozyme variation: its determinants in space and time. Pp. 60–77 *in:* Molecular Evolution, F.J. Ayala (ed.), Sinauer, Sunderland, MA.

Soulé, M. 1980. Thresholds for survival: Maintaining fitness and evolutionary potential. Pp. 111–124 *in:* Conservation Biology: An Evolutionary-Ecological Perspective, M.E. Soulé and B.A. Wilcox (eds.), Sinauer, Sunderland, MA.

Soulé, M. and B.R. Stewart. 1970. The "niche-variation" hypothesis: a test and alternatives. Amer. Natur. 104:85–97.

Soumalainen, E., A. Saura, and J. Lokki. 1976. Evolution of parthenogenetic insects. Evol. Biol. 9:209–257.

Sourdis, J. and C. Krimbas. 1987. Accuracy of phylogenetic trees estimated from DNA sequence data. Mol. Biol. Evol. 4:159–166.

Southern, E.M. 1975. Detection of specific sequences among DNA fragments separated by gel electrophoresis. J. Mol. Biol. 98:503–517.

Southern, S.O., P.J. Southern, and A.E. Dizon. 1988. Molecular characterization of a cloned dolphin mitochondrial genome. J. Mol. Evol. 28:32–42.

Sparrow, A.H., H.J. Price, and A.G. Underbrink. 1972. A survey of DNA content per cell and per chromosome of prokaryotic and eukaryotic organisms: Some evolutionary considerations. Brookhaven Symp. Biol. 23:451–494.

Spiess, E.B. and C.M. Wilke. 1984. Still another attempt to achieve assortative mating by disruptive selection in *Drosophila*. Evolution 38:505–515.

Spilliaert, R., G. Vikingsson, Ú. Árnason, Á. Pálsdóttir, J. Sigurjónsson, and A. Árnason. 1991. Species hybridization between a female blue whale (*Balaenoptera musculus*) and a male fin whale (*B. physalus*): Molecular and morphological documentation. J. Heredity 82:269–274.

Spolsky, C.M., C.A. Phillips, and T. Uzzell. 1992. Antiquity of clonal salamander lineages revealed by mitochondrial DNA. Nature 356:706–708.

Spolsky, C.M. and T. Uzzell. 1984. Natural interspecies transfer of mitochondrial DNA in amphibians. Proc. Natl. Acad. Sci. USA 81:5802–5805.

Spolsky, C.M. and T. Uzzell. 1986. Evolutionary history of the hybridogenetic hybrid frog *Rana esculenta* as deduced from mtDNA analyses. Mol. Biol. Evol. 3:44–56.

Sprague, G.F., Jr. 1991. Genetic exchange between kingdoms. Curr. Opin. Genet. Dev. 1:530–533.

Springer, M., E.H. Davidson, and R.J. Britten. 1992. Calculation of sequence divergence from the thermal stability of DNA duplexes. J. Mol. Evol. 34:379–382.

Springer, M. and J.A.W. Kirsch. 1989. Rates of single-copy DNA evolution in phalangeriform marsupials. Mol. Biol. Evol. 6:331–341.

Springer, M. and J.A.W. Kirsch. 1991. DNA hybridization, the compression effect, and the radiation of diprotodontian marsupials. Syst. Zool. 40:131–151.

Springer, M., J.A.W. Kirsch, K. Aplin, and T. Flannery. 1990. DNA hybridization, cladistics, and the phylogeny of phalangerid marsupials. J. Mol. Evol. 30:298–311.

Springer, M. and C. Krajewski. 1989. DNA hybridization in animal taxonomy: a critique from first principles. Quart. Rev. Biol. 64:291–318.

Ståhl, G. 1987. Genetic population structure of Atlantic salmon. Pp. 121–140 *in:* Population Genetics and Fisheries Management, N. Ryman and F. Utter (eds.), Univ. Washington Press, Seattle.

Stallings, R.L., A.F. Ford, D. Nelson, D.C. Torney, C.E. Hildebrand, and R.K. Moyzis. 1991. Evolution and distribution of (GT)n repetitive sequences in mammalian genomes. Genomics 10:807–815.

Stangel, P.W., M.R. Lennartz, and M.H. Smith. 1992. Genetic variation and population structure of red-cockaded woodpeckers. Cons. Biol. 6:283–292.

Stangel, P.W., J.A. Rodgers, Jr., and A.L. Bryan. 1990. Genetic variation and population structure of the Florida wood stork. Auk 107:614–619.

Stanley, S.M. 1975. A theory of evolution above the species level. Proc. Natl. Acad. Sci. USA 72:646–650.

Stapel, S.O., J.A.M. Leunissen, M. Versteeg, J. Wattel, and W.W. deJong. 1984. Ratites as oldest offshoot of avian stem—evidence from α-crystallin A sequences. Nature 311:257–259.

Stebbins, G.L. 1950. Variation and Evolution in Plants. Columbia Univ. Press, New York.

Stebbins, G.L. 1970. Adaptive radiation in angiosperms. I. Pollination mechanisms. Annu. Rev. Ecol. Syst. 1:307–326.

Stebbins, G.L. 1989. Plant speciation and the founder principle. Pp. 113–125 *in:* Genetics, Speciation and the Founder Principle, L.V. Giddings, K.Y. Kaneshiro, and W.W. Anderson (eds.), Oxford Univ. Press, New York.

Steffens, G.J., J.V. Bannister, W.H. Bannister, L. Flohe, W.A. Gunzler, S.M. Kim, and F. Otting. 1983. The primary structure of Cu-Zn superoxide dismutase from *Photobacterium leiognathi:* evidence for a separate evolution of Cu-Zn superoxide dismutase in bacteria. Hoppe-Seyler's Z. Physiol. Chem. 364:675–690.

Stephan, W. and C.H. Langley. 1992. Evolutionary consequences of DNA mismatch inhibited repair opportunity. Genetics 132:567–574.

Stephens, J.C. 1985. Statistical methods of DNA sequence analysis: detection of intragenic recombination or gene conversion. Mol. Biol. Evol. 2:539–556.

Stephens, J.C., M.L. Cavanaugh, M.I. Gradie, M.L. Mador, and K.K. Kidd. 1990a. Mapping the human genome: current status. Science 250:237–244.

Stephens, J.C. and M. Nei. 1985. Phylogenetic analysis of polymorphic DNA sequences at the Adh locus in *Drosophila melanogaster* and its sibling species. J. Mol. Evol. 22:289–300.

Stephens, J.C., J. Rogers, and G. Ruano. 1990b. Theoretical underpinning of the single-molecule-dilution (SMD) method of direct haplotype resolution. Amer. J. Human Genet. 46:1149–1155.

Stern, B.R. and D.G. Smith. 1984. Sexual behaviour and paternity in three captive groups of rhesus monkeys (*Macaca mulatta*). Anim. Behav. 32:23–32.

Stern, D.B. and J.D. Palmer. 1984. Extensive and widespread homologies between mitochondrial DNA and chloroplast DNA in plants. Proc. Natl. Acad. Sci. USA 81:1946–1950.

Stevens, P.F. 1980. Evolutionary polarity of character states. Annu. Rev. Ecol. Syst. 11:333–358.

Stille, M., B. Stille, and P. Douwes. 1991. Polygyny, relatedness and nest founding in the polygynous myrmecine ant *Leptothorax acervorum* (Hymenoptera; Formicidae). Behav. Ecol. Sociobiol. 28:91–96.

Stine, G.J. 1989. The New Human Genetics. Wm. C. Brown. Publ., Dubuque, Iowa.

Stock, D.W., K.D. Moberg, L.R. Maxson, and G.S. Whitt. 1991. A phylogenetic analysis of the 18S ribosomal RNA sequence of the coelacanth *Latimeria chalumnae*. Env. Biol. Fishes 32:99–117.

Stock, D.W. and D.L. Swofford. 1991. Coelacanth's relationships. Nature 353:217–218.

Stock, D.W. and G.S. Whitt. 1992. Evidence from 18S ribosomal RNA sequences that lampreys and hagfishes form a natural group. Science 257:787–789.

Stoddart, J.A. 1983a. Asexual production of planulae in the coral *Pocillopora damicornis*. Mar. Biol. 76:279–288.

Stoddart, J.A. 1983b. The accumulation of genetic variation in a parthenogenetic snail. Evolution 37:546–554.

Stoddart, J.A. 1984a. Genetic structure within populations of the coral *Pocillopora damicornis*. Mar. Biol. 81:19–30.

Stoddart, J.A. 1984b. Genetic differentiation amongst populations of the coral *Pocillopora damicornis* off southwestern Australia. Coral Reefs 3:149–156.

Stoddart, J.A., R.C. Babcock, and A.J. Heyward. 1988. Self-fertilization and maternal enzymes in the planulae of the coral *Goniastrea favulus*. Mar. Biol. 99:489–494.

Stoneking, M., K. Bhatia, and A.C. Wilson. 1986. Mitochondrial DNA variation in eastern highlanders of Papua New Guinea. Pp. 87–100 *in:* Genetic Variation and Its Maintenance, D.F. Roberts and G.F. DeStefano (eds.), Cambridge Univ. Press, Cambridge, England.

Stoneking, M., D. Hedgecock, R.G. Higuchi, L. Vigilant, and H.A. Erlich. 1991. Population variation of human mtDNA control region sequences detected by enzymatic amplification and sequence-specific oligonucleotide probes. Am. J. Human Genet. 48:370–382.

Straney, D.O. 1981. The stream of heredity: genetics in the study of phylogeny. Pp. 100–138 *in:* Mammalian Population Genetics, M.H. Smith and J. Joule (eds.), Univ. of Georgia Press, Athens.

Strassmann, J.E., C.R. Hughes, D.C. Queller, S. Turillazzi, R. Cervo, S.K. Davis, and K.F. Goodnight. 1989. Genetic relatedness in primitively eusocial wasps. Nature 342:268–269.

Strauss, S.H., J.D. Palmer, G.T. Howe, and A.H. Doerksen. 1988. Chloroplast genomes of two conifers lack a large inverted repeat and are extensively rearranged. Proc. Natl. Acad. Sci. USA 85:3898–3902.

Stringer, C.B. and P. Andrews. 1988. Genetic and fossil evidence for the origin of modern humans. Science 239:1263–1268.

Studier, J.A. and K.J. Keppler. 1988. A note on the neighbor-joining algorithm of Saitou and Nei. Mol. Biol. Evol. 5:729–731.

Sturmbauer, C. and A. Meyer. 1992. Genetic divergence, speciation and morphological stasis in a lineage of African cichlid fishes. Nature 358:578–581.

Swift, C.C., C.R. Gilbert, S.A. Bortone, G.H. Burgess, and R.W. Yerger. 1985. Zoogeography of the southeastern United States: Savannah River to Lake Ponchartrain. Pp. 213–265 in: Zoogeography of the North American Freshwater Fishes, C.H. Hocutt and E.O Wiley (eds.), John Wiley and Sons, New York.

Swofford, D.L. 1981. On the utility of the distance Wagner procedure. Pp. 25–43 *in:* Advances in Cladistics. Proceedings of the First Meeting of the Willi Hennig Society, V.A. Funk and D.R. Brooks (eds.), New York Botanical Gardens, Bronx, NY.

Swofford, D.L, and S.H. Berlocher. 1987. Inferring evolutionary trees from gene frequency data under the principle of maximum parsimony. Syst. Zool. 36:293–325.

Swofford, D.L. and G.J. Olsen. 1990. Phylogeny reconstruction. Pp. 411–501 *in:* Molecular Systematics, D.M. Hillis and C. Moritz (eds.), Sinauer, Sunderland, MA.

Swofford, D.L. and R.K. Selander. 1981. BIOSYS-1: A FORTRAN program for the comprehensive analysis of electrophoretic data in population genetics and systematics. J. Heredity 72:281–283.

Syren, R.M. and P. Luykx. 1977. Permanent segmental interchange complex in the termite *Incisitermes schwarzi*. Nature 266:167–168.

Sytsma, K.J. and J.F. Smith. 1992. Molecular systematics of Onagraceae: examples from *Clarkia* and *Fuchsia*. Pp. 295–323 *in:* Molecular Systematics of Plants, P.S. Soltis, D.E. Soltis, and J.J. Doyle (eds.), Chapman and Hall, New York.

Szmidt, A.E., T. Alden, and J.-E. Hallgren. 1987. Paternal inheritance of chloroplast DNA in *Larix*. Plant Mol. Biol. 9:59–64.

Szymura, J.M., C. Spolsky, and T. Uzzell. 1985. Concordant change in mitochondrial and nuclear genes in a hybrid zone between two frog species (genus *Bombina*). Experientia 41:1469–1470.

Tabachnick, W.J., L.E. Munstermann, and J.R. Powell. 1979. Genetic distinctness of sympatric forms of *Aedes aegypti* in east Africa. Evolution 33:287–295.

Tabachnick, W.J. and J.R. Powell. 1978. Genetic structure of the East African domestic population of *Aedes aegypti*. Nature 272:535–537.

Taberlet, P. and J. Bouvet. 1991. A single plucked feather as a source of DNA for bird genetic studies. Auk 108:959–960.

Taberlet, P. and J. Bouvet. 1992. Bear conservation genetics. Nature 358:197.

Tajima, F. 1983. Evolutionary relationships of DNA sequences in finite populations. Genetics 105:437–460.

Takahata, N. 1985. Population genetics of extranuclear genomes: a model and review. Pp. 195–212 *in:* Population Genetics and Molecular Evolution, T. Ohta and K. Aoki (eds.), Springer-Verlag, Berlin.

Takahata, N. 1988. More on the episodic clock. Genetics 118:387–388.

Takahata, N. 1989. Gene genealogy in three related populations: consistency probability between gene and population trees. Genetics 122:957–966.

Takahata, N. 1993. Allelic genealogy and human evolution. Mol. Biol. Evol. 10:2–22.

Takahata, N. and M. Nei. 1990. Allelic genealogy under overdominant and frequency-dependent selection and polymorphism of major histocompatibility complex loci. Genetics 124:967–978.

Takahata, N. and S.R. Palumbi. 1985. Extranuclear differentiation and gene flow in the finite island model. Genetics 109:441–457.

Tamarin, R.H., M. Sheridan, and C.K. Levy. 1983. Determining matrilineal kinship in natural populations of rodents using radionuclides. Can. J. Zool. 61:271–274.

Tateno, Y., M. Nei, and F. Tajima. 1982. Accuracy of estimated phylogenetic trees from molecular data. I. Distantly related species. J. Mol. Evol. 18:387–404.

Tauber, C.A. and M.J. Tauber. 1989. Sympatric speciation in insects: perception and perspective. Pp. 307–344 *in:* Speciation and Its Consequences, D. Otte and J.A. Endler (eds.), Sinauer, Sunderland, MA.

Tautz, D. 1989. Hypervariability of simple sequences as a general source for polymorphic DNA markers. Nucleic Acids Res. 17:6463–6471.

Tegelström, H. 1987a. Genetic variability in mitochondrial DNA in a regional population of the great tit (*Parus major*). Biochem. Genet. 25:95–110.

Tegelström, H. 1987b. Transfer of mitochondrial DNA from the northern red-backed vole (*Clethrionomys rutilus*) to the bank vole (*C. glareolus*). J. Mol. Evol. 24:218–227.

Tegelström, H. and H.P. Gelter. 1990. Haldane's rule and sex biased gene flow between two hybridizing flycatcher species (*Ficedula albicollis* and *F. hypoleuca*, Aves: Muscicapidae). Evolution 44:2012–2021.

Tegelström, H., J. Searle, J. Brookfield, and S. Mercer. 1991. Multiple paternity in wild common shrews (*Sorex araneus*) is confirmed by DNA-fingerprinting. Heredity 66:373–379.

Tegelström, H., P.-I. Wyoni, H. Gelter, and M. Jaarola. 1988. Concordant divergence in proteins and mitochondrial DNA between two vole species in the genus *Clethrionomys*. Biochem. Genet. 26:223–237.

Templeton, A.R. 1980a. Modes of speciation and inferences based on genetic distances. Evolution 34:719–729.

Templeton, A.R. 1980b. The theory of speciation via the founder principle. Genetics 94:1011–1038.

Templeton, A.R. 1983. Phylogenetic inference from restriction endonuclease cleavage site maps with particular reference to the humans and apes. Evolution 37:221–244.

Templeton, A.R. 1987. Nonparametric inference from restriction cleavage sites. Mol. Biol. Evol. 4:315–319.

Templeton, A.R. 1989. The meaning of species and speciation: a genetic perspective. Pp. 3–27 *in:* Speciation and Its Consequences, D. Otte and J.A. Endler (eds.), Sinauer, Sunderland, MA.

Templeton, A.R. 1992. Human origins and analysis of mitochondrial DNA sequences. Science 255:737.

Templeton, A.R., K.A. Crandall, and C.F. Sing. 1992. A cladistic analysis of phenotypic associations with haplotypes inferred from restriction endonuclease mapping and DNA sequence data. III. Cladogram estimation. Genetics 132:619–633.

Theisen, B.F. 1978. Allozyme clines and evidence of strong selection in three loci in *Mytilus edulis* (Bivalvia) from Danish waters. Ophelia 17:135–142.

Thoday, J.M. and J.B. Gibson. 1962. Isolation by disruptive selection. Nature 193:1164–1166.

Thomas, R.H. and J.A. Hunt. 1991. The molecular evolution of the alcohol dehydrogenase locus and the phylogeny of Hawaiian *Drosophila*. Mol. Biol. Evol. 8:687–702.

Thomas, R.H., W. Schaffner, A.C. Wilson, and S. Pääbo. 1989. DNA phylogeny of the extinct marsupial wolf. Nature 340:465–467.

Thomas, W.K. and A.T. Beckenbach. 1989. Variation in salmonid mitochondrial DNA: evolutionary constraints and mechanisms of substitution. J. Mol. Evol. 29:233–245.

Thomas, W.K., S. Pääbo, F.X. Villiblanca, and A.C. Wilson. 1990. Spatial and temporal continuity of kangaroo rat populations shown by sequencing mitochondrial DNA from museum specimens. J. Mol. Evol. 31:101–112.

Thompson, J.D., E.A. Herre, J.L. Hamrick, and J.L. Stone. 1991. Genetic mosaics in strangler fig trees: implications for tropical conservation. Science 254:1214–1216.

Thorpe, J.P. 1982. The molecular clock hypothesis: biochemical evolution, genetic differentiation and systematics. Annu. Rev. Ecol. Syst. 13:139–168.

Thorson, G. 1961. Length of pelagic larval life in marine bottom invertebrates as related to larval transport by ocean currents. Pp. 455–474 *in:* Oceanography, M. Sears (ed.), American Association for the Advancement of Science, Washington, D.C.

Thresher, R.E. and E.B. Brothers. 1985. Reproductive ecology and biogeography of Indo-West Pacific angelfishes (Pisces: Pomacentridae). Evolution 39:878–887.

Throckmorton, L. 1975. The phylogeny, ecology and geography of *Drosophila*. Pp. 421–469 *in:* Handbook of Genetics, Vol. 3, R.C. King (ed.), Plenum, New York.

Tibayrenc, M. and F.J. Ayala. 1987. Forte correlation entre classification isoenzymatique et variabilite de l'ADN kinetoplastique chez *Trypanosoma cruzi*. Comp. Rend. Acad. des Sci. 304:89–92.

Tibayrenc, M. and F.J. Ayala. 1988. Isozyme variability in *Trypanosoma cruzi*, the agent of Chagas' disease: Genetical, taxonomical, and epidemiological significance. Evolution 42:277–292.

Tibayrenc, M., F. Kjellberg, J. Arnaud, B. Oury, S.F. Breniere, M.-L. Darde, and F.J. Ayala. 1991a. Are eukaryotic microorganisms clonal or sexual? A population genetics vantage. Proc. Natl. Acad. Sci. USA 88:5129–5133.

Tibayrenc, M., F. Kjellberg, and F.J. Ayala. 1990. A clonal theory of parasitic protozoa: The population structures of *Entamoeba, Giardia, Leishmania, Naegleria, Plasmodium, Trichomonas,* and *Trypanosoma* and their medical and taxonomical consequences. Proc. Natl. Acad. Sci. USA 87:2414–2418.

Tibayrenc, M., F. Kjellberg, and F.J. Ayala. 1991b. The clonal theory of parasitic protozoa. BioScience 41:767–774.

Tibayrenc, M., P. Ward, A. Moya, and F.J. Ayala. 1986. Natural populations of *Trypanosoma cruzi*, the agent of Chagas disease, have a complex multiclonal structure. Proc. Natl. Acad. Sci. USA 83:115–119.

Tilley, S.G. and J.S. Hausman. 1976. Allozymic variation and occurrence of multiple inseminations in populations of the salamander *Desmognathus ochrophaeus*. Copeia 1976:734–741.

Torroni, A., T.G. Schurr, C.-C. Yang, E.J.E. Szathmary, R.C. Williams, M.S. Schanfield, G.A. Troup, W.C. Knowler, D.N. Lawrence, K.M. Weiss, and D.C. Wallace. 1992. Native American mitochondrial DNA analysis indicates that the Amerind and the Nadene populations were founded by two independent migrations. Genetics 130:153–162.

Tracey, M.L., K. Nelson, D. Hedgecock, R.A. Shleser, and M.L. Pressick. 1975. Biochemical genetics of lobsters: genetic variation and the structure of American lobster (*Homerus americanus*) populations. J. Fish. Res. Bd. Can. 32:2091–2101.

Travis, J., J.C. Trexler, and M. Mulvey. 1990. Multiple paternity and its correlates in female *Poecilia latipinna* (Poeciliidae). Copeia 1990:722–729.

Triggs, S.J., R.G. Powlesland, and C.H. Daugherty. 1989. Genetic variation and conservation of kakapo (*Strigops habroptilus:* Psittaciformes). Cons. Biol. 3:92–96.

Triggs, S.J., M.J. Williams, S.J. Marshall, and G.K. Chambers. 1992. Genetic structure of blue duck (*Hymenolaimus malacorhynchos*) populations revealed by DNA fingerprinting. Auk 109: 80–89.

Trivers, R.L. 1972. Parental investment and sexual selection. Pp. 136–179 *in:* Sexual Selection and the Descent of Man, 1871–1971, B. Campbell (ed.), Aldine Press, Chicago, IL.

Turner, B.J. 1974. Genetic divergence of Death Valley pupfish species: Biochemical versus morphological evidence. Evolution 28:281–294.

Turner, B.J. 1982. The evolutionary genetics of a unisexual fish, *Poecilia formosa*. Pp. 265–305 *in:* Mechanisms of Speciation, C. Barigozzi (ed.), Alan R. Liss, New York.

Turner, B.J. 1983. Genic variation and differentiation of remnant natural populations of the desert pupfish, *Cyprinodon macularius*. Evolution 37:690–700.

Turner, B.J., J.F. Endler, Jr., T.F. Laughlin, and W.P. Davis. 1990. Genetic variation in clonal vertebrates detected by simple-sequence DNA fingerprinting. Proc. Natl. Acad. Sci. USA 87: 5653–5657.

Turner, B.J., J.F. Endler, Jr., T.F. Laughlin, W.P. Davis, and D.S. Taylor. 1992. Extreme clonal diversity and divergence in populations of a selfing hermaphroditic fish. Proc. Natl. Acad. Sci. USA 89:10643–10647.

Turner, B.J. and D.J. Grosse. 1980. Trophic differentiation in *Ilyodon,* a genus of stream-dwelling goodeid fishes: speciation versus ecological polymorphism. Evolution 34:259–270.

Turner, M.E., J.C. Stephens, and W.W. Anderson. 1982. Homozygosity and patch structure in plant populations as a result of nearest-neighbor pollination. Proc. Natl. Acad. Sci. USA 79:203–207.

Turner, T.R., M.L. Weiss, and M.E. Pereira. 1992. DNA fingerprinting and paternity assessment in Old World monkeys and ringtailed lemurs. Pp. 96–112 *in:* Paternity in Primates: Genetic Tests and Theories, R.D. Martin, A.F. Dixson, and E.J. Wickings (eds.), Karger, Basel, Switzerland.

Upholt, W.B. 1977. Estimation of DNA sequence divergence from comparison of restriction endonuclease digests. Nucleic Acids Res. 4:1257–1265.

Utter, F. 1991. Biochemical genetics and fishery management: an historical perspective. J. Fish Biol. 39A:1–20.

Uzzell, T. and C. Spolsky. 1981. Two data sets: alternative explanations and interpretations. Ann. N.Y. Acad. Sci. 361:481–499.

Valdés, A.M. and D. Piñero. 1992. Phylogenetic estimation of plasmid exchange in bacteria. Evolution 46:641–656.

Valdes, A.M., M. Slatkin, and N.B. Freimer. 1993. Allele frequencies at microsatellite loci: the stepwise mutation model revisited. Genetics 133:737–749.

Valentine, J.W. 1976. Genetic strategies of adaptation. Pp. 78–94 *in:* Molecular Evolution, F.J. Ayala, (ed.), Sinauer, Sunderland, MA.

Valentine, J.W. and F.J. Ayala. 1974. Genetic variation in *Frieleia halli,* a deep-sea brachiopod. Deep-Sea Res. 22:37–44.

VandeBerg, J.L., M.J. Aivaliotis, L.E. Williams, and C.R. Abee. 1990. Biochemical genetic markers of squirrel monkeys and their use for pedigree validation. Biochem. Genet. 28:41–56.

van Delden, W. 1982. The alcohol dehydrogenase polymorphism in *Drosophila melanogaster:* selection at an enzyme locus. Evol. Biol. 15:187–211.

Van den Eynde, H., R. De Baere, E. De Roeck, Y. Van de Peer, A. Vandenberghe, P. Willekens, and R. de Wachter. 1988. The 5S RNA sequences of a red algal rhodoplast and a gymnosperm chloroplast. Implications for the evolution of plastids and cyanobacteria. J. Mol. Evol. 27:126–132.

Van de Peer, Y., J.-M. Neefs, and R. De Wachter. 1990. Small ribosomal subunit RNA sequences, evolutionary relationships among different life forms, and mitochondrial origins. J. Mol. Evol. 30:463–473.

Vane-Wright, R.I., C.J. Humphries, and P.H. Williams. 1991. What to protect—systematics and the agony of choice. Biol. Cons. 55:235–254.

Vanlerberghe, F., P. Boursot, J.T. Nielsen, and F. Bonhomme. 1988. A steep cline for mitochondrial DNA in Danish mice. Genet. Res. Camb. 52:185–193.

Vanlerberghe, F., B. Dod, P. Boursot, M. Bellis, and F. Bonhomme. 1986. Absence of Y-chromosome introgression across the hybrid zone between *Mus musculus domesticus* and *Mus musculus musculus*. Genet. Res. Camb. 48:191–197.

van Pijlen, I.A., B. Amos, and G.A. Dover. 1991. Multilocus DNA fingerprinting applied to population studies of the Minke whale *Balaenoptera acutorostrata*. Pp. 245–254 *in:* Genetic Ecology of Whales and Dolphins, A.R. Hoelzel (ed.), Black Bear Press, Cambridge, England.

Van Valen, L. 1962. A study of fluctuating asymmetry. Evolution 16:125–142.

Van Wagner, C.E. and A.J. Baker. 1990. Association between mitochondrial DNA and morphological evolution in Canada geese. J. Mol. Evol. 31:373–382.

Varvio, S.-L., R. Chakraborty, and M. Nei. 1986. Genetic variation in subdivided populations and conservation genetics. Heredity 57:189–198.

Vassart, G., M. Georges, R. Monsieur, H. Brocas, A.S. Lequarre, and D. Christophe. 1987. A sequence in M13 phage detects hypervariable minisatellites in human and animal DNA. Science 235:683–684.

Vawter, A.T., R. Rosenblatt, and G.C. Gorman. 1980. Genetic divergence among fishes of the eastern Pacific and the Caribbean: support for the molecular clock. Evolution 34:705–711.

Vawter, L. and W.M. Brown. 1986. Nuclear and mitochondrial DNA comparisons reveal extreme rate variation in the molecular clock. Science 234:194–196.

Verspoor, E. and J. Hammar. 1991. Introgressive hybridization in fishes: the biochemical evidence. J. Fish Biol. 39A:309–334.

Victor, B.C. 1986. Duration of the planktonic larval stage of one hundred species of Pacific and Atlantic wrasses (family Labridae). Mar. Biol. 90:317–326.

Vigilant, L., R. Pennington, H. Harpending, T.D. Kocher, and A.C. Wilson. 1989. Mitochondrial DNA sequences in single hairs from a southern African population. Proc. Natl. Acad. Sci. USA 86:9350–9354.

Vigilant, L., M. Stoneking, H. Harpending, K. Hawkes, and A.C. Wilson. 1991. African populations and the evolution of human mitochondrial DNA. Science 253:1503–1507.

Vigneault, G. and E. Zouros. 1986. The genetics of asymmetrical male sterility in *Drosophila mojavensis* and *Drosophila arizonensis* hybrids: interactions between the Y-chromosome and autosomes. Evolution 40:1160–1170.

Villanueva, E., K.R. Luehrsen, J. Gibson, N. Delihas, and G.E. Fox. 1985. Phylogenetic origins of the plant mitochondrion based on a comparative analysis of 5S ribosomal RNA sequences. J. Mol. Evol. 22:46–52.

Visscher, P.K. 1986. Kinship discrimination in queen rearing by honey bees (*Apis mellifera*). Behav. Ecol. Sociobiol. 18:453–460.

Vrijenhoek, R.C. 1985. Homozygosity and interstrain variation in the self-fertilizing hermaphroditic fish, *Rivulus marmoratus*. J. Heredity 76:82–84.

Vrijenhoek, R.C., R.M. Dawley, C.J. Cole, and J.P. Bogart. 1989. A list of known unisexual vertebrates. Pp. 19–23 *in:* Evolution and Ecology of Unisexual Vertebrates, R. Dawley and J. Bogart (eds.), New York State Museum, Albany, NY.

Vrijenhoek, R.C., M.E. Douglas, and G.K. Meffe. 1985. Conservation genetics of endangered fish populations in Arizona. Science 229:400–402.

Vrijenhoek, R.C. and M.A. Graven. 1992. Population genetics of Egyptian *Biomphalaria alexandrina* (Gastropoda, Planorbidae). J. Heredity 83:255–261.

Vrijenhoek, R.C. and P.L. Leberg. 1991. Let's not throw the baby out with the bathwater: a comment on management for MHC diversity in captive populations. Cons. Biol. 5:252–254.

Vrijenhoek, R.C. and R.J. Schultz. 1974. Evolution of a trihybrid unisexual fish (*Poeciliopsis*, Poeciliidae). Evolution 28:306–319.

Vulliamy, T.J., A. Othman, M. Town, A. Nathwani, A.G. Falusi, P.J. Mason, and L. Luzzatto. 1991. Polymorphic sites in the African population detected by sequence analysis of the glucose-6-phosphate dehydrogenase gene outline the evolution of the variants A and A-. Proc. Natl. Acad. Sci. USA 88:8568–8571.

Vuorinen, J. and O.K. Berg. 1989. Genetic divergence of anadromous and nonanadromous Atlantic salmon (*Salmo salar*) in the River Namsen, Norway. Can. J. Fish. Aquat. Sci. 46:406–409.

Vyas, D.K., C. Moritz, D.M. Peccinini-Seale, J.W. Wright, and W.M. Brown. 1990. The evolutionary history of parthenogenetic *Cnemidophorus lemniscatus* (Sauria: Teiidae). II. Maternal origin and age inferred from mitochondrial DNA analyses. Evolution 44:922–932.

Wada, H., K.W. Makabe, M. Nakauchi, and N. Satoh. 1992. Phylogenetic relationships between solitary and colonial ascidians, as inferred from the sequence of the central region of their respective 18S rDNAs. Biol. Bull. 183:448–455.

Wada, S., T. Kobayashi, and K.-I. Numachi. 1991. Genetic variability and differentiation of mitochondrial DNA in Minke whales. Pp. 203–215 *in:* Genetic Ecology of Whales and Dolphins, A.R. Hoelzel (ed.), Black Bear Press, Cambridge, England.

Wade, M.J. 1982. The effect of multiple inseminations on the evolution of social behaviors in diploid and haplo-diploid organisms. J. Theoret. Biol. 95:351–368.

Wade, M.J. and D. McCauley. 1984. Group selection: the interaction of local deme size and migration in the differentiation of small populations. Evolution 38:1047–1058.

Wagner, D.B., G.R. Furnier, M.A. Saghai-Maroof, S.M. Williams, B.P. Dancik, and R.W. Allard. 1987. Chloroplast DNA polymorphisms in lodgepole and jack pines and their hybrids. Proc. Natl. Acad. Sci. USA 84:2097–2100.

Wain, R.P. 1983. Genetic differentiation during speciation in the *Helianthus debilis* complex. Evolution 37:1119–1127.

Wainscoat, J.S., A.V.S. Hill, A.L. Boyce, J. Flint, M. Hernandez, S.L. Thein, J.M. Old, J.R. Lynch, A.G. Falusi, D.J. Weatherall, and J.B. Clegg. 1986. Evolutionary relationships of human populations from an analysis of nuclear DNA polymorphisms. Nature 319:491–493.

Wake, D.B. and K.P. Yanev. 1986. Geographic variation in allozymes in a "ring species," the plethodontid salamander *Ensatina eschscholtzii* of western North America. Evolution 40:702–715.

Wake, D.B., K.P. Yanev, and C.W. Brown. 1986. Intraspecific sympatry in a "ring species," the plethodontid salamander *Ensatina eschscholtzii*, in southern California. Evolution 40:866–868.

Wake, D.B., K.P. Yanev, and M.M. Frelow. 1989. Sympatry and hybridization in a "ring species": the plethodontid salamander *Ensatina eschscholtzii*. Pp. 134–157 *in:* Speciation and Its Consequences, D. Otte and J.A. Endler (eds.), Sinauer, Sunderland, MA.

Waldman, B. 1988. The ecology of kin recognition. Annu. Rev. Ecol. Syst. 19:543–571.

Waldman, B. 1991. Kin recognition in amphibians. Pp. 162–219 *in:* Kin Recognition, P.G. Hepper (ed.), Cambridge Univ. Press, Cambridge, England.

Waldman, B., J.E. Rice, and R.L. Honeycutt. 1992. Kin recognition and incest avoidance in toads. Amer. Zool. 32:18–30.

Wallace, B. 1958. The role of heterozygosity in *Drosophila* populations. Proc. 10th Int. Congr. Genet. 1:408–419.

Wallace, B. 1970. Genetic Load. Prentice-Hall, Englewood Cliffs, NJ.

Wallace, B. 1991. Fifty Years of Genetic Load. An Odyssey. Cornell Univ. Press, Ithaca, NY.

Wallace, D.C. 1982. Structure and evolution of organelle genomes. Microbiol. Rev. 46:208–240.

Wallace, D.C. 1986. Mitochondrial genes and disease. Hospital Practice 21:77–92.

Wallace, D.C. 1992. Mitochondrial genetics: a paradigm for aging and degenerative diseases? Science 256:628–632.

Wallace, D.C., G. Singh, M.T. Lott, J.A. Hodge, T.G. Schurr, A.M.S. Lezza, L.J. Elsas II, and E.K. Nikoskelainen. 1988. Mitochondrial DNA mutation associated with Leber's hereditary optic neuropathy. Science 242:1427–1430.

Waller, D.M., D.M. O'Malley, and S.C. Gawler. 1987. Genetic variation in the extreme endemic *Pedicularis furbishiae* (Scrophulariaceae). Cons. Biol. 1:335–340.

Wallis, G.P. and J.W. Arntzen. 1989. Mitochondrial-DNA variation in the crested newt superspecies: limited cytoplasmic gene flow among species. Evolution 43:88–104.

Walters, M. 1992. A Shadow and a Song: Extinction of the Dusky Seaside Sparrow. Chelsea Green Publ. Co., Post Mills, VT.

Waples, R.S. 1987. A multispecies approach to the analysis of gene flow in marine shore fishes. Evolution 41:385–400.

Waples, R.S. 1991. Heterozygosity variation in bony fishes: an alternative view. Evolution 45:1275–1280.

Ward, B.L., R.S. Anderson, and A.J. Bendich. 1981. The mitochondrial genome is large and variable in a family of plants (Cucurbitaceae). Cell 25:793–803.

Ward, P.S. 1983. Genetic relatedness and colony organization in a species complex of ponerine ants, I: Phenotypic and genotypic composition of colonies. Behav. Ecol. Sociobiol. 12:285–299.

Ward, P.S. and R.W. Taylor. 1981. Allozyme variation, colony structure and genetic relatedness in the primitive ant *Nothomyrmecia macrops* Clark (Hymenoptera: Formicidae). J. Austral. Entomol. Soc. 20:177–183.

Ward, R.D. 1977. Relationship between enzyme heterozygosity and quaternary structure. Biochem. Genet. 15:123–135.

Ward, R.D., D.O.F. Skibinski, and M. Woodwark. 1992. Protein heterozygosity, protein structure, and taxonomic differentiation. Evol. Biol. 26:73–159.

Ward, R.H., B.L. Frazier, K. Dew-Jager, and S. Pääbo. 1991. Extensive mitochondrial diversity within a single Amerindian tribe. Proc. Natl. Acad. Sci. USA 88:8720–8724.

Waterman, M.S., T.F. Smith, M. Singh, and W.A. Beyer. 1977. Additive evolutionary trees. J. Theoret. Biol. 64:199–213.

Watson, J.D. and F.H.C. Crick. 1953. A structure for DNA. Nature 171:736–738.

Watson, J.D., M. Gilman, J. Witkowski, and M. Zoller. 1992. Recombinant DNA (2nd ed.). W.H. Freeman Co., New York.

Watt, W.B. 1977. Adaptation at specific loci. I. Natural selection on phosphoglusose isomerase of *Colias* butterflies: biochemical and population aspects. Genetics 87:177–194.

Watt, W.B., P.A. Carter, and S.M. Blower. 1985. Adaptation at specific loci. IV. Differential mating success among glycolytic allozyme genotypes of *Colias* butterflies. Genetics 109:157–194.

Watt, W.B., R.C. Cassin, and M.S. Swan. 1983. Adaptation at specific loci, III. Field behavior and survivorship differences among *Colias* PGI genotypes are predictable from in vitro biochemistry. Genetics 103:725–739.

Watts, R.J., M.S. Johnson, and R. Black. 1990. Effects of recruitment on genetic patchiness in the urchin *Echinometra mathaei* in Western Australia. Mar. Biol. 105:145–151.

Wayne, R.K. 1992. On the use of morphologic and molecular genetic characters to investigate species status. Cons. Biol. 6:590–592.

Wayne, R.K. and S.M. Jenks. 1991. Mitochondrial DNA analysis implying extensive hybridization of the endangered red wolf *Canis rufus*. Nature 351:565–568.

Wayne, R.K., N. Lehman, M.W. Allard, and R.L. Honeycutt. 1992. Mitochondrial DNA variability of the gray wolf: genetic consequences of population decline and habitat fragmentation. Cons. Biol. 6:559–569.

Wayne, R.K., N. Lehman, D. Girman, P.J.P. Gogan, D.A. Gilbert, K. Hansen, R.O. Peterson, U.S. Seal, A. Eisenhawer, L.D. Mech, and R.J. Krumenaker. 1991b. Conservation genetics of the endangered Isle Royale gray wolf. Cons. Biol. 5:41–51.

Wayne, R.K., B. Van Valkenburgh, and S.J. O'Brien. 1991a. Molecular distance and divergence time in carnivores and primates. Mol. Biol. Evol. 8:297–319.

Weatherhead, P.J. and R.D. Montgomerie. 1991. Good news and bad news about DNA fingerprinting. Trends Ecol. Evol. 6:173–174.

Weber, J.L. and P.E. May. 1989. Abundant class of human DNA polymorphisms which can be typed using the polymerase chain reaction. Amer. J. Human Genet. 44:388–396.

Weir, B.S. 1990. Genetic Data Analysis. Sinauer, Sunderland, MA.

Weir, B.S. 1992. Independence of VNTR alleles defined as fixed bins. Genetics 130:873–887.

Weir, B.S. and C.C. Cockerham. 1973. Mixed self and random mating at two loci. Genet. Res. 21:247–262.

Weir, B.S. and C.C. Cockerham. 1984. Estimating F-statistics for the analysis of population structure. Evolution 38:1358–1370.

Weissenbach, J., G. Gyapay, C. Dib, A. Vignal, J. Morissette, P. Millasseau, G. Vaysseix, and M. Lathrop. 1992. A second-generation linkage map of the human genome. Nature 359:794–801.

Weller, J., M. Sober, and T. Brody. 1988. Linkage analysis of quantitative traits in an interspecific cross of tomato (*Lycopersicon esculentum* × *Lycopersicon pimpinellifolium*) by means of genetic markers. Genetics 118:329–339.

Weller, R. and D.M. Ward. 1989. Selective recovery of 16S rRNA sequences from natural microbial communities in the form of cDNA. Appl. Environ. Microbiol. 55:1818–1822.

Weller, R., J.W. Weller, and D.M. Ward. 1991. 16S rRNA sequences of uncultivated hot spring cyanobacterial mat inhabitants retrieved as randomly primed cDNA. Appl. Environmen. Microbiol. 57:1146–1151.

Welsh, J., C. Petersen, and M. McClelland. 1991. Polymorphisms generated by arbitrarily primed PCR in the mouse: application to strain identification and genetic mapping. Nucleic Acids Res. 19:303–306.

Wendel, J.F. and V.A. Albert. 1992. Phylogenetics of the cotton genus (*Gossypium* L.): Character-state weighted parsimony analysis of chloroplast-DNA restriction site data and its systematic and biogeographic implications. Syst. Bot. 17:115–143.

Wendel, J.F., J.M. Stewart, and J.H. Rettig. 1991. Molecular evidence for homoploid reticulate evolution among Australian species of *Gossypium*. Evolution 45:694–711.

Wenink, P.W., A.J. Baker, and M.G.J. Tilanus. 1993. Hypervariable-control-region sequences reveal global population structuring in a long-distance migrant shorebird, the Dunlin (*Caladris alpina*). Proc. Natl. Acad. Sci. USA 90:94–98.

Werman, S.D., M.S. Springer, and R.J. Britten. 1990. Nucleic acids I: DNA-DNA hybridization. Pp. 204–249 *in:* Molecular Systematics, D.M. Hillis and C. Moritz (eds.), Sinauer, Sunderland, MA.

Werth, C.R., S.I. Guttman, and W.H. Eshbaugh. 1985. Recurring origins of allopolyploid species in *Asplenium*. Science 228:731–733.

West-Eberhard, M.J. 1990. The genetic and social structure of polygynous social wasp colonies (Vespidae: Polistinae). Pp. 254–255 *in:* Social Insects and the Environment, G.K. Veeresh, B. Mallik, and C.A. Viraktamath (eds.), Oxford and IBH, New Delhi, India.

Westneat, D.F. 1987. Extra-pair fertilizations in a predominantly monogamous bird: genetic evidence. Anim. Behav. 35:877–886.

Westneat, D.F. 1990. Genetic parentage in the indigo bunting: a study using DNA fingerprinting. Behav. Ecol. Sociobiol. 27:67–76.

Westneat, D.F., P.C. Frederick, and R.H. Wiley. 1987. The use of genetic markers to estimate the frequency of successful alternative reproductive tactics. Behav. Ecol. Sociobiol. 21:35–45.

Westneat, D.F., P.W. Sherman, and M.L. Morton. 1990. The ecology and evolution of extra-pair copulations in birds. Pp. 331–369 *in:* Current Ornithology, Vol. 7, D.M. Power (ed.), Plenum Press, New York.

Wetton, J.H., R.E. Carter, D.T. Parkin, and D. Walters. 1987. Demographic study of a wild house sparrow population by DNA fingerprinting. Nature 327:147–149.

Wetton, J.H., D.T. Parkin, and R.E. Carter. 1992. The use of genetic markers for parentage analysis in *Passer domesticus* (house sparrow). Heredity 69:243–254.

Wheat, T.E., G.S. Whitt, and W.F. Childers. 1973. Linkage relationships of six enzyme loci in interspecific sunfish hybrids (genus *Lepomis*). Genetics 74:343–350.

Wheeler, M.R. 1986. Additions to the catalog of the world's Drosophilidae. Pp. 395–409 *in:* The Genetics and Biology of *Drosophila*, Vol. 3e, M. Ashburner, H.L. Carson, and J.N. Thompson, Jr. (eds.), Academic Press, New York.

Wheeler, N.C. and R.P. Guries. 1982. Population structure, genetic diversity and morphological variation in *Pinus contorta* Dougl. Can. J. Forest Res. 12:595–606.

Wheelis, M.L., O. Kandler, and C.R. Woese. 1992. On the nature of global classification. Proc. Natl. Acad. Sci. USA 89:2930–2934.

Whisenant, E.C., B.K.A. Rasheed, H. Ostrer, and Y.M. Bhatnagar. 1991. Evolution and sequence analysis of a human Y-chromosomal DNA fragment. J. Mol. Evol. 33:133–141.

White, M.J.D. 1978a. Modes of Speciation. W.H. Freeman and Co., San Francisco.

White, M.J.D. 1978b. Cytogenetics of the parthenogenetic grasshopper *Warramaba* (formerly *Moraba*) *virgo* and its bisexual relatives III. Meiosis of male "synthetic *virgo*" individuals. Chromosoma 67:55–61.

White, T.J., N. Arnheim, and H.A. Erlich. 1989. The polymerase chain reaction. Trends Genet. 5:185–188.

Whitt, G.S. 1983. Isozymes as probes and participants in developmental and evolutionary genetics. Isozymes X:1–40.

Whitt, G.S. 1987. Species differences in isozyme tissue patterns: Their utility for systematic and evolutionary analysis. Isozymes XV:1–26.

Whitt, G.S., W.F. Childers, J.B. Shaklee, and J. Matsumoto. 1976. Linkage analysis of the multilocus glucosephosphate isomerase isozyme system in sunfish (Centrarchidae, Teleostii). Genetics 82:35–42.

Whitt, G.S., J.B. Shaklee, and C.L. Markert. 1975. Evolution of the lactate dehydrogenase isozymes of fishes. Isozymes IV:381–400.

Whittaker, R.H. 1959. On the broad classification of organisms. Quart. Rev. Biol. 34:210–226.

Whittam, T.S., A.G. Clark, M. Stoneking, R.L. Cann, and A.C. Wilson. 1986. Allelic variation in human mitochondrial genes based on patterns of restriction site polymorphism. Proc. Natl. Acad. Sci. USA 83:9611–9615.

Whittam, T.S., H. Ochman, and R.K. Selander. 1983a. Geographic components of linkage disequilibrium in natural populations of *Escherichia coli*. Mol. Biol. Evol. 1:67–83.

Whittam, T.S., H. Ochman, and R.K. Selander. 1983b. Multilocus genetic structure in natural populations of *Escherichia coli*. Proc. Natl. Acad. Sci. USA 80:1751–1755.

Whittemore, A.T. and B.A. Schaal 1991. Interspecific gene flow in oaks. Proc. Natl. Acad. Sci. USA 88:2540–2544.

Wildt, D.E., M. Bush, K.L. Goodrowe, C. Packer, A.E. Pusey, J.L. Brown, P. Joslin, and S.J. O'Brien. 1987. Reproductive and genetic consequences of founding isolated lion populations. Nature 329:328–331.

Wiley, E.O. 1981. Phylogenetics. John Wiley and Sons, New York.

Wiley, E.O. 1988. Vicariance biogeography. Annu. Rev. Ecol. Syst. 19:513–542.

Willers, B. 1992. Toward a science of letting things be. Cons. Biol. 6:605–607.

Williams, G.C. 1975. Sex and Evolution. Princeton Univ. Press, Princeton, NJ.

Williams, G.C. and R.K. Koehn. 1984. Population genetics of North Atlantic catadromous eels (*Anguilla*). Pp. 529–560 *in:* Evolutionary Genetics of Fishes, B.J. Turner (ed.), Plenum Press, New York.

Williams, G.C., R.K. Koehn, and J.B. Mitton. 1973. Genetic differentiation without isolation in the American eel, *Anguilla rostrata*. Evolution 27:192–204.

Williams, G.C. and D.C. Williams. 1957. Natural selection of individually harmful social adaptations among sibs with special reference to social insects. Evolution 11:32–39.

Williams, J.G.K., A.R. Kubelik, J. Livak, J.A. Rafalski, and S.V. Tingey. 1990. DNA polymorphisms amplified by arbitrary primers are useful as genetic markers. Nucleic Acids Res. 18:6531–6535.

Williams, S.A. and M. Goodman. 1989. A statistical test that supports a human/chimpanzee clade based on noncoding DNA sequence data. Mol. Biol. Evol. 6:325–330.

Williams, S.M., R. DeSalle, and C. Strobeck. 1985. Homogenization of geographical variants at the nontranscribed spacer of rDNA in *Drosophila mercatorum*. Mol. Biol. Evol. 2:338–346.

Williams, S.M., G.R. Furnier, E. Fuog, and C. Strobeck. 1987. Evolution of the ribosomal DNA spacers of *Drosophila melanogaster:* different patterns of variation on X and Y chromosomes. Genetics 116:225–232.

Williams, S.M. and C. Strobeck. 1986. Measuring the multiple insemination frequency of *Drosophila* in nature: use of a Y-linked molecular marker. Evolution 40:440–442.

Wilson, A.C. 1976. Gene regulation in evolution. Pp. 225–234 *in:* Molecular Evolution, F.J. Ayala (ed.), Sinauer, Sunderland, MA.

Wilson, A.C. 1985. The molecular basis of evolution. Sci. Amer. 253(4):164–173.

Wilson, A.C., R.L. Cann, S.M. Carr, M. George, Jr., U.B. Gyllensten, K.M. Helm-Bychowski, R.G. Higuchi, S.R. Palumbi, E.M. Prager, R.D. Sage, and M. Stoneking. 1985. Mitochondrial DNA and two perspectives on evolutionary genetics. Biol. J. Linn. Soc. 26:375–400.

Wilson, A.C., S.S. Carlson, and T.J. White. 1977. Biochemical evolution. Annu. Rev. Biochem. 46:473–639.

Wilson, A.C., L.R. Maxson, and V.M. Sarich. 1974a. Two types of molecular evolution. Evidence from studies of interspecific hybridization. Proc. Natl. Acad. Sci. USA 71:2843–2847.

Wilson, A.C., H. Ochman, and E.M. Prager. 1987. Molecular time scale for evolution. Trends Genet. 3:241–247.

Wilson, A.C., V.M. Sarich, and L.R. Maxson. 1974b. The importance of gene rearrangement in evolution: evidence from studies on rates of chromosomal, protein, and anatomical evolution. Proc. Natl. Acad. Sci. USA 71:3028–3030.

Wilson, A.C., M. Stoneking, and R.L. Cann. 1991. Ancestral geographic states and the peril of parsimony. Syst. Zool. 40:363–365.

Wilson, E.O. 1971. The Insect Societies. Harvard Univ. Press, Cambridge, MA.

Wilson, E.O. 1975. Sociobiology. Harvard Univ. Press, Cambridge, MA.

Wilson, E.O. 1984. Biophilia. Harvard Univ. Press, Cambridge, MA.

Wilson, E.O. 1987. Kin recognition: an introductory synopsis. Pp. 7–18 *in:* Kin Recognition in Animals, D.J.C. Fletcher and C.D. Michener (eds.), John Wiley and Sons, New York.

Wilson, G.M., W.K. Thomas, and A.T. Beckenbach. 1987. Mitochondrial DNA analysis of Pacific Northwest populations of *Oncorhynchus tshawytscha*. Can. J. Fish. Aquat. Sci. 44:1301–1305.

Wimpee, C.F., T.-L. Nadeau, and K.H. Nealson. 1991. Development of species-specific hybridization probes for marine luminous bacteria by using *in vitro* DNA amplification. Appl. Environ. Microbiol. 57:1319–1324.

Winans, G.A. 1980. Geographic variation in the milkfish *Chanos chanos*. I. Biochemical evidence. Evolution 34:558–574.

Winkler, C., A. Schultz, S. Cevario, and S.J. O'Brien. 1989. Genetic characterization of *FLA*, the cat major histocompatibility complex. Proc. Natl. Acad. Sci. USA 86:943–947.

Witter, M.S. and G.D. Carr. 1988. Adaptive radiation and genetic differentiation in the Hawaiian silversword alliance (Compositae: Madiinae). Evolution 42:1278–1287.

Woese, C.R. 1987. Bacterial evolution. Microbiol. Rev. 51:221–271.

Woese, C.R. and G.E. Fox. 1977. Phylogenetic structure of the prokaryotic domain: the primary kingdoms. Proc. Natl. Acad. Sci. USA 74:5088–5090.

Woese, C.R., O. Kandler, and M.L. Wheelis. 1990. Towards a natural system of organisms: proposal for the domains Archaea, Bacteria, and Eucarya. Proc. Natl. Acad. Sci. USA 87:4576–4579.

Woese, C.R., O. Kandler, and M.L. Wheelis. 1991. A natural classification. Nature 351:528–529.

Wolfe, K.H., M. Gouy, Y.-W. Yang, P.M. Sharp, and W.-H. Li. 1989. Date of the monocot-dicot divergence estimated from chloroplast DNA sequence data. Proc. Natl. Acad. Sci. USA 86:6201–6205.

Wolfe, K.H., W.-H. Li, and P.M. Sharp. 1987. Rates of nucleotide substitution vary greatly among plant mitochondrial, chloroplast, and nuclear DNAs. Proc. Natl. Acad. Sci. USA 84:9054–9058.

Wolpoff, M.H. 1989. Multiregional evolution: the fossil alternative to Eden. Pp. 62–108 *in:* The Human Revolution: Behavioural and Biological Perspectives on the Origins of Modern Humans, P. Mellars and C. Stringer (eds.), Princeton Univ. Press, Princeton, NJ.

Wolpoff, M.H., X.Z. Yu, and A.G. Thorne. 1984. Modern *Homo sapiens* origins: a general theory of hominid evolution involving the fossil evidence from East Asia. Pp. 411–483 *in:* The Origins of Modern Humans: A World Survey of the Fossil Evidence, F.H. Smith and F. Spencer (eds.), Alan R. Liss, New York.

Wolstenholme, D.R. 1992. Animal mitochondrial DNA: structure and evolution. Int. Rev. Cytol. 141:173–216.

Wolstenholme, D.R., D.O. Clary, J.L. MacFarlane, J.A. Wahleithner, and L. Wilcox. 1985. Organization and evolution of invertebrate mitochondrial-genomes. Pp. 61–69 *in:* Achievements and Perspectives of Mitochondrial Research Vol. II: Biogenesis, E. Quagliariello, E.C. Slater, F. Palmieri, C. Saccone, and A.M. Kroon (eds.), Elsevier, Amsterdam.

Wolstenholme, D.R., J.L. MacFarlane, R. Okimoto, D.O. Clary, and J.A. Wahleithner. 1987. Bizarre tRNAs inferred from DNA sequences of mitochondrial genomes of nematode worms. Proc. Natl. Acad. Sci. USA 84:1324–1328.

Wong, Z., V. Wilson, A.J. Jeffreys, and S.L. Thein. 1986. Cloning a selected fragment from a human DNA fingerprint: Isolation of an extremely polymorphic minisatellite. Nucleic Acids Res. 14:4605–4616.

Wong, Z., V. Wilson, I. Patel, S. Povey, and A.J. Jeffreys. 1987. Characterization of a panel of highly variable minisatellites cloned from human DNA. Ann. Human Genet. 51:269–288.

Woodin, S.A. 1986. Settlement of infauna: larval choice? Bull. Mar. Sci. 39:401–407.

Woodruff, D.S. and S.J. Gould. 1987. Fifty years of interspecific hybridization: genetics and morphometrics of a controlled experiment on the land snail *Cerion* in the Florida Keys. Evolution 41:1022–1045.

Woodruff, D.S., M. Mulvey, and M.W. Yipp. 1985. Population genetics of *Biomphalaria straminea* in Hong Kong. J. Heredity 76:355–360.

Woolfenden, G.E. and J.W. Fitzpatrick. 1984. The Florida Scrub Jay. Princeton Univ. Press, Princeton, NJ.

Wooten, M.C., K.T. Scribner, and M.H. Smith. 1988. Genetic variability and systematics of *Gambusia* in the southeastern United States. Copeia 1988:283–289.

Wrege, P.H. and S.T. Emlen. 1987. Biochemical determination of parental uncertainty in white-fronted bee-eaters. Behav. Ecol. Sociobiol. 20:153–160.

Wright, C.A. 1974. Biochemical and Immunological Taxonomy of Animals. Academic Press, New York.

Wright, J.W. 1983. The evolution and biogeography of the lizards of the Galapagos archipelago: evolutionary genetics of *Phyllodactylus* and *Tropidurus* populations. Pp. 123–156 *in:* Patterns of Evolution in Galapagos Organisms, R.I. Bowman, M. Berson, and A.E. Leviton (eds.), American Association for the Advancement of Science, San Francisco, CA.

Wright, J.W., C. Spolsky, and W.M. Brown. 1983. The origin of the parthenogenetic lizard *Cnemidophorus laredoensis* inferred from mitochondrial DNA analysis. Herpetologica 39:410–416.

Wright, S. 1931. Evolution in Mendelian populations. Genetics 16:97–159.

Wright, S. 1951. The genetical structure of populations. Ann. Eugen. 15:323–354.

Wrischnik, L.A., R.G. Higuchi, M. Stoneking, H.A. Erlich, N. Arnheim, and A.C. Wilson. 1987. Length mutations in human mitochondrial DNA: direct sequencing of enzymatically amplified DNA. Nucleic Acids Res. 15:529–542.

Wu, C.-I. 1991. Inferences of species phylogeny in relation to segregation of ancient polymorphisms. Genetics 127:429–435.

Wu, C.-I. and W.-H. Li. 1985. Evidence for higher rates of nucleotide substitution in rodents than in man. Proc. Natl. Acad. Sci. USA 82:1741–1745.

Wyatt, R. 1988. Phylogenetic aspects of the evolution of self-pollination. Pp. 109–131 *in:* Plant Evolutionary Biology, L.D. Gottlieb and S.K. Jain (eds.), Chapman and Hall, New York.

Wyatt, R., E.A. Evans, and J.C. Sorenson. 1992. The evolution of self-pollination in granite outcrop species of *Arenaria* (Caryophyllaceae). VI. Electrophoretically detectable genetic variation. Syst. Bot. 17:201–209.

Wyatt, R., I.J. Odrzykoski, A. Stoneburner, H.W. Bass, and G.A. Galau. 1988. Allopolyploidy in bryophytes: multiple origins of *Plagiomnium medium*. Proc. Natl. Acad. Sci. USA 85:5601–5604.

Wykoff, R.W.G. 1972. The Biochemistry of Animal Fossils. Scientechnica, Bristol.

Wyles, J.S., J.G. Kunkel, and A.C. Wilson. 1983. Birds, behavior, and anatomical evolution. Proc. Natl. Acad. Sci. USA 80:4394–4397.

Wyles, J.S. and V.M. Sarich. 1983. Are the Galapagos iguanas older than the Galapagos? Pp. 177–199 *in:* Patterns of Evolution in Galapagos Organisms, R.I. Bowman, M. Berson, and A.E. Leviton (eds.), American Association for the Advancement of Science, San Francisco, CA.

Xia, X. and J.S. Millar. 1991. Genetic evidence of promiscuity in *Peromyscus leucopus*. Behav. Ecol. Sociobiol. 28:171–178.

Xiong, W., W.-H. Li, I. Posner, T. Yamamura, A. Yamamoto, A.M. Gotto., Jr., and L. Chan. 1991. No severe bottleneck during human evolution: evidence from two apolipoprotein C-II deficiency alleles. Amer. J. Human Genet. 48:383–389.

Xiong, Y. and T.H. Eickbush. 1990. Origin and evolution of retroelements based upon their reverse transcriptase sequences. The EMBO J. 9:3353–3362.

Yang, D., Y. Oyaizu, H. Oyaizu, G.J. Olsen, and C.R. Woese. 1985. Mitochondrial origins. Proc. Natl. Acad. Sci. USA 82:4443–4447.

Yang, S.Y. and J.L. Patton. 1981. Genic variability and differentiation in the Galapagos finches. Auk 98:230–242.

Yokoyama, S. and T. Gojobori. 1987. Molecular evolution and phylogeny of the human AIDS viruses LAV, HTLV-III, and ARV. J. Mol. Evol. 24:330–336.

Yuhki, N. and S.J. O'Brien. 1990. DNA variation of the mammalian major histocompatibility complex reflects genomic diversity and population history. Proc. Natl. Acad. Sci. USA 87:836–840.

Zambryski, P., J. Tempe, and J. Schell. 1989. Transfer and function of T-DNA genes from *Agrobacterium* Ti and Ri plasmids in plants. Cell 56:193–201.

Zeh, D.W., J.A. Zeh, M.A. Coffroth, and E. Bermingham. 1992. Population-specific DNA fingerprints in a neotropical pseudoscorpion (*Cordylochernes scorpioides*). Heredity 69:201–208.

Zera, A.J. 1981. Genetic structure of two species of waterstriders (Gerridae: Hemiptera) with differing degrees of winglessness. Evolution 35:218–225.

Zhang, Q., M. Tibayrenc, and F.J. Ayala. 1988. Linkage disequilibrium in natural populations of *Trypanosoma cruzi* (Flagellate), the agent of Chagas' disease. J. Protozool. 35:81–85.

Zimmer, E.A., S.L. Martin, S.M. Beverley, Y.W. Kan, and A.C. Wilson. 1980. Rapid duplication and loss of genes coding for the α chains of hemoglobin. Proc. Natl. Acad. Sci. USA 77:2158–2162.

Zimmerman, E.G., C.W. Kilpatrick, and B.J. Hart. 1978. The genetics of speciation in the rodent genus *Peromyscus*. Evolution 32:565–579.

Zink, R.M. 1991. Geography of mitochondrial DNA variation in two sympatric sparrows. Evolution 45:329–339.

Zink, R.M. and N.K. Johnson. 1984. Evolutionary genetics of flycatchers. I. Sibling species in the genera *Empidonax* and *Contopus*. Syst. Zool. 33:205–216.

Zink, R.M., M.F. Smith, and J.L. Patton. 1985. Associations between heterozygosity and morphological variance. J. Heredity 76:415–420.

Zinkham, W.H., H. Isensee, and J.H. Renwick. 1969. Linkage of lactate dehydrogenase B and C loci in pigeons. Science 164:185–187.

Zouros, E. 1976. Hybrid molecules and the superiority of the heterozygote. Nature 262:227–229.

Zouros, E., K.R. Freeman, A.O. Ball, and G.H. Pogson. 1992. Direct evidence for extensive paternal mitochondrial DNA inheritance in the marine mussel *Mytilus*. Nature 359:412–414.

Zouros, E., S.M. Singh, and M.E. Miles. 1980. Growth rate in oysters: an overdominant phenotype and its possible explanations. Evolution 34:856–867.

Zuckerkandl, E. and L. Pauling. 1965. Evolutionary divergence and convergence in proteins. Pp. 97–166 *in:* Evolving Genes and Proteins, V. Bryson and H.J. Vogel (eds.), Academic Press, New York.

Zurawski, G. and M.T. Clegg. 1987. Evolution of higher-plant chloroplast DNA-encoded genes: implications for structure-function and phylogenetic studies. Annu. Rev. Plant Physiol. 38:391–418.

Index to Taxonomic Genera

General Index

507